Handbook of Research on Monitoring and Evaluating the Ecological Health of Wetlands

Ashok K. Rathoure
Akone Services, India

IGI Global
PUBLISHER of TIMELY KNOWLEDGE

A volume in the Practice, Progress, and Proficiency in Sustainability (PPPS) Book Series

Published in the United States of America by
IGI Global
Engineering Science Reference (an imprint of IGI Global)
701 E. Chocolate Avenue
Hershey PA, USA 17033
Tel: 717-533-8845
Fax: 717-533-8661
E-mail: cust@igi-global.com
Web site: http://www.igi-global.com

Library of Congress Cataloging-in-Publication Data

Names: Rathoure, Ashok K., 1983- editor.
Title: Handbook of research on monitoring and evaluating the ecological health of wetlands / Ashok K. Rathoure, editor.
Description: Hershey : Engineering Science Reference, 2021. | Includes bibliographical references and index. | Summary: "This book highlights the challenges of wetland conservation and the current scenario of existing wetlands including inland wetland and discusses Inventory, assessment and monitoring are as vital components of effective wetland management"-- Provided by publisher.
Identifiers: LCCN 2021037621 (print) | LCCN 2021037622 (ebook) | ISBN 9781799894988 (hardcover) | ISBN 9781799895008 (ebook)
Subjects: LCSH: Wetland conservation. | Wetland management.
Classification: LCC QH75 .M655 2021 (print) | LCC QH75 (ebook) | DDC 333.91/816--dc23
LC record available at https://lccn.loc.gov/2021037621
LC ebook record available at https://lccn.loc.gov/2021037622

This book is published in the IGI Global book series Practice, Progress, and Proficiency in Sustainability (PPPS) (ISSN: 2330-3271; eISSN: 2330-328X)

British Cataloguing in Publication Data
A Cataloguing in Publication record for this book is available from the British Library.

All work contributed to this book is new, previously-unpublished material. The views expressed in this book are those of the authors, but not necessarily of the publisher.

For electronic access to this publication, please contact: eresources@igi-global.com.

Practice, Progress, and Proficiency in Sustainability (PPPS) Book Series

Ayman Batisha
International Sustainability Institute, Egypt

ISSN:2330-3271
EISSN:2330-328X

Mission

In a world where traditional business practices are reconsidered and economic activity is performed in a global context, new areas of economic developments are recognized as the key enablers of wealth and income production. This knowledge of information technologies provides infrastructures, systems, and services towards sustainable development.

The **Practices, Progress, and Proficiency in Sustainability (PPPS) Book Series** focuses on the local and global challenges, business opportunities, and societal needs surrounding international collaboration and sustainable development of technology. This series brings together academics, researchers, entrepreneurs, policy makers and government officers aiming to contribute to the progress and proficiency in sustainability.

Coverage

- Global Content and Knowledge Repositories
- Global Business
- Green Technology
- Intellectual Capital
- Environmental informatics
- Socio-Economic
- Knowledge clusters
- Technological learning
- Strategic Management of IT
- Sustainable Development

IGI Global is currently accepting manuscripts for publication within this series. To submit a proposal for a volume in this series, please contact our Acquisition Editors at Acquisitions@igi-global.com or visit: http://www.igi-global.com/publish/.

The Practice, Progress, and Proficiency in Sustainability (PPPS) Book Series (ISSN 2330-3271) is published by IGI Global, 701 E. Chocolate Avenue, Hershey, PA 17033-1240, USA, www.igi-global.com. This series is composed of titles available for purchase individually; each title is edited to be contextually exclusive from any other title within the series. For pricing and ordering information please visit http://www.igi-global.com/book-series/practice-progress-proficiency-sustainability/73810. Postmaster: Send all address changes to above address.

Titles in this Series

For a list of additional titles in this series, please visit: www.igi-global.com/book-series

Achieving Sustainability Using Creativity, Innovation, and Education A Multidisciplinar Approach
Ziska Fields (University of Johannesburg, South Africa)
Information Science Reference • © 2022 • 260pp • H/C (ISBN: 9781799879633) • US $215.00

Handbook of Research on Sustainable Development Goals, Climate Change, and Digitalization
Rui Alexandre Castanho (WSB University, Poland)
Engineering Science Reference • © 2022 • 684pp • H/C (ISBN: 9781799884828) • US $315.00

Sociological Perspectives on Sustainable Development and Poverty Reduction in Rural Populations
Ladislaus M. Semali (Pennsylvania State University, USA)
Information Science Reference • © 2022 • 303pp • H/C (ISBN: 9781799846468) • US $195.00

Handbook of Research on Novel Practices and Current Successes in Achieving the Sustainable Development Goals
Cristina Raluca Gh. Popescu (University of Bucharest, Romania & The Bucharest University of Economic Studies, Romania & University of Craiova, Romania & The National Institute for Research and Development in Environmental Protection (I.N.C.D.P.M.), Romania & National Research and Development Institute for Gas Turbines COMOTI, Romania)
Engineering Science Reference • © 2021 • 461pp • H/C (ISBN: 9781799884262) • US $295.00

Advancements in Sustainable Architecture and Energy Efficiency
Roberto A. González-Lezcano (Universidad CEU San Pablo, Spain)
Engineering Science Reference • © 2021 • 337pp • H/C (ISBN: 9781799870234) • US $195.00

Management and Conservation of Mediterranean Environments
Rui Alexandre Castanho (WSB University, Poland) and José Martín Gallardo (Universidad de Extremadura, Spain)
Engineering Science Reference • © 2021 • 336pp • H/C (ISBN: 9781799873914) • US $195.00

Handbook of Research on Institution Development for Sustainable and Inclusive Economic Growth in Africa
Evans S. Osabuohien (Covenant University, Nigeria) Emmanuel A. Oduntan (Covenant University, Nigeria) Obindah Gershon (Covenant University, Nigeria) Olaronke Onanuga (Covenant University, Nigeria) and Oluyomi Ola-David (Covenant University, Nigeria)
Business Science Reference • © 2021 • 507pp • H/C (ISBN: 9781799848172) • US $235.00

IGI Global
PUBLISHER of TIMELY KNOWLEDGE

701 East Chocolate Avenue, Hershey, PA 17033, USA
Tel: 717-533-8845 x100 • Fax: 717-533-8661
E-Mail: cust@igi-global.com • www.igi-global.com

The book is dedicated to Ms. Anika A. Rathoure.

Editorial Advisory Board

List of Contributors

Ali, Tagorti Mohamed / *Univeristy of Sousse, Tunisia* ... 267
Banerjee, Debolina / *Kidderpore College, India* ... 128
Banerjee, Priya / *Centre for Distance and Online Education, Rabindra Bharati University, India* ... 280
Bedoui, Sana / *Higher Institute of Water Sciences and Techniques of Gabes, University of Gabes, Tunisia* ... 215
Bhat, Sami Ullah / *Department of Environmental Science, University of Kashmir, India* ... 51
Bhattacharjee, Chandani / *H.R. College of Commerce and Economics, India* ... 171
Bhushan, Brij / *Graphic Era University, Dehradun, India* ... 1
Chaudhary, Vidhi / *Daulat Ram College, India* ... 27, 128
Dar, Sajad Ahmad / *Department of Environmental Science, Uttarakhand Technical University, India* ... 51
Dar, Shahid Ahmad / *Department of Environmental Science, University of Kashmir, India* ... 51
Datta, Soumi / *Jamia Hamdard, India* ... 27, 128
Essefi, Elhoucine / *University of Gabes, Tunisia* ... 197, 215, 230, 252, 267
Ganguli, Sayak / *St. Xavier's College (Autonomous), Kolkata, India* ... 71
Ghosh, Mahashweta Mitra / *St. Xavier's College (Autonomous), Kolkata, India* ... 71
Gupta, Rashmi / *Jindal Global Law School, Sonipat, India* ... 92
Hajji, Soumaya / *Faculty of Sciences of Sfax, Tunisia* ... 197
Haldar, Anwesha / *East Calcutta Girls' College, India* ... 280
Hamed, Younes / *University of Gafsa, Tunisia* ... 215
Kar, Somnath / *Holy Cross College, India* ... 27
Karmakar, Rupsha / *Vivekananda College for Women, India* ... 71
Karnena, Manoj Kumar / *Department of Environmental Science, GITAM Institute of Science, GITAM University (Deemed), India* ... 182
Khatik, Nisha / *Maharshi Dayanand Saraswati University, Ajmer, India* ... 17
Konni, Madhavi / *Department of Basic Sciences and Humanities, Dadi Institute of Engineering and Technology, India* ... 182
Kundu Chaudhuri, Rituparna / *Krishnagar Government College, India* ... 128
Madhuri, Pulavarthi / *Department of Basic Sciences and Humanities, Dadi Institute of Engineering and Technology, India* ... 182
Maji, Nibedita / *Chandernagore College, India* ... 106
Mitra, Pallavi / *Centre for Distance and Online Education, Rabindra Bharati University, India* ... 280
Mukhopadhyay, Aniruddha / *University of Calcutta, India* ... 150
Nayak, Arunima / *Graphic Era University, Dehradun, India* ... 1

Roy Goswami, Moumit / *Netaji Nagar College for Women, Kolkata, India* 150
Saritha, Vara / *Department of Environmental Science, GITAM Institute of Science, GITAM University (Deemed), India* 182
Sarkar (Paria), Dipanwita / *Chandernagore College, India* 106
Siddiqui, Nabeela / *CHRIST University (Deemed), India* 92
Singh, Anjana / *Deshbandhu College, India* 27
Singh, Meesha / *St. Xavier's College (Autonomous), Kolkata, India* 71
Sinha, Dwaipayan / *Government General Degree College, Mohanpur, India* 27, 128
Soma Sekhar, K. / *Department of Basic Sciences and Humanities, Dadi Institute of Engineering and Technology, India* 182
Tari, Vinaya Satyawan / *University of Mumbai, India* 92

Table of Contents

Foreword xx

Preface xxi

Acknowledgment xxviii

Introduction xxix

Chapter 1
Wetland Ecosystems and Their Relevance to the Environment: Importance of Wetlands 1
Arunima Nayak, Graphic Era University, Dehradun, India
Brij Bhushan, Graphic Era University, Dehradun, India

Chapter 2
Wetland Creation, Restoration, and Conservation: Wetland Protection 17
Nisha Khatik, Maharshi Dayanand Saraswati University, Ajmer, India

Chapter 3
Water Pollution of Wetlands: A Global Threat to Inland, Wetland, and Aquatic Phytodiversity 27
Soumi Datta, Jamia Hamdard, India
Dwaipayan Sinha, Government General Degree College, Mohanpur, India
Vidhi Chaudhary, Daulat Ram College, India
Somnath Kar, Holy Cross College, India
Anjana Singh, Deshbandhu College, India

Chapter 4
Wetland Ecosystem Monitoring Through Water Quality Perspectives 51
Shahid Ahmad Dar, Department of Environmental Science, University of Kashmir, India
Sami Ullah Bhat, Department of Environmental Science, University of Kashmir, India
Sajad Ahmad Dar, Department of Environmental Science, Uttarakhand Technical University, India

Chapter 5
Metagenomics-Guided Assessment of Water Quality and Predicting Pathogenic Load 71
Sayak Ganguli, St. Xavier's College (Autonomous), Kolkata, India
Rupsha Karmakar, Vivekananda College for Women, India
Meesha Singh, St. Xavier's College (Autonomous), Kolkata, India
Mahashweta Mitra Ghosh, St. Xavier's College (Autonomous), Kolkata, India

Chapter 6
Impact of Climate Change on Upper Ganga Ramsar Site of UP, India: Sustainable Restoration of Wetland Ecosystem 92
Vinaya Satyawan Tari, University of Mumbai, India
Rashmi Gupta, Jindal Global Law School, Sonipat, India
Nabeela Siddiqui, CHRIST University (Deemed), India

Chapter 7
Status and Threats of Wetland Change in Land Use Pattern and Planning: Impact of Land Use Patterns and Urbanization 106
Dipanwita Sarkar (Paria), Chandernagore College, India
Nibedita Maji, Chandernagore College, India

Chapter 8
Ecological Succession of Wetlands: A Review of the Current Scenario 128
Dwaipayan Sinha, Government General Degree College, Mohanpur, India
Soumi Datta, Jamia Hamdard, India
Vidhi Chaudhary, Daulat Ram College, University of Delhi, India
Debolina Banerjee, Kidderpore College, India
Rituparna Kundu Chaudhuri, Krishnagar Government College, India

Chapter 9
An Overview of the Determination of Biodiversity of Wetlands 150
Moumit Roy Goswami, Netaji Nagar College for Women, Kolkata, India
Aniruddha Mukhopadhyay, University of Calcutta, India

Chapter 10
Solid Waste Disposal on Inland Water Wetlands: A Sustainable Restoration of Wetland Ecosystems 171
Chandani Bhattacharjee, H.R. College of Commerce and Economics, India

Chapter 11
Global Climate Change by Wetland Greenhouse Gas Fluxes: Mechanisms, Effects, and Control 182
Madhavi Konni, Department of Basic Sciences and Humanities, Dadi Institute of Engineering and Technology, India
Vara Saritha, Department of Environmental Science, GITAM Institute of Science, GITAM University (Deemed), India
Pulavarthi Madhuri, Department of Basic Sciences and Humanities, Dadi Institute of Engineering and Technology, India
K. Soma Sekhar, Department of Basic Sciences and Humanities, Dadi Institute of Engineering and Technology, India
Manoj Kumar Karnena, Department of Environmental Science, GITAM Institute of Science, GITAM University (Deemed), India

Chapter 12
Variation of the Descriptive Grain Size Distribution Along Cores From the Sebkha of Sidi El Hani, Eastern Tunisia 197
Elhoucine Essefi, University of Gabes, Tunisia
Soumaya Hajji, Faculty of Sciences of Sfax, Tunisia

Chapter 13
Multidisciplinary Study of Cores From the Wetland of El Melah, Southwestern Tunisia: Climatic and Environmental Significance 215
Sana Bedoui, Higher Institute of Water Sciences and Techniques of Gabes, University of Gabes, Tunisia
Elhoucine Essefi, University of Gabes, Tunisia
Younes Hamed, University of Gafsa, Tunisia

Chapter 14
Millennial to Centennial Cyclicity Within the Exoreic Saline System of Boujmel, Southern Tunisia........ 230
Elhoucine Essefi, University of Gabes, Tunisia

Chapter 15
Record of the Sedimentary Dynamics and Climatic Variability Within the Sedimentary Filling of Mchiguig Wetland During the Late Holocene: Central Tunisia – Astronomic, Atmospheric, and Oceanographic Forces........ 252
Elhoucine Essefi, University of Gabes, Tunisia

Chapter 16
The Geobiology of Bacteria Within the Extreme Conditions of Sidi El Hani Wetland, Eastern Tunisia: Environmental and Astrobiological Significance........ 267
Tagorti Mohamed Ali, Univeristy of Sousse, Tunisia
Elhoucine Essefi, University of Gabes, Tunisia

Chapter 17
Public Participation in Restoration and Sustainable Use of Wetland Ecosystem Services in India ... 280
Pallavi Mitra, Centre for Distance and Online Education, Rabindra Bharati University, India
Anwesha Haldar, East Calcutta Girls' College, India
Priya Banerjee, Centre for Distance and Online Education, Rabindra Bharati University, India

Compilation of References ... 304

About the Contributors ... 360

Index ... 368

Detailed Table of Contents

Foreword xx

Preface xxi

Acknowledgment xxviii

Introduction xxix

Chapter 1
Wetland Ecosystems and Their Relevance to the Environment: Importance of Wetlands 1
Arunima Nayak, Graphic Era University, Dehradun, India
Brij Bhushan, Graphic Era University, Dehradun, India

As compared to the tropical evergreen forest, wetlands are also one of the most productive ecosystems in the biosphere and make a major contribution to the ecological sustainability of a region. The wetlands maintain biologically diverse communities having both ecological and economic value. Based on the immense environmental and sustainability benefits, wetlands have been demarcated as essential for the future of human existence. The future challenges pertaining to food, clean water and energy security, well-being of humans, natural disaster risk reduction, and climate change resilience can be met by preserving the wetlands. The chapter has an aim to provide insight on the fundamentals like the classifications, major functions, as well as the various factors affecting the wetland ecosystem. Other important aspects like the major threats leading to the loss of the wetlands, consequences of the loss or degradation of wetlands, and ways to preserve the wetlands are discussed.

Chapter 2
Wetland Creation, Restoration, and Conservation: Wetland Protection 17
Nisha Khatik, Maharshi Dayanand Saraswati University, Ajmer, India

Water-dominated landscapes include wetland areas. The term "wetland" has not been commonly used until quite recently. It is believed to be a euphemistic equivalent of "swamp." Every year on the second day of February, World Wetlands Day marks the adoption of the convention on Wetland by Ramsar, the Iranian city that has a special place in Iranian history. Flood protection, water quality improvement, shoreline erosion control, natural products, recreation, and aesthetics are some of the many advantages of wetlands, as well as the fact that they are vital habitats for a variety of animals and plants. Several studies illustrated the importance of wetlands in reducing carbon emissions and regulating climate on a global scale. In recent years, these advantages of wetlands have been recognized by governments worldwide and have led to legislation, regulations, and management plans creating wetlands for conservation, protection, and restoration. Unfortunately, the destruction of wetlands is a concern since they are among the planet's most productive areas.

Chapter 3
Water Pollution of Wetlands: A Global Threat to Inland, Wetland, and Aquatic Phytodiversity......... 27
Soumi Datta, Jamia Hamdard, India
Dwaipayan Sinha, Government General Degree College, Mohanpur, India
Vidhi Chaudhary, Daulat Ram College, India
Somnath Kar, Holy Cross College, India
Anjana Singh, Deshbandhu College, India

Pollution has become a matter of grave concern at present with all the components of the environment laden with pollutants largely from anthropogenic sources and unplanned urbanization. Inland wetlands are very delicate ecosystems and encompass a variety of water bodies, namely ponds, rivers, swamps, etc. They house some unique floristic patterns that are crucial in the primary productivity and maintaining a balance of the wetland ecosystem. In addition to it, the inland water bodies are also productive and are of immense importance to humans. The inland wetlands are also an integral part of boosting the economy of the region as they support a number of industries including fishing and recreation. Thus pollution of water bodies has impacted the human race in a deleterious manner. This chapter is an attempt to overview the inland water bodies, their biodiversity pattern, pollution, and their effect on flora at large.

Chapter 4
Wetland Ecosystem Monitoring Through Water Quality Perspectives.. 51
Shahid Ahmad Dar, Department of Environmental Science, University of Kashmir, India
Sami Ullah Bhat, Department of Environmental Science, University of Kashmir, India
Sajad Ahmad Dar, Department of Environmental Science, Uttarakhand Technical University, India

Water quality monitoring is an important tool in determining the safety and suitability of water for various desired and intended uses. The procedures involved in the evaluation of water quality are numerous and multifaceted. Therefore, taking into consideration the specific objectives of water quality monitoring, sampling design is of vital importance. Most of the physical parameters of water quality are determined via in-situ measurements using modern testing equipment/field testing kits. Although there are some good field-based sensors that are being used for evaluation of water quality, the chemical parameters traditionally are mostly analyzed through laboratory-based experiments. This chapter is aimed to offer an inclusive knowledge and insights on the importance and assessment of physico-chemical parameters that are of high priority for monitoring the water quality of wetlands.

Chapter 5
Metagenomics-Guided Assessment of Water Quality and Predicting Pathogenic Load 71
Sayak Ganguli, St. Xavier's College (Autonomous), Kolkata, India
Rupsha Karmakar, Vivekananda College for Women, India
Meesha Singh, St. Xavier's College (Autonomous), Kolkata, India
Mahashweta Mitra Ghosh, St. Xavier's College (Autonomous), Kolkata, India

Antibiotic-resistant bacteria (ARB) are becoming more prevalent in the environment and are efficiently disseminating through contaminated wastewater resulting in resistome cycling. This chapter compares the bacterial profile of hospital effluents collected from rural, urban, and delta regions of West Bengal,

India. Comparative metagenomics analysis identified pathogenic bacterial genera like pseudomonas, escherichia, staphylococcus, lactobacillus, prevotella, acinetobacter across the samples. Delta sample showed highest abundance of pseudomonas whereas rural sample had lower titre of all the common bacterial genera. Urban sample reflected more diversity of different genera in terms of abundance. Pathogenic load prediction revealed significant occurrence of diarrhea, irritable bowel syndrome, liver cirrhosis, ulcerative colitis in the disease network. This chapter proposes a monitoring programme for assessing wastewater health using a combination of culture independent and culture-dependent molecular techniques in order to prevent the spread of pollutants in tropical environments.

Chapter 6
Impact of Climate Change on Upper Ganga Ramsar Site of UP, India: Sustainable Restoration of Wetland Ecosystem 92
Vinaya Satyawan Tari, University of Mumbai, India
Rashmi Gupta, Jindal Global Law School, Sonipat, India
Nabeela Siddiqui, CHRIST University (Deemed), India

According to the IPCC Second Assessment Report, climate change will lead to an alteration of the hydrological cycle and could have major impacts on regional water resources. India features a diverse range of wetlands, including high-altitude alpine lakes, littoral swamps in the form of mangroves and corals, and inland wetlands of various sorts. The Upper Ganga Ramsar Site is Uttar Pradesh's only Ramsar Site geographical distribution and may fluctuate as a result of climate change. Wetland reactions to climate change are frequently left out of global climate change models. The climate change adaptations must be incorporated into the economic development, planning, and implementation process.

Chapter 7
Status and Threats of Wetland Change in Land Use Pattern and Planning: Impact of Land Use Patterns and Urbanization 106
Dipanwita Sarkar (Paria), Chandernagore College, India
Nibedita Maji, Chandernagore College, India

Wetland-related studies documented the loss of native species diversity and promotion of the biotic homogenization due to wetland loss. Excessive withdrawals of water from wetlands for residential, agricultural, or industrial use are responsible for wetland degradation. Constructions of dams impedes water flow and replenishment of wetlands, and it also creates a hazard to aquatic living organisms. Climate change causing some wetlands to disappear under rising sea levels, while others are severely impacted by changing climatic conditions, including drought. So necessary steps such as increase wetlands and prevention of the illegal swamping of wetlands, etc. should be taken for conserving the wetland biodiversity from the threatening of unplanned urbanization, purifying the environment and mainlining the sustainable development. Though the Ramsar Convention policy exists to persevere wetlands and achieve sustainable development throughout the world, mass consciousness, greater participation of local people, use of indigenous knowledge in the management strategies are needed to protect wetlands.

Chapter 8
Ecological Succession of Wetlands: A Review of the Current Scenario... 128
Dwaipayan Sinha, Government General Degree College, Mohanpur, India
Soumi Datta, Jamia Hamdard, India
Vidhi Chaudhary, Daulat Ram College, University of Delhi, India
Debolina Banerjee, Kidderpore College, India
Rituparna Kundu Chaudhuri, Krishnagar Government College, India

The change in community structure induced by environmental perturbances over time is called ecological succession and results in modification of community structure in a particular ecosystem. However, with climatic changes and pollution, wetlands are most likely to have modified their successional trails, especially as these ecosystems are exposed to various fluctuating water rhythms enhanced carbon dioxide, salinity invasion, and climatic temperatures. The chapter is an attempt to review the present condition of wetlands with respect to pollution and its impact on successional patterns.

Chapter 9
An Overview of the Determination of Biodiversity of Wetlands 150
Moumit Roy Goswami, Netaji Nagar College for Women, Kolkata, India
Aniruddha Mukhopadhyay, University of Calcutta, India

Wetland ecosystems support rich and unique biodiversity. Biodiversity of a given ecosystem in general and wetlands in particular provide important insights to the ecological health of an area. The Ramsar Convention 1971 identified nine criteria for identifying wetlands of international importance. Out of the nine criteria, eight are linked to biodiversity of which three are based on sites of international importance for conserving biological diversity, two are specific for water birds, two are specific for fish, and one criterion for other taxa. Hence, determination of biodiversity of wetlands is of utmost importance. In order to understand that birds, fishes, amphibians, odonates, mammals, and aquatic plants were particularly selected as indicators of wetland biodiversity, the chapter discusses the different methodologies about determination of each of these taxa under different criteria as mentioned above. These methodologies will help various stakeholders in appropriate determination of biodiversity of wetlands of a particular area.

Chapter 10
Solid Waste Disposal on Inland Water Wetlands: A Sustainable Restoration of Wetland Ecosystems.. 171
Chandani Bhattacharjee, H.R. College of Commerce and Economics, India

Generation of solid waste precedes the surge of urbanization. The earliest waste dumping is recorded in Greece as early as 500 BCE, the conservancy workers in France, wastewater treatment in London, and aqueduct systems in oriental civilizations. The magnitude of waste has been compounding annually with the rise of global population, urbanization, and economic growth. Waste has been overtly and irresponsibly dumped in inland water bodies and the wetlands around it causing inherent damage to the fluvial, pond, or riverine ecosystems. The United Nations has declared this decade to be for ecosystem restoration, and hence, this chapter intends to ponder and establish the concerns of health, species modification, ecosystem endangering, pollution of the surface and subsurface water, impact on the vegetation along the water stretches, to name a few. The objective of this chapter is to evaluate the impact on the ecospheres while arriving at sustainable restoration options.

Chapter 11
Global Climate Change by Wetland Greenhouse Gas Fluxes: Mechanisms, Effects, and Control 182
Madhavi Konni, Department of Basic Sciences and Humanities, Dadi Institute of Engineering and Technology, India
Vara Saritha, Department of Environmental Science, GITAM Institute of Science, GITAM University (Deemed), India
Pulavarthi Madhuri, Department of Basic Sciences and Humanities, Dadi Institute of Engineering and Technology, India
K. Soma Sekhar, Department of Basic Sciences and Humanities, Dadi Institute of Engineering and Technology, India
Manoj Kumar Karnena, Department of Environmental Science, GITAM Institute of Science, GITAM University (Deemed), India

Wetlands (WLs) in the landscapes are important for the GHGs production, ingesting, and exchange with the atmosphere. In this chapter, the authors illustrated how the WLs influence climate change, even though it is typical for determining the climatic role of WLs in the broader perspective. The conclusions might be wary based on the radiative balance as the radiative forcing since the 1750s or climatic roles are continuously changing in the wetlands. Degradation of WLs leads to reducing their functioning, and GHG fluxes might change and alter the climatic roles of the WLs. The chapter demonstrated that WL disturbances might cause global warming for a longer duration even though the WLs are restored or managed by replacing them with the mitigation WLs. Thus, activities that cause disturbance in the WLs leading to carbon oxidation in the soils should be avoided. Regulating the climate is an ecosystem service in the WLs; during the planning of the WLs, protection, restoration, and creation, environmental management should be considered.

Chapter 12
Variation of the Descriptive Grain Size Distribution Along Cores From the Sebkha of Sidi El Hani, Eastern Tunisia 197
Elhoucine Essefi, University of Gabes, Tunisia
Soumaya Hajji, Faculty of Sciences of Sfax, Tunisia

In this chapter, cores were the object of descriptive classifications of the grain size distribution, which were meant to describe the grain size continuous variability within cores and to correlate between them. The statistical treatment of the crude data was done on the basis of two different methods (the method of moments statistics and the method of inclusive graphic statistics) to compute statistical parameters of the grain size distribution such as mean and median. The correlations between cores were done on the basis of sand/silt/clay percentages. Even though it has given special care to test different methods of studying the grain size distribution, this study has not deviated from its primary purpose of investigating the filling of the playa; correlations between different cores were meant to infer their sedimentary dynamics.

Chapter 13
Multidisciplinary Study of Cores From the Wetland of El Melah, Southwestern Tunisia: Climatic and Environmental Significance 215
Sana Bedoui, Higher Institute of Water Sciences and Techniques of Gabes, University of Gabes, Tunisia
Elhoucine Essefi, University of Gabes, Tunisia
Younes Hamed, University of Gafsa, Tunisia

Saline systems are candidates to be threatened by climatic change. In terms of methodology and materials, color identification, geochemistry, and mineralogy analysis were used. The spectral analysis of data of the amount of active K and Na firstly shows the individualization of a cycle stretching along with 1000 yr. Then, a less pronounced cycle of 2500 years is marked with a lower intensity. Also, on the spectral analysis of K data, the drowned cycles of 650 and 500 years are marked by weak intensities not reaching the threshold level. These two comparable cycles may be artifacts due to analysis errors or variability in the rate of sedimentation resulting in the bifurcation of two hybrid cycles from one real cycle. This work confirmed the cycles found by color studies through real analyses such as geochemical and magnetic measurements. As a matter of fact, the majority of cycles found out by the spectral analysis of colors data are confirmed through analyses.

Chapter 14
Millennial to Centennial Cyclicity Within the Exoreic Saline System of Boujmel, Southern Tunisia........ 230
Elhoucine Essefi, University of Gabes, Tunisia

Previous works proposed different age models of sedimentation in Sebkha Lagoon of Boujmel leading to the setting of controversial interpretations of eustatic and climatic phases. The aim of this work is carrying out a geological correlation and an astrochronological calibration based on the Holocene cyclostratigraphy leading to the setting of an age model satisfying dates of climatic and eustatic phases identified in southern Tunisia, including the Anthropocene and the Great Acceleration. Along a 130 cm core, four major climatic phases were upward recorded.

Chapter 15
Record of the Sedimentary Dynamics and Climatic Variability Within the Sedimentary Filling of Mchiguig Wetland During the Late Holocene: Central Tunisia – Astronomic, Atmospheric, and Oceanographic Forces........ 252
Elhoucine Essefi, University of Gabes, Tunisia

This work aimed to study the cyclicity of the geochemical chemical parameters and the carbonate percentages along a 59 cm core from the sebkha of Mchiguig, Central Tunisia. In fact, from the bottom upwards, six climatic phases were recorded including the Warming Present (Great Acceleration), the Late Little Ice Age (Anthropocene), the Early Little Ice Age, the Medieval Climatic Anomaly, the Dark Age, and the Roman Warm Period. In fact, the spectral analysis of the studied parameters visualized many cycles. Those cycles are related to sun activity, oceanographic, and atmospheric factors. Solar activity generated 500 yr cycles; however, the oceanographic circulation generated other cycles of 1500 yr and 700-800 yr. The 1500 yr cycle may be the result of the solar activity and NAO-like circulation.

Chapter 16
The Geobiology of Bacteria Within the Extreme Conditions of Sidi El Hani Wetland, Eastern Tunisia: Environmental and Astrobiological Significance.. 267
Tagorti Mohamed Ali, Univeristy of Sousse, Tunisia
Elhoucine Essefi, University of Gabes, Tunisia

Extreme conditions in wetlands may be the niche of development of colonies of bacteria. In this chapter, the authors study the gnotobiology of moderate bacteria within the saline wetland of Sidi El Hani. In terms of geology, the coring within the sedimentary filling of the wetland shows color variation from white, grey, black to red related to the variability of bacteria species. On the other hand, in terms of microbiological investigations, isolates of bacteria show a variability from Gram+ to Gram-, from oxidase+ to oxidase-, from catalase+ to catalase-. This geobiological variability is related to radical change in climatic conditions. In doing so, the wetland of Sidi El Hani may record the climatic variability during the Late Holocene. On the other hand, it may be considered as terrestrial analogue with a development of extremophiles.

Chapter 17
Public Participation in Restoration and Sustainable Use of Wetland Ecosystem Services in India ... 280
Pallavi Mitra, Centre for Distance and Online Education, Rabindra Bharati University, India
Anwesha Haldar, East Calcutta Girls' College, India
Priya Banerjee, Centre for Distance and Online Education, Rabindra Bharati University, India

Ecosystem services include conditions and processes that make up natural ecosystems and the species present therein. According to recent studies, wetland ecosystems provide the maximum service value per area by playing a significant role in regulating and purifying water supplies, controlling flood, acting as carbon-sinks, and sustaining human lives and livelihoods. Challenges like wetland loss and degradation, declining freshwater resources, and probable consequences of climate change have attracted significant scientific and public attention towards wetland conservation and restoration. Despite diligent conservation efforts, the global status of wetland security is still alarming. Long-term sustainable management and use of wetlands necessitate active public participation from all sectors. This study reviews the current status of different wetlands in India. It also provides a detailed discussion of different aspects of economic evaluation of ecosystem services, wetland restoration, and public participation for improving wetland policies and governance.

Compilation of References 304

About the Contributors 360

Index 368

Foreword

Wetlands are vital for human survival. They are among the world's most productive environments; cradles of biological diversity that provide the water and productivity upon which countless species of plants and animals depend for survival. The Convention on Wetlands, signed in Ramsar, Iran, in 1971, is an intergovernmental treaty which provides the framework for national action and international cooperation for the conservation and wise use of wetlands and their resources. Wetlands provide habitat for thousands of species of aquatic and terrestrial plants and animals. Wetlands are valuable for flood protection, water quality improvement, shoreline erosion control, natural products, recreation, and aesthetics. Wetlands provide many societal benefits: food and habitat for fish and wildlife, including threatened and endangered species; water quality improvement; flood storage; shoreline erosion control; economically beneficial natural products for human use; and opportunities for recreation, education, and research. Wetland also help to reduce the greenhouse gases.

Wetlands are vital sponges in the city. They prevent flood and recharge groundwater. Along with being an aesthetic entity, they provide social, economic, and environmental beneficiaries. They improve the quality of water and also sustain the surrounding diverse flora and fauna (aquatic and wildlife habitat). Being rich in nutrients, they provide a diverse and productive ecosystem for the natural environment to survive. Originally emphasizing the conservation and wise use of wetlands primarily to provide a habitat for waterbirds, the Convention has subsequently broadened its scope to address all aspects of wetland conservation and wise use. This shift in focus reflects the increasing recognition of the importance of wetlands as ecosystems that contribute to both biodiversity conservation and human well-being. Wetlands cover an estimated nine percent of the Earth's land surface and contribute significantly to the global economy in terms of water supply, fisheries, agriculture, forestry, and tourism.

The current book is highlighting the challenges of wetland conservation and current scenario of existing wetland with importance of inland wetland. Inventory, assessment, and monitoring are vital components of effective wetland management. The conservation of inland wetland is critical and most importantly to retain the wetland and protect them in situ. Other focus are are - promote the inventory, assessment and monitoring of wetlands, through discussion of practical approaches, methodologies and techniques.

Here, in the current book, Dr. Ashok K. Rathoure has done a great job for the same in the area of wetland conservation and its monitoring. This will be a milestone in the current era where the wetland is degrading, and its conservation is a major global problem.

Vishnu D. Rajput
Southern Federal University, Russia

Preface

The Convention on Wetlands of International Importance especially as Waterfowl Habitat (the Ramsar Convention) was signed in Ramsar, Iran, on 2 February 1971 and came into force on 21 December 1975. The Convention provides a framework for national action and international cooperation for the conservation and wise use of wetlands and their resources.

The Ramsar Convention is the only environmental treaty dealing with a particular ecosystem, and as of December 2008 had 158 parties. Also as of December 2008, a total of 1822 wetland sites, covering 168 million hectares, were included in the Ramsar List of Wetlands of International Importance. Parties to the Convention commit themselves to: designate at least one site that meets the Ramsar Criteria for inclusion in the Ramsar List and ensure maintenance of the ecological character of each Ramsar site; include wetland conservation within national land-use planning in order to promote the wise use of all wetlands within their territory; establish nature reserves on wetlands and promote training in wetland research and management; and consult with other parties about Convention implementation, especially with regard to transboundary wetlands, shared water systems, shared species and development projects affecting wetlands.

Contracting parties meet every three years to assess progress in implementing the Convention and wetland conservation, share knowledge and experience on technical issues, and plan for the next triennium. In addition to the Conference of the Parties (COP), the Convention's work is supported by a Standing Committee, a Scientific and Technical Review Panel, and the Ramsar Bureau, which carries out the functions of a Secretariat. There have been ten meetings of the COP since the Convention's entry into force: COP1 in Cagliari, Italy (24-29 November 1980); COP2 in Gröningen, the Netherlands (7-12 May 1984); COP3 in Regina, Canada (27 May-5 June 1987); COP4 in Montreux, Switzerland (27 June-4 July 1990); COP5 in Kushiro, Japan (9-16 June 1993); COP6 in Brisbane, Australia (19-27 March 1996); COP7 in San José, Costa Rica (10-18 May 1999); COP8 in Valencia, Spain (18-26 November 2002); COP9 in Kampala, Uganda (8-15 November 2005); and COP10 in Changwon, Republic of Korea (28 October-4 November 2008).

COP7: At COP7, delegates focused on the interrelations between human societies and wetland habitats. They considered the Convention's implementation in each region and adopted 30 resolutions and four recommendations on policy, programme and budgetary issues. In addition, a Memorandum of Understanding was signed between the Ramsar Bureau and the UNESCO World Heritage Center.

COP8: With over 1000 participants, COP8 was the largest COP in the Convention's history. Focusing on Water, wetlands, life and culture, delegates adopted more than 40 resolutions addressing policy, technical, programme and budgetary matters, including wetlands and agriculture, climate change, cultural issues, mangroves, water allocation and management, and the Report of the World Commission on Dams. They also approved the Convention's budget and Work Plan for 2003-2005, and its Strategic Plan for 2003-2008.

COP9: The theme of COP9, the first COP held in Africa, was Wetlands and water: supporting life, sustaining livelihoods. Delegates adopted 25 resolutions on a wide range of policy, programme and budgetary matters, including: additional scientific and technical guidance for the implementation of the Ramsar wise use concept; engagement of the Convention in ongoing multilateral processes dealing with water; the Convention's role in disaster prevention, mitigation and adaptation; wetlands and poverty reduction; cultural values of wetlands; and the emergence of avian flu. The COP also reviewed its Strategic Plan 2003-2008, and held two technical sessions on the wise use of wetlands in integrated water management and culture and knowledge in wetland management. Participating ministers adopted the Kampala Declaration, which emphasizes the role of the Convention in arresting continuing loss and degradation of wetland ecosystems.

COP10: COP10 was organized around the theme Healthy Wetlands, Healthy People, and convened from 28 October 2008 - 4 November 2008, in Changwon, Republic of Korea. COP10 adopted 33 resolutions, including on: wetlands and climate change; wetlands and biofuels; wetlands and extractive industries; wetlands and poverty eradication; wetlands and human health and well-being; enhancing biodiversity in rice paddies as wetland systems; and promoting international cooperation on the conservation of waterbird flyways. The COP also adopted the Convention's budget for 2009-2013 and Strategic Plan 2009-2014. The main issue of institutional reform – the legal status of the Convention's Secretariat – was deferred to an intersessional working group.

The target audience of this book is professionals and researchers working in the field of ecological management and wetland conservation in various disciplines, *e.g.* life sciences, environmental sciences, ecology & biodiversity, aquatic environment pollution, thermal pollution, environmental pollution, industries, etc. Moreover, the book provides insights and support executives concerned with the management of wetland, status, monitoring information, trophic status index, productivity, NPV, ecological health and strategy development in different types of wetland conservation practices. It covers the most cost-effective techniques in conservation of wetland technologies and the most cutting-edge research on monitoring of wetland health and its applications.

ORGANIZATION OF THE BOOK

The book is organized into 17 chapters. A brief description of each chapter follows:

Chapter 1 states about wetland ecosystem and relevance to environment. As compared to the tropical evergreen forest, wetlands are also one of the most productive ecosystems in the biosphere and have a major contribution to the ecological sustainability of a region. he wetlands maintain biologically diverse communities having both ecological and economic value. Based on the immense environmental and sustainability benefits, wetlands have been demarcated as essential for the future of human existence. The future challenges pertaining to food, clean water and energy security, well-being of humans, natural disaster risk reduction and climate change resilience can be met by preserving the wetlands. The chapter has an aim to provide insight on the fundamentals like the classifications, major functions as well as the various factors affecting the wetland ecosystem. Other important aspects like the major threats leading to the loss of the wetlands, consequences of the loss or degradation of wetlands and ways to preserve the wetlands are discussed.

Chapter 2 is about wetland creation, restoration and conservation. Water-dominated landscapes include wetland areas. The term wetland has not been commonly used until quite recently. It is believed to be a euphemistic equivalent of swamp. Every year on the second day of February, World Wetlands Day marks the adoption of the convention on Wetland by Ramsar, the Iranian city that has a special place in Iranian history. Flood protection, water quality improvement, shoreline erosion control, natural products, recreation and aesthetics are some of the many advantages of wetlands, as well as the fact that they are vital habitats for a variety of animals and plants. Several studies illustrated the importance of wetlands in reducing carbon emissions and regulating climate on a global scale. In recent years, these advantages of wetlands have been recognized by governments worldwide and have led to legislation, regulations and management plans creating wetlands for conservation, protection and restoration. Unfortunately, the destruction of wetlands is a concern since they are among the planet's most productive areas.

Chapter 3 describes the water pollution of wetland. Pollution has become a matter of grave concern at present with all the components of the environment have become laden with pollutants largely from anthropogenic sources and unplanned urbanization. Inland wetlands are very delicate ecosystems and encompass varieties of water bodies namely ponds, rivers, swamps, etc. They house some unique floristic patterns that are crucial in the primary productivity and maintaining a balance of the wetland ecosystem. In addition to it, the inland water bodies are also productive and are of immense importance to humans. The inland wetlands are also an integral part of boosting the economy of the region as they support a number of industries including fishing and recreation. Thus, pollution of water bodies has impacted the human race in a deleterious manner. This chapter is an attempt to overview the inland water bodies, their biodiversity pattern, pollution, and their effect on flora at large.

Chapter 4 deliberates about wetland ecosystem monitoring through water quality perspectives: water quality monitoring. Water quality monitoring is an important tool in determining the safety and suitability of water for various desired and intended uses. The procedures involved in the evaluation of water quality are numerous and multifaceted. Therefore, taking into consideration the specific objectives of water quality monitoring, sampling design is of vital importance. Most of the physical parameters of water quality are determined via in-situ measurements using modern testing equipment's/field testing kits. Although, there are some good field based sensors that are being used for evaluation of water quality, the chemical parameters traditionally are mostly analysed through laboratory-based experiments. This chapter is aimed to offer an inclusive knowledge and insights on the importance and assessment of physico-chemical parameters that are of high priority for monitoring the water quality of wetlands.

Chapter 5 provides metagenomics guided assessment of water quality and predicting pathogenic load: wastewater analysis for water quality monitoring using metagenomics. Antibiotic Resistant Bacteria (ARB) are becoming more prevalent in the environment and are efficiently disseminating through contaminated wastewater resulting in resistome cycling. This chapter compares the bacterial profile of hospital effluents collected from rural, urban and delta regions of West Bengal, India. Comparative metagenomics analysis identified pathogenic bacterial genera like Pseudomonas, Escherichia, Staphylococcus, Lactobacillus, Prevotella, Acinetobacter across the samples. Delta sample showed highest abundance of Pseudomonas whereas Rural sample had lower titer of all the common bacterial genera. Urban sample reflected more diversity of different genera in terms of abundances. Pathogenic load prediction revealed significant occurrence of Diarrhea, Irritable Bowel Syndrome, Liver Cirrhosis, Ulcerative Colitis in the disease network. This chapter proposes a monitoring programme for assessing wastewater health using a combination of culture independent and culture dependent molecular techniques in order to prevent the spread of pollutants in tropical environments.

Chapter 6 discusses impact of climate change on Upper Ganga Ramsar Site, UP, India. Wetlands are defined as Areas of the marsh, fen, bog, or water, whether natural or artificial, permanent or temporary, with water that is static or flowing, fresh, brackish, or salt, including areas of marine water the depth of which at low tide does not exceed six meters according to the Ramsar Convention (Ramsar Convention, 1971, Article 1.1). According to the IPCC Second Assessment Report, climate changes will lead to an alteration of the hydrological cycle and could have major impacts on regional water resources. India features a diverse range of wetlands, including high-altitude alpine lakes, littoral swamps in the form of mangroves and corals, and inland wetlands of various sorts. The Upper Ganga Ramsar Site is Uttar Pradesh's only Ramsar Site geographical distribution may fluctuate as a result of climate change. Wetland reactions to climate change are frequently left out of global climate change models. The climate change adaptations are must be incorporated into the economic development, planning, and implementation process.

Chapter 7 highlights the status and threats of wetlands change in land use pattern and planning. Wetland related studies documented the loss of native species diversity and promotion of the biotic homogenization due to wetland loss. Excessive withdrawals of water from wetlands for residential, agricultural or industrial use responsible for wetland degradation. Constructions of dams impedes water flow and replenishment of wetlands and it also creates a hazard to aquatic living organisms. Climate change causing some wetlands to disappear under rising sea levels, while others are severely impacted by changing climatic conditions, including draught. So necessary steps such as increase wetlands and prevent the illegal swamping of wetlands etc. should be taken for conserving the wetland biodiversity from the threatening of unplanned urbanization, purifying the environment and mainlining the sustainable development. Though the Ramsar Convention policy to persevere wetlands and achieve sustainable development throughout the world, yet mass consciousness, greater participation of local people, use of indigenous knowledge in the management strategies are needed to protect wetlands.

Chapter 8 provides the review of current scenario of ecological succession of wetlands. The change in community structure induced by environmental perturbances over time is called ecological succession and results in modification of community structure in a particular ecosystem. However, with climatic changes and pollution, wetlands are most likely to have modified their successional trails, especially as these ecosystems are exposed to various fluctuating water rhythms enhanced carbon dioxide, salinity invasion, and climatic temperatures. The Chapter is an attempt to review the present condition of wetlands with respect to pollution and its impact on successional patterns.

Chapter 9 delivers an overview of determination of biodiversity of wetland. Wetland ecosystem supports rich and unique biodiversity. Biodiversity of a given ecosystem in general and wetlands in particular provide important insights to ecological health of an area. The Ramsar Convention 1971 identified nine criteria's for identifying wetlands of international importance. Out of the nine criteria, eight are linked to biodiversity of which three are based on sites of international importance for conserving biological diversity, two are specific for water birds, two specifics for fish and one criterion for other taxa. Hence determination of biodiversity of wetlands is of utmost importance. In order to understand that birds, fishes, amphibians, odonates, mammals and aquatic plants were particularly selected as indicators of wetland biodiversity. The chapter discusses the different methodologies about determination of each of these taxa under different criteria as mentioned above inhabiting in different types of wetlands. These methodologies will help various stakeholders in appropriate determination of biodiversity of wetland of a particular area.

Chapter 10 highlights the impact of solid waste disposal on inland water wetland. Generation of solid waste precedes the surge of urbanization. The earliest waste dumping is recorded in Greece as early as 500 BCE, the conservancy workers in France, waste water treatment in London and aqueduct systems in oriental civilizations. The magnitude of waste has been compounding, 11.2 billion, annually (UNEP) with the rise of global population, urbanization and economic growth. Waste has been overtly and irresponsibly dumped in inland water bodies and the wetlands around it causing inherent damage to the fluvial, pond or riverine ecosystems. The United Nations has declared this decade to be for ecosystem restoration and hence this paper intends to ponder and establish the concerns of health, specie modification, ecosystem endangering, pollution of the surface and subsurface water, impact on the vegetation along the water stretches, to name a few. The objective of this paper is to evaluate the impact on the ecospheres while arriving at sustainable restoration options.

In Chapter 11, the authors have illustrated how the WLs influencing climate change, even though it is typical for determining the climatic role of WLs in the broader perspective. The conclusions might be wary based on the radiative balance as the radiative forcing since the 1750s or climatic roles are continuously changing in the wetlands. Degradation of WLs leads to reducing their functioning, and GHGs fluxes might change and alter the climatic roles of the WLs. The current chapter demonstrated that WLs disturbances might cause global warming for a longer duration even though the WLs are restored or managed by replacing them with the mitigation WLs. Thus, activities that cause disturbance in the WLs lead to carbon oxidation in the soils should be avoided. Regulating the climate is an ecosystem service in the WLs; during the planning of the WLs, protection, restoration, and creation, environmental management should be considered.

In Chapter 12, cores were the object of descriptive classifications of the grain size distribution, which were meant to describe the grain size continuous variability within cores and to correlate between them. The statistical treatment of the crude data was done on the basis of two different methods (the method of Moments Statistics and the method of Inclusive Graphic Statistics) to compute statistical parameters of the grain size distribution such as mean and median. The correlations between cores were done on the basis of sand/silt/clay percentages. Even though it has given special care to test different methods of studying the grain size distribution, this study was not deviated from its primary purpose of investigating the filling of the playa; correlations between different cores were meant to infer their sedimentary dynamics.

In Chapter 13, the author presented climatic and environmental significance of wetland of El Melah, Southwestern Tunisia. Saline systems are more candidate to be threatened by climatic change. In terms of methodology and materials, color identification, geochemistry, and mineralogy analysis were used. The spectral analysis of data of the amount of active K and Na firstly shows the individualization of a cycle stretching along with 1000 yr. Then, a less pronounced cycle of 2500 years is marked with a lower intensity. Also, on the spectral analysis of K data, the drowned cycles of 650 and 500 years are marked by weak intensities not reaching the threshold level. These two comparable cycles may be artifacts due to analysis errors or variability in the rate of sedimentation resulting in the bifurcation of two hybrid cycles from one real cycle. This work confirmed the cycles found by color studies through real analyses such as geochemical and magnetic measurements. As a matter of fact, the majority of cycles found out by the spectral analysis of colors data are confirmed through analyses.

Chapter 14 highlights different age models of sedimentation in sebkha-lagoon of Boujmel leading to the setting of controversial interpretations of eustatic and climatic phases. The aim of this work is carrying out a geological correlation and an astrochronological calibration based on the Holocene cyclostratigraphy leading to the setting of an age model satisfying dates of climatic and eustatic phases identified in southern Tunisia, including the Anthropocene and the Great Acceleration. Along a 130 cm core, four major climatic phases were upward recorded.

Chapter 15 presents record of the sedimentary dynamics and climatic variability within the sedimentary filling of Mchiguig Wetland during the Late Holocene: Central Tunisia. This work aimed to study the cyclicity of the geochemical chemical parameters, and the carbonate percentages along a 59 cm core from the Sebkha of Mchiguig, Central Tunisia. In fact, from the bottom upwards, six climatic phases were recorded including the Warming Present (Great Acceleration), the Late Little Ice Age (Anthropocene), the Early Little Ice Age, the Medieval Climatic Anomaly, the Dark Age and the Roman Warm Period. In fact, the spectral analysis of the studied parameters visualized many cycles. Those cycles are related to sun activity, oceanographic and atmospheric factors. Solar activity generated 500 yr cycles, however, the oceanographic circulation generated other cycles of 1500 yr and 700-800 yr. The 1500 yr cycle may be the result of the solar activity and NAO-like circulation.

Chapter 16 presents the gnotobiology of moderate bacteria within the saline wetland of Sidi El Hani. In terms of geology, the coring within the sedimentary filling of the wetland shows color variation from white, grey, black to red related to the variability of bacteria specie. On the other hand, in terms of microbiological investigations, isolates of bacteria show a variability from Gram+ to Gram-, from oxidase+ to oxidase-, from catalase+ to catalase-. This geobiological variability is related to radical change in climatic conditions. In doing so, the wetland of Sidi El Hani may record the climatic variability during the Late Holocene. On the other hand, it may be considered as terrestrial analogue with a development of extremophiles.

In Chapter 17, the authors have emphasized on the public participation in restoration and sustainable use of wetland ecosystem services. Ecosystem services include conditions and processes that make up natural ecosystems and the species present therein. According to recent studies, wetland ecosystems provide the maximum service value per area by playing a significant role in regulating and purifying water supplies, controlling flood, acting as carbon-sinks and sustaining human lives and livelihoods. Challenges like wetland loss and degradation, declining freshwater resources, and probable consequences of climate change has attracted significant scientific and public attention towards wetland conservation and restoration. Despite diligent conservation efforts, the global status of wetland security is still alarming. Long-term sustainable management and use of wetlands necessitate active public participation from all sectors. This study reviews the current status of different wetlands in India. It also provides a detailed discussion of different aspects of economic evaluation of ecosystem services, wetland restoration and public participation for improving wetland policies and governance.

CONCLUSION

Wetlands are highly productive ecosystems that provide the world with nearly two-thirds of its fish harvest. Wetlands play an integral role in the ecology of the watershed. The combination of shallow water, high levels of nutrients is ideal for the development of organisms that form the base of the food web and feed many species of fish, amphibians, shellfish and insects. Wetlands have several benefits for the environment. With the frequent addition of new names to the list of wetlands, the concern for wetlands is increasing simultaneously all over the world. Climate activists and climate organizations raise their voices from time immemorial to protect wetlands. Wetlands prove to be a playing a very important part in maintaining the ecological balance in the environment. They harbor several plant and animal species and help in their prevention from extinction. They are also useful for human beings as they prevent floods and serve as a great source for recycling organic waste. Given the importance of wetlands, it has become important for us to look into various issues involved in wetland conservation and suggest measures to protect them. The conservation of wetlands needs management of environmental needs with a holistic approach. There should be people's participation and developmental activities should consider the immediate stress on the biota and the local community residing there. Legal instruments will not merely help in serving the purpose rather we need to involve all the stakeholders to conserve such ecosystems. The management programs should include sustainable measures so that those who are dependent on it do not suffer and biodiversity is protected. There is a need to adopt scientific, technical, and socio-economic cooperation to conserve the wetlands. Integrated management with a collaborative approach can pave the way for attaining the goal.

Ashok K. Rathoure
Akone Services, India

Acknowledgment

This book is the result of the dedicated effort of numerous individuals. We would like to acknowledge the help of all the people involved in this project and, more specifically, to the authors and reviewers that took part in the review process. Without their support, this book would not have become a reality.

First, we would like to thank each one of the authors for their contributions. Our sincere gratitude goes to the chapter's authors who contributed their time and expertise to this book.

Second, we wish to acknowledge the valuable contributions of the reviewers regarding the improvement of quality, coherence, and content presentation of chapters. Most of the authors also served as referees; we highly appreciate their double task.

Additionally, we would like to thank our Editorial Advisory Board.

Our Production Coordinator skillfully managed the production of the textbook. We wish to thank the publisher for the sustained interest shown by him during the entire work. A sincere expression of the thanks goes to the entire staff of IGI Global.

Generous participation of faculties and students to eliminate errors in the text and to eliminate errors in the text and to refine the presentation will be greatly acknowledged.

Last and not least, we beg forgiveness of all those who have been with us over the course of the years and whose names we have failed to mention.

Ashok K. Rathoure
Akone Services, India

Introduction

WETLAND CONSERVATION AND MANAGEMENT

India has nearly 4.6% of its land as wetlands, covering an area of 15.26 million hectares and has 42 sites designated as Wetlands of International Importance (Ramsar Sites), with a surface area of 1.08 million hectares. The year 2021 also commemorates the 50th anniversary of the signing of the Ramsar Convention on Wetlands on 2nd February 1971 in Ramsar, Iran, celebrated annually as World Wetlands Day.

The Ministry of Environment, Forest, and Climate Change (MoEF&CC), India has notified Wetlands (Conservation and Management) Rules, 2017 (hereinafter Wetlands Rules) under the provisions of the Environment (Protection) Act, 1986 as regulatory framework for conservation and management of wetlands in India. These guidelines have been drafted to support the State Governments / Union Territory (UT) Administrations in the implementation of the Rules by providing guidance on:

1. Preparing a list of wetlands in the State / UT
2. Identifying wetlands for notification under Wetlands (Conservation and Management) Rules, 2017
3. Preparation of Brief Document
4. Determining 'wise use' and ecological character
5. Developing a list of activities to be regulated and permitted
6. Developing an Integrated Management Plan
7. Constitution and operational matters of the Wetlands Authorities
8. Overlapping provisions.

The new rules, notified by the environment ministry, decentralise wetlands management by giving states powers to not only identify and notify wetlands within their jurisdictions but also keep a watch on prohibited activities. Seeking to protect over 2 Lakhs wetlands across the country, the Centre has come out with rules to identify and manage these ecologically fragile areas which play an important role in flood control, groundwater recharge, preserving plant varieties, supporting migratory birds and protecting coastlines. The new rules, notified by the environment ministry, decentralise wetlands management by giving states powers to not only identify and notify wetlands within their jurisdictions but also keep a watch on prohibited activities. It also indirectly widens the ambit of permitted activities by inserting the 'wise use' principle, giving powers to state-level wetland authorities to decide what can be allowed in larger interest. As per this notification; the wetlands shall be conserved and managed in accordance with the principle of wise use as determined by the Wetlands Authority. The Centre's role under the Wetlands (Conservation and Management) Rules, 2017, is restricted to monitoring its implementation by states/UTs, recommending trans-boundary wetlands for notification and reviewing integrated management of selected wetlands under the Ramsar Convention — an international arrangement to preserve identified wetlands.

RAMSAR CONVENTION

The Ramsar Convention is an international treaty signed in Ramsar, Iran for conservation and wise use of wetlands. The agreement was signed on February 2, 1971 and came into force from December 21, 1975. It is one of the oldest specific conventions that deal not only with the conservation of the wetlands but also its wise use. There are at present 158 contracting parties for this convention. About 1831 wetlands of international importance have been listed as Ramsar sites. To undertake immediate remedial measures against pollution or ecological degradation, wetlands are included under Montraux Record. In India, the convention on wetlands came into force on February 1st 1982. Initially three lakes namely Chilika (Orissa), Loktak (Manipur) and Keoladeo lakes (Rajasthan) were included in the Montreux Record for remedial measures and monitoring. Later on Chilika was removed, as the Government had claimed an improvement in the ecology.

CATEGORIES OF WETLAND

There are three categories of wetland:

Category A

1. Wetlands listed under the Ramsar Convention
2. Wetlands recognized as or lying within a world heritage site or a national heritage site
3. Transboundary wetlands
4. Inter-state wetlands which do not fall under category B or C;
5. Wetlands with an area equivalent to or more than 1000 ha in arid regions, 5000 ha in semi-arid regions, 10,000 ha in sub-humid and 1,00,000 ha in humid tropic regions;
6. Wetland which is a major source of drinking water for 'Class A' cities.

Category B

1. Wetlands recognized as, or lying within, a state heritage site;
2. Wetlands with an area of 25 ha but below 1000 ha in arid regions, 100 ha but below 5000 ha in semi-arid regions, 500 ha but below 10,000 ha in sub-humid; and 2,500 ha but below 1, 00,000 in humid tropic regions;
3. High altitude wetland at 2,500 metres or more above mean sea level;
4. Wetland which is a major source of drinking water for 'Class B' towns.

Category C

1. Wetlands other than those covered under category A and B;
2. Wetland with an area less than 25 ha in arid regions, less than 100 ha in semi-arid region, less than 500 ha in sub-humid and less than 2,500 ha in humid tropic regions;
3. Wetland which is a major source of drinking water for local communities involving at least 100 households;
4. Wetland which is socially and/or culturally important to the local communities

Chapter 1
Wetland Ecosystems and Their Relevance to the Environment:
Importance of Wetlands

Arunima Nayak
https://orcid.org/0000-0002-1762-8019
Graphic Era University, Dehradun, India

Brij Bhushan
https://orcid.org/0000-0002-9360-3083
Graphic Era University, Dehradun, India

ABSTRACT

As compared to the tropical evergreen forest, wetlands are also one of the most productive ecosystems in the biosphere and make a major contribution to the ecological sustainability of a region. The wetlands maintain biologically diverse communities having both ecological and economic value. Based on the immense environmental and sustainability benefits, wetlands have been demarcated as essential for the future of human existence. The future challenges pertaining to food, clean water and energy security, well-being of humans, natural disaster risk reduction, and climate change resilience can be met by preserving the wetlands. The chapter has an aim to provide insight on the fundamentals like the classifications, major functions, as well as the various factors affecting the wetland ecosystem. Other important aspects like the major threats leading to the loss of the wetlands, consequences of the loss or degradation of wetlands, and ways to preserve the wetlands are discussed.

INTRODUCTION

As compared to the tropical evergreen forest, wetlands are also one of the most productive ecosystems in the biosphere and have a major contribution to the ecological sustainability of a region (ten Brink et al., 2012). Wetlands support a diversity of biotic life ranging from diverse aquatic plants (floating pond lilies, cattails, cypress, tamarack, and blue spruce); algae; fishes, shell fishes; insects; carnivorous ani-

DOI: 10.4018/978-1-7998-9498-8.ch001

mals like dragonflies, otters, alligators and osprey; diverse migratory birds, waterfowl etc. (Ghermandi et al., 2008). Thus, the wetlands maintain biologically diverse communities having both ecological and economic value. The global area of wetland extends from 917 million hectares to more than 1275 million hectares which is a meager 6% of the Earth's surface (Finlayson and Spiers, 1999; Turner et al., 2000). Despite this, wetlands are a crucial aspect of life on earth as they provide majority of services related to sustainability of the ecosystem (Smith et al., 1995). For instance, besides, maintaining biodiversity, wetlands mitigate floods, protect the coastal areas from erosion, provide as nurseries for aquatic and terrestrial life, improve and preserve water quality, recharge groundwater aquifers, serve as sinks, sources, or transformers of materials, and produce food and energy for human use (Boutwell and Westra, 2016). While evaluating the economic value, Costanza et al. (1997) concluded that the economic value provided by wetland ecosystems was second only to that provided by coastal estuaries and the same exceeded that provided by lakes, streams, forests, and grasslands. Wetlands across the globe have exhibited diversified characteristics ranging from inland freshwater swamps to brackish water coastal marshes and mangroves, peatlands, riparian wetlands, and marshes (Costanza, 2021). Also, at the transitional areas, wetlands can exhibit characteristics of either terrestrial or aquatic ecosystems while some wetlands can possess characteristics common to both. But, despite the diversified characteristics, a wetland must possess an area (a) that results in wet or flooded soils (b) having soils undergoing anaerobic processes and (c) which has a biological entity (plant or animal) that are adapted to life in flooded, anaerobic environments. The diversity in wetlands has resulted in scientists and technocrats giving different definitions. Basically, wetlands are geographical areas where water bodies are near land or soil throughout the year or for varying periods of time during the year. The earliest definition was given by Cowardian et al. who proposed that wetlands form the transitional zone between land and water, where saturation with water is the dominant factor determining the nature of soil development and the types of plant and animal communities living in and on it (Cowardian et al., 1979). The Ramsar Convention on Wetland in 1971, in Iran (http://www.ramsar.org) characterized wetlands as "... areas of marsh, fen, peatland, whether natural or artificial, permanent or temporary, with water that is static or flowing, fresh, brackish or salt, including areas of marine water the depth of which at low tide does not exceed six meters." The National Research Council in 1995 (NRC 1995) proposed that "the minimum essential characteristics of a wetland are recurrent, sustained inundation or saturation at or near the surface and the presence of physical, chemical, and biological features reflective of recurrent, sustained inundation or saturation. Common diagnostic features of wetlands are hydric soils and hydrophytic vegetation". Based on the immense environmental and sustainability benefits, wetlands have been demarcated as essential for the future of human existence. The future challenges pertaining to food, clean water and energy security, well-being of humans, natural disaster risk reduction and climate change resilience can be met by preserving the wetlands (Sun and Carson, 2020). But literature data has revealed the steady decline and disappearance of the existence of wetlands by 64% since 1900 (www.ramsar.org). This loss has been particularly higher in Asian countries. The major causes identified as responsible for the decline of the wetlands are rapid urbanization and infrastructure development; excessive use of land; drainage to agricultural use; pollution due to agriculture, industrial and municipal sources; climate change and variability. The negligence towards maintaining and preserving the wetlands has resulted in adverse effects in the form of fewer access to fresh water for people worldwide, scarcity of food and energy and more occurrences of floods thereby causing irreparable damage to coastland ecosystem and livelihood. A steady decline in the terrestrial and aquatic species by 76% between 1970 and 2010 according to WWF's Living Planet Index has caused an adverse effect on the biodiversity (Cherry, 2011). Many of their resources are considered under risk as a result of anthropic impacts related to water management, damming, fishing, farming, oil exploitation, agriculture and forestry, etc. The losses are larger and faster on inland than on coastal natural areas.

The damaging consequences of the declining wetlands as well as the increased awareness of the importance of wetland ecosystem to the environment and to the biosphere has led to the creation of governing bodies, laws, regulations to restore and protect the wetlands across the globe. The protection of wetlands falls under the Clean Water Act of 1972 in the United States of America. Under this Act, activities include dredging and filling of the wetlands as well as monitoring of water quality. The National Wetlands Policy Forum created in 1988, had an aim to restrict further wetland loss in the US, and further requiring creation, restoration and mitigation to offset or replace wetland damages done due to human activity. The Ramsar Convention, which is an international treaty aimed at conserving wetlands, requires all member countries to develop their own national wetland policies, to establish wetland reserves, and to designate one or more wetlands as an area of international importance. Such activities in the form of convention, laws etc. have an aim to protect or conserve wetlands and the ecosystem services they provide.

The chapter thus, has an aim to provide insight to the fundamentals like the classifications, major functions as well as the various factors affecting the wetland ecosystem. Other important aspects like the major threats leading to the loss of the wetlands, consequences of the loss or degradation of wetlands and ways to preserve the wetlands will be discussed.

CLASSIFICATION OF WETLANDS

Wetlands have different features and characteristics. Irrespective of the variability, the feature that is common to all wetlands is the presence of a water table that is near the soil surface (Conner & Day, 1982). All wetlands are characterized by the combination of salinity of water, type of soil and the associated biota. There is a basic requirement to classify the different types of wetlands. But because of the variability, till date there is no single wetland classification type (Scott and Jones 1995). An attempt has been made to classify wetland ecosystem based on traditional methods and based on scientific background.

Traditional Wetland Classification

Traditional method of wetland classification is based on the mineral and organic soil types (Richardson, 1994; www.epa.gov/wetlands). Typical examples of mineral soil-based wetland types are marsh and swamp.

Marsh is a type of wetland ecosystem which is found near the shallow edges of rivers, lakes and streams or near the deltas of rivers. It exhibits poorly drained mineral soils and is dominated by deep-rooted herbaceous plants like the grasses, sedges, rushes and other herbs. Deep rooted herbaceous plants inhibit the flow of water and hence the soil nutrients; thereby favoring conditions for the further development of marshy conditions. Marshes provide habitat to invertebrates, fishes, amphibians, waterfowl and other aquatic mammals. Marshes are of three types: tidal salt marshes, tidal freshwater marshes and inland freshwater marshes. Both tidal salt and freshwater marshes are rich in biodiversity and offer other functions like inhibit soil and shoreline erosion, protect the aquatic life, offer nesting sites for migratory aquatic birds and absorb excess nutrients which are otherwise detrimental to aquatic and wildlife. Inland freshwater marshes exist near the surface of lakes and rivers where the water level is very high. Size variations of such wetlands range from bowl-shaped depressions called prairie potholes to the watery grasslands of the Florida Everglades national Park. The Everglades is an example of the largest freshwater marsh in the United States and is drowned throughout the year in a shallow layer of water. The Everglades (also called the "River of grass") support plants like sawgrass, cypress, and mangrove forests. They offer shelter to animals like ducks, geese, raccoons, turtles and frogs. Predatory animals like the alligators and panthers also thrive the Everglades. An example of the largest freshwater marsh is in the Okavango Delta in Botswana. The Okavango delta in an arid region consisting of marshes that span about 15,000 square kilometers. The marshy area is made up of papyrus, water lilies, and underwater plants such as bladderworts. The Okavango Delta is a source of freshwater to diverse animal species like the hippopotamuses, crocodiles, giraffes and elephants in the middle of the dry Kalahari Desert. Pallikaranai wetland is a freshwater marsh situated near the Bay of Bengal in the city of Chennai, India. The marsh has been home to several endangered migratory birds from various places within and outside the country. The plant species include water hyacinth and water lettuce. Various birds like the glossyibis, grey-headed lapwings, pheasant-tailed jacana, cormorants, darters, herons, egrets, open-billed storks, spoonbills, white ibis, little grebe, Indian moorhen, black-winged stilts, purple moorhens, warblers, coots and dabchicks have been spotted in large numbers in the marshland.

A swamp, like the marsh is a land area that is permanently filled with water. But unlike the marsh, swamps are dominated by trees. Depending on the characteristics of the water, swamps are classified into freshwater and salty swamps. Freshwater swamps are found inland while the saltwater swamps are found in the coastal areas. The Amazon basin of Brazil and the "Fertile Crescent" between the Tigris and Euphrates Rivers in the Middle East are examples of freshwater swamps. Ratargul freshwater swamp forest located in Gowainghat of Sylhet district has a rich growth of the tree species *Dalbergia reniformis*. The swamp forest has a rich and diverse existence of plants and animals. The Indian Sundarbans is the freshwater swamp and is comprised of hundreds of islands and a network of rivers, tributaries, and creeks in the delta of the Ganga and the Brahmaputra at the mouth of the Bay of Bengal in India and Bangladesh. The Sundarbans has over 60% of India's total mangrove forest area. The mangrove forests protect the land from storms, cyclones, prevent soil erosion and seepage of saltwater inland and into the waterways. Besides, the wetland provides a nursery to different types of fishes. Other functions include providing freshwater, food and serve as a shock absorber for various natural calamities. Louisiana swamp in New Orleans has a rich biotic life housing plants like cypress trees, Spanish moss and native animals like alligators, brown pelicans, the Louisiana state bird etc.

The organic soil-based wetlands are referred to as peatland because of the ability of soil to form peat. Peatlands are of two types: bogs and fens; both are adapted to same climatic and geographic regions. Bog is characterized by poorly drained wet and spongy soil and acidic waters. The soil has a rich growth of mosses, Sphagnum, and heaths. The source of water is exclusively rainwater and hence the soil is less in nutrients; thereby plant growth is restricted. Bogs have served as home to various mammals like caribou, moose, and beavers as well as for species of nesting shorebirds, such as Siberian cranes and yellowlegs. Various reptiles and insects like lizards, yellow fly, butterfly etc. have thrived the bogs.

Fens, unlike the bogs are characterized by the presence of water that is less acidic; the source of such water is the surface and ground water. The soil is peat based and is dominated by brown mosses and herbaceous plants like grasses, sedges, rushes, wildflowers, and reeds. Less acidic nature of water causes the growth of woody plants, shrubs and small trees. Wider variety of plant species help support the nurture and thriving of animal species like the deer, moose, black bear and various amphibians. The nutrient rich shallow water provides home to various fishes, reptiles, insects and wading birds.

Figure 1. (a) Pallikaranai marsh in Chennai, India (b)marsh in Jyväskylä, Finland (c) salt marsh in Scotland (d) Sawgrass prairie in Everglades National Park

Ramsar Classification System

1. Marine/Coastal Wetlands
    a. **Permanent shallow marine waters** in most cases less than six metres deep at low tide; includes sea bays and straits.
    b. **Marine subtidal aquatic beds**; includes kelp beds, sea-grass beds, tropical marine meadows.
    c. **Coral reefs**.
    d. **Rocky marine shores**; includes rocky offshore islands, sea cliffs.
    e. **Sand, shingle or pebble shores**; includes sand bars, spits and sandy islets; includes dune systems and humid dune slacks.
    f. **Estuarine waters**; permanent water of estuaries and estuarine systems of deltas.
    g. **Intertidal mud, sand or salt flats.**
    h. **Intertidal marshes**; includes salt marshes, salt meadows, saltings, raised salt marshes; includes tidal brackish and freshwater marshes.
    i. **Intertidal forested wetlands**; includes mangrove swamps, nipah swamps and tidal freshwater swamp forests.
    j. **Coastal brackish/saline lagoons**; brackish to saline lagoons with at least one relatively narrow connection to the sea.

Figure 2. (a) Ratargul Swamp Forestin Gowainghat of Sylhet district, Bangladesh (b) Louisiana swamp, New Orleans (c) Freshwater Swamp Forests of Lake Sentarum, Indonesia (d) Asmat Swamp, Indonesia (e) Myristica swamp in Maharashtra, India (f) Sundarbans, India

k. **Coastal freshwater lagoons**; includes freshwater delta lagoons.
   i. **Karst and other subterranean hydrological systems**, marine/coastal

2. Inland Wetlands
   a. **Permanent inland deltas.**
   b. **Permanent rivers/streams/creeks**; includes waterfalls.
   c. **Seasonal/intermittent/irregular rivers/streams/creeks.**
      i. **Permanent freshwater lakes (over 8 ha)**; includes large oxbow lakes.
   d. **Seasonal/intermittent freshwater lakes (over 8 ha)**; includes floodplain lakes.
   e. **Permanent saline/brackish/alkaline lakes.**
   f. **Seasonal/intermittent saline/brackish/alkaline lakes and flats.**
   g. **Permanent saline/brackish/alkaline marshes/pools.**
   h. **Seasonal/intermittent saline/brackish/alkaline marshes/pools.**

Figure 3. (a)wet Sphagnum bog in Frontenac National Park, Quebec, Canada (b)Carbajal Valley peat bogs, Tierra del Fuego Province, Argentina(c)bog in Duck Lake, Oregon (d) Ponemah Kettle Hole, an example of a peat bog, Amherst, New Hampshire

Figure 4. Fen in the Butterfly Valley in spring, Plumas National Forest

i. **Permanent freshwater marshes/pools**; ponds (below 8 ha), marshes and swamps on inorganic soils; with emergent vegetation water-logged for at least most of the growing season.
j. **Seasonal/intermittent freshwater marshes/pools on inorganic soils**; includes sloughs, potholes, seasonally flooded meadows, sedge marshes.
k. **Non-forested peatlands**; includes shrub or open bogs, swamps, fens.

l. **Alpine wetlands**; includes alpine meadows, temporary waters from snowmelt.
m. **Tundra wetlands**; includes tundra pools, temporary waters from snowmelt.
n. **Shrub-dominated wetlands**; shrub swamps, shrub-dominated freshwater marshes, shrub carr, alder thicket on inorganic soils.
o. **Freshwater, tree-dominated wetlands**; includes freshwater swamp forests, seasonally flooded forests, wooded swamps on inorganic soils.
p. **Forested peatlands**; peatswamp forests.
q. **Freshwater springs**; oases.
r. **Zg - Geothermal wetlands**
s. **Zk(b)- Karst and other subterranean hydrological systems, inland**

Note: "Floodplain" is a broad term used to refer to one or more wetland types, which may include examples from the R, Ss, Ts, W, Xf, Xp, or other wetland types. Some examples of floodplain wetlands are seasonally inundated grassland (including natural wet meadows), shrublands, woodlands and forests. Floodplain wetlands are not listed as a specific wetland type herein.

3. Human-made wetlands
    a. **Aquaculture** (e.g., fish/shrimp) **ponds**
    b. **Ponds**; includes farm ponds, stock ponds, small tanks; (generally below 8 ha).
    c. **Irrigated land**; includes irrigation channels and rice fields.
    d. **Seasonally flooded agricultural land** (including intensively managed or grazed wet meadow or pasture).
    e. **Salt exploitation sites**; salt pans, salines, etc.
    f. **Water storage areas**; reservoirs/barrages/dams/impoundments (generally over 8 ha).
    g. **Excavations**; gravel/brick/clay pits; borrow pits, mining pools.
    h. **Wastewater treatment areas**; sewage farms, settling ponds, oxidation basins, etc.
    i. **Canals and drainage channels, ditches**.
    j. **Karst and other subterranean hydrological systems**, human-made

MAJOR FUNCTIONS OF WETLAND ECOSYSTEM

Wetlands serve various functions that are essential to the society and to the environment. The functions are diverse and may vary among the wetlands. The extent to which a wetland may serve a function depends on the conditions of water, soil, vegetation, size and its geographical location. Some of the ecological benefits of wetlands are providing habitats for wildlife, aquatic species and plants; carbon storage; water quality improvement; flood water storage and protection from floods (Erwin, 2009; Nyman, 2011).

Water Storage

Continuous rain, thunder storm and/or melting of ice or glacier may result in increased water level flowing over the surface of land and may result in flooding of plains, washing away of habitats of aquatic life or erosion of surface soil. Many wetlands have the capacity to temporarily store surface waters or run-off waters or flood waters during any of the events as stated above (Bassi et al., 2014). Wetlands are hence

Figure 5. Water storage function by wetlands

referred to as natural sponges or natural tubs. Trees, natural vegetation of wetlands impedes the flow of surface flood water and help in distributing them over the plains. The rate of flow of the excess water from the wetlands into the natural streams or into the plains is very slow. Thus, the wetlands have the capacity to hold back some of the flood waters or excess waters and can reduce the damage that could have resulted from flooding and erosion. Because of the capacity to control the rate and volume of runoff, wetlands have made economic contribution like increased public safety, increased safety to property, have prevented water logging of agricultural lands and have enhanced the stability of habitat of aquatic life.

Surface and Ground Water Protection

The quality of water is deteorating due to increased agricultural, industrial and man-made activities. Run-off or flooding has resulted in washing away of the pollutants from urban and agricultural lands to the water bodies. The pollutants usually encountered as a result of agricultural and industrial activities are soil particles, pesticides, fertilizers, dyes, metal ions, oil etc. (Arslan and Gamal El-Din, 2021). In this respect, wetlands have helped in removal of pollutants thereby helping in improving surface water quality. The roots of trees and vegetation in a wetland trap the sediments along with the heavy metal ions as well as bind to the organic pollutants. Thereby wetlands help in removing as much as 90% of the pollutants present in the run-off (Odum, 1979; Mulder et al., 2020). Nutrients like nitrates and phosphates present in fertilizers and washed away from agricultural land promote the growth of algae, cyanobacteria and plants; which in turn generates chemicals that are toxic to the plant and aquatic life. Such nutrients are trapped by the wetland vegetation in less toxic forms and hence restrict their flow to water bodies. The decay and death of the vegetation result in recycling of the nutrients within the wetlands. Thus, wetlands help in recharging a drinking water source, reduce concentration of pollutants in surface water and enhance the quality of water via sediment trapping, nutrient removal and chemical detoxification.

Stabilization of Habitat for Wildlife, Fish Community and Protection of Endangered Species

Wetlands have exhibited high rates for growth of plants and trees; the dense vegetation provides not only food but also shelter for different animal and bird species (Men´endez et al., 2020). Migratory birds often build their nests in the dense vegetation of wetlands for breeding, rearing or feeding of their

Figure 6. Surface and ground water protection of wetlands

The roots of wetland plants can bind and remove as much as 90 percent of sediments present in runoff or streamflow.

Nutrients such as nitrogen and phosphorus threaten water quality in lakes and rivers, and wetlands act as a natural water filter. The wetland plants take up excess nutrients and convert them into less harmful forms, and over time, the nutrients are recycled within the wetland.

young. Among the different wildlife, mammals, amphibians and reptiles have found their home for food, shelter, reproduction and protection from their predators. Because of the dense vegetation both on land and in water which serve as source of food, wetlands have served as farmlands for aquatic environment. Wetlands have provided spawning, nursery, feeding habitat for the fish community. They have also provided ambient aquatic environment by moderation of temperature.

Many plant and animal species are endangered and require safe habitats for their survival and growth. Wetlands provide a safe haven to such species.

Erosion Control

The dense vegetation of the wetlands along the shores of water bodies protects the erosion of soil (Costanza et al., 2008). Soil erosion is mainly caused by waves along the shorelines in the event of floods and storms. The vegetation prevents the erosion by binding to the soil particles and stabilizing the soil

WETLANDS AND RELEVANCE TO ENVIRONMENT

Wetlands are productive ecosystems supporting a wide range of biodiversity. Wetlands play a constructive role in water purification and are responsible for feeding the fishes, birds, amphibians, reptiles etc. Besides, they are responsible for protecting the water bodies, coastal areas and help in mitigating climate change. They play a vital role in protecting the environment. With respect to water bodies, wetlands help to preserve ground and surface water, help to regulate the impact of floods and storms and also help to preserve water quality. Fresh water is a vital natural resource and is slowly diminishing in its availability. Yet humans need 20-30 liters of water a day for production of food, cooking, drinking and for cleaning purposes. Wetlands help to preserve both the quantity as well as the quality of water. Wetlands also serve as absorbing medium for rainwater and provide pools for storage of floodwater; which in turn can help to safeguard against droughts. Wetlands help to control soil erosion and transport of sediments; thereby helping in the restoration of land. Wetlands are the major providers of support to all types of biota. Various species of birds, endangered species of birds, migratory birds etc. have found their home

Figure 7. Stabilization of habitat for wildlife and fish

in such wetlands. Such birds have not only found their feeding support but also have found their shelter and breeding grounds in such wetlands. Wetlands are also an important source of food to diverse types of species of birds, reptiles, fishes, amphibians etc. Finally, wetlands provide sustainable livelihoods in the fields of fishery, timber industry, agriculture, horticulture, vegetable oil, natural products, and medicinal plants. Summarily, it can be said that the fate of wetlands is dependent on the future of humanity.

FACTORS AFFECTING THE WETLAND ECOSYSTEM

A wetland ecosystem is complex comprising of both the biotic and abiotic components. Biotic components include plants, animals, microbes, fishes etc. while the abiotic components refer to the water, availability of oxygen, soil, minerals, nutrients etc. present within the ecosystem.

Water is an essential component of the wetland ecosystem (Gosselink and Turner,1978). The balance between the biotic and abiotic components as well as the survival/existence of the biotic components depends on the availability, distribution, flow and quality of water in a wetland. Water can enter the ecosystem via precipitation, flooding, storms and/or coastal tides. Surface water run-off, sub-surface water outflow or evapo-transpiration is some of the causes for water outlet from a wetland. Water logging leads to the formation of hydric soils in which the soil remains flooded or inundated for an extended period of time leading to anaerobic conditions. The level of water and the distribution or flow pattern within an ecosystem may vary with respect to the input of water, timing and duration of soil inundation and flooding patterns due to coastal tides. The duration of flooding, soil inundation by surface or ground water has a determining effect on whether the wetland is marshy or swampy. The nature and type of vegetation or the animal/fish/amphibian species also depend on the water characteristics in a wetland. Thus the nature of a wetland depends on its water level which again is decided by the amount of inflow and outflow of water, geomorphology, and soil composition. The hydrochemistry i.e the oxygen and carbon-dioxide content in water in a wetland depend on the temperature and atmospheric pressure of the area; soil characteristics like pH, conductivity, salinity, nutrient levels, hardness etc. as well as the landscape characteristics.

Amount of oxygen availability is another important abiotic factor affecting the characteristics of a wetland ecosystem. Decreased oxygen availability may occur due to prolonged inundation of wetland soils resulting in anaerobic conditions. Other conditions like temperature of soil, rate of microbial respiration etc. may also result in anaerobic conditions.

Decreased oxygen levels because of the presence of anaerobic conditions have important implication on the biogeochemistry of a wetland like affecting the carbon, nitrogen or sulphur cycle. The nitrogen cycle of a wetland is altered as a result of anaerobic conditions. For example, the nitrogen gas or nitrous oxide via bacterial activities get converted to nitrates because of de-nitrification process brought about by anaerobic conditions. The nitrates in turn are lost to the atmosphere.

Other abiotic factors affecting the characteristics of a wetland are salinity, nutrient availability, sedimentation, temperature, landscape, climatic conditions and also human intervention. Human intervention could be due to agricultural, industrial or urban developmental activities. In many such wetlands, nutrient availability is greatly altered due to agricultural activities and hence contributing to the change in structure and function of the ecosystem.

MAJOR THREAT LEADING TO LOSS OF WETLAND ECOSYSTEM

A well-maintained wetland creates a healthy ecosystem and provides habitat for both aquatic/terrestrial species as well as plants, trees, ferns and other vegetation. Besides, the wetland ensures other important functions like natural floodwater storage, enhanced water pollution abatement, reduced soil, coastal and surface water erosion etc. Despite the importance of wetlands, approximately 50% of world's wetlands (in North America, Europe, Australia, and New Zealand) have been lost because of the rapid urbanization, industrialization, and agricultural activities. Approximately 5000km^2 of wetland area is lost annually in Asia alone mainly on account of development activities. India has lost 38% of its wetlands in the accounting decade 1991-2001(Davidson, 2014).

On the other hand, majority of such useful functions are inhibited in a degraded wetland. A rapidly degrading wetland implies loss of preservation of biotic species, extinction of birds/fishes/reptiles/amphibians as well as well as an impaired ecosystem. As many as 21% of bird species; 37% of mammal species and 20% of freshwater fish species have become either endangered or globally threatened. Some of the other serious consequences of rapid loss of wetlands are increased flooding, decline in water quality and preservation. With an aim to minimize or eliminate such consequences, it has been imperative to preserve such ecosystems. For this, the factors that have contributed to the loss of wetland ecosystem needs to be identified and corrective actions implemented.

Human activities are the only cause for wetland degradation and loss. Such activities have resulted in alteration in the quality, quantity and flow rate of water. Various pollutants have been introduced to various extents and introduction of various non-native species have altered the composition of species existing in a wetland. The major characteristic of a wetland is decided by its hydrological conditions. Human activities have caused a major change in the hydrology of wetlands resulting in major alterations in the soil chemistry as well as plant and animal biotic communities. Some of the human activities identified for initiating the change in the hydrology condition of a wetland are

1. Construction and developmental work like deposition of fill material within a wetland premises.
2. Agricultural and farming activities leading to drainage of water table from a wetland.
3. Dredging activities for navigation, development, and flood control.
4. Construction of dams to form ponds and lakes.
5. Diverting water flow to or from wetlands.
6. Installing rocks, stones, pebbles and other impervious objects in the watershed, causing enhanced water and pollutant runoff into the wetlands.

Modern methods of agricultural activities, mining activities, construction/developmental activities, industrial activities, automobile emissions, lifestyle changes etc. have resulted in enhanced input of pollutant into wetlands via the surface run-off water. The wetlands have a limited capacity to absorb pollutants from aqueous medium; thus, excess pollutants have posed a danger to the biotic species of a wetland ecosystem. Major pollutants identified to have caused a degrading effect on the wetlands leading to their loss are heavy metal ions, pesticides, fertilizers, pharmaceuticals, sediments, human wastes etc.

Other activities that have been identified for wetlands degradation are grazing by domestic animals, removal of vegetation for mining or construction activities, planting of non-native plants that have been found to be competitive with native vegetation. Such activities have adversely affected the native vegetation causing wetlands degradation and their eventual loss.

WAYS TO PRESERVE WETLANDS

The Ramsar convention was the first international convention that was adopted in 1971 as a result of rising environmental awareness on nature conservation and on sustainable development. The philosophies practiced under the Ramsar Convention which include identification of endangered wetlands, need for restoration and management of wetlands has been adopted and practiced in other international and national conventions. It is imperative for all to preserve the wetlands via different initiatives like reducing the use of fertilizers, pesticides, herbicides for agricultural activities; protecting the wetlands of neighboring areas; helping the local bodies in preserving the wetlands in terms of money, time and labor; educating people about the importance and preservation of wetlands; working with local bodies in framing laws for protection and restoration of wetlands; minimizing any activities that can degrade or alter wetlands; participate in Clean Water Act program and state or local regulating bodies.

CONCLUSION

Wetland ecosystems are known to provide habitats to diverse species of plants, terrestrial and aquatic animals. Wetlands are contributory to the preservation of many endangered biotic species. They support diverse biological species and are responsible for providing a wide variety of ecosystem goods and services. Services provided include fresh and clean water supply, surface water recharge, flood control, carbon sequestration and pollution control. In their natural and fully operational form, wetlands are highly productive and constitute a vital part of the water cycle. But the preservation and maintenance of the wetlands have received least attention by the national and international government agencies while drafting the water and environment protection agenda. This has resulted in indiscriminate and undue pressures on the wetlands of urban and rural areas in the form of changes in land use in the catchment areas, encroachments, excessive pollution load from industries, mining, automobiles, lifestyle changes and modern agricultural activities. The Ramsar convention drafted in the 1996 was the first global intergovernmental treatise on the conservation and wise use of natural wetlands. Various wetland sites across the globe were identified and major restoration work was started with a focus for restore and maintenance. But till date no significant progress could be achieved mainly on account of only few selected sites getting attention by means of financial and technical assistance under the wetland conservation programs. Major research work on wetlands is focused on the limnological aspects and ecological or environmental aspects of wetland management. While work on the physical and socio-economic activities causing changes in wetlands has been largely ignored. Also, governmental policies, rules, regulations and setup of organizations for conservation and maintenance of wetlands has largely received little attention. Thus, more effort is anticipated from researchers, environmentalists, scientists for research on the physical, socio-economic and institutional factors that have a role in determining the condition of wetlands. The research outcomes will help in better management strategies for various such wetlands that are non-functional and stressed out due to various anthropogenic and climatic factors.

REFERENCES

Arslan, M., & Gamal El-Din, M. (2021). Removal of per- and poly-fluoroalkyl substances (PFASs) by wetlands: Prospects on plants, microbes and the interplay. *The Science of the Total Environment*, *800*, 149570. doi:10.1016/j.scitotenv.2021.149570 PMID:34399352

Badamfirooz, J., Mousazad, R., & Hamid, S. (2021). A proposed framework for economic valuation and assessment of damages cost to national wetlands ecosystem services using the benefit-transfer approach. *Environmental Challenges, 5.*

Bassi, N., Kumar, M. D., Sharma, A., & Saradhi, P. P. (2014). Status of wetlands in India: A review of extent, ecosystem benefits, threats and management strategie. *Journal of Hydrology. Regional Studies*, *2*, 1–19. doi:10.1016/j.ejrh.2014.07.001

Boutwell, J. L., & Westra, J. V. (2016). The role of wetlands for mitigating economic damage from Hurricanes. *Journal of the American Water Resources Association*, *52*(6), 1472–1481. doi:10.1111/1752-1688.12473

Brown. (2021). Opportunities for improving recognition of coastal wetlands in global ecosystem assessment frameworks. *Ecological Indicators, 126.*

Cherry, J. A. (2011). Ecology of Wetland Ecosystems: Water, Substrate, and Life. *Nature Education Knowledge*, *3*(10), 16.

Conner, W. H., & Day, J. W. (1982). The ecology of forested wetlands in the southeastern United States. In *Wetlands: Ecology and Management* (pp. 69–87). National Institute of Ecology and International Scientific Publications.

Costanza, R. (2021). The global value of coastal wetlands for storm protection. *Global Environmental Change*, 70.

Costanza, R., Anderson, S. J., Sutton, P., Mulder, K., Mulder, O., Kubiszewski, I., Wang, X., Liu, X., Pérez-Maqueo, O., Luisa Martinez, M., Jarvis, D., & Dee, G. (2021). The global value of coastal wetlands for storm protection. *Global Environmental Change*, *70*, 102328. doi:10.1016/j.gloenvcha.2021.102328

Costanza, R., P'erez-Maqueo, O., Martinez, M. L., Sutton, P., Anderson, S. J., & Mulder, K. (2008). The value of coastal wetlands for hurricane protection. *Ambio*, *37*(4), 241–248. doi:10.1579/0044-7447(2008)37[241:TVOCWF]2.0.CO;2 PMID:18686502

Costanza, R. W., Costanza, R., d'Arge, R., de Groot, R., Farber, S., Grasso, M., Hannon, B., Limburg, K., Naeem, S., O'Neill, R. V., Paruelo, J., Raskin, R. G., Sutton, P., & van den Belt, M. (1997). The value of the world's ecosystem services and natural capital. *Nature*, *387*(6630), 253–260. doi:10.1038/387253a0

Cowardin, L. M., Carter, V., Golet, F. C., & LaRoe, E. T. (1979). *Classification of Wetlands and Deepwater Habitats of the United States. FWS/OBS-79/31.* US Fish and Wildlife Service.

Davidson, N. C. (2014). How much wetland has the world lost? Long-term and recent trends in global wetland area. *Marine and Freshwater Research*, *65*(10), 934. doi:10.1071/MF14173

Erwin, K. L. (2009). Wetlands and global climate change: The role of wetland restoration in a changing world. *Wetlands Ecology and Management*, *17*(1), 71–84. doi:10.100711273-008-9119-1

Finlayson, C. M., & Spiers, A. G. (Eds.). (1999). *Global Review of Wetland Resources and Priorities for Wetland Inventory*. Supervising Scientist.

Ghermandi, A., van den Bergh, J. C. J. M., Brander, L. M., & Nunes, P. A. L. D. (2008). *The Economic Value of Wetland Conservation and Creation: A Meta-Analysis* [Working Paper 79]. Fondazione Eni Enrico Mattei.

Gosselink, J. G., & Turner, R. E. (1978). The role of hydrology in freshwater wetland ecosystems. In *Freshwater wetlands Ecological processes and management potential.* Academic Press. https://www.epa.gov/sites/default/files/2016-02/documents/wetlandfunctionsvalues.pdf

Kumar, R., Sharma, P., & Bandyopadhyay, S. (2021). Evidence of microplastics in wetlands: Extraction and quantification in Freshwater and coastal ecosystems. *Journal of Water Process Engineering, 40.*

Li, D., Tian, P., Luo, Y., Dong, B., Cui, Y., & Khan, S. (n.d.). Importance of stopping groundwater irrigation for balancing agriculture and wetland ecosystem. *Ecological Indicators, 127.* http:// doi:10.1016/j.ecolind.2021.107747

Men'endez, P., Losada, I. J., Torres-Ortega, S., Narayan, S., & Beck, M. W. (2020). The global flood protection benefits of mangroves. *Scientific Reports*, *10*(1), 1–11. doi:10.103841598-020-61136-6 PMID:32157114

Mulder, O. J., Mulder, K. P., Kubiszewski, I., Anderson, S. J., Costanza, R., & Sutton, P. (2020). The value of coastal wetlands for storm protection in Australia. *Ecosystem Services*, *46*, 101205. doi:10.1016/j.ecoser.2020.101205

National Research Council (NRC). (1995). *Wetlands: Characteristics and Boundaries*. National Academy Press.

Nyman, J. A. (2011). Ecological Functions of Wetlands. *Wetlands*, 115–128. doi:10.1007/978-94-007-0551-7_6

Odum, E. P. (1979). The value of wetlands: a hierarchical approach. *Vegetatio*, *118*, 3–16.

Qian, J., Tang, S., Wang, P., Lu, B., Li, K., Jin, W., & He, X. (2020). From source to sink: Review and prospects of microplastics in wetland ecosystems. *The Science of the Total Environment*.

Richardson, C. J. (1994). Ecological functions and human values n wetlands: A framework for assessing forestry impacts. *Wetlands*, *14*(1), 1–9. doi:10.1007/BF03160616

Smith, R. D., Ammann, A., Bartoldus, C., & Brinson, M.M. (1995). *An approach for assessing wetland functions using hydrogeomorphic classification, reference wetlands, and functional indices: Vicksburg, Miss., U.S. Army Engineers Waterways Experiment Station.* Technical Report TRWRP-DE, 10, 100.

Sun, F., & Carson, R. T. (2020). Coastal wetlands reduce property damage during tropical cyclones. *Proceedings of the National Academy of Sciences of the United States of America*, *117*(11), 5719–5725. doi:10.1073/pnas.1915169117 PMID:32123063

Ten Brink, P., Badura, T., Farmer, A., & Russi, D. (2012). *The Economics of Ecosystem and Biodiversity for Water and Wetlands: A Briefing Note*. Institute for European Environmental Policy.

Turner, R. K., van der Bergh, J. C. J. M., Soderqvist, T., Barendregt, A., van der Straaten, J., Maltby, E., & van Ierland, E. C. (2000). Ecological-economic analysis of wetlands: Scientific integration for management and policy. *Ecological Economics*, *35*(1), 7–23. doi:10.1016/S0921-8009(00)00164-6

Wallace, J., & Waltham, N. J. (2021). On the potential for improving water quality entering the Great Barrier Reef lagoon using constructed wetlands. *Marine Pollution Bulletin*, *170*, 112627. doi:10.1016/j.marpolbul.2021.112627 PMID:34182303

Zedler, J. B., & Kercher, S. (2005). Wetland resources: Status, trends, ecosystem services, and restorability. *Annual Review of Environment and Resources*, *30*(1), 39–74. doi:10.1146/annurev.energy.30.050504.144248

Chapter 2
Wetland Creation, Restoration, and Conservation:
Wetland Protection

Nisha Khatik
https://orcid.org/0000-0002-4417-8105
Maharshi Dayanand Saraswati University, Ajmer, India

ABSTRACT

Water-dominated landscapes include wetland areas. The term "wetland" has not been commonly used until quite recently. It is believed to be a euphemistic equivalent of "swamp." Every year on the second day of February, World Wetlands Day marks the adoption of the convention on Wetland by Ramsar, the Iranian city that has a special place in Iranian history. Flood protection, water quality improvement, shoreline erosion control, natural products, recreation, and aesthetics are some of the many advantages of wetlands, as well as the fact that they are vital habitats for a variety of animals and plants. Several studies illustrated the importance of wetlands in reducing carbon emissions and regulating climate on a global scale. In recent years, these advantages of wetlands have been recognized by governments worldwide and have led to legislation, regulations, and management plans creating wetlands for conservation, protection, and restoration. Unfortunately, the destruction of wetlands is a concern since they are among the planet's most productive areas.

INTRODUCTION

Wetlands are one of the most expensive ecosystems on the planet (Mitsch & Gosselink, 2007). Wetlands are important for a wide range of processes, and they are key locations in the landscape for the creation, consumption, and maintenance and preservation of many natural cycles along with supporting biodiversity. Waterfowl, shorebirds, wading birds, raptors, loons, grebes, cranes, woodcock, kingfishers, and many songbirds rely upon wetlands during all or part of their life cycles. Wetlands related with springs and leaks might be as small as a few square feet while some Great Lakes swamps or peat lands cover a large number of sections of land. India has a long history and custom of preservation of natural assets.

DOI: 10.4018/978-1-7998-9498-8.ch002

As a feature of strict custom, individuals adore and worship numerous streams and wetlands all through the country. For giving protection to delegate wetlands in India, Ministry of Environment and Forests, Govt. of India has attempted a National Plan for Conservation of Aquatic Ecosystems (NPCA) in the nation and brought National Wetland Conservation Program (NWCP) and National Lake Conservation Plan (NLCP) under its umbrella. A huge number of wetlands including Ramsar destinations, public parks, bird sanctuaries, advised wetlands and lakes get subsidizing for their preservation and action plan readiness under this program. Ministry has additionally sanctioned a few legislations for their protection also, the executives of wetlands including mangroves and coral reefs.

Ramsar convention (1971) on wetlands, the first intergovernmental convention for protection and 'wise utilization' of any environment, gone into force in 1975 and has 168 contracting parties from everywhere the world. So far the convention has 2169 wetland destinations under its aegis covering 206.64 million ha region (Ramsar Convention, 2013). India marked the Convention on February 1, 1982 and has recognized 26 wetlands (Chilka, Keoladev Public Park, Harike, Loktak, Chilka, Wular, Sambhar, Ashtamudi, Bhitarkanika, Bhoj, Deepor Beel, East Calcutta Wetlands, Kolleru, Point Calimere natural life and bird safe-haven, Pong dam lake, Chandratal, Renuka, Sasthamkotta, Tso Moriri, Hokera, Surinsar-Mansar, Vembanad-Kol, Kanjali, Ropar, Rudrasagar, Upper Ganga River-Brijghat to Narora, Bhoj and Nal Sarovar) for bearing the cost of need preservation in the country (Garg, 2013). In fact, more than one-third of the species listed as threatened or endangered in the United States live exclusively in wetlands and almost half use wetlands eventually in their lives (USEPA, 1995).

Wetland creation and Restoration can assist with keeping up with the advantages of wetlands and their encompassing biological systems and simultaneously accommodate the human requirement for advancement. Wetland creation is that the development of a wetland on a site that never was a wetland, while wetland restoration involves returning a wetland to its unique or past wetland state. Consequently, creation is more difficult than restoration. Restoration is the manipulation of the physical, chemical, or biological characteristics of a site with the goal of returning natural/historic functions to former or degraded wetland. For Restoration takes place on land that has been, or still is, a wetland. A term related with restoration is "enhanced." An upgraded wetland is a current wetland that has been adjusted to upgrade a particular function, normally to the detriment of different functions.

Wetland restoration and conservation is crucial to give natural surroundings to a large number of types of aquatic and terrestrial plants and animals. Before the mid-1970s, waste and obliteration of wetlands were acknowledged practices all throughout the world and were even empowered by explicit government policies. Wetlands were supplanted by agricultural fields and by business and private turn of events. The rate at which wetlands are being lost on a worldwide scale is just now getting clear, in part with the use of most recent advances related with satellite symbolism. Thus, in the event that we secure the wetlands and their resources, we additionally ensure our planet and ourselves.

BACKGROUND

Wetland impacts have included both wetland alteration and wetland destruction. In earlier times, wetland waste was considered the only arrangement for overseeing wetlands inside the western world. With over 70% of the world's population living on or close to coastlines, coastal wetlands have long been destroyed through a combination of excessive harvesting, hydrologic adjustment and seawall construction, coastal development, pollution, and other human activities. In like manner, inland wetlands are persistently influenced, eapecially through hydrologic alteration and agrarian and substantial turn of events. By 1985, 56–65% of wetlands in North America and Europe, 27% in Asia, 6% in South America, and 2% in Africa had been depleted for escalated horticulture. Other human activities like ranger service, transfer channelization, hydroponics, dam, and seawall development, mining, water contamination, and ground water withdrawal all had impacts, some extreme, on wetlands. Wetlands are debased and annihilated in a roundabout way likewise through shift or silt designs in waterways, hydrologic variation, roadway development, and land subsidence. Wetlands are additionally overseen near their natural state for specific goals like fish and wildlife upgrade, agrarian and hydroponics creation, water-quality improvement, and flood control. Management of wetlands for coastal conservation has now taken on more significance (Mitch, 2013)

WETLAND CREATION

Wetland creation worries to the change of a diligent upland or shallow water region in to a wetland by human activity. This might be done distinctly on a site where conditions exist which will create and support a wetland. A term usually connected with wetland creation is "constructed." A built wetland could be a wetland made explicitly for the point of treating wastewater, storm water, corrosive mine seepage, or farming spill over (Hammer, 1989). Creation is that the "development of a wetland in a section that was not a wetland inside the new past (inside the last 100-200 years) which is detached from existing wetlands (i.e., not straightforwardly neighbouring)" (Gwin, et al., 1999). At the end of the day, creation happens when a wetland is pot on the scene by some follow up on a non-wetland site (Lewis, 1989). Normally, a wetland is framed by excavation of upland soils to elevations which will uphold the development of wetland species through the foundation of an appropriate hydrology.

CHALLENGES AND RECOMMENDATIONS FOR IMPROVING WETLAND RESTORATION

Wetland restoration mostly incorporates the restoration or reconstruction of degraded or vanished wetlands through natural innovation and environmental designing (Cui &Yan, 2006). During the time spend wetland restoration; consideration ought to be paid to regarding the natural attributes of wetland ecosystems, keeping away from excessive human obstruction with the crude wetlands, all together that they will take advantage their self-recuperation capacity (Mitsch & Gosselink, 2015; Cui & Yan, 2006). The Society for Ecological Restoration (SER) characterizes ecological restoration as "the process of assisting the recovery and management of ecological integrity. Ecological integrity includes a critical range of variability in biodiversity, ecological processes and structures, regional and historical context, and sustainable cultural practices." Ecological issues and actual limits are significant variables to consider when getting ready for wetland restoration or creation. As of late, there has been an expanding utilization of wetlands as supplement traps (e.g. Jansson et al., 1994; Leonardson et al., 1994; Vymazal et al., 1998; Trepel, 2000).

Restoration will be a mind boggling undertaking that incorporates a wide scope of disciplines including ecology, aquatic biology, hydrology and hydraulics, geomorphology, engineering, planning, communications and scientific discipline. Structure and function are firmly connected in river corridors, lakes, wetlands, estuaries and other aquatic resources. Restoration requires a plan upheld the total watershed, not simply a piece of the water body that will be the first debased site. Activities throughout the watershed can have adverse effects on the aquatic resource that is being restored. Exercises all through the watershed can effectsly affect the oceanic asset that is being re-established. A large number of the issues concerning eutrophication of oceanic frameworks are credited to heightening of farming, including expanded treatment and fixing of streams and bringing down of groundwater tables.

A portion of the difficulties distinguished in reports recently distributed are summed up beneath:

- Practical experience and furthermore the accessible science base on restoration and creation are restricted for a few kinds and fluctuate territorially.
- Many wetland restoration and creation projects do not have explicit and quantifiable objectives, entangling endeavours to survey progress towards accomplishing execution and "achievement."
- Monitoring of deliberate wetland restoration and creation projects has been less thorough than for relief projects.
- Knowledge is missing on the way to re- make a thoroughly working wetland that is indistinguishable from the one being lost (despite the fact that the reasons of guidelines are that it is feasible to attempt to do as such).
- Performance assumptions in Section 404 government licenses have regularly been hazy or potentially depended intensely on guideline, and consistence has frequently not been guaranteed, accomplished or very much reported.
- Wetlands are frequently re-established or created without thinking about the more extensive watershed setting.
- Support for administrative choosing is inadequate and a shortage of assets handicaps endeavours to survey projects and recognize approaches to require changes to upgrade execution.
- Lack of satisfactory thoughtfulness regarding soils and hydrology has obstructed our comprehension of an approach to re-establish wetlands. (Kusler & Kentula, 1989; NRC, 2001)

Numerous aquatic resources in need of restoration have issues that started with adjustment the physical, substance or organic attributes, which progressively may have prompted living space debasement, changes in stream systems and siltation. As one answer for the present issue, the turn of events, reconstruction or restoration of wetlands has been started (for example Fleischer et al., 1994; Hoffmann, 1998; Trepel, 2000).Wetland restoration rehabilitates a degraded wetland or re-establishes a wetland that has been destroyed. Restoring the reasonable normal design can bring back useful capacities. Re-foundation winds up in modifying a previous wetland and prompts an addition in wetland sections of land. For instance, restoring the underside rise in a really wetland might be basic for restoring the hydrological system, regular aggravation cycles and supplement motions. Restoration prompts an increase in wetland work, however doesn't end in an addition in wetland sections of land.

TARGETS FOR WETLAND PROTECTION, RESTORATION AND CONSTRUCTION

Wetland restoration alludes to the arrival of a wetland from an upset or modified condition brought about by act to a formerly existing condition. The wetland may are debased or hydrological changed and rebuilding then, at that point may include restoring hydrologic condition store-set up past vegetation networks. The necessity for protection and restoration of wetlands has been perceived for a significant time frame in ecological strategy and nature protection. Beforehand the principle target has for the most part been on wetlands as environments for untamed life. During the last decade the significance of wetlands as sinks and source or potentially hotspots for supplements and ozone harming substances has gotten progressively apparent (Mitsch, 1994). Today, four fundamental focuses for wetland protection, restoration and construction are frequently recognized:

- Improving water quality (catching and change of supplements and toxins)
- Restore or secure wetlands as living spaces for plants and creatures
- Management of Water (flood control, water supply)
- Recreation

In a few, yet not through and through cases the different targets can correspond, for example in wetlands built for supplement maintenance which may likewise expand biodiversity of farming scenes (Kiehl & Weisner, 1997).

PRESENT STATUS OF WETLANDS IN INDIA

In India, a various demonstrations and legal provisions have been applied to conserve the delicate wetland environment, and because of which the Ministry of Environment, Forest and Climate Change, Government of India has pronounced 17 destinations as told wetlands, while 26 have as of now been pronounced as Wetlands of International Importance under Ramsar Convention (Tables 1 and 2). Aside from this, the few wetlands covered by the National Wetland Conservation Program (NWCP) and National Lake Conservation Plan (NLCP) have too expanded to 115 and 61, individually (MoEF 2012). In the valuation of wetlands in India, a sum of 4.63% of the geographic region has been confirmed under wetlands (Fig. 1) (MoEF 2011). India has 63% normal wetland (66, 23,067 ha), and the staying 37% have human-made inland wetlands (39, 41,832 ha).

Table 1. Present status (numeric) of wetlands in India

S. No.	Particulars	Status (Total No.)
1.	Total no. of wetlands in the country as per the latest National Wetland Atlas	55,862
2.	No. of natural wetlands under conservation:	
	a. Wetlands designated as Ramsar sites and under NWCP	119
	b. Wetlands under NLCP	61
	c. Wetlands under NRCP 39	39
	d. Overall no. of wetlands under conservation	219

Sources: Data compiled from MoEF (2007, 2012) and Ramsar Convention on Wetlands (2012)

Table 2. List of wetlands designated as Ramsar sites in India and their criteria

S. No.	Wetland's Name	State	Ramsar Criteria
1.	Ashtamudi Wetland	Kerala	1,2,3,8
2.	Bhitarkanika Mangroves	Orissa	2,4,6,8,9
3.	Bhoj Wetland	Madhya Pradesh	2,4,5,6
4.	Chandertal Wetland	Himachal Pradesh	2,3
5.	Chilika Lake	Orissa	2,4,5,6,8,9
6.	Deepor Beel	Assam	2,5
7.	East Calcutta Wetlands	West Bengal	1
8.	Harike Lake	Punjab	2,5,6
9.	Hokera Wetland	Jammu and Kashmir	2,5,6
10.	Kanjli	Punjab	3
11.	Keoladeo National Park	Rajasthan	2,5,6
12.	Kolleru Lake	Andhra Pradesh	2,4,5,6
13.	Loktak Lake	Manipur	2,5,6
14.	Nal Sarovar Bird Sanctuary	Gujarat	2,5,6
15.	Point Calimere Wildlife and Bird Sanctuary	Tamil Nadu	2,4,5
16.	Pong Dam Lake	Himachal Pradesh	2,5,6
17.	Renuka Wetland	Himachal Pradesh	3,4
18.	Ropar	Punjab	5,6
19.	Rudrasagar Lake	Tripura	2,3,8
20.	Sambhar Lake	Rajasthan	2,5,6
21.	Sasthamkotta Lake	Kerala	1,2,7,8
22.	Surinsar-Mansar Lakes	Jammu and Kashmir	2,3,4
23.	Tso Moriri	Jammu and Kashmir	2,6
24.	Upper Ganga River (Brijghat to Narora Stretch)	Uttar Pradesh	2,5
25.	Vembanad-Kol Wetland	Kerala	4,5,6
26.	Wular Lake	Jammu and Kashmir	2,5,6

Source: Ministry of Environment and Forest, Govt. of India

WETLAND PROTECTION AND RESTORATION ACT

Wetlands do confront the awfulness of lodge, as is obvious from present quality and steep decrease in their numbers. The excellent justification this state is primarily a direct result of absence of coordination among numerous organizations required inside the administration and suitable legitimate measures to protect these biological systems. As on today, Wetlands are not outlined under a particular managerial locale. A few wetlands are ensured after the detailing of the Wildlife Protection Act. In any case, it is insufficient and most are in grave peril of extinction. Compelling coordination between the different services, energy, industry, fisheries income, horticulture, transport and water assets, is fundamental for the assurance of those biological systems. Prevailing laws are incapable to the extent the protection or preservation of amphibian environments are worried as a large portion of them by implication contact wetland protection (fragmented approach);

- The Indian Fisheries Act - 1857
- The Indian Forest Act - 1927
- Wildlife (Protection) Act - 1972
- Water (Prevention and Control of Pollution) Act - 1974
- Territorial Water, Continental Shelf, Exclusive Economic Zone and other Marine Zones Act - 1976
- Water (Prevention and Control of Pollution) Act - 1977
- Maritime Zone of India (Regulation and fishing by foreign vessels) Act - 1980
- Forest (Conservation) Act - 1980
- Environmental (Protection) Act - 1986
- Coastal Zone Regulation Notification - 1991
- Wildlife (Protection) Amendment Act – 1991
- National Conservation Strategy and Policy Statement on Environment and Development -1992

India, in spite of being a signatory to the Ramsar Convention on Wetlands and furthermore the Convention of Biological Diversity, there is no critical improvement towards supporting these environments, either because of absence of coordination among organizations included or absence of consciousness of the upsides of wetlands among the arrangement producers and execution offices. The powerful administration of those wetlands requires an intensive evaluation of the current laws, foundations and practices. The inclusion of grouped individuals from various areas is urgent inside the supportable administration of those wetlands (Ramachandra & Ahalya, 2001).

WETLAND CONSERVATION METHODS AND STRATEGIES

Eighteen years after the term 'wetland' was officially utilized, wetlands turned into the main natural surroundings to be the focal point of a worldwide show, presently known as the Ramsar Convention, perceiving their significance particularly as waterfowl environments (Carp, 1972; Matthews, 1993). On understanding their value, individuals and state run administrations began laying weight on conservation of the wetlands. An exceptional spray of interest in wetlands was seen at all levels among scientists, engineers, administrators, traditionalists, sociologists, financial specialists, different NGOs, or more every one of the Governments. A great deal of studies is being led to comprehend wetland environments prompting many examination papers and books. Various associations are working for preservation and feasible administration of wetlands across the world. Policy makers additionally understand the worth of wetlands. However, despite this load of positive drives, a significant number of our wetlands don't stop to be out of danger. One interesting example is Dal Lake in Kashmir valley, where spending reserves summing huge number of crores doesn't appear to be working. Lake Wular has a similar story as well. Three of our best wetlands, Keoladeo National Park, Bharatpur, Rajasthan and Loktak Lake in Manipur and Chilika Lake in Orissa had been set on Montreux Record long back. Chilika has been eliminated of the register following improvement in its condition yet Loktak and Keoladeo keep on being on it. Why, disregarding investing in such a large amount energy and money, is their condition not improving, is a state of discussion. It is necessitated that different holes in protection measure be recognized so these significant environments can be moderated. The methodologies for preservation and the executives of a wetland should be chosen keeping in see the neighbourhood topography, hydrology, environment, financial state of nearby networks and strategy system. It is no insight to settle on choices without thinking about the ground-level situation. The protection techniques for a wetland should be founded on the sound logical examination (Sharma et al., 2014).

WETLAND CONSERVATION PROGRAMS

The Ramsar Convention, a global deal pointed toward conserving wetlands, requires member countries to develop national wetland policies, to find out wetland holds, and to assign at least one wetland as a space of worldwide significance (Cherry, 2011). Environmental Protection Agency (EPA), states, and tribes share an expansive objective for their wetlands programs: to protect and restore wetland assets for their economic, environmental and human medical advantages. The Enhancing State and Tribal Programs (ESTP) Initiative is made upon a system of 4 Core Elements that are the establishment of a state or tribal wetlands management and protection plan. The Core Elements Framework (CEF) was intended to help states and tribes develop Wetland Program Plans (WPP).

The Core Elements Framework (CEF) serves as an establishment for improving state and tribal program improvement. It is upheld state and tribal wetlands program insight in the course of the last 10-15 years and was created with state and tribal information. EPA distinguishes four core elements that involve and reinforce viable wetlands programs. The core components are fundamental program works that form the inspiration of wetlands management and protection during a state or tribe. These include:

1. Monitoring and Assessment;
2. Regulatory activities including 401 certification;
3. Voluntary Restoration and Protection;
4. Water Quality Standards for wetlands.

Previously, EPA has embraced six core elements of state and tribal wetlands programs: Regulation, Monitoring and Assessment, Restoration, Waters Quality Standards, Public Private Partnerships, and Coordination among State and Federal Agencies.

Wetland Program Plans (WPPs) are voluntary plans created and carried out by state organizations and tribes which articulate what these substances need to achieve with their wetland programs over time. WPPs describe overall program objectives along with wide based activities and more explicit exercises which will assist with accomplishing the objectives

FUTURE CHALLENGES

Wetlands are presumably the world's most essential freshwater asset and they perform complex capacities inside the upkeep of the environmental equilibrium of the district like water stockpiling, security from tempests and floods, re-energize of ground water, water decontamination, fisheries, horticulture, untamed life asset, transport, entertainment, and so forth Late worries over wetlands misfortunes have produced a prerequisite for making of wetlands, which are expected to copy the capacities and upsides of regular wetlands that are annihilated. Restoration is subsequently a brilliant chance to oversee wetlands for expansive untamed life objectives, as re-established wetlands give upgraded natural life benefits, also to different advantages, simultaneously. The fate of humankind relies upon wetlands. Wetlands are number of the principal significant bio-assorted regions inside the world. A significant number of the difficulties of things to come are regularly met through moderating and economically utilizing wetlands, similar to food and water security, human wellbeing, calamity hazard decrease and environmental change flexibility.

REFERENCES

Babu, M. S. U., & Nautiyal, S. (2015). Conservation and Management of Forest Resources in India: Ancient and Current Perspectives. *Natural Resources*, *6*(04), 256–272. doi:10.4236/nr.2015.64023

Carp, E. (Ed.). (1972). *Proceedings of the International Conference on the Conservation of Wetlands and Waterfowl.* Ramsar, Iran: IWRB.

Cherry, J. A. (2011). Ecology of Wetland Ecosystems: Water, Substrate, and Life. *Nature Education Knowledge*, *3*(10), 16.

Cui, B. S., & Yang, Z. F. (2006). Wetland Science. Beijing Normal University Press.

Fleischer, S., Gustafson, A., Joelsson, A., Pansar, J., & Stibe, L. (1994). Nitrogen removal in created ponds. *Ambio*, *6*, 349–357.

Garg, J. K. (2013). Wetland assessment, monitoring and management in India using geospatial techniques. *Journal of Environmental Management*, 1–12. doi:10.1016/j.jenvman.2013.12.018 PMID:24486190

Gwin, S. E., Kentula, M. E., & Shaffer, P. W. (1999). Evaluating the Effects of Wetland Regulation through Hydrogeomorphic Classification and Landscape Profiles. *Wetlands*, *19*(3), 477–489. doi:10.1007/BF03161687

Hoffmann, C. C. H. (1998). *Nutrient retention in wet meadows and fens* (PhD thesis). Ministry of Environment and Energy, National Environmental Research Institute, Silkeborg, Denmark.

Jansson, M., Andersson, R., Berggren, H., & Leonardson, L. (1994). Wetlands and lakes as nitrogen traps. *Ambio*, *23*, 320–325.

Kusler, J. A., & Kentula, M. E. (1989a). Wetland Creation and Restoration: The Status of The Science. Vol. I. Regional review. EPA/600/3–89/038. USEPA, Environmental Research Laboratory.

Leonardson, L., Bengtson, L., Davidsson, T., Person, T., & Emanuelsson, U. (1994). Nitrogen retention in artificially flooded meadows. *Ambio*, *23*, 332–341.

Lewis, R. R., III. (1989). Wetland restoration/creation/enhancement terminology: Suggestions for standardization. Wetland Creation and Restoration: The Status of the Science, Vol. II. EPA 600/3/89/038B. U.S. Environmental Protection Agency.

Ministry of Environment and Forests. (2007). *Conservation of wetlands in India: a profile (approach and guidelines). MoEF*. Government of India.

Mitsch, W. J. (1994). *Global wetlands - old world and new*. Elsevier.

Mitsch, W. J., & Gosselink, J. G. (2007). *Wetlands* (4th ed.). John Wiley and Sons.

Mitsch, W. J., & Gosselink, J. G. (2015). *Wetlands*. John Wiley & Sons.

Mitsch, W. L. (2013). Wetland creation and restoration. Encyclopedia of Biodiversity, 7, 367-383. doi:10.1016/B978-0-12-384719-5.00221-5

National Research Council (NRC). (2001). *Compensating for Wetland Losses Under the Clean Water Act*. National Academy Press.

New Delhi Ministry of Environment and Forests. (2011). *National wetland inventory and assessment report. MoEF*. Government of India.

New Delhi Ministry of Environment and Forests. (2012). *Annual report 2011–12. MoEF*. Government of India.

Ramachandra, T. V., & Ahalya, N. (2001). Wetlands Restoration and Conservation. In *Proceedings of National Conference on 'Control of Industrial Pollution and Environmental Degradation'*. Department of Civil Engineering, PSG College of Technology.

Ramsar Convention. (2013). *The Ramsar Convention Bureau*. www.ramsar.org

Ramsar Convention on Wetlands. (2012). *The annotated Ramsar list: India*. The Secretariat of the Convention on Wetlands.

Sharma, S., Rattan, R., Jamwal, P. S., Kumar, R., & Shukla, S. (2014). *Wetlands: Strategy for Natural Resource Management through Wetlands Conservation in India. Bharti Publications.*

Trepel, M. (2000). Quantifizierung der Stickstoffdynamik von Ökosystemen auf Niedermoorböden mit dem Modellsystem WASMOD. *EcoSys*, *29*, 1–140.

United States Environmental Protection Agency (USEPA). (1995). *America's Wetlands: Our Vital Link Between Land and Water*. Office of Water, Office of Wetlands, Oceans, and Watersheds.

Vymazal, J., Brix, H., Cooper, P. F., Green, M. B., & Haberl, R. (1998). *Constructed wetlands for wastewater treatment in Europe*. Backhuys Publishers.

Chapter 3
Water Pollution of Wetlands:
A Global Threat to Inland, Wetland, and Aquatic Phytodiversity

Soumi Datta
https://orcid.org/0000-0003-0360-4910
Jamia Hamdard, India

Dwaipayan Sinha
https://orcid.org/0000-0001-7870-8998
Government General Degree College, Mohanpur, India

Vidhi Chaudhary
Daulat Ram College, India

Somnath Kar
Holy Cross College, India

Anjana Singh
Deshbandhu College, India

ABSTRACT

Pollution has become a matter of grave concern at present with all the components of the environment laden with pollutants largely from anthropogenic sources and unplanned urbanization. Inland wetlands are very delicate ecosystems and encompass a variety of water bodies, namely ponds, rivers, swamps, etc. They house some unique floristic patterns that are crucial in the primary productivity and maintaining a balance of the wetland ecosystem. In addition to it, the inland water bodies are also productive and are of immense importance to humans. The inland wetlands are also an integral part of boosting the economy of the region as they support a number of industries including fishing and recreation. Thus pollution of water bodies has impacted the human race in a deleterious manner. This chapter is an attempt to overview the inland water bodies, their biodiversity pattern, pollution, and their effect on flora at large.

DOI: 10.4018/978-1-7998-9498-8.ch003

BACKGROUND

A distinct ecosystem flooded by water, either permanently or seasonally, where oxygen-free processes prevail, is known as '*wetlands*' and the main factor that differentiates it from other water bodies is its characteristic vegetation of aquatic plants, adapted to distinctive hydric soil (Fagorite et al., 2019). Inland wetlands consist of inland water bodies like lakes, rivers, ponds, streams, springs, cave waters, floodplains, as well as bogs, marshes, and swamps (U.S.Fish & Wildlife Service, 2021). Since the inland aquatic ecosystems are ecologically dynamic, therefore cannot be differentiated by artificial conceptual boundaries, and are best studied from an ecosystem or landscape perspective (Keddy, 2010). Thus, we consider a holistic approach towards water pollution of 'Inland Water Bodies (IWBs)' as'wetlands' are a part of IWBs.

These IWBs possess several characteristics like physical, chemical, and biological. IWBs are broadly categorized into fresh, saline, and brackish water (Radwan, 2020) however the terminology is often perplexing like IWBs may include freshwater (Caspian Sea) and hyper-saline (Dead Sea), and exclude largely freshwater (Baltic Sea). In the deep interiors of Australia, several wetlands are saline; while others, just a few meters of the ocean are freshwater; thus, it is the location that matters and not their ecological nature. The most diversified ecosystems arethe 'wetlands' as they shelter a varied array of animal and plant species that live permanently in wetlands or rely on wetlands for at least part of their life cycle (EPA, 2021); especially significant for migratory water bird species (NPS, 2016). Therefore, we should view the IWBs from a broader perspective concerning flora and fauna, and landscape rather than narrowing them to superficial boundaries (Keddy, 2010).

The concept of "aquatic biodiversity" embrace variety at three different levels, namely, species, genetic and ecosystem commonly comprises of all forms of life that are dependent on marine habitat except drinking or transpiration in plants (WHO, UN, 2015); however, since the entire terrestrial flora and fauna depend on freshwater, the territory between water and land are indistinct. Apart from the aquatic life (e.g. fish); it may also include other forms of "terrestrial" species of fauna (e.g. water birds), semi-aquatic creatures (e.g. crocodiles, hippopotamus, beaver), and flora (e.g. mangroves, flooded forest, water-bound foliage, etc.); whereas the majority of amphibians breed in freshwater. There are immovable species (e.g., freshwater fish), which are restricted to inland waters only (CBD, 2008), and thus such water bodies are endemic to freshwater species; and these species often possess high genetic diversity, like, higher reaches of rivers sub-catchments areas or various lakes, or located at proximity.

FLORISTIC DIVERSITY AND SUCCESSIONAL PATTERNS IN WETLANDS

Wetlands represent the land areas that are covered by water bodies and formed due to flood i.e., flood-water (perpetual or periodic), strong tides (in coastal areas), percolation of water from a spring or an aquifer, etc. (Kundu, 2018). According to Cherry (2011), they are the transitional areas situated between land-based or terrestrial ecosystems and water-based or aquatic ecosystems.

Wetlands are the habitat for aquatic plants which are also called hydrophytes (Cronk and Fennessy, 2009). Hydrophytes or water-loving plants adapted to living in hydric soil throughout the year. In wetlands, plants are recognized as an indicator (Ling et al., 2019). Even plants also performa vital part in the wetlands classification (Shelef et al., 2013).

In wetlands, circumstances vary extensively resulting frequent alteration of the population to a wide array of the water quality. It includes turbidity, salinity, oxygen availability, pollution level, amount and availability of water, temperature etc (Bhateria and Jain, 2016).

Wetland floras are generally considered obligate aquatic species or facultative aquatic species (Sieben et al., 2021). Mitsch et al., (2012) remarked that wetland plants have to be partly or entirely submerged in water, at least have to remain in that condition for one of their developmental phases. Wetlands essentially have to endure anaerobic soils (Tiner, 1991; Cronk and Fennessy, 2009).

Wetlands plants are usually categorized as emergent, submergent, floating, and riparian. Emergent plants are rooted in the soil and some parts are outspread above the water level. Submerged plants live in submerged conditions. Floating plants having leaves that float on the surface on the other hand roots are either drape down in the water or fixed in the soil. While Riparian plant types are found at the wetland edges. According to Tiner,2009, wetlands may be classified into four categories based on their dominant flora:

1. Emergent wetlands are inhabited by herbaceous plants.
2. Shrub wetlands possess multi-branched woody plants characteristically having height > 6 m.
3. Forested wetlands which are tree dominated
4. Aquatic wetlands which possess floating or submerged plants.

Emergent wetlands can be divided again into three broad categories (Tiner, 2009) such as

1. Marshes.
2. Fens.
3. Wet meadows.
    a. Marshes- Marshes are those wetlands that are frequently flooded (Keddy, 2008) and they are generally inhabited by *Pontederia crassipes*, *Lemna* sp, *Naja* sp., *Nymphaea* sp., *Wolffia* sp., *Pistia* sp., *Salvinia* sp., *Chara* sp., *Ceratophyllum* sp., *Vallisneria* sp., *Hydrilla* sp., *Azolla* sp., etc.
    b. Fens- Fens are those wetlands which are having slightly mineral-rich peat soil (Keddy, 2010) and generally inhabited by *Carex* sp., *Scirpus cespitosus*, *Juncus* sp., *Scheuch zeriapalustris*, *Pinguicula vulgaris*, *Selaginella* sp., *Myrica gale*, *Salix* sp., etc.
    c. Wet meadows- These wetlands are having mineral-rich soil (Keddy, 2008) that promotes *Eupatoriumsp, Mimulussp, Bidenssp, Juncus* sp., *Agrostis* sp., *Poa* sp., *Carex* sp., *Polygonum* sp., *Equisetum* sp. to grow there.

Shrub wetlands are very frequent and again two kinds(Tiner, 2009) such as a) shrub swamps having mineral-rich soils and b) shrub bogs having low nutrient and acidic content. *Alnus* sp., *Salix* sp., *Syzygium* sp., *Dalbergia* sp., *Mimosa* sp. are the common plants found in Shrub swamps wetlands whereas Shrub bog wetlands possess *Andromeda* sp., *Kalmia* sp., *Empetrum* sp., *Betula* sp., *Rhodora* sp., *Ilex* sp., *Drosera* sp., *Utricularia* sp., *Eriophorum* sp., *Rhynchospora* sp., *Sphagnum* sp., *Cladonia* sp., etc.

Fens are generally predominated tree species or shrub species mainly *Populus* sp., *Quercus* sp., *Betula* sp., *Magnolia* sp., *Thuja* sp., *Eucalyptus* sp., *Shorea* sp., *Litea* sp., *Bombax* sp., *Cassia* sp., *Ficus* sp., *Phoenix* sp., *Mitragyna* sp., *Croton* sp., *Terminalia* sp., *etc.* Forested wetlands may include *Picea* sp., *Tsuga* sp., *Nothofagus* sp., *Sphagnum* sp., etc.

Succession denotes the alteration in floras due to interruption as well as purulence of plant species (Van der Valk, 1981; Emery, 2010). It refers to the change of species in the case of wetlands, these disturbances or interruptions comprise water quality particularly change in the salinity level in addition to other aspects such as herbivores, natural calamities, fire, etc that can modify vegetation. Clements (1916) stated that the furtherance of succession in the direction of a climax state is occurred due to natural disturbances. Each several stages are having a set of intertwined species. Generally, wetland succession characteristically ensues in four typical stages such as:

1. Stage 1- these include deep water with diminutive foliage levels.
2. Stage 2- these include low water levels subjugated by submerged as well as floating water plants.
3. Stage 3- these include very low water levels or wet ground subjugated by emergent aquatic floras.
4. Stage 4- these include parched ground and dominated by adjacent highland floras.

However, Instabilities in water availability and water heights can root for the revert of a former phase instead of the final succession stage. It has been also observed that the wetlands which possess ample deep-water level or lithic base show a slow succession rate.

Now a day, the disturbance initiates successional deviations in plant communities that ensure species activities withinthe community perform autonomously. Gleason (1926) provided an idea on succession. Van der Valk (1981), also observed succession in the North American wetlands i.e."Prairie pothole wetlands". Succession also observed wetlands formed during the rainy season in Australia (Finlayson, 1991) and India (Middleton, 1999). There is another model of wetland succession given by van der Valk(1981) that ensues sequentially which sustains wetlands over ages. This continues as stagnated freshwater marshes are devastated by instabilities due to inflowing tide and herbivory resulting in the formation of open marsh due to vegetation destruction. These conditions remain upheld for ages as the nascent foliage cannot sprout in water-logged situations.

Middleton (1999) and Parolin and Wittmann (2010) observed that the Indian, African, Australian and South American wetlandsthat are formed during rainy seasons are repeatedly held yearly of drought and flood cycles, resulting in the occurrence of successional cycle ensues in a short period.

According to Niering (1987), several forms are occasionally used to manifest wetland though several stages. Vegetation studies along with palynology, and other interrelated tactics displayed that the conditional several phases are not correlated to vegetation transition over time (Twiddle, 2012). By eradicating the dominant plants; disturbances perform on the vegetation resulting upkeep of wetlands phytodiversity by providing a space of contingency to the subsidiary species to amplify.

POLLUTION IN INLAND WETLANDS: A GLOBAL SCENARIO

The global wetlands are at greater risk from pollution, threatening the drinking water supply for people and decreasing biodiversity, but at times it may provide the solution to the problem.The inland waters not only include lakes, rivers, and freshwaters; but it is also inclusive of saline and brackish water; and these are usually contaminated by two major ways, firstly by pollutants deposited within the nearby terrain, that drains into the river basin; and secondly by direct discard into water. The pollution of inland water is caused by several other factors like pesticides, water sanitization, and pollution of underground water, lake, stagnant canals, rivers, and accumulation of ferminatization of male fish. Accordingly to

Figure 1. Types of impact factors affecting wetlands with their respective percentage are pollution (54%), use of biological resources (53%), natural system modification (53%), and agriculture and aquaculture (42%).

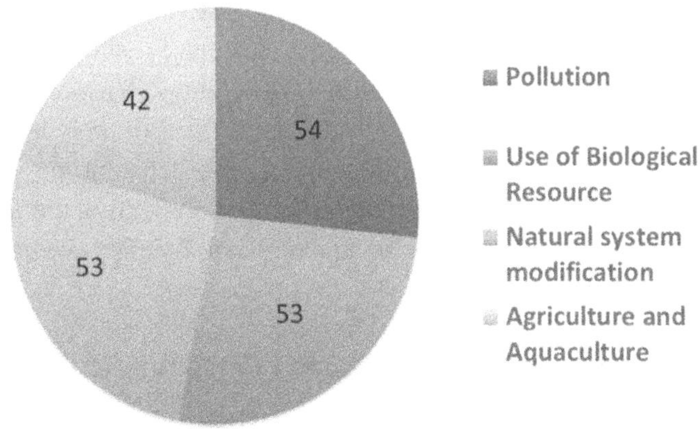

United Nations, wastewater is approximately comprised of 99% water and 1% suspended, dissolved, and colloidal solids (UN-Water, 2015a).

A total of 2303 wetlands unevenly distributed in different continents are of international importance; where Africa owns the largest area of sites and Europe largest number of sites. About fifty-five percent of sites are majorly affected by four impact factors-i.e. pollution, use of biological resources, modification of natural system, agriculture and aquaculture (Fig.1) The type of impact on various wetlands varies, i.e. river and lake are mostly influenced by land area and faces maximum impact on phytodiversity; marshes by regulation of water sources and marine/coastal wetlands by climate change (Xu et al., 2019).

With the increasing technologies, modernization, escalating population, and rapid rise of the economy over decades now, discarded andwaste materials are growingly been disposed into marine bodies, and such toxic products are highly hazardous to nature and have serious implications; especially in scanty water bodies like streams, canals,etc as decomposition of invertebrate species may provide real indications to aquatic pollution. In many coastal regions like India and China, the issue of water pollution is highlighted due to aquatic saline imposition into underground water due to over-exploitationof aquifers and the rise of sea level (Post VEA, 2005; Schwarzenbach et al., 2010). In the last 30 years, the rising 'Persistent Organic Pollutants (POPs)' like PBDEs have resulted in an exponential increase by a factor of approximately a hundred in concentration among human tissues, and have multiplied twice over five years in different locations around the globe covering various exposure channels (mainly terrestrial); and such traits are also seen in aquatic mammals of Northern America and Northern Europe (Hites, 2004). An application of pesticide in city areas by an enhanced overflow of rainwater containing these chemicals over sealed surfaces like roads and roofs have resulted in water pollution in drainage systems (Kristoffersen et al., 2008); while the effect of such an exposure mechanism is still questionable in Europe and North America (Calvert et al., 2008; Steerenberg et al., 2008), but unintentional exposure-conscious misuse of pesticides is more common in developing countries (Oluwole and Cheke, 2009; Williamson et al., 2008); and have resulted in the poisoning of approximately three millionpeople with around twenty thousand accidental deaths annually (UNEP,2007). Pollution by geogenic contaminants like the disposition of arsenic in underground water is reported to affect around seventy-five million people in Bangladesh (Chen, Parvez and Gamble, 2009), six million in India (West Bengal) (Haque et al., 2003); and various other parts of the world especially in thickly populated river deltas like Vietnam-and Cambodia (Berg et al., 2007), United States (Frost et al., 2003; Peters, 2008); and Eastern Europe. Chemical spills (hazardous or radioactive material) have resulted in contamination of groundwater, as hundreds of milliontons of wastes are discarded in landfills (EEA, 2000; EEA, 2000; EPA, 2008; Giusti, 2009). Mining operations like gold miningwith mercury have resulted in contamination of water in around fifty countries including Indonesia, Tanzania, Brazil,and Vietnam (Hinton et al., 2003). A study reported in Brazil (Eisler, 2003) anticipates over hundred tons of mercury been discarded into the ecosystem annually, half of it mobilized into surface waters and helps biomagnification of mercury to 106-fold inpredatory fish and subsequently enters the food chain affecting the native population. In the next phase, the threats and impacts of pollution to aquatic phytodiversity would be discussed in detail.

GLOBAL THREATS TO INLAND AQUATIC PHYTODIVERSITY

Natural Threats

"Thousands have lived without love, not one without water." W. H. Auden

Water is very important for life, yet we pollute! It is reported that around eighty percent of the global wastewater is discarded to the environment, mostly untreated, thus polluting the water bodies like oceans, rivers, and lakes (UNESCO, WWAP, 2017). Water is a "universal solvent", so its capacity to dissolve substances is highest than any other solvent; and thus is exclusively susceptible to contamination. The previous century (the 1780s-1980s) witnessed a steady and rapid disappearance of wetlands across the globe, for example-approximately, twenty-seven percentage of wetlands was lost in Montana, USA (Votteler and Muir, 2002). Some of the major causes of wetland loss are the introduction of non-native species or alien or alien invasive species like Hydrilla, Gypsy Moth, Kudzu Nutria, Sea Lamprey and Zebra Mussel ; air and water pollution, logging, and runoff.

The natural causes of water pollution are mainly volcanoes, animal waste, algal blooms, and deposits from storms and floods; while natural contamination of water is of two types, namely-aquifers get contaminated by dissolving various elements found in different layers of soil with underground water; fluorides been most important element naturally found, followed by arsenic (frequent metalloid present in the atmosphere, soil and the hydrosphere) and other natural contaminants like Mn, Rn, Cs, and Li (Sharma and Bhattacharya, 2017). The epidemic caused by fluorides (called 'fluorosis') has escalated manifold, as a result of over exploitation of aquifers that compelled the localities to dig to greater depths, sometimes may reach up to 200 meters, have resulted in the contact of underground water zones with soils for more than thousands of years, and this "*old water*'' often contain higher fluoride and arsenic concentrations (Jha et al., 2013). For example, the aquifers at Guanajuato (Mexico) have been over exploited by commercial, agricultural, and tourist activities due to unplanned hydrology management. Toxic materials from various sources like urban areas, agricultural farms, and industries dissolve these pollutants and contaminate the water bodies. The various types of water pollution are illustrated in Fig.2 .

There is clear scientific evidence that the changing climate will continue to change, thus affecting the masses through the water as it will influence the quality, quantity, and availability required for primary human needs.

Figure 2. Types of water pollution

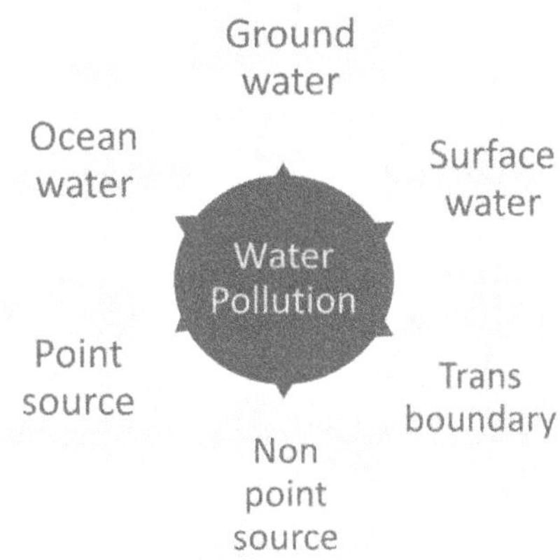

Anthropogenic Threats

Unplanned Urbanization

A report states that in the next one or two decades, the greatest urbanization will take place in smaller city centers (between 500,000 and 1 million inhabitants) and this will immensely impact the production of wastewater (UN-Habitat, 2016). Unplanned urbanization is often linked to the pollution of water bodies in urban areas. The increasing pollution of the wetland is greatly influenced by human activities like hydrologic alterations, pollution inputs, and vegetation damage (EPA, 2001).

The key nonpoint and point pollutants areroad salts, transportation, commercial and residential waste, heavy metals, trash, microbes, petroleum hydrocarbons, sediment, oxygen-requiring materials, contaminated rainstorm water, nutrients, etc. overspills from city landscapes, and getdeposited into nearby water bodies. Roadside construction, deficittownship planning, sanitation lead, and supply of watercater to sediment erosion. Extensive urbanization and human settlements are the major causes of water pollution; and have severe consequences on the hydrological cycle, namely flooding (caused by more soil sealing), alteration in groundwater, lakes, river systems, and aquatic pollution (Strohschon et al., 2013) and scarcity of water (as a result of increased consumption).Reports have shown that lack of proper groundwater withdrawal, correct analysis of water resources, and loss of precious potable water-through seepages, inadequate data analysis of possible existing aquifers, ineffective sewage and water supply system account for pollution in water bodies (Wakode et al., 2018).

Deposition of Industrial Wastes

Water pollution due to urbanization and industrialization is a serious global concern influencing the biosphere (Mathuthu et al., 1997). Various toxic substances and high concentration of elemental nutri-entsare present in untreated wastewaters discharged from industrial effluents into aquatic bodies, causing algal blooms and pathogenic organisms which deteriorate the aquatic environment (Cornish et al., 1999). Many of these compounds are carcinogenic (Kinuthia et al., 2020; Ozturk and Arici, 2021), and they can build up in the tissues of organisms as they progress through the food web's trophic levels (Raamsdonk et al., 2020).The hydrogeological and hydrological conditions in a contaminated zone are substantially altered by the inflow of polluted deposits and wastewater lakes and streams which ultimately results in deterioration of ground and surface water quality (Khatri and Tyagi, 2015; McGrane et al., 2016; Bi et al., 2018).The impact of the wetland environment varies with the wetland locations as well as industries type.

The waste discharged from the industries may contain the following types of contaminants:

- **Biodegradable Organic Substances:** The predominant pollutant load in industrial effluents processing plant or animal products, as well as wastes, is a mixture of complex organic compounds such as carbohydrates, proteins, and lipids (Alves et al., 2009; Alibardi and Cossu, 2016; Slavov, 2017).They can be decomposed by the natural decomposing microorganisms when they are introduced into the natural milieu (Wan et al., 2021).
- **Plant Nutrients:** Nutrient availability is a key regulator of aquatic ecosystem biological productivity (Zehr and Ward, 2002).The majority of young lakes are oligotrophic or nutrient-poor (Nyanda, 2000), but they naturally gets enriched with time as a result deposition of industrial effluents (Yan and Ma, 2010).

Waste from animals and humans, effluents from industry, runoffs from agricultureadds up excess nutrients to waterways (Manuel, 2014). Excess nutrients are harmful and cause algal blooms (Christensen et al., 2019), oxygen deficiencies (Malone and Newton, 2020), and an increase in turbidity (Järvenpääand Lindström, 2004). Anaerobic conditions result from a lack of dissolved oxygen and general deterioration of the water body's ecological and aesthetic aspects is observed.

- **Pathogenic Organisms:** Untreated sewage is a major cause of water-borne disease transmission (Tulchinsky et al., 2018; Zahedi et al., 2021). They are caused by water-borne pathogenic organisms which may be derived from various anthropogenic as well as industrial activities with inadequate wastewater treatment facilities (Pandey et al., 2014; Alegbeleye and Sant'Ana, 2020). The untreated effluents being discharged into the streams may result in a significant build-up of pollutants and could have a significant negative impact on the quality of water in receiving streams (Preisner et al., 2020).
- **Heavy Metals:** High heavy metal concentration in wetlands is a major threat to the protection and sustainable utilization of water resources (Jaishankar et al., 2014; Hon et al., 2020). Metals are generally dissolved in waterbodies by natural or man-made sources. Metal ions are dispersed-during their passagethrough various biotic or abiotic compartments of aquatic ecosystems such as fish, water, and sediment, plants (Ho et al., 2012). Heavy metals that persist in contaminated sediments can accumulate in bacteria, which can then enter the food chain ultimately causing toxicity in molecular levels in the cell of an organism (Agarwal and Khan, 2020) which in turn gets manifested by several physiological symptoms (Shakeri and Moore, 2010).

Deposition of Fertilizers, Pesticides, and Insecticides (Run-Offs)

Maintaining the quality of wetlands and other aquatic water bodiesis the most petrifying challenge human beings are facing universally in the twenty-first century. Anthropogenic activities threaten the healthy life of not only human beings but also result in the defilement of aquatic water bodies, particularly wetlands, resulting in the unloading of countless pollutants into the aquatic environments, thereby, making it unsuitable. Aquatic environments are gatherers forwastes like detergents, industrial solvents, fertilizers, volatile organic compounds, heavy metals, radioactive elements, microfibers, plastics, fuel combustion, nanoparticles, antibiotics, personal care products, drugs, veterinary medicines, fertilizers, insecticides,and pesticides (Bhat et al., 2017; Hampel et al., 2015) leading to water pollution and resulting in the unfit water body. These pollutants have the potential to inflict primary or secondary damage, as well as have a direct effect on the ecosystem or in the form of noticeableflutter in the delicate balance of the inter-related biological food chains that is perceivable after a long time (Ghani, 2015; Naggar et al., 2014). In 2017, Mateo-Sagasta et al., observed that globally, 80% of untreated wastewater from the municipality is released into water bodies,and industry is accountable for throwing away millions of tons of heavy metals, toxic sludge, solvents, and other wastes every year into these aquatic environs, whereas agriculture plays a key role in water contamination and enterprise about 70% of water (Mateo-Sagasta et al., 2017). Although there is a natural tendency of dilution of mostof the aquatic ecosystems, severe contamination mayaffect the alteration in the flora and fauna of the community due to the massive release of drug residues, organic matter, saline drainage, sediments, and agrochemicals from agricultural lands (Mateo-Sagasta et al., 2017). Furthermore, until a threshold level is reached, pollution of these water bodies does not receive significant consideration, ultimately posing serious threats to human health, activities, and the aquatic ecosystem (Halpern et al., 2008). In advanced countries, the major factor in the degradation of water reservoirs is agriculture. According to Connor (2015), 38% of water bodies in the European Union are significantly under agricultural pressure. As per USEPA 2016, agriculture is the paramount source of pollution in streams and rivers, wetlands are the next source and the third main source is the lakes. According to FAO 2013, China is suffering from groundwater pollutionby the release of a huge amount of agriculture products like nitrogen. In developing nations, limitless untreated municipal and industrial waste waters are causing harmful menace. In the world, China (1,806 million Kg/year), US (386 million Kg/year), Argentina (265 million Kg/year), Thailand (87 million Kg/year), Brazil (76 million Kg/year), Italy (63 million Kg/year), France (62 million Kg/year), Canada (54 million Kg/year), Japan (52 million Kg/year) and India (40 million Kg/year) are amongst the top pesticide consuming countries (Carvalho,2017). On one hand, pesticides improve the crop productivity in agricultural land andpromote faster growth to plants, therefore, improve the economy of the world by meeting the demands. On the other hand, the higher exposure to pesticides may influence the soil quality, water purity, air quality, ultimately leading to bio-magnification (Carvalho,2017). As an alternative, eco-friendly pesticides may be usedwhich better service to the environment and human race such as, *Eucalyptus* oil, Neem oil, citrus oil, onion and garlic spray, mineral oil, Chile pepper, and Chrysanthemum flower tea (Pandya, 2018). Increasing awareness about contamination, regulations, mitigation measures and appropriate monitoring system is an urgent need to adaptto respond to these challenges and to combat problems arising out of agricultural run-offs.

GLOBAL IMPACT OF WATER POLLUTION ON AQUATIC PHYTODIVERSITY

The pollution of water bodies not only affects humans but also is equally harmful to plants. Most of the pollutants of water cause detrimental effects on the physiological process of plants. Since aquatic plants are in direct contact with water, there lies a high probability of direct absorption of pollutants in their bodies. A study reported contamination of aquatic plants in the lakes of Taihu basin of China. It was observed that the enrichment factors of heavy metals such as gas copper, chromium, manganese, zinc, cobalt, lead, and nickel were greater than 2 which indicate significant contamination of heavy metals in the sediments of the lake.

Impact on Non-Vascular Plants

The marine and rheophilous bryophyte community structure, reduced species, and richness changes with concentration of pollutants in water; as a result these non-vascular plants lichens and bryophytes are often used as indicators of water pollution (Martinez et al., 2002; Ruiz and Aguirre, 2003; Davies, 2007; Tipping et al., 2008; Vuori and Helisten, 2010; Gecheva et al., 2011; Vargas and Morales, 2014; Debén et al., 2015; Gecheva et al., 2017; Kováˇcik et al., 2017; Favas et al., 2018; Vasquez et al., 2019) and such results are mostly consequence of higher degree of urbanization and wastewater deposition in the wetlands. Vasquez et al., (2019) reported sixteen indicator species of control zones (most effective species-*Fissidens serratus* and *Monoclea gottschei*); three in south zone (best indicator *Noteroclada conflens*) and four in north and central zones (*Campylopus pauper* and *Marchantia polymorpha*; and *Fissidens elegans* in north) of Zamora river, Loja City (Ecuador).

Impact on Vascular Plants

A study reported the accumulation of heavy metals by *Kandelia obovate* and *Sonneratia apetala* in the Pearl River estuary of China (Chai et al., 2019); and *Hydrocharis morsus-ranae* and *Myriophyllum verticillatum* growing in kizilirmak delta of Turkey which was more than the permissible levels (Engin et al., 2015). Polluted inland waters also cause damage to crop plants. In a study, it was reported that irrigation with waters from a lake contaminated with microcystins resulted in its accumulation in rapeseed, ryegrass, and lettuce and the probability of getting transmitted in the food chain (Crush et al., 2008).In another study, it was observed that crops such as rice and maize irrigated with contaminated water of Zefta drain in El-Kobraredion of Egypt resulted in significant bioaccumulation of copper, zinc, lead, cadmium, and manganese in their bodies. A similar accumulation of pattern was also observed in aquatic plants namely *Cynodon dactylon*, *Phragmite* spp., and *Typha* inhabiting the deltaic region of Zefta (Mahmoud and Ghoneim, 2016).

Impact on Aquatic Plants

It was observed that the heavy metal accumulation in submerged plants was more than that of the floating plants (Bai et al., 2018). It is also reported that aquatic plants namely *Vallisneria natans, Potamogeton pectinatus, Hydrilla verticillata, Myriophyllum spicatum*, and *Potamogeton crispus* growing in the waters of Yuxi River in China exhibited a high accumulation of heavy metals (Lu et al., 2014). Contamination of water bodies is intimately associated with the process of eutrophication. In this process, there is a nutrient over abundance in aquatic habitats, resulting in algal blooms and anoxic events (Carpenter, 2005). Several studies indicate that the eutrophication process is closely linked to the accumulation of heavy metals (Xing et al., 2013; Yuan et al., 2015; Jia et al., 2018). Agricultural activities also result in the eutrophication of lakes. A study reports the presence of substantial quantities of phosphorus in sediments of Poyang Lake in China and is speculated to be one of the reasons for eutrophication (Wang and Liang, 2015). Other reports state about eutrophication in Lake Erie due to prolonged agricultural practices accompanied by loading of phosphorus in the water (Michalak et al., 2013). Eutrophication also results due to the accumulation of nitrates in the water bodies. A study reported eutrophication accompanied by water pollution in Samica Stęszewska River and lakes Tomickie and Niepruszewskie due to nitrate enrichment from agricultural and livestock practices (Lawniczak et al., 2016). The most important characteristic feature of eutrophication is the harmful algal bloom. It is a drastic increase in the population of algae in an aquatic ecosystem and is generally triggered by nutrient enrichment of water. Algal bloom poses threat to the aquatic ecosystem. The dense growth of algal mat on the surface of water bodies prevents the access of sunlight and oxygen thereby inhibiting photosynthesis and cause hypoxia (Liu et al., 2019) to the plants growing in the bottom of the water bodies. As a result, the plants of the littoral zone often experience reduced growth and ultimately results in death (Chislock et al., 2013). A simulation experiment was performed to study the death of aquatic plants during algal bloom using water hyacinth as a model system. It was observed that there was a reduction in photosynthetic rate along with an increase in the concentration of ammonium ions. In addition to it, there was damage to new and old roots (Liu et al., 2015). In another study, it was reported that there was a reduction of biomass of *Elodea nuttallii* by 30% in presence of floating *Cladophora* sp. This was accompanied by an increase in the content of malondialdehyde and the activities of catalase and peroxidase (Guo et al., 2018). It is reported in another study that algal blooms result in the release of microcystins in the waters of Meiliang bay of Lake Taihu in China. The algal bloom along with secreted microcystins also results in ammonium stress which results in an imbalance in carbon-nitrogen metabolism and oxidative stress. There is also a reduction in the activity of superoxide dismutase among the submerged plants which is speculated to be the reason for the weakening of the antioxidant defense system (Zhang et al., 2011). Microcystin from cyanobacterial bloom also had an inhibitory effect on the growth of *Iris pseudacorus*. This was accompanied by damage of cell membrane and reduction in chlorophyll content which in turn affected photosynthesis. Microcystin also inducted oxidative stress and resulted in the accumulation of hydrogen peroxide and increased the activities of antioxidant enzymes (Wang and Wang,

Table 1. Effect of water pollutants on wetland floras

S.No	Environmental Pollutants	Organism Affected	Effect	Overall Outcome	Reference
1.	Chemical pollutants, pesticides, Complex mixtures of organic pollutants, industrials sludges etc.	Phytoplankton	Population increases in some species as well as decrease in some species, chlorophyll degradation, increase in BOD, bioaccumulation, increase in growth rate and algal bloom	Population increases in case of some dominant species resulting eutrophication. On the other hand, lethal effects are also observed in some species.	D'Costa *et al.*, 2017; Echeveste *et al.*, 2010
2.	Chemical pollutants, pesticides, Complex mixtures of organic pollutants, Heavy metals, pesticides, eutrophication, industrials sludges etc.	Plants	Population decreases,Sensetive species loss, Bioaccumulation.	Decreases in the population occurred and causes a number of physiological, behavioural problem and also reduces reproductive potentiality and also causes lethal effect.	Lawniczak et al., 2016; Bai et al., 2018; Jia et al., 2018; Guo et al., 201; Wang and Wang, 2018; Liu et al., 2019

2018).The effect of cyanobacterial toxins namely microcystin-LR and cylindrospermopsin on *Lactuca sativa* have been studied in an experiment. It was observed that the toxins resulted in an increase in activity of GST along with alterations in mineral contents of the plant (Freitas et al., 2015). Table 1 illustrates the effect of water pollutants on wetland floras.

REMEDIAL PLANS FOR CONSERVATION AND DECONTAMINATION

Wetlands act as natural bio-filters that help to remove an array of contaminants (like heavy metals, nutrients, organic matter, solid matters, and pathogens) from water as they possess the capacity to clean up to sixty percent of metals, trap and retain up to ninety percent of sediment runoff and eradicate up to ninety percent of nitrogen (Maiga et al., 2017). The wetland management measures include (EHSO, 2021):-

1. Wetland and riparian areas should be protected: It decreases non-source pollution by NPS-related pollutant filtration and this role is attributed to its characteristic geochemistry, and composition of vegetation and fauna; any modulation to these features affects its filtering capacities.
2. Pre-existing wetland and riparian areas are restored and are vital for subsiding nonpoint source pollution; as they also provide to attenuate flows from higher storm events, and there by safeguarding downstream areas from effects of erosion of stream-bank, scouring of the channel, and various chemical and temperature fluctuations.

Establishment of engineered vegetated treatment systems to serve as abatement function for nonpoint source pollution, as vegetated filter helps to remove sediment and other runoff pollutants and sewage water by absorption, adsorption, decomposition, deposition, filtration, infiltration, and volatilization, thus declining the pollution entering adjacent marine bodies. Several seminal remedial recommendations towards control of wetland pollution were suggested during the an International Symposium on Wetland Pollutant Dynamics and Control (WETPOL) was held at Nantes, France in 2013 (Chazarenc et al., 2015); some of them are listed as:-

- Specialized wetlands of various designs and operational variables to combat TSS removal and phenomena of clogging to treat city sewage water were constructed (Al- Isawi et al., 2015).
- CWs were constructed with combined biological and chemical instruments for the management of landfill leachate in polyculture CWs (Madera-Parra et al., 2015) and surfactant-polluted water by cascade CWs (Tamiazzo et al., 2015).
- Microbiological mechanism with importance on N_2 cycle to compare several CW models on NH removal from polluted underground water (Coban et al., 2015); monitor the effect of N_2 removal rates by evaluating performances of various multistage CWs, and thus linking oxidized and anoxic conditions (Gajewska et al., 2015); and partial nitrification and de-nitrification models in hybrid bio-film reactor (Zeng et al., 2015).

Various models of CWs were used to treat sewage water flows like hybrid CWs to treat effluent from dairy milking parlor (Harada et al., 2015), swimming pool (Van Oirschot et al., 2015); and city sludge wastewater (Nielsen and Bruun, 2015). Thus, the wetland systems are instrumental in sustainability and resiliency, key components of future environment and water.

FUTURE PROSPECTS AND CONCLUSION

Wetlands, freshwater, and aquatic environments are facing a major problem of water pollution as it is harmful to human health, flora and fauna.; and with time such diverse phytodiversities are becoming highly endangered in nature due to the distractions caused by the modification of water flow, invasion of non-native species, overexploitation, water contamination, degradation of habitat, and climate change. Therefore, a strategic and scientifically justifiable framework by systematic conservation planning is required for the conservation of phytodiversity. The global wetlands can be protected and restored by funding, knowledge, policy and management, monitoring, and restoration (Dudgeon, 2012).The worldwide gamut of threats to the phytodiversity of wetlands, scenario for sustaining normal function of ecosystems, maintaining the goods and services that support human livelihoods appear to be limited. Routine commercial cleaning up of polluted water from industrial, mining, urbanization, and underground water contamination from various sources require scientific reforms and decisions to tackle the particular hydrological conditions, geochemical and microbial transformation pathways, and suitable remediation mechanisms to develop the most efficient strategies and these must be implemented in long run by a practical approach based on lifecycle evaluations and cradle-to-grave stewardship for toxic compounds.

There are several constraints to conversation aspects like insufficient knowledge of wetland phytodiversity, insensitivity of awareness towards its significance to humans; and such gaps in systems may result ininadvertence where the human interventions on phytodiversity are often overlooked. The possible impacts are ignored due to technical, economic, or politicalconveniencefavoring the development. The readily available data may be misunderstood by policymakers or misinterpreted as the subject is sector site-specific. The problem of misinterpretation and misunderstanding is particularly pertinent in the case of wetland ecosystem, as it requires integrated management, incorporation, and integration of several types of information is collected at different areas covering a broad spectrum of sources including data on socio-economics, hydrology, a cover of vegetation, quality of water,terrain-use,ecology, etc. Therefore, it would be an enormous challengetoreconcile the basic human needs for water without compromising the supply of goods and services, livelihoods support resulting from functioning ecosystems and its phytodiversity sustenance; and hereby 'legitimize'policies for freshwater ecosystems as water users (Arthington et al., 2006) to take responsibility of trade-off among environmental and human requirement for water through proper planning, decision-making processes, and management.

REFERENCES

Agarwal, S., & Khan, S. (2020). Heavy Metal Phytotoxicity: DNA Damage. In Cellular and Molecular Phytotoxicity of Heavy Metals. Springer.

Al-Isawi, R., Scholz, M., Wang, Y., & Sani, A. (2015). Clogging of vertical-flow constructed wetlands treating urban wastewater contaminated with a diesel spill. *Environmental Science and Pollution Research International*, *22*(17), 12779–12803. doi:10.100711356-014-3732-8 PMID:25339533

Alegbeleye, O. O., & Sant'Ana, A. S. (2020). Manure-borne pathogens as an important source of water contamination: An update on the dynamics of pathogen survival/transport as well as practical risk mitigation strategies. *International Journal of Hygiene and Environmental Health*, *227*, 113524. doi:10.1016/j.ijheh.2020.113524 PMID:32298989

Alibardi, L., & Cossu, R. (2016). Effects of carbohydrate, protein and lipid content of organic waste on hydrogen production and fermentation products. *Waste Management*, *47*(Pt A), 69–77. doi:10.1016/j.wasman.2015.07.049

Alves, M. M., Pereira, M. A., Sousa, D. Z., Cavaleiro, A. J., Picavet, M., Smidt, H., & Stams, A. J. (2009). Waste lipids to energy: How to optimize methane production from long-chain fatty acids (LCFA). *Microbial Biotechnology*, *2*(5), 538–550. doi:10.1111/j.1751-7915.2009.00100.x PMID:21255287

Arthington, A.H., Bunn, S.E., Poff, N.L., & Naiman, R.J. (2006). The challenge of providing environmental flow rules to sustain river ecosystems. *Ecological Applications*, *16*(4), 1311–1318.

Bai, L., Liu, X. L., Hu, J., Li, J., Wang, Z. L., Han, G., Li, S. L., & Liu, C. Q. (2018). Heavy metal accumulation in common aquatic plants in rivers and lakes in the Taihu basin. *International Journal of Environmental Research and Public Health*, *15*(12), 2857. doi:10.3390/ijerph15122857 PMID:30558148

Berg, M., Stengel, C., Pham, T. K., Pham, H. V., Sampson, M. L., Leng, M., Samreth, S., & Fredericks, D. (2007). Magnitude of arsenic pollution in the Mekong and Red River Deltas—Cambodia and Vietnam. *The Science of the Total Environment*, *372*(2-3), 413–425. doi:10.1016/j.scitotenv.2006.09.010 PMID:17081593

Bhat, R. A., Shafiq-ur-Rehman, M. M., Dervash, M. A., Mushtaq, N., Bhat, J. I. A., & Dar, G. H. (2017). Current status of nutrient load in Dal Lake of Kashmir Himalaya. *Journal of Pharmacognosy and Phytochemistry*, *6*(6), 165–169.

Bhateria, R., & Jain, D. (2016). Water quality assessment of lake water: A review. *Sustainable Water Resources Management*, *2*(2), 161–173. doi:10.100740899-015-0014-7

Bi, W., Weng, B., Yuan, Z., Ye, M., Zhang, C., Zhao, Y., Yan, D., & Xu, T. (2018). Evolution characteristics of surface water quality due to climate change and LUCC under scenario simulations: A Case study in the Luanhe River Basin. *International Journal of Environmental Research and Public Health*, *15*(8), 1724. doi:10.3390/ijerph15081724 PMID:30103482

Calvert, G. M., Karnik, J., Mehler, L., Beckman, J., Morrissey, B., Sievert, J., Barrett, R., Lackovic, M., Mabee, L., Schwartz, A., Mitchell, Y., & Moraga-McHaley, S. (2008). Acute pesticide poisoning among agricultural workers in the United States, 1998-2005. *American Journal of Industrial Medicine*, *51*(12), 883–898. doi:10.1002/ajim.20623 PMID:18666136

Carpenter, S. R. (2005). Eutrophication of aquatic ecosystems: Biostability and soil phosphorus. *Proceedings of the National Academy of Sciences of the United States of America*, *102*(29), 10002–10005. doi:10.1073/pnas.0503959102 PMID:15972805

Carvalho, F. P. (2017). Pesticides, environment, and food safety. *Food and Energy Security*, *6*(2), 48–60. doi:10.1002/fes3.108

CBD-Convention of Biological Diversity. (2008). *Inland water biodiversity*. https://www.cbd.int/waters/inland-waters

Chai, M., Li, R., Tam, N., & Zan, Q. (2019). Effects of mangrove plant species on accumulation of heavy metals in sediment in a heavily polluted mangrove swamp in Pearl River Estuary, China. *Environmental Geochemistry and Health*, *41*(1), 175–189. doi:10.100710653-018-0107-y PMID:29876676

Chazarenc, F., Gagnon, V., & Korkusuz, E. A. (2015). Treatment of pollution in constructed wetlands: From the fundamental mechanisms to the full-scale applications. *Environmental Science and Pollution Research International*, *22*(17), 12775–12778. doi:10.100711356-015-4969-6 PMID:26160123

Chen, Y., Parvez, F., Gamble, M., Islam, T., Ahmed, A., Argos, M., Graziano, J. H., & Ahsan, H. (2009). Arsenic exposure at low-to-moderate levels and skin lesions, arsenic metabolism, neurological functions, and biomarkers for respiratory and cardiovascular diseases: Review of recent findings from the Health Effects of Arsenic Longitudinal Study (HEALS) in Bangladesh. *Toxicology and Applied Pharmacology*, *239*(2), 184–192. doi:10.1016/j.taap.2009.01.010 PMID:19371619

Cherry, J. A. (2011). Ecology of Wetland Ecosystems: Water, Substrate, and Life. *Nature Education Knowledge*, *3*(10), 16.

Chislock, M. F., Doster, E., Zitomer, R. A., & Wilson, A. E. (2013). Eutrophication: Causes, consequences, and controls in aquatic ecosystems. *Nature Education Knowledge*, *4*(4), 10.

Christensen, V. G., Maki, R. P., Stelzer, E. A., Norland, J. E., & Khan, E. (2019). Phytoplankton community and algal toxicity at a recurring bloom in Sullivan Bay, Kabetogama Lake, Minnesota, USA. *Scientific Reports*, *9*(1), 1–11. doi:10.103841598-019-52639-y PMID:31695119

Clements, F. E. (1916). *Plant succession analysis of the development of vegetation*. Carnegie Institution of Washington Publication Sciences. doi:10.5962/bhl.title.56234

Coban, O., Kuschk, P., Wells, N. S., Strauch, G., & Knoeller, K. (2015). Microbial nitrogen transformation in constructed wetlands treating contaminated groundwater. *Environmental Science and Pollution Research International*, *22*(17), 12829–12839. doi:10.100711356-014-3575-3 PMID:25233917

Connor, R. (2015). *The United Nations world water development report 2015: water for a sustainable world* (Vol. 1). UNESCO Publishing.

Cornish, G., Mensah, E., & Ghesquire, P. (1999). *Water quality and peri-urban irrigation: an assessment of surface water quality for irrigation and its implications for human health in the peri-urban zone of Kumasi, Ghana*. Project Report. HR Wallingford Ltd.

Cronk, J. K., & Fennessy, M. S. (2009). Wetland Plants. In G. E. Likens (Ed.), *Encyclopedia of Inland Waters* (pp. 590–598). Academic Press. doi:10.1016/B978-012370626-3.00060-0

Crush, J. R., Briggs, L. R., Sprosen, J. M., & Nichols, S. N. (2008). Effect of irrigation with lake water containing microcystins on microcystin content and growth of ryegrass, clover, rape, and lettuce. *Environmental Toxicology*, *23*(2), 246–252. doi:10.1002/tox.20331 PMID:18214908

D'Costa, P. M., D'Silva, M. S., & Naik, R. K. (2017). Impact of pollution on Phytoplankton and implications for marine econiches. In M. Naik & S. Dubey (Eds.), *Marine Pollution and Microbial Remediation*. Springer. doi:10.1007/978-981-10-1044-6_13

Davies, T. D. (2007). Sulphate toxicity to the aquatic moss, *Fontinalis antipyretica. Chemosphere, 66*(3), 444–451. doi:10.1016/j.chemosphere.2006.06.021 PMID:16899275

Debén, S., Aboal, J., Carballeira, A., Cesa, M., Real, C., & Fernandez, J. (2015). Inland water quality monitoring with native bryophytes: A methodological review. *Ecological Indicators, 53*, 115–124. doi:10.1016/j.ecolind.2015.01.015

Dudgeon, D. (2012). Threats to freshwater biodiversity globally and in the Indo-Burma Biodiversity *Hotspot.*Chair of Ecology and Biodiversity, School of Biological Sciences, The University of Hong Kong, Pok Fu Lam Road, Hong Kong SAR, China. *Environmental Science.*

Echeveste, P., Dachs, J., Galbán-Malagón, C., Berrojalbiz, N., & Agust, S. (2010). Decrease in the abundance and viability of oceanic phytoplankton due to trace levels of complex mixtures of organic pollutants. *Chemosphere, 81*(2), 161–168. doi:10.1016/j.chemosphere.2010.06.072 PMID:20673958

EHSO. (2021). *Causes and Control of Water Pollution in Wetlands*. Retrieved June 25, 2021, from http://www.ehso.com/WaterPollution_Wetlands.htm

Eisler, R. (2003). Health risks of gold miners: A synoptic review. *Environmental Geochemistry and Health, 25*(3), 325–345. doi:10.1023/A:1024573701073 PMID:12971253

Emery, S. (2010). Succession: A Closer Look. *Nature Education Knowledge, 3*(10), 45.

Engin, M. S., Uyanik, A., & Kutbay, H. G. (2015). Accumulation of heavy metals in water, sediments and wetland plants of Kizilirmak delta (Samsun, Turkey). *International Journal of Phytoremediation, 17*(1-6), 66–75. doi:10.1080/15226514.2013.828019 PMID:25174426

Environmental Protection Office of Wetlands. (2001). *Agency Oceans and Watersheds* (4502T). Author.

EPA. (2001). Treats to wetlands. United States Office of Water. EPA 843-F-01-002d.

EPA. (2021). *Why are wetlands important?* https://www.epa.gov/wetlands/why-are-wetlands-important

Eur. Environ. Agency. (EEA). (2000). *Management of contaminated sites in Western Europe: Topic report No.13/1999*. EEA. Retrieved June 24, 2021, from https://www.eea.europa.eu/publications/Topic_report_No_131999

Fagorite, V. I., Odundun, O. A., Iwueke, L. E., Nwaigbo, U. N., & Okeke, O. C. (2019). Wetlands: A review of their classification, significance and management for sustainable development. *International Journal of Advanced Academic Research, Sciences, Technology and Engineering, 5*(3), 2488–9849.

FAO. (2013). Guidelines to control water pollution from agriculture in China. *Water Reports*, 40.

Favas, P. J., Pratas, J., Rodrigues, N., D'Souza, R., Varun, M., & Paul, M. S. (2018). Metal (loid) accumulation in aquatic plants of a mining area: Potential for water quality biomonitoring and biogeochemical prospecting. *Chemosphere, 194*, 158–170. doi:10.1016/j.chemosphere.2017.11.139 PMID:29202268

Finlayson, C. M. (1991). Plant ecology and management of an internationally important wetland in monsoonal Australia. In *Proceedings of an international symposium on wetlands and river corridor management*. New York: Association of State Wetland Managers.

Freitas, M., Azevedo, J., Pinto, E., Neves, J., Campos, A., & Vasconcelos, V. (2015). Effects of microcystin-LR, cylindrospermopsin and a microcystin-LR/cylindrospermopsin mixture on growth, oxidative stress and mineral content in lettuce plants (*Lactucasativa* L.). *Ecotoxicology and Environmental Safety*, *116*, 59–67. doi:10.1016/j.ecoenv.2015.02.002 PMID:25768423

Frost, F. J., Muller, T., Petersen, H. V., Thomson, B., & Tollestrup, K. (2003). Identifying US populations for the study of health effects related to drinking water arsenic. *Journal of Exposure Science & Environmental Epidemiology*, *13*(3), 231–239. doi:10.1038j.jea.7500275 PMID:12743617

Gajewska, M., Jóźwiakowski, K., Ghrabi, A., & Masi, F. (2015). Impact of influent wastewater quality on nitrogen removal rates in multistage treatment wetlands. *Environmental Science and Pollution Research International*, *22*(17), 12840–12848. doi:10.100711356-014-3647-4 PMID:25300181

Gecheva, G., Pall, K., & Hristeva, Y. (2017). Bryophyte communities' responses to environmental factors in highly seasonal rivers. *Botany Letters*, *164*(1), 79–91. doi:10.1080/23818107.2016.1263238

Gecheva, G., Yurukova, L., & Ganeva, A. (2011). Assessment of pollution with aquatic bryophytes in Maritsa River (Bulgaria). *Bulletin of Environmental Contamination and Toxicology*, *87*(4), 480–485. doi:10.100700128-011-0377-x PMID:21814844

Ghani, S. A. A. (2015). Trace metals in seawater, sediments and some fish species from MarsaMatrouh Beaches in north-western Mediterranean coast, Egypt. *Egyptian Journal of Aquatic Research*, *41*(2), 145–154. doi:10.1016/j.ejar.2015.02.006

Giusti, L. (2009). A review of waste management practices and their impact on human health. *Waste Management (New York, N.Y.)*, *29*(8), 2227–2239. doi:10.1016/j.wasman.2009.03.028 PMID:19401266

Gleason, H. A. (1926). The individualistic concept of the plant association. *Bulletin of the Torrey Botanical Club*, *53*(1), 7–26. doi:10.2307/2479933

Guo, L., Dai, L., Yang, K., Li, D., & Li, G. (2018). Physiological changes of submerged macrophytes in response to a floating filamentous green algae bloom in clear-water conditions. *Journal of Oceanology and Limnology*, *36*(5), 1604–1614. doi:10.100700343-018-7139-0

Halpern, B. S., Walbridge, S., Selkoe, K. A., Kappel, C. V., Micheli, F., D'Agrosa, C., & Watson, R. (2008). A global map of human impact on marine ecosystems. *Science*, *319*(5865), 948–952. doi:10.1126cience.1149345 PMID:18276889

Hampel, M., Blasco, J., & Segner, H. (2015). Molecular and cellular effects of contamination in aquatic ecosystems. *Environmental Science and Pollution Research International*, *22*(22), 17261–17266. doi:10.100711356-015-5565-5 PMID:26507731

Haque, R., Mazumder, D. G., Samanta, S., Ghosh, N., Kalman, D., Smith, M. M., & Smith, A. H. (2003). Arsenic in drinking water and skin lesions: Dose-response data from West Bengal, India. *Epidemiology (Cambridge, Mass.)*, *14*(2), 174–182. doi:10.1097/01.EDE.0000040361.55051.54 PMID:12606883

Harada, J., Inoue, T., Kato, K., Uraie, N., & Sakuragi, H. (2015). Performance evaluation of hybrid treatment wetland for six years of operation in cold climate. *Environmental Science and Pollution Research International*, *22*(17), 12861–12869. doi:10.100711356-014-3843-2 PMID:25424033

Hinton, J. J., Veiga, M. M., & Veiga, A. T. C. (2003). Clean artisanal gold mining: A utopian approach? *Journal of Cleaner Production*, *11*(2), 99–115. doi:10.1016/S0959-6526(02)00031-8

Hites, R. A. (2004). Polybrominateddiphenyl ethers in the environment and in people: A meta-analysis of concentrations. *Environmental Science & Technology*, *38*(4), 945–956. doi:10.1021/es035082g PMID:14998004

Ho, Y. C., Show, K. Y., Guo, X. X., Norli, I., Abbas, F. A., & Morad, N. (2012). Industrial discharge and their effect to the environment. In *Industrial waste*. Intech.

Hong, Z., Zhao, Q., Chang, J., Peng, L., Wang, S., Hong, Y., & Ding, S. (2020). Evaluation of water quality and heavy metals in wetlands along the Yellow River in Henan Province. *Sustainability*, *12*(4), 1300. doi:10.3390u12041300

Jaishankar, M., Tseten, T., Anbalagan, N., Mathew, B. B., & Beeregowda, K. N. (2014). Toxicity, mechanism and health effects of some heavy metals. *Interdisciplinary Toxicology*, *7*(2), 60–72. doi:10.2478/intox-2014-0009 PMID:26109881

Järvenpää, M., & Lindström, K. (2004). Water turbidity by algal blooms causes mating system breakdown in a shallow-water fish, the sand go by *Pomatos chistusminutus*. *Proceedings. Biological Sciences*, *271*(1555), 2361–2365. doi:10.1098/rspb.2004.2870 PMID:15556888

Jha, S. K., Singh, R. K., Damodaran, T., Mishra, V. K., Sharma, D. K., & Rai, D. (2013). Fluoride in groundwater: Toxicological exposure and remedies. *Journal of Toxicology and Environmental Health. Part B, Critical Reviews*, *16*(1), 52–66. doi:10.1080/10937404.2013.769420 PMID:23573940

Jia, Y., Chen, W., Zuo, Y., Lin, L., & Song, L. (2018). Heavy metal migration and risk transference associated with cyanobacterial blooms in eutrophic freshwater. *The Science of the Total Environment*, *613-614*, 1324–1330. doi:10.1016/j.scitotenv.2017.09.180 PMID:28968935

Kansiime, F., & Maimuna, N. (1999). *Wastewater treatment by a natural wetland: The Nakivubo swamp*. CRC Press.

Keddy, A. P. (2008). Freshwater Marshes. In *Encyclopaedia of Ecology*. Academic Press; doi:10.1017/CBO9780511778179

Keddy, P. A. (2010). *Wetland ecology: Principles and conservation*. Cambridge University Press.

Khatri, N., & Tyagi, S. (2015). Influences of natural and anthropogenic factors on surface and groundwater quality in rural and urban areas. *Frontiers in Life Science*, *8*(1), 23–39. doi:10.1080/21553769.2014.933716

Kinuthia, G. K., Ngure, V., Beti, D., Lugalia, R., Wangila, A., & Kamau, L. (2020). Levels of heavy metals in wastewater and soil samples from open drainage channels in Nairobi, Kenya: Community health implication. *Scientific Reports*, *10*(1), 1–13. doi:10.103841598-020-65359-5 PMID:31913322

Kováčik, J., Babula, P., & Hedbavny, J. (2017). Comparison of vascular and non-vascular aquatic plant as indicators of cadmium toxicity. *Chemosphere*, *180*, 86–92. doi:10.1016/j.chemosphere.2017.04.002 PMID:28391156

Kristoffersen, P., Rask, A. M., Grundy, A. C., Franzen, I., Kempenaar, C., Raisio, J., & Zarina, L. (2008). A review of pesticide policies and regulations for urban amenity areas in seven European countries. *Weed Research, 48*(3), 201–214. doi:10.1111/j.1365-3180.2008.00619.x

Kundu, P. (2018). Review article on the assessment of wetland. *International Journal of Research and Analytical Reviews, 5*(3), 343–346.

Lawniczak, A. E., Zbierska, J., Nowak, B., Achtenberg, K., Grześkowiak, A., & Kanas, K. (2016). Impact of agriculture and land use on nitrate contamination in groundwater and running waters in central-west Poland. *Environmental Monitoring and Assessment, 188*(3), 172. doi:10.100710661-016-5167-9 PMID:26887311

Ling, J. E., Casanova, M. T., Shannon, I., & Powell, M. (2018). Development of a wetland plant indicator list to inform the delineation of wetlands in New South Wales. *Marine and Freshwater Research, 70*(3), 322–344. doi:10.1071/MF18114

Liu, G.F., He, J., Yang, Y.Z., & Han, S.Q. (2015). Article. *Huanjingkexue= Huanjingkexue, 36*(8), 2887–2894.

Liu, X.Q., Huang, T.L., Li, N., Yang, S.Y., Li, Y., Xu, J., &Wang, H.Y. (2019). Article. *Huanjingkexue= Huanjingkexue,*40(5), 2258–2264.

Lu, X., Zhang, X., Li, L. Y., & Chen, H. (2014). Assessment of metals pollution and health risk in dust from nursery schools in Xi'an, China. *Environmental Research, 128*, 27–34. doi:10.1016/j.envres.2013.11.007 PMID:24407476

Madera-Parra, C. A., Peña, M. R., Peña, E. J., & Lens, P. N. L. (2015). Contamination status of *Salmonella* spp., *Shigella* spp. and *Campylobacter* spp. in surface and groundwater of the Kelani river basin, Sri Lanka. *Water (Basel), 12*(8), 2187.

Mahmoud, E. K., & Ghoneim, A. M. (2016). Effect of polluted water on soil and plant contamination by heavy metals in El-Mahla El-Kobra, Egypt. *Solid Earth, 7*(2), 703–711. doi:10.5194e-7-703-2016

Maiga, Y., von Sperling, M., & Mihelcic, J. R. (2017). Constructed Wetlands. In Water and Sanitation for the 21st Century: Health and Microbiological Aspects of Excreta and Wastewater Management (Global Water Pathogen Project). Part 4: Management Of Risk from Excreta and Wastewater - Section: Sanitation System Technologies, Pathogen Reduction in Sewered System Technologies). Michigan State University. doi:10.14321/waterpathogens.66

Malone, T. C., & Newton, A. (2020). The globalization of cultural eutrophication in the coastal ocean: Causes and consequences. *Frontiers in Marine Science, 7*, 670. doi:10.3389/fmars.2020.00670

Manuel, J. (2014). Nutrient pollution: A persistent threat to waterways. *Environmental Health Perspectives, 122*(11), A304–A309. doi:10.1289/ehp.122-A304 PMID:25360879

Martinez, J., Garcia, A., Beaucourt, N., & Nuñez, E. (2002). Combined seasonal and longitudinal variations of element concentrations in two aquatic mosses (*Fontinalis antipyretica* and *Fontinalis squamosa*). *Nova Hedwigia, 74*(3-4), 349–364. doi:10.1127/0029-5035/2002/0074-0349

Mateo-Sagasta, J., MarjaniZadeh, S., Turral, H., & Burke, J. (2017). *Water pollution from agriculture: a global review.* Food and Agriculture Organization of the United Nations. Retreived from: https://www.fao.org/3/i7754e/i7754e.pdf

Mathuthu, A. S., Mwanga, K., & Simoro, A. (1997). *Impact assessment of industrial and sewage effluents on water quality of the receiving Marimba River in Harare. Lake Chivero: A Polluted Lake*. University of Zimbabwe Publications.

McGrane, S. J. (2016). Impacts of urbanisation on hydrological and water quality dynamics, and urban water management: A review. *Hydrological Sciences Journal*, *61*(13), 2295–2311. doi:10.1080/02626667.2015.1128084

Michalak, A. M., Anderson, E. J., Beletsky, D., Boland, S., Bosch, N. S., Bridgeman, T. B., Chaffin, J. D., Cho, K., Confesor, R., Daloglu, I., Depinto, J. V., Evans, M. A., Fahnenstiel, G. L., He, L., Ho, J. C., Jenkins, L., Johengen, T. H., Kuo, K. C., Laporte, E., ... Zagorski, M. A. (2013). Record-setting algal bloom in Lake Erie caused by agricultural and meteorological trends consistent with expected future conditions. *Proceedings of the National Academy of Sciences of the United States of America*, *110*(16), 6448–6452. doi:10.1073/pnas.1216006110 PMID:23576718

Middleton, B. A. (1999). Succession and herbivory in monsoonal wetlands. *Wetlands Ecology and Management*, *6*(4), 189–202. doi:10.1023/A:1008495121557

Mitsch, W. J., Zhang, L., Stefanik, K. C., Nahlik, A. M., Anderson, C. J., Bernal, B., Hernandez, M., & Song, K. (2012). Creating wetlands: Primary succession, water quality changes, and self-design over 15 years. *Bioscience*, *62*(3), 237–250. doi:10.1525/bio.2012.62.3.5

Naggar, Y. A., Naiem, E., Mona, M., Giesy, J. P., & Seif, A. (2014). Metals in agricultural soils and plants in Egypt. *Toxicological and Environmental Chemistry*, *96*(5), 730–742. doi:10.1080/02772248.2014.984496

National Park Service. (2016). *Why are wetlands important.* https://www.nps.gov/subjects/wetlands/why.htm

Nielsen, S., & Bruun, E. W. (2015). Sludge quality after 10–20 years of treatment in reed bed systems. *Environmental Science and Pollution Research International*, *22*(17), 12885–12891. doi:10.100711356-014-3815-6 PMID:25422113

Niering, W. A. (1987). Vegetation dynamics (succession and climax) in relation to plant community management. *Conservation Biology*, *1*(4), 287–295. doi:10.1111/j.1523-1739.1987.tb00049.x

Nyanda, M. (2000). Report on the study of agrochemical use and handling in the Lake Zone. Academic Press.

Oluwole, O., & Cheke, R. A. (2009). Health and environmental impacts of pesticide use practices: A case study of farmers in Ekiti State, Nigeria. *International Journal of Agricultural Sustainability*, *7*(3), 153–163.

Ozturk, A., & Arici, O. K. (2021). Carcinogenic-potential ecological risk assessment of soils and wheat in the eastern region of Konya (Turkey). *Environmental Science and Pollution Research International, 28*(12), 15471–15484. https://doi.org/10.1007/s11356-020-11697-w

Pandey, P. K., Kass, P. H., Soupir, M. L., Biswas, S., & Singh, V. P. (2014). Contamination of water resources by pathogenic bacteria. *AMB Express, 4*, 51. https://doi.org/10.1186/s13568-014-0051-x

Pandya, I. Y. (2018). Pesticides and their applications in agriculture. *Asian Journal of Applied Science and Technology, 2*(2), 894–900.

Parolin, P., & Wittmann, F. (2010). Struggle in the flood: Tree responses to flooding stress in four tropical floodplain systems. *AoB Plants.*

Peters, S. C. (2008). Arsenic in groundwaters in the Northern Appalachian Mountain belt: A review of patterns and processes. *Journal of Contaminant Hydrology, 99*(1-4), 8–21.

Post, V. E. A. (2005). Fresh and saline groundwater interaction in coastal aquifers: Is our technology ready for the problems ahead? *Hydrogeology Journal, 13*(1), 120–123.

Preisner, M. (2020). Surface water pollution by untreated municipal wastewater discharge due to a sewer failure. *Environmental Processes, 7*(3), 767–780.

Radwan, G. A. E. (2020). Physical properties of inland lakes and their interaction with global warming: A case study of Lake Nasser, Egypt. *The Egyptian Journal of Aquatic Research, 46*(2), 103–115.

Ruiz, C. A., & Aguirre, J. (2003). Las comunidades de briófitos y su relación con la estructura de la vegetación fanerogamica, en el gradiente altitudinal de la Serranía del Perijá (Cesar–Colombia). *Bryophyte Diversity and Evolution, 24*, 101–113.

Schwarzenbach, R. P., Egli, T., Hofstetter, T. B., Von Gunten, U., & Wehrli, B. (2010). Global water pollution and human health. *Annual Review of Environment and Resources, 35*, 109–136.

Shakeri, A., & Moore, F. (2010). The impact of an industrial complex on freshly deposited sediments, Chener Rahdar river case study, Shiraz, Iran. *Environmental Monitoring and Assessment, 169*(1), 321–334.

Sharma, S., & Bhattacharya, A. (2017). Drinking water contamination and treatment techniques. *Applied Water Science, 7*, 1043–1067.

Sharma, S. D. (2020). Risk assessment via oral and dermal pathways from heavy metal polluted water of Kollerulake - A Ramsar wetland in Andhra Pradesh, India. *Environmental Analysis. Health and Toxicology, 35*(3), e2020019.

Shelef, O., Gross, A., & Rachmilevitch, S. (2013). Role of Plants in a Constructed Wetland: Current and New Perspectives. *Water (Basel), 5*, 405–419.

Sieben, E. J. J., Glen, R. P., Deventer, H. V., & Dayaram, A. (2021). The contribution of wetland flora to regional floristic diversity across a wide range of climatic conditions in southern Africa. *Biodiversity and Conservation, 30*(3), 1–22.

Slavov, A. K. (2017). General Characteristics and Treatment Possibilities ofDairy Wastewater-A Review. *Food Technology and Biotechnology, 55*(1), 14–28. https://doi.org/10.17113/ftb.55.01.17.4520

Steerenberg, P., Van Amelsvoort, L. G. P. M., Colosio, C., Corsini, E., Fustinoni, S., Vergieva, T., & Van Loveren, H. (2008). Toxicological evaluation of the immune function of pesticide workers, a European wide assessment. *Human and Experimental Toxicology*, *27*(9), 701–707.

Strohschön, R., Wiethoff, K., Baier, K., Lu, L., Bercht, A.L., Wehrhahn, R., & Azzam, R. (2013). Land use and Water Quality in Guangzhou, China: A survey of ecological and Social Vulnerability in Four Urban Units of the Rapidly Developing Megacity. *International Journal of Environmental of Research*, *7*, 343–358.

Tamiazzo, J., Breschigliaro, S., Salvato, M., & Borin, M. (2015). Performance of a wall cascade constructed wetland treating surfactant-polluted water. *Environmental Science and Pollution Research International*, *22*(17), 12816–12828.

Tiner, R. W. (1991). The concept of a hydrophyte for wetland identification. *Bioscience*, *41*, 236–247.

Tiner, R. W. (2009). *Ecology of Wetlands: Classification Systems. In Encyclopedia of Inland Waters*. Academic Press.

Tipping, E., Vincent, C. D., Lawlor, A. J., & Lofts, S. (2008). Metal accumulation by stream bryophytes, related to chemical speciation. *Environmental Pollution*, *156*, 936–943.

Todd, M. J., Smith, K., & Bannister, P. (2006). Supervising a Social Science Dissertation: Staff Experiences and Perceptions. *Teaching in Higher Education*, *11*(2), 161–173. doi:10.1080/13562510500527693

Tulchinsky, T. H. (2018). John Snow, Cholera, the Broad Street Pump; Waterborne Diseases Then and Now. *Case Studies in Public Health*, 77–99. doi:10.1016/B978-0-12-804571-8.00017-2

Twiddle, C. (2012). Pollen Analysis: Not Just a Qualitative Tool. *Geo-Morphological Techniques,* 1-8.

UN-Habitat. (2016). https://unhabitat.org/un-habitat-launches-the-world-cities-report 2016#:~:text=With%20the%20United%20Nations%20Conference,to%20prosperity%20and%20equitable%20growth

UN-Water. (2015a). https://www.unwater.org/publications/wastewater-management-un-water-analytical-brief/

UNESCO World Water Assessment Programme (WWAP). (2017). *The United Nations world water development report 2017: wastewater: the untapped resource; facts and figures*. https://unesdoc.unesco.org/ark:/48223/pf0000247553

United Nations Environment Program (UNEP). (2007). *Global Environment Outlook GEO.* Retrieved June 24, 2021, from https://www.unep.org/resources/global-environment-outlook-4

U.S. Fish & Wildlife Service. (2021). *Classification and types of wetlands*. https://www.epa.gov/wetlands/classification-and-types-wetlands#undefined

USEPA. (2016). *Water quality assessment and TMDL information*. United States Environmental Protection Agency.

Van der Valk, A. (1981). Succession in wetlands: A Gleasonian approach. *Ecology*, *62*, 688–696.

Van Oirschot, D., Wallace, S., & Van Deun, R. (2015). Wastewater treatment in a compact intensified wetland system at the Badboot: A floating swimming pool in Belgium. *Environmental Science and Pollution Research International*, *22*(17), 12870–12878.

vanRaamsdonk, L., van der Zande, M., Koelmans, A. A., Hoogenboom, R., Peters, R., Groot, M. J., Peijnenburg, A., & Weesepoel, Y. (2020). Current insights into monitoring, bioaccumulation, and potential health effects of microplastics present in the food chain. *Foods*, *9*(1), 72. doi:10.3390/foods9010072

Vargas, R., & Morales, E. (2014). Hepáticas del Parque Natural Municipal "Robledales de Tipacoque", Boyacá-Colombia. *Universitas Scientiarum*, *19*, 201–211.

Vásquez, C., Calva, J., Morocho, R., Donoso, D. A., & Benítez, Á. (2019). Bryophyte communities along a tropical urban river respond to heavy metal and arsenic pollution. *Water (Basel)*, *11*, 813.

Votteler, T. H., & Muir, T. A. (2002). *Wetland Protection Legislation*. Wetland Management and Research. https://water.usgs.gov/nwsum/WSP2425/legislation.html

Vuori, K. M., & Helisten, H. (2010). The use of aquatic mosses in assessment of metal pollution: Appraisal of type-specific background concentrations and inter-specific differences in metal accumulation. *Hydrobiologia*, *656*, 99–106.

Wakode, H. B., Baier, K., Jha, R., & Azzam, R. (2018). Impact of urbanization on groundwater recharge and urban water balance for the city of Hyderabad, India. *International Soil and Water Conservation Research*, *6*, 51–6261.

Wan, J., Wang, X., Yang, T., Wei, Z., Banerjee, S., Friman, V. P., Mei, X., Xu, Y., & Shen, Q. (2021). Livestock manure type affects microbial community composition and assembly during composting. *Frontiers in Microbiology*, *12*, 621126. https://doi.org/10.3389/fmicb.2021.621126

Wang, L., & Liang, T. (2015). Distribution characteristics of phosphorus in the sediments and overlying water of Poyang lake. *PLoS One*, *10*(5), e0125859.

Wang, N., & Wang, C. (2018). Effects of microcystin-LR on the tissue growth and physiological responses of the aquatic plant *Iris pseudacorus* L. *Aquatic Toxicology (Amsterdam, Netherlands)*, *200*, 197–205.

WHO. (2015). *Connecting Global Priorities: Biodiversity and Human Health*. https://apps.who.int/iris/bitstream/handle/10665/174012/9789241508537_eng.pdf

Williamson, S., Ball, A., & Pretty, J. (2008). Trends in pesticide use and drivers for safer pest management in four African countries. *Crop Protection (Guildford, Surrey)*, *27*(10), 1327–1334.

Xie, P. (2011). Ammonium, microcystins, and hypoxia of blooms in eutrophic water cause oxidative stress and C-N imbalance in submersed and floating-leaved aquatic plants in Lake Taihu, China. *Chemosphere*, *82*(3), 329–339.

Xing, W., Wu, H., Hao, B., Huang, W., & Liu, G. (2013). Bioaccumulation of heavy metals by submerged macrophytes: Looking for hyperaccumulators in eutrophic lakes. *Environmental Science & Technology*, *47*, 4695–4703. doi:10.1021/es303923w

Yan, C. P., & Ma, Y. J. (2010). Research progress on impacts of human industrial activities on wetland environment. *Wetland Science*, *8*(1), 98–104.

Yuan, H., Liu, E., & Shen, J. (2015). The accumulation and potential ecological risk of heavy metals in microalgae from an eutrophic lake (Taihulake, China). *Environmental Science and Pollution Research International*, *22*(21), 17123–17134.

Zahedi, A., Monis, P., Deere, D., & Ryan, U. (2021). Wastewater-based epidemiology—Surveillance and early detection of waterborne pathogens with a focus on SARS-CoV-2, *Cryptosporidium* and *Giardia*. *Parasitology Research*, 1–22.

Zehr, J. P., & Ward, B. B. (2002). Nitrogen cycling in the ocean: New perspectives on processes and paradigms. *Applied and Environmental Microbiology*, *68*(3), 1015–1024. https://doi.org/10.1128/AEM.68.3.1015-1024.2002

Zeng, M., Soric, A., & Roche, N. (2015). Modeling partial nitrification and denitrification in a hybrid biofilm reactor: Calibration by retention time distribution and respirometric tests. *Environmental Science and Pollution Research International*, *22*(17), 12849–12860.

ADDITIONAL READING

The Encyclopedia of Earth. (n.d.). http://www.eoearth.org/article/Global_Environment_Outlook_%28GEO4%29:_Environment_for_Development_%28e-book%29

United States Environmental Protection Agency. (2018). National Air Toxics Assessment. Available from: https://www.epa.gov/national-air-toxics-assessment/nata-glossary-terms#hq

We Are Water Foundation. (2019). Natural water pollutants, a silent threat. Retrieved from https://www.wearewater.org/en-IN/natural-water-pollutants-a-silent-threat_320991

Chapter 4
Wetland Ecosystem Monitoring Through Water Quality Perspectives

Shahid Ahmad Dar
https://orcid.org/0000-0002-1086-2312
Department of Environmental Science, University of Kashmir, India

Sami Ullah Bhat
Department of Environmental Science, University of Kashmir, India

Sajad Ahmad Dar
Department of Environmental Science, Uttarakhand Technical University, India

ABSTRACT

Water quality monitoring is an important tool in determining the safety and suitability of water for various desired and intended uses. The procedures involved in the evaluation of water quality are numerous and multifaceted. Therefore, taking into consideration the specific objectives of water quality monitoring, sampling design is of vital importance. Most of the physical parameters of water quality are determined via in-situ measurements using modern testing equipment/field testing kits. Although there are some good field-based sensors that are being used for evaluation of water quality, the chemical parameters traditionally are mostly analyzed through laboratory-based experiments. This chapter is aimed to offer an inclusive knowledge and insights on the importance and assessment of physico-chemical parameters that are of high priority for monitoring the water quality of wetlands.

INTRODUCTION

Wetlands are amongst the highly valuable ecosystems of the earth surface, delivering a wide variety of ecosystem services (Mitsch et al. 2015; Dar et al. 2021a, b). They help in the purification of water resources, recharge of groundwater, and play a significant role in water supplies, livelihoods, and irri-

DOI: 10.4018/978-1-7998-9498-8.ch004

gation of farmlands (Keddy et al. 2009; Mitsch and Gosselink 2015). Wetlands harbor a huge diversity of aquatic species including macrophytes, fishes, macroinvertebrates, phytoplankton, and zooplankton (Blackwell and Pilgrim 2011). Besides, they provide breeding and nesting grounds for a variety of bird species. However, during the last few decades, growing human population, urbanization, land use land cover changes, degradation of natural landscapes, and pollution have resulted in degradation and deterioration of water quality of wetlands (Ahmad et al. 2019; Dar et al. 2020a). The anthropogenic activities have deprived the desired uses of wetland water with major effects on the biotic communities living in the wetlands. As a result, many species dependent on wetlands including mammals (37%), birds (21%), and fishes (20%) are either threatened or extinct (MEA 2005). The increased runoff from urban areas and agricultural fields has intensified the loadings of nutrients (nitrogen and phosphorus), resulting in cultural eutrophication, algal blooms, subsequent depletion of dissolved oxygen, and fish kills (Dar et al. 2020b). Human activities such as ingress of raw sewage (Parvez and Bhat 2014), unplanned urbanization, over-exploitation of water, encroachments, and catchment degradation (Romshoo and Rashid 2014), has significantly degraded the water quality of wetlands, thereby seriously affecting their use for navigation, recreational activities, industrial and agricultural purposes.

Globally, the massive degradation and loss of wetland ecosystems has led to the efforts for conservation and restoration of these ecologically important ecosystems (Galatowitsch 2018). Monitoring water quality of wetlands provides important information about the possible contaminants of wetland ecosystems and help in identification of pollution sources (Osei et al. 2010; Dar et al. 2021c). Because of the dynamic nature of wetland ecosystems, wetland management authorities are significantly challenged in their efforts to monitor and manage the water quality of wetlands for multiple uses (Ramachandra 2009). Physical and chemical characterization of wetland water quality provides valuable inputs about the health of wetland ecosystems (Dar et al. 2021c). Therefore, continuous monitoring of water quality of wetland ecosystems is crucial for developing strategies for management of these ecologically important ecosystems.

MONITORING OF WATER QUALITY

Water quality may be defined as the state or condition of water with respect to the various biological, chemical, and physical characteristics (Meybeck et al. 1996). It provides a magnitude of the quality of water with regard to certain attributes in relation to the necessities of human beings, biotic communities, and for other desired purposes. Most of the physical parameters of water quality can be monitored and determined via in-situ measurements using modern testing equipment's/field testing kits or via laboratory-based experiments (Table 1). While, the chemical parameters of water quality are mostly laboratory based. Both the methods (In-situ and laboratory based) have their advantages and disadvantages.

PHYSICAL PARAMETERS

Physical parameters of water quality are those characteristics that are sensed by means of sight, touch, smell, and taste. These include color, turbidity, taste and odor, temperature, foam, transparency, and total suspended solids.

Table 1. Methodology for assessment of physico-chemical parameters of water quality

Parameters	Unit	Method	Reference
Color	Hazen	Tintometer/ platinum cobalt	APHA 2005
Odor and taste	TON/TTN	Odor intensity	APHA 2005
Foam	Pounds/ft^3	Aeration	APHA 2005
Turbidity	NTU	Nephelometric/Turbidimeter	APHA 2005
Water temperature	°C	Thermometer/ probe	APHA 2005
Transparency	cm	Secchi disc	APHA 2005
Depth	cm	Graduated Rod	APHA 2005
pH	-	Potentiometric	APHA 2005
Electrical conductivity	µS cm^{-1}	Potentiometric	APHA 2005
Solids	mg L^{-1}	Gravimetric filtration	APHA 2005
Salinity	mg L^{-1}	Probe	APHA 2005
Dissolved oxygen	mg L^{-1}	Winkler's method/ probe	APHA 2005
Free carbon dioxide	mg L^{-1}	Titrimetric method	APHA 2005
Total alkalinity	mg L^{-1}	Titrimetric method	APHA 2005
Total hardness	mg L^{-1}	EDTA Titrimetric method	APHA 2005
Calcium hardness	mg L^{-1}	EDTA Titrimetric method	APHA 2005
Magnesium	mg L^{-1}	Titrimetric method	APHA 2005
Chloride	mg L^{-1}	Argentometric method	APHA 2005
Ammoniacal nitrogen	µg L^{-1}	Phenate Spectrophotometric method	APHA 2005
Nitrite	µg L^{-1}	Sulphanilamide Spectrophotometric method	APHA 2005
Nitrate	µg L^{-1}	Salicylate method	APHA 2005
Total Kjeldahl nitrogen	µg L^{-1}	Semi-Micro Kjeldahl method	APHA 2005
Ortho-phosphorus	µg L^{-1}	Stannous chloride method	APHA 2005
Total phosphorus	µg L^{-1}	Stannous chloride method	APHA 2005
Chlorophyll	µg m^{-3}	Acetone extraction method	APHA 2005
Biological oxygen demand	mg L^{-1}	5-day incubation method	APHA 2005
Chemical oxygen demand	mg L^{-1}	Open reflux method	APHA 2005

Color

The natural unpolluted water is colorless. The water gets its color due to the presence of impurities or any other foreign materials which is aesthetically objectionable (Sharma and Bhattacharya 2017) details in Table 2. High color of water in wetlands is dangerous as it hampers the photosynthetic activity by reducing the penetration of sunlight. Therefore, appearance of any color in the water of wetlands indicates pollution. Color is determined by an instrument known as tintometer or by comparing the shade of water with a standard color solution of platinum cobalt (APHA 2005). One color unit equals to 1 mg L^{-1} solution of platinum. Color of water is reported either as apparent color that is the color of water consisting of both the suspended and dissolved materials, and true color which is the color measured after removing all the suspended materials through filtration.

Table 2. Sources of impurities giving color to water of wetlands

Color Type	Sources
Reddish	Presence of precipitated or dissolved iron, iron reducing bacteria
Blackish	Presence of iron, manganese, and anaerobic bacteria
Yellowish	Presence of humid substances, aerobic and anaerobic bacteria
Milky	Presence of excessive air, suspended solids, and precipitation of carbonates

Odor and Taste

Pure water is odorless and doesn't give any taste. In wetlands odor and taste are due to metabolic products of algae, actinomycetes, and cyanobacteria (Watson 2003). Therefore, the wetland water giving any odor/smell or having any taste is aesthetically objectionable and indicates pollution of water. Odor and taste are measured by a term known as odor intensity (US EPA 1979). To determine the odor of wetland water, the water sample is diluted with distilled water, and the amount of the water at which the detection of odor by human observation is lost is measured. The threshold number is used to define the unit of odor and taste as

$$TON\ or\ TTN = \frac{(A+B)}{A} \tag{1}$$

where TON is threshold odor number, TTN is threshold taste number, A is volume of water, and B is volume of distilled water or odor free water.

Foam

Foam in wetlands is due to sewage, detergents, and effluents from sewage treatment plants (Fig. 1a).

Foam is viewed as an aesthetically unpleasant and a dangerous substance for aquatic life in wetlands because it creates anoxic conditions (Schilling and Zessner 2011). An overview of methods used in measurement of foam is provided in Table 3.

Table 3. Overview of methods used for measurement of foam

Method	Parameter Measured	Reference
Shaking	Foaming factor	Ruzicka et al. 2009
Webcam	Foam index	Ruzicka et al. 2009
Mixing	Height of foam column as function of time	Vikingstad et al. 2006
Alka-Seltzer	Bubble size and stability, foam height	Paris 2004
Aeration	Scum index	Hladikova et al. 2002
Handshaking	Foam height, foam stability	Madrange et al. 1993
Aeration	Foam generation, foam stability	Blackall and Marshall 1989

Figure 1. (a) Foam in wetland caused due to discharge of effluents from sewage treatment plants, (b) Turbidity of water due to clay, silt, and sediments, (c) On-site measurement of temperature using multi-parameter probe (PCS TESTR 35), and (d) Measurement of transparency using Secchi disc

Turbidity

Turbidity refers to the cloudiness or murkiness of water and is a measurement of amount of light scattered by the material present in water (Lambrou et al. 2009). Turbidity in water is aesthetically unpleasant and is due to the occurrence of suspended materials (Fig. 1b) like plankton, organic materials, silt, and clay particles (Alley 2007). Turbidity in the wetland can be measured using a turbidity rod or in the laboratory using an instrument called nephelometric/microprocessor turbidimeter. The turbidimeter is first calibrated with a standard solution of Hydrazine sulfate and Hexamethylenetetramine. Turbidity is expressed as NTU or TU, a TU is equal to 1 mg L^{-1} of silica in suspension.

Solids

Solids exist in wetlands either as dissolved or as suspended materials (Tchobanoglous et al. 2003). The total solids are a measurement of both the total suspended solids (TSS) and total dissolved solids (TDS) in a water sample. Water is classified into several uses on the basis of TDS values (Table 4).

$$Total\ Solids(TS) = TSS + TDS \quad (2)$$

The suspended and dissolved solids from the water sample can be measured by filtering the water sample using a filter paper. The materials that remain on the surface of filter paper are TSS, and those that pass through the filter paper in water are TDS.

Procedure for Estimation of TSS

Whatman's filter paper is washed by filtering through it 20 mL of deionized water followed by drying in a hot air oven at 104 ± 1 °C for 1 h and weighed. About 250 mL of the sample is filtered and allowed to dry at room temperature after which the filter paper is weighed again (APHA 2005). The quantity of total suspended solids is calculated as

$$TSS(mgL^{-1}) = \frac{A - B \times 1000}{mL\,of\,sample} \tag{3}$$

where A is the weight of filter with dried residue (mg), and B is the weight of filter (mg).

Procedure for Estimation of TDS

Lab-based measurement: The filtered sample is evaporated in a water bath. The solids collected in the dish as residue are called total dissolved solids.

On-site measurement: The TDS of water samples is recorded on-site by using a portable multi-parameter probe (PCS Testr 35). The probe is immersed into the water column for five minutes and the readings are recorded and expressed as mg L^{-1}.

Water Temperature

Water temperature is an important parameter for maintaining the thermal regime, stratification, circulation, nutrient cycling, distribution of dissolved gases, growth, and reproduction of biota in the aquatic ecosystems (Wetzel and Likens 2000). It is essential to have an on-site measurement of temperature of a waterbody, otherwise the water temperature may change. The temperature of water is measured on-site using a thermometer. The mercury bulb of the thermometer is immersed in water and after two minutes temperature is recorded. The readings are mainly expressed in þC. Additionally, the temperature of water

Table 4. Classification of water on the basis of TDS values

TDS mg L^{-1}	Classification
<500	Desirable for drinking
500-1000	Permissible for drinking
1000-3000	Useful for irrigation
>3000	unfit for drinking and irrigation

Source: (Davis and DeWiest 1966)

samples is recorded on-site by using a portable multi-parameter probe (PCS Testr 35) (Fig. 1c). The probe is immersed into the water column for two minutes and the readings are recorded and expressed as þC.

Transparency

The transparency of water is a function of reflection of light, reflected from the surface of the Secchi disc. Transparency is a measure of how deep the sunlight penetrates through the water column and has important implications for the diversity and productivity of aquatic biota (Liu et al. 2020). The transparency of water column is measured on-site using a Secchi disc (Fig. 1d). The Secchi disc is dropped into the water column and the depth at which the disc disappears is measured. The disc is then raised until it reappears and that depth is recorded again. The average of the two readings is recorded as the transparency of the water and results are expressed as cm or m.

Depth

Water depth is an important parameter regulating the distribution of macrophytes, fishes, and other forms of aquatic life in a wetland ecosystem (Angeler et al. 2005). The depth of the water column is measured within the wetland using a graduated metallic rod. The rod is immersed into the water column until it touches the bottom sediments. The depth is recorded and the results are expressed as cm or m.

pH

pH is the negative logarithm of the hydrogen ion concentration, and a measure showing the alkaline or acidic nature of water. pH is an important parameter for the optimal growth and survival of organisms in a water body. Most of the aquatic organisms live within the pH range between 6.5-8.0, the pH value both above and below this range is not favorable for organisms as they become physiologically stressed and some may die. Both low and high pH reduces the hatching of eggs and irritates the fishes and insects, damages their membranes, and kills them (Wurts and Durborow 1992). pH can also alter the behavior of other chemicals such as ammonia and chlorine in water that may affect the plant and animal life in wetlands (Shah et al. 2019).

Determination: The pH of water samples is recorded on-site by using a portable multi-parameter probe (PCS Testr 35). The instrument is first calibrated with standard solutions having pH 4, 7, and 9.2. After that the probe is immersed into the water column for five minutes and the readings are recorded. Additionally, the pH can be determined in the laboratory using a digital pH meter (Fig. 2a).

Electrical Conductivity (EC)

Conductivity is a measure of the ability of water to carry an electric current (Hamid et al. 2020). Conductivity of water is due to the presence of ions, and is dependent on their total concentration, mobility, valency, and on the water temperature during the measurement. The higher values of conductivity indicate higher concentration of ions mainly due to anthropogenic activities, geological, and weathering activities in the catchments (Costello et al. 1984). The unit of conductivity is mho per centimeter and is often reported as micromho/cm. The EC of water samples is recorded on-site by using a portable multi-parameter probe (PCS Testr 35). The probe is first calibrated using standard solution of 0.1M potassium

Figure 2. (a) Laboratory based pH measurement using digital pH meter, (b) On-site fixing of dissolved oxygen in a water sample, orange color shows the presence of dissolved oxygen, (c) White color depicts that the oxygen is not present in a water sample, and (d) A multiparameter probe (Sonde EXO 2) used for on-site measurement and continuous monitoring of water quality

chloride. After that the probe is immersed into the water column for five minutes and the readings are recorded and expressed as $\mu S\ cm^{-1}$.

Salinity

Salinity is a measure of the amount of salt dissolved in water. Increased salinity affects fishes, microorganisms, and macrophytes (Hart et al. 1990). Higher levels result in degradation of wetlands, decreasing wetland species and endangering wetland biodiversity.

The salinity of water samples is recorded on-site by using a portable multi-parameter probe (PCS Testr 35). The probe is immersed into the water column for five minutes and the readings are recorded and expressed as $mg\ L^{-1}$.

CHEMICAL PARAMETERS

Dissolved Oxygen (DO)

Dissolved oxygen is a measure of the quantity of oxygen dissolved in a unit volume of water. Measurement of dissolved oxygen provides an estimate of the capability of water to sustain life in aquatic ecosystems. Aquatic organisms such as fish and other species consume substantial amount of oxygen dissolved in water (Shoko et al. 2014). Therefore, dissolved oxygen is essential for maintenance of aquatic life in wetland ecosystems.

The amount of oxygen in a water sample is determined as per Winkler's method (APHA 2005). The DO is fixed on-site by adding 1 mL of manganous sulphate followed by the addition of 1 mL of alkali-iodide to the water sample in a 300 mL reagent bottle. Stopper is tightly placed over the bottle to exclude air bubbles after which it is inverted few times so as to completely mix the chemicals. The added reagents result in the formation of a floc that is dissolved by the addition of 2 mL concentrated sulfuric acid. The presence and absence of DO in water is visualized by the presence of orange and white color respectively (Fig. 2b, c). From the fixed sample, 50 mL is taken in a titration flask and titrated against standard 0.025N sodium thiosulphate solution using starch as indicator. The amount of DO present in water sample is calculated from the formula given below and the results are expressed in mg L^{-1}.

$$DO(mgL^{-1}) = \frac{V_1 \times N \times e \times 1000}{V_2} \tag{4}$$

where, V_1 is the volume of sodium thiosulphate used (mL), V_2 is the volume of sample taken for titration (mL), N is the normality of sodium thiosulphate solution, and e is the equivalent weight of oxygen.

Free Carbon Dioxide

Free carbon dioxide is an important raw material for photosynthetic activity of primary producers of an aquatic ecosystem. The measurement of free carbon dioxide in wetlands is important as it changes the pH of the water by forming H_2CO_3 (Wurts 2003).

Determination: The water sample is collected in a sampling bottle and the cap is tightly placed to avoid any exchange of gas. From the collected sample, 50 mL is taken in a titration flask to which 2 drops of phenolphthalein indicator are added after which it is titrated with 0.02N NaOH till color change. Free carbon dioxide concentration is calculated using the following formula:

$$F - CO_2(mgL^{-1}) = \frac{V_1 \times 1000}{V_2} \tag{5}$$

Where, V_1 is the volume of sodium hydroxide used (mL), and V_2 is the volume of the sample taken for titration (mL).

Total Alkalinity

Alkalinity is a measure of the capacity of the water to neutralize acids (Moyle 1946). It is a sum of all titratable bases including carbonates, bicarbonates, and hydroxide content (Gibbs 1970). Alkalinity plays a central role in determining the capacity of water to neutralize acidic pollution, supports the growth of algae, and other aquatic organisms. Alkalinity in excess of the concentration of alkaline earth metals is important in determining the aptness of water for irrigational purposes.

Measurement: Take a suitable volume (50 mL) of sample in a conical flask, and add 0.2 mL of methyl orange. Titrate the contents of the flask with 0.02N H_2SO_4 till color change. Total alkalinity is calculated as

$$Total\ Alkalinity(mgL^{-1}) = \frac{V_1 \times N \times 50000}{V_2} \tag{6}$$

Where, V_1 is volume of titrant used for the sample (mL), N is normality of H_2SO_4, and V_2 is volume of sample taken for titration.

Total Hardness

Hardness of water is caused due to cations, anions, and by a variety of other metals (Mairs 1966). Besides, calcium and magnesium are the other important contributors of hardness to water. Based on the values of hardness, water is classified as soft, moderately hard, and very hard (Table 5).

Determination: Complexometric method (APHA 2005). Take a suitable volume of sample (50 mL) is taken in an Erlenmeyer flask, and add 1 mL standard buffer solution to raise the pH of sample to about 10 followed by addition of 1 mL of sodium sulphide. Add a pinch of indicator Erichrome Black-T indicator that turns the color of solution to wine red. Titrate the contents of the flask with standard 0.01 M EDTA solution. The end point is detected by color change from wine red to blue. Total hardness is calculated as

$$Total\ hardness(as\ CaCO_3)(mgL^{-1}) = \frac{T \times 1000}{V} \tag{7}$$

Where, T is the volume of EDTA (mL) used in the titration, and V is volume of sample used.

Table 5. Classification of water on the basis of Hardness values

Total Hardness mg L^{-1}	Classification
<75	Soft
75-150	Moderately hard
150-300	Hard
>300	Very hard

Source: (Sawyer and McCarthy 1967)

Calcium Hardness

Calcium in the form of Ca^{2+} is one of the major inorganic cations presents in water. It originates from the dissolution of salts of calcium, chloride, and calcium sulphate in water (Bhateria and Jain 2016). Most of the calcium in wetlands originates from the streams and rivers flowing over limestone, gypsum, $CaCO_3$, and other calcium containing rocks and minerals.

Determination: EDTA Titrimetric (APHA 2005). Take 50 mL of sample in a flask and add 2 mL of 1N NaOH followed by a pinch of murexide indicator. Titrate the contents of the flask with 0.01N EDTA till color changes from pink to purple. Calcium (mg L^{-1}) is estimated using the formula as

$$Calcium\ hardness(as\ CaCO_3)(mgL^{-1}) = \frac{V_1 \times 1.05 \times 1000}{V_2} \tag{8}$$

Where, V_1 is the volume of EDTA (mL) used in the titration of sample and V_2 is the volume of sample used.

Magnesium Hardness

The most common aquatic species imparting hardness is Mg^{2+} (Bhateria and Jain 2016). Magnesium hardness is calculated as a difference between total hardness and calcium as $CaCO_3$. The results are expressed in mg L^{-1}.

$$Magnesium\ hardness(mgL^{-1}) = Total\ hardness - Calcium\ hardness \tag{9}$$

Chloride

Chloride is an essential micronutrient for wetland plants but it is toxic in excessive amounts (Schraufnagel 1967). Chloride in water is due to influx of domestic sewage and salts of calcium, sodium, and potassium (Kaushik and Saksena 1999). However, the high concentration of chloride is harmful to aquatic organisms as it interferes with the osmoregulation.

Determination: Argentometric Titration (APHA 2005). Take 50 mL of water sample in a flask and add 2 mL of potassium dichromate indicator. Titrate the mixture using 0.0141N silver nitrate till the attainment of brick red color. Concentration of chloride is calculated using the formula

$$Chloride(mgL^{-1}) = \frac{(A - B) \times N \times 35 \times 450}{V} \tag{10}$$

Where, A is volume of titrant used for sample (mL), B is the volume of titrant used for blank (mL), N is the normality of the silver nitrate solution, and V is the volume of sample used (mL).

Nitrogen

Nitrogen is an important nutrient element required to support primary production in wetland ecosystems. Nitrogen in wetlands is present in various forms such as particulate nitrogen, inorganic (ammonium, nitrate), and dissolved organic compounds (amino acids and urea) (Anita et al. 1991; Peterls and Paerl 1997).

Ammoniacal Nitrogen

Ammonia is the second most important parameter after DO that affects the growth and production of fish and aquatic foods (Leoni et al. 2018). Ammoniacal nitrogen in wetlands is produced as an end product of the decomposition of organic matter by heterotrophic bacteria. Ammonia produced is readily assimilated by aquatic plants in the trophogenic zone (Wetzel 2001). Ammonia even in low concentration causes stress and susceptibility of fish to bacterial infections, and damage to gills, and tissues (Francis-Floyd et al. 2009).

Determination: Phenate Spectrophotometric (APHA 2005)

Take 25 mL of sample in a flask and add 1 mL phenol solution, 1 mL sodium nitroprusside solution, and 2.5 mL of oxidizing solution (100 mL alkaline citrate and 25 mL sodium hypochlorite). Avoid the direct exposure of light by covering the flasks with foil. Place the samples in dark for 1 hr till color development takes place. Measure the intensity of blue color at 640 nm wavelength using a spectrophotometer. Prepare a standard curve by dissolving 3.891 g of anhydrous ammonium chloride in water and dilute to 1000 mL (1 mL = 1 mg N = 1.22 mg NH_3). Compute sample concentration by comparing absorbance with the standard curve.

Nitrite Nitrogen (NO_2^--N)

Nitrite is formed by the oxidation of ammonia and is readily available for plant uptake. In wetlands, nitrite supports the production of both the algae and plants (Durand et al. 2011). However, higher concentration of nitrite is of great concern to wetland managers as it is highly toxic to aquatic flora, and fauna (Philips et al. 2002). Nitrite is a known bacteriostatic molecule as due to its affinity with metal ions, therefore, it influences the enzymatic systems of microorganisms and their processes (Wild et al. 1995).

Determination: Sulphanilamide Spectrophotometric (Golterman and Clymo 1978). Take a suitable volume (45 mL) of sample in a 50 mL Erlenmeyer flask. Add 1 mL of Sulfanilamide reagent, mix well and after 5 minutes add 1 mL of N-1-napthylenediamine dihydrochloride. Raise the volume to 50 mL, and after 10 minutes, measure the absorbance at 543 nm. Prepare a standard curve by dissolving 1.232 g sodium nitrite ($NaNO_2$) in water and dilute to 1000 mL (1 mL = 250 μg N). Compute sample concentration by comparing absorbance with the standard curve.

Nitrate Nitrogen (NO_3^-N)

Nitrate is formed in wetlands through the oxidation of ammonia, nitrite, and other nitrogen compounds such as amino acids. Besides nitrate is also introduced through sewage, septic systems, landfills, garbage dumps, and fertilizer inputs (Dar et al. 2021d, e). In wetlands, the occurrence of some fixed amount of nitrogen isn't problematic, however, the higher concentration of NO_3^-N can cause serious water problems. In water, the proteins, amino acids, urea, and organic matter are broken down by microorganisms, the nitrogen formed in the process combines with oxygen and forms NO_3^-. This decreases the oxygen dissolved in water thereby forming anoxic zones that has severe detrimental effects for all organisms living in water.

Determination: Salicylate Method (APHA 2005). Take 50 mL of sample in an Erlenmeyer flask and add 1 mL of sodium salicylate solution. Evaporate the contents of the flask on a water bath till the appearance of white color residue. To the residue, add 1 mL of concentrated sulfuric acid followed by mixing. Allow to stand for 10 minutes and add cautiously 6 mL of distilled water and 7 mL of 30% NaOH solution. Measure the intensity of yellow color at 410 nm using a spectrophotometer. Prepare a standard curve by dissolving anhydrous 0.7218 g potassium nitrate (KNO_3) in water and dilute to 1000 mL (1 mL = 100 µg NO_3-N). Compute sample concentration by comparing absorbance with the standard curve.

Total Kjeldahl Nitrogen

Kjeldahl nitrogen is a term that reflects the technique used for determination of organic nitrogen together with the ammonia. Organic nitrogen is organically bound nitrogen in tri-negative oxidation state and includes proteins, peptides, nucleic acids, urea, and synthetic organic materials. Typical values of organic nitrogen range from few µg L^{-1} to >20 mg L^{-1} in raw sewage.

Determination: Kjeldahl method (APHA 2005). Take a suitable volume of sample (50 mL) in a digestion tube, add catalyst mixture potassium sulphate and copper sulphate in the ratio of 5:1, followed by 10 mL of concentrated H_2SO_4. Preheat the digestion unit to 350 þC. Place the digestion tubes in the digestion unit and continue digestion of the sample at 420 þC for about 1 hr, the presence of green color indicates the digestion is complete. After cooling the sample, dilute the sample with distilled water (10 mL) and place the digestion tube (sample) in distillation unit on one end and titration flask laden with 25 mL of 4% boric acid and mixed indicators (bromocresol green and methyl red) at the receiver end. Start adding alkali (40% NaOH) until dark brown color appears. Start the process for 9 minutes, during the process liquid ammonia will collect in the boric acid that is reflected by the change in color of boric acid. After completion of the process, titrate the contents of the flack with 0.1N H_2SO_4 till color change. Calculate the concentration of TKN as

$$TKN(mgL^{-1}) = \frac{(A-B)\times 280}{mL\ Sample}$$

where A is the volume used for titration of sample, and B is the volume used for titration of blank.

Phosphorus

Phosphorus is one of the vital nutrient elements for the growth of plant and animal communities representing the aquatic food web. Phosphorus is considered as a limiting nutrient for phytoplankton production in wetland ecosystems (Rabalais 2002). Phosphorus is a critical nutrient available in low concentration in wetlands whose modest increase leads to undesirable effects such as eutrophication, algal blooms, and accelerated growth of algae.

1. Orthophosphate phosphorus

Orthophosphates also known as reactive phosphates are the main constituents in fertilizers used in horticulture and agriculture production. These are carried to wetlands and other aquatic ecosystems through surface runoff. In wetlands, orthophosphates are readily utilized and provide a good estimate of the amount of phosphorus available for plant and algal growth.

Determination: Stannous chloride (APHA 2005). Take 50 mL of sample in an Erlenmeyer flask and add 4 mL of ammonium molybdate and 0.5 mL of stannous chloride and mix thoroughly. After

10 minutes, measure the absorbance of each sample at 690 nm. Prepare a standard curve by dissolving anhydrous 219.5 mg of anhydrous KH_2PO_4 in water and dilute to 1000 mL (1 mL = 50 µg PO_4^{3-}-P). Compute sample concentration by comparing absorbance with the standard curve.

2. Total phosphorus

Total phosphorus in wetlands includes various forms of phosphorus such as orthophosphates, condensed phosphates (pyro, meta, and other polyphosphates), and organically bound phosphates. They occur in solution, in particles or detritus, or in the bodies of aquatic ecosystems. The discharge of raw or treated sewage, and agricultural drainage to the wetlands may stimulate the growth of photosynthetic aquatic micro and macro-organisms in nuisance quantities.

Method: Stannous chloride (APHA 2005)

Take 50 mL of water sample in an Erlenmeyer flask and add 1 mL of sulfuric acid and 5 mL of nitric acid. Digest the contents of the flask on a hot plate till the volume of sample reduces to 1 mL. After digestion, allow the sample to cool and then add 20 mL of distilled water. After this add 0.05 (1 mL) of phenolphthalein and titrate the sample with 0.2N NaOH till faint pink color is obtained. Raise the contents to 50 mL using distilled water and add 10N H_2SO_4 dropwise to discharge the color. Add 4 mL of ammonium molybdate and 0.5 mL of stannous chloride. After 10 minutes measure the intensity of blue color at 690 nm using a spectrophotometer. Prepare a standard curve by dissolving anhydrous 219.5 mg of anhydrous KH_2PO_4 in water and dilute to 1000 mL (1 mL = 50 µg PO_4^{3-}-P). Compute sample concentration by comparing absorbance with the standard curve.

Chlorophyll

Chlorophyll-a is an important water quality parameter that is extensively used for monitoring of phytoplankton biomass and assessment of trophic status of wetlands and other aquatic ecosystems (Datti et al. 2017). The important degradation products of chlorophyll found in aquatic ecosystems include chlorophyllides, pheophorbides, and pheophytins. The presence or absence of the various photosynthetic pigments is used among other features to separate the major algal groups.

Determination: Acetone extraction (Spectrophotometric) (APHA 2005). Filter a suitable volume (500 mL) of water sample through a 0.45 µm pore size filter. Add 2 mL of 90% acetone solution and disrupt the cells in a tissue grinder. Transfer the sample to a centrifugation tube and make the volume 10 mL by adding 90% acetone solution. Continue the centrifugation for 15 minutes. Transfer 3 mL of clarified extract to a 1 cm cuvette and read the optical density in a spectrophotometer at 750 and 664 nm. Acidify the extract in the cuvette by adding 0.1 mL of 0.1N HCL. Shake the extract and after 90 seconds read the optical density again at 750 and 665 nm. The concentration of chlorophyll is calculated as

$$Chlorophyll\ a\,(mg\,m^3) = \frac{26.7(664_b - 665_a)\times V_1}{V_2 \times L} \quad (11)$$

Where, 26.7 is absorbance correction factor and equals A×K, A is the absorbance coefficient for chlorophyll-a at 664 nm =11.0, and K is the ratio expressing correction for acidification. V_1 is the volume of extract (L), V_2 is the volume of sample (m^3), and L is the light path length or width of cuvette (cm).

Biochemical Oxygen Demand (BOD)

BOD is an estimate of the amount of dissolved oxygen consumed by microorganisms during the oxidation of organic matter (Bhateria and Jain 2016). Measurements of BOD in wetlands is vital as it determines the quantity of organic matter in water. Higher values of BOD mean that more rapidly is the oxygen depleted from water, making less oxygen available to aquatic organisms. High BOD results in anoxic conditions in wetlands leading to death of fish, macroinvertebrates, and other aerobic organisms.

Determination: 5 Day Incubation (APHA 2005). Take 1000 mL of distilled water and thoroughly aerate it. Add 1 mL of phosphate buffer solution, magnesium sulphate ($MgSO_{4.}7H_2O$) solution, calcium chloride ($CaCl_2$) solution, and ferric chloride ($FeCl_3$.6H2O) solution. Fill two sets of BOD bottles each of 300 mL capacity with required volume of sample and dilution water at a ratio of 5:100 (5%). Determine the amount of dissolved oxygen in one set immediately following the Winkler's method and incubate the another set for 5 days in a BOD incubator at 27 C. After 5 days determine the dissolved oxygen content of the incubated bottle following Winkler's method. Calculate the BOD mg L^{-1} as

$$BOD_5\left(mgL^{-1}\right) = \frac{D_I - D_F}{N} \quad (12)$$

where D_I is the initial DO, D_F is final DO, and N is the dilution factor

Chemical Oxygen Demand (COD)

COD is a measure of the amount of O_2 needed for the chemical oxidation of organic matter using strong chemical oxidants. COD as compared to the BOD is a quick and inexpensive means for estimation of organics in wetlands. COD is always higher than BOD because there are some inert materials in addition to biodegradable compounds that are degraded by strong chemical oxidants in the process.

Selection of Method: The open reflux method is suitable for a wide range of wastes where a large sample size is preferred. The closed reflux methods are more economical in the use of metallic salt reagents and generate smaller quantities of hazardous waste, but require homogenization of samples containing suspended solids to obtain reproducible results.

Determination: Open reflux method (APHA 2005). Take 25 mL of sample and add 25 mL of distilled water, 5 mL of sulfuric acid reagent (sulfuric acid + silver sulphate), 25 mL of potassium dichromate ($K_2Cr_2O_7$) solution, 30 mL of sulfuric acid reagent, mercuric sulphate powder and add few glass beads. The contents are then refluxed for digestion in COD digester at 150 þC for 2 hours. Cool the contents and titrate against 0.25N ferrous ammonium sulphate (Fe (NH_4)2 $SO_{4.}7H_2O$) using ferroin as an indicator till color changes from green to reddish brown (end point). Calculate the COD using the following formula:

$$COD\left(mgL^{-1}\right) = \frac{\left(B - A\right) \times N \times 1000 \times 8}{mL\ of\ sample} \quad (13)$$

where, B is volume of titrant used for blank, A is volume of titrant used for sample, and M is normality.

PROBES

While, the lab-based analysis provides a favorable environment for testing large number of samples, there is an additional cost associated with lab-based analysis such as transportation of samples, storage, and also the quality of water may change with respect to time. Modern sophisticated equipment's/multi-parameter probes such as Sonde EXO 2 have the advantages that they can be used to assess the water quality of wetlands on-site with better precision and can be carried to far-flung remote locations. The only disadvantages with the modern testing equipment's being the high cost and ability to determine only some select parameters. The German made multi-parameter probe such as Sonde EXO 2 (Fig. 2d) have the advantage that it can be used to assess the water quality of wetlands on-site with better precision and can be carried to far-flung remote locations. The disadvantages with the instrument being the high cost and ability to determine only some select parameters such as pH, conductivity, TDS, salinity, chlorophyll, nitrate, free dissolved organic matter, depth, ammonium, and chloride.

CONCLUSION

Water quality monitoring has proved to be one of the robust tools to offer the insights on the status of the wetlands for policy makers and the interventions thereof. Water quality of wetland ecosystems can be measured via both the In-situ and the laboratory measurements. In-situ measurements via sophisticated instruments are less time consuming and make results readily available. The main elements of laboratory analysis are the collection of samples, transportation, preservation, and analysis. Both the methods finally involve the interpretation and reporting of results. The results of analysis performed on a single water sample are only valid for the particular location and time at which that sample was taken. One purpose of a continuous monitoring program is therefore, to gather sufficient data by means of regular or intensive sampling and analysis to assess spatial and/ or temporal variations in water quality.

REFERENCES

Ahmad, Z., Hussain, A., & Shakeel, A. (2019). *Economics Importance of Wetlands, their Benefits and Values Case of Pakistan*. Academic Press.

American Public Health Association (APHA). (2005). *Standard methods for the examination of water and waste water* (21st ed.). APHA.

Angeler, D. G., Sanchez-Carrillo, S., Rodrigo, M. A., Viedma, O., & Alvarez-Cobelas, M. (2005). On the importance of water depth, macrophytes and fish in wetland picocyanobacteria regulation. *Hydrobiologia, 549*(1), 23–32. doi:10.100710750-005-1700-5

Antia, N. J., Harrison, P. J., & Oliveira, L. (1991). The role of dissolved organic nitrogen in phytoplankton nutrition, cell biology and ecology. *Phycologia, 30*(1), 1–89. doi:10.2216/i0031-8884-30-1-1.1

Assessment, M. E. (2005). *Ecosystems and human well-being: wetlands and water*. World Resources Institute.

Bhateria, R., & Jain, D. (2016). Water quality assessment of lake water: A review. *Sustainable Water Resources Management*, *2*(2), 161–173. doi:10.100740899-015-0014-7

Blackall, L. L., & Marshall, K. C. (1989). The mechanism of stabilization of actinomycete foams and the prevention of foaming under laboratory conditions. *Journal of Industrial Microbiology*, *4*(3), 181–187. doi:10.1007/BF01574075

Blackwell, M. S., & Pilgrim, E. S. (2011). Ecosystem services delivered by small-scale wetlands. *Hydrological Sciences Journal*, *56*(8), 1467–1484. doi:10.1080/02626667.2011.630317

Dar, S. A., Bhat, S. U., Aneaus, S., & Rashid, I. (2020b). A geospatial approach for limnological characterization of Nigeen Lake, Kashmir Himalaya. *Environmental Monitoring and Assessment*, *192*(2), 1–18. doi:10.100710661-020-8091-y PMID:31953600

Dar, S. A., Bhat, S. U., & Rashid, I. (2021a). The Status of Current Knowledge, Distribution and Conservation Challenges of Wetland Ecosystems in Kashmir Himalaya, India. In Sharma, S. & Singh, P. (eds) Wetlands Conservation: Current Challenges and Future Strategies. https://doi.org/10.1002/9781119692621.ch10

Dar, S. A., Bhat, S. U., & Rashid, I. (2021d). Landscape transformations, morphometry and water quality of Anchar wetland in Kashmir Himalaya, India: Implications for urban wetland management. *Water, Air, and Soil Pollution*, *232*(462) 1-19. doi:10.100711270-021-05416-5

Dar, S. A., Bhat, S. U., Rashid, I., & Dar, S. A. (2020a). Current status of wetlands in Srinagar City: Threats, management strategies, and future perspectives. *Frontiers in Environmental Science*, *7*, 199. doi:10.3389/fenvs.2019.00199

Dar, S. A., Hamid, I., Bhat, S. U., & Rashid, I. (2021c). Identification of anthropogenic contribution to wetland degradation: Insights from the Environmetric Techniques. *Stochastic Environmental Research and Risk Assessment*, 1–15. doi:10.100700477-021-02121-x

Dar, S. A., Rashid, I., & Bhat, S. U. (2021b). Land system transformations govern the trophic status of an urban wetland ecosystem: Perspectives from remote sensing and water quality analysis. *Land Degradation & Development*, *32*(14), 1–18. doi:10.1002/ldr.3924

Dar, S. A., Rashid, I., & Bhat, S. U. (2021e). Linking land system changes (1980-2017) with the trophic status of an urban wetland: Implications for wetland management. *Environmental Monitoring and Assessment*, 193(710), 1–17. https://doi.org/10.1007/s10661-021-09476-2

Datti, A. A., Nanda, S., Annadurai, R., & Sengupta, M. (2017). Geospatial estimation of chlorophyll-a in urban lakes. *Rasayan Journal of Chemistry*, *10*(1), 263–270.

Davis, S. N., & DeWiest, R. J. (1966). *Hydrogeology*. Wiley.

Durand, P., Breuer, L., Johnes, P. J., Billen, G., Butturini, A., Pinay, G., ... Wright, R. (2011). *Nitrogen processes in aquatic ecosystems*. Academic Press.

Francis-Floyd, R., Watson, C., Petty, D., & Pouder, D.B. (2009). *Ammonia in aquatic systems*. UF/IFAS University of Florida (UF)/Institute of Food and Agricultural Sciences (IFAS).

Galatowitsch, S. M. (2018). Natural and anthropogenic drivers of wetland change. In C. Finlayson, G. Milton, R. Prentice, & N. Davidson (Eds.), *The Wetland Book II: Distribution, Description, and Conservation* (pp. 359–367). Springer Nature. doi:10.1007/978-94-007-4001-3_217

Gibbs, R. J. (1970). Mechanisms controlling world water chemistry. *Science, 170*(3962), 1088–1089. doi:10.1126cience.170.3962.1088 PMID:17777828

Hladikova, K., Ruzickova, I., Klucova, P., & Wanner, J. (2002). An investigation into studying of the activated sludge foaming potential by using physicochemical parameters. *Water Science and Technology, 46*(1-2), 525–528. doi:10.2166/wst.2002.0529 PMID:12216681

Kaushik, S., & Saksena, D. N. (1999). Physico-chemical limnology of certain water bodies of central India. In K. Vijay Kumar (Ed.), *Freshwater ecosystem of India*. Daya Publishing House Delhi.

Keddy, P. A., Fraser, L. H., Solomeshch, A. I., Junk, W. J., Campbell, D. R., Arroyo, M. T., & Alho, C. J. (2009). Wet and wonderful: The world's largest wetlands are conservation priorities. *Bioscience, 59*(1), 39–51. doi:10.1525/bio.2009.59.1.8

Lambrou, T. P., Anastasiou, C. C., & Panayiotou, C. G. (2009, September). A nephelometric turbidity system for monitoring residential drinking water quality. In *International Conference on Sensor Applications, Experimentation and Logistics* (pp. 43-55). Springer.

Leoni, B., Patelli, M., Soler, V., & Nava, V. (2018). Ammonium transformation in 14 lakes along a trophic gradient. *Water (Basel), 10*(3), 265. doi:10.3390/w10030265

Liu, D., Duan, H., Loiselle, S., Hu, C., Zhang, G., Li, J., Yang, H., Thompson, J. R., Cao, Z., Shen, M., Ma, R., Zhang, M., & Han, W. (2020). Observations of water transparency in China's lakes from space. *International Journal of Applied Earth Observation and Geoinformation, 92*, 102187. doi:10.1016/j.jag.2020.102187

Madrange, L., Chaboury, P., Ferrandon, O., Mazet, M., & Rodeaud, J. (1993). Study of the formation and stability of chemical foam on the Vienne river between Limoges and Confolens [Etude de la formation et de la stabilite des mousses chimiques de surface de la Vienne]. *Revue des sciences de l'eau/journal of water science, 6*(3), 315-334.

Mairs, D. F. (1966). A total alkalinity atlas for marine lake waters. *Limnology and Oceanography, 11*(1), 68–72. doi:10.4319/lo.1966.11.1.0068

Meybeck, M., Kuusisto, E., Makela, A., & Malkki, E. (1996). *Water quality monitoring: a practical guide to the design and implementation of freshwater quality studies and monitoring programmes* (J. Bartram & R. Ballance, Eds.). CRC Press.

Mitsch, W. J., Bernal, B., & Hernandez, M. E. (2015). Ecosystem services of wetlands. *The International Journal of Biodiversity Science, Ecosystem Services & Management, 11*(1), 1–4. doi:10.1080/21513732.2015.1006250

Mitsch, W. J., & Gosselink, J. G. (2015). *Wetlands*. John Wiley & Sons.

Moyle, J. B. (1946). Some chemical factors infl uencing the distribution of aquatic plants in Minnesota. *American Midland Naturalist, 34*(2), 402–420. doi:10.2307/2421128

Osei, J., Nyame, F. K., Armah, T. K., Osae, S. K., Dampare, S. B., Fianko, J. R., Adomako, D., & Bentil, N. (2010). Application of multivariate analysis for identification of pollution sources in the Densu delta wetland in the vicinity of a landfill site in Ghana. *Journal of Water Resource and Protection*, *2*(12), 1020–1029. doi:10.4236/jwarp.2010.212122

Paris, S. (2004). *Bekämpfung von Schwimmschlamm, verursacht durch Microthrix parvicella* (Doctoral dissertation). Technische Universität München.

Parvez, S., & Bhat, S. U. (2014). Searching for water quality improvement in Dal lake, Srinagar, Kashmir. *Journal of Himalayan Ecology and Sustainable Development*, *9*, 51–64.

Peierls, B. L., & Paerl, H. W. (1997). Bioavailability of atmospheric organic nitrogen deposition to coastal phytoplankton. *Limnology and Oceanography*, *42*(8), 1819–1823. doi:10.4319/lo.1997.42.8.1819

Philips, S., Laanbroek, H. J., & Verstraete, W. (2002). Origin, causes and effects of increased nitrite concentrations in aquatic environments. *Reviews in Environmental Science and Biotechnology*, *1*(2), 115–141. doi:10.1023/A:1020892826575

Rabalais, N. N. (2002). Nitrogen in aquatic ecosystems. *Ambio*, *31*(2), 102–112. doi:10.1579/0044-7447-31.2.102 PMID:12077998

Ramachandra, T. V. (2009). *Conservation and management of urban wetlands: Strategies and challenges*. Academic Press.

Romshoo, S. A., & Rashid, I. (2014). Assessing the impacts of changing land cover and climate on Hokersar wetland in Indian Himalayas. *Arabian Journal of Geosciences*, *7*(1), 143–160. doi:10.100712517-012-0761-9

Ruzicka, K., Gabriel, O., Bletterie, U., Winkler, S., & Zessner, M. (2009). Cause and effect relationship between foam formation and treated wastewater effluents in a transboundary river. *Physics and Chemistry of the Earth Parts A/B/C*, *34*(8-9), 565–573. doi:10.1016/j.pce.2009.01.002

Sawyer, C., & McCarthy, P. (1967). *Chemical and Sanitary Engineering* (2nd ed.). McGraw-Hill.

Schilling, K., & Zessner, M. (2011). Foam in the aquatic environment. *Water Research*, *45*(15), 4355–4366. doi:10.1016/j.watres.2011.06.004 PMID:21757217

Schraufnagel, F.H. (1967). Pollution associated with chemical de-icing. *Highway Research Record*, *193*, 22–33.

Shah, J. A., Pandit, A. K., & Shah, G. M. (2019). Physico-chemical limnology of a shallow lake in the floodplains of western Himalaya from last four decades: Present status. *Environmental Systems Research*, *8*(1), 1–10. doi:10.118640068-019-0136-2

Sharma, S., & Bhattacharya, A. (2017). Drinking water contamination and treatment techniques. *Applied Water Science*, *7*(3), 1043–1067. doi:10.100713201-016-0455-7

Shoko, A. P., Limbu, S. M., Mrosso, H. D., & Mgaya, Y. D. (2014). A comparison of diurnal dynamics of water quality parameters in Nile tilapia (Oreochromis niloticus, Linnaeus, 1758) monoculture and polyculture with African sharp tooth catfish (Clarias gariepinus, Burchell, 1822) in earthen ponds. *International Aquatic Research.*, *6*(1), 1–13. doi:10.100740071-014-0056-8

Tchobanoglous, G., Burton, F. L., & Stensel, H. D. (2003). Metcalf & Eddy wastewater engineering: treatment and reuse. International Edition. McGraw-Hill.

United States Environmental Protection Agency. (1979). *National Secondary Drinking Water Regulations EPA 570/9-76-000*. US EPA.

Vikingstad, A. K. (2006). *Static and dynamic studies of foam and foam-oil interactions. Department of Chemistry*. University of Bergen.

Watson, S. B. (2003). Cyanobacterial and eukaryotic algal odour compounds: Signals or by-products? A review of their biological activity. *Phycologia, 42*(4), 332–350. doi:10.2216/i0031-8884-42-4-332.1

Wetzel, R. G. (2001). *Limnology: lake and river ecosystems*. Academic Press.

Wetzel, R. G., & Likens, G. E. (2000). Light and Temperature. In *Limnological Analyses*. Springer., doi:10.1007/978-1-4757-3250-4_2

Wild, D., Von Schulthess, R., & Gujer, W. (1995). Structured modelling of denitrification intermediates. *Water Science and Technology, 31*(2), 45–54. doi:10.2166/wst.1995.0070

Wurts, W. A. (2003). Daily pH cycle and ammonia toxicity. *World Aquaculture, 34*(2), 20–21.

Wurts, W. A., & Durborow, R. M. (1992). Interactions of pH, carbon dioxide, alkalinity and hardness in fish ponds. Southern Regional Aquaculture Centre. *Fact Sheet*, (464), 1–4.

Chapter 5
Metagenomics-Guided Assessment of Water Quality and Predicting Pathogenic Load

Sayak Ganguli
https://orcid.org/0000-0001-5481-7436
St. Xavier's College (Autonomous), Kolkata, India

Rupsha Karmakar
Vivekananda College for Women, India

Meesha Singh
St. Xavier's College (Autonomous), Kolkata, India

Mahashweta Mitra Ghosh
St. Xavier's College (Autonomous), Kolkata, India

ABSTRACT

Antibiotic-resistant bacteria (ARB) are becoming more prevalent in the environment and are efficiently disseminating through contaminated wastewater resulting in resistome cycling. This chapter compares the bacterial profile of hospital effluents collected from rural, urban, and delta regions of West Bengal, India. Comparative metagenomics analysis identified pathogenic bacterial genera like pseudomonas, escherichia, staphylococcus, lactobacillus, prevotella, acinetobacter across the samples. Delta sample showed highest abundance of pseudomonas whereas rural sample had lower titre of all the common bacterial genera. Urban sample reflected more diversity of different genera in terms of abundance. Pathogenic load prediction revealed significant occurrence of diarrhea, irritable bowel syndrome, liver cirrhosis, ulcerative colitis in the disease network. This chapter proposes a monitoring programme for assessing wastewater health using a combination of culture independent and culture-dependent molecular techniques in order to prevent the spread of pollutants in tropical environments.

INTRODUCTION

DOI: 10.4018/978-1-7998-9498-8.ch005

A Well-developed sewage infrastructure, efficient wastewater treatment, and availability of drinking water are just a few of the basic requirements for maintaining and improving the public health (Bürgmann et al., 2018). Water management systems around the world are confronted with a huge issue of contamination of water supplies. One of the few reasons behind this contamination is lack of proper sanitary infrastructure which pollutes the environment and pervades all societal processes leading to human health risks that in turn can reduce overall developmental potential. According to the United Nations, about 80 percent of sewage is discharged in the adjoining aquatic bodies without any treatment (UN WWAP (United Nations World Water Assessment Programme), 2017). Animal excreta, variety of biomolecules, decomposition products, sands, grits, natural and synthetic organic substances from industrial sources, various micronutrients, macronutrients, heavy metals, traces of antibiotics are the basic contents of domestic, industrial and hospital effluents (Holmberg, 2019; Ajiboye et al., 2021; Hassan et al., 2020). Hospital wastes are in constant limelight as these include antibiotic residues such as Oxytetracycline, Tetracycline, Ciprofloxacin, Fluoroquinolones, Sulfonamides, Macrolides, Trimethoprim, disinfectants, anaesthetics, lipid regulators, radioactive markers, organic matters, excreta from patients and many more (Hassan et al., 2020; Ngigi et al., 2019; Thai et al., 2018). Additionally, various kinds of pathogenic microbial strains can be found in wastewater (Yasir, 2021).

Moreover, domestic sewage, hospital effluents and drainage from animal husbandry act as reservoirs of Antibiotic resistant bacteria (ARB) and Antibiotic resistance genes (ARGs) (He et al., 2020; Khan et al., 2019). ARBs have increased their population in the environment due to severe use of antibiotics for medical purposes and their continuous release into the environment (Serwecińska, 2020; Kraemer et al., 2019). World Health Organization has classified ARGs as a new contaminant due to their increasing incidence and widespread distribution. Extensive use of antibiotics in medicine and agriculture has been recognized as one of the major causes of rising ARGs. Similarly, heavy metal exposure from various sources can influence the spread of Metal resistant genes (MRGs) and ARGs. Heavy metal resistance and antibiotic resistance can be selected simultaneously in the heavy metal contamination environment. Two mechanistic interpretations of this event are co-selection (selection of two or more genetically linked resistant genes) or cross-selection (an event where a single genetic element offers tolerance to many antimicrobial drugs) (Chen et al., 2019). In agriculture, animal husbandry, wastewater treatment systems and sediment, the co-selection of MRGs and ARGs has been documented (Wang et al., 2021; Li et al., 2021; Yuan et al., 2018). Bacterial communities, on the other hand, are influenced by a diverse set of evolutionary, ecological, and environmental factors which includes the concentration and distribution of MRGs and ARGs also (Chen et al., 2019). Studies have proven that various constituents of waste water, such as antibiotic residues, disinfectants, and heavy metals exert selection pressure for antibiotic resistance (Karkman et al., 2018). Several studies have recently revealed that antibiotic residues, resistant bacteria, and resistance genes are well abundant in the clinical sewages as well as in the recipient aquatic ecosystems which clearly indicates the sites of the development of antibiotic resistance in the environment (Voigt et al., 2020). Being a hotspot of drug resistant organisms, hospital wastes elevate the numbers of resistant bacteria in the recipient sewers facilitating the resistant gene transfer in the bacterial strain through recombination, horizontal gene transfer, mutation etc (Mondal et al., 2021; Abe et al., 2020). In due course of time these Antibiotic Resistant bacteria get the chance to enter the food chain when wastewaters are used for irrigation and sewage sludges are applied as fertilizer (Lamastra et al., 2018). Untreated human sewage reaches streams by point source discharges or combined sewer overflows from wastewater treatment plants (Mansfeldt et al., 2020; Quintela-Baluja et al., 2019). These streams are in turn utilized to provide drinking water to the community. Wastewater treatment plants (WWTPs) have been proved to curtail the release of antibiotics, pathogenic microorganisms into the environment successfully (Aghalari et al., 2020). But the efficiency is only 10 to 100 folds in terms of microbial reduction (Karkman et al., 2018). Multiple studies have revealed the presence of several antibiotic resistance genes (ARGs) and metal resistance genes (MRGs) from metagenomes collected from influent, activated sludge (AS) and digested sludge (DS) of Urban and Municipal WWTPs (Wang et al., 2020; Sun et al., 2021; Yoo et al., 2020). The above claims can be correlated through the widespread use of antibiotics in human and veterinary medicine resulting in antibiotic resistance and spreading it across the wastewater network. A study in Göttingen, Germany showed dominance of pathogenic bacteria in the samples collected from hospital effluent at different steps of wastewater treatment (Schneider et al., 2020). Hence, discharge of treated effluents is equally responsible for contaminating adjoining natural water bodies with ARGs and MRGs.

The emergence of novel antibiotic-resistant bacterial strains that are resistant to the majority of currently available antimicrobial drugs has become a serious challenge in modern medicine, especially with the rise in immune compromised patients. They are vigorously harming public health with economic and social consequences around the world. According to the World Health Organization, diseases like diarrhoea, cholera, dysentery, typhoid can all be spread by contaminated water. As per WHO, each year, 4,85,000 diarrhoeal fatalities are estimated to be caused by contaminated drinking water. Antibiotic resistance chart shows increasing resistance of pathogens toward the antibiotics like Amikacin, Carbapenems, aminoglycosides, Cephalosporins etc. From 2008 to 2020, consistent increase in the resistance events is noticeable (The Center for Disease, 2021). A report from the European Medicines Agency and the European Centre for Disease Prevention and Control has revealed the death of about 25000 patients annually in the European Union from an infection with the Multi Drug Resistant (MDR) bacteria. They have warned that very few antibacterial agents are left to deal with this drug resistant catastrophes. For decades no unique mechanism of action and novel target agents are in the trail to deal with the situation (European Medicines Agency and the European Centre for Disease Prevention and Control, 2009). A list of Global Priority Pathogens (GPP) containing 12 species of bacteria categorized under 3 priority ranks was published by World Health Organization (WHO) in 2017 with the objective to prioritize the need of new therapeutic agents (WHO, 2017).

BACKGROUND

Characterization of wastewater microbial composition and potential pathogenic load is mandatory to predict the overexpressed resistance pathways along with prevalent disease network. Wastewater-based epidemiology (WBE) that means wastewater screening as a tool of public health monitoring is now widely accepted (Daughton, 2018; Xagoraraki & O'Brien, 2020). WBE entails collection of wastewaters, analysis of drug or metabolite residues in wastewater, computational outputs of drug consumption and drug excretion rates which enables to estimate drug prevalence and consumption by a particular population. It has the advantages of being less costly and producing more reliable, near-real-time data. It augments a lot of promise as a supplement to current drug monitoring systems. Wastewater-based epidemiology can be used to create large-scale monitoring networks that reveal spatial patterns and temporal trends in drug misuse (Feng et al., 2018). For the last few years WBE have been used to treat a diverse of waterborne, foodborne, faecal-oral pathogenic bacteria and viruses, which are usually excreted through the stool of the infected people (Bisseux et al., 2018). Frequent analysis of municipal wastewater helps to evaluate the viral circulation pattern in the population as they act as ideal sampling station concentrating discharges from the entire community.

Sudden emergence of COVID-19 pandemic has forced the world to reconsider the matter of wastewater hazards. Multiple studies have reported the traces of SARS-CoV-2 coronavirus in municipal wastes (Ahmed et al., 2020a), hospital sewage (Zhang et al., 2020), activated sludges derived from wastewater treatment plants (Kocamemi et al., 2020) and even in the river water (Rimoldi et al., 2020). Usually, excreta of patients which are expected to run away in sewers systems eventually ends up into wastewater and sewage. Thus, sanitary plumbing system is considered to be the prime route of viral transmission from community into the water and wastewater system (Ahmed et al., 2020a; Barcelo, 2020). Besides from faeces and urine of the patients, casual disposal of surgical face masks and medical equipment containing viral traces is another way that add viral load in sewer systems (Chin et al., 2020). Guerrero-Latorre et al., 2020 and Rimoldi et al., 2020 reported the existence of SARS-CoV-2 virus in the river water of Ecuador and Italy respectively which receives effluents from all over the city. Rosa et al. (2020) collected multiple sewage samples from various sites of Milan and Rome, Italy between the time span of 3rd February 2020 to 2nd April 2020. Molecular analysis of samples confirmed the occurrence of SARS-CoV-2 RNA fragments in the test samples. Most interesting point to be noted is that the first Italian SARS-CoV-2 positive case was reported on February 21st. Significantly, COVID-19 infections were still rare in Italy by the end week of February 2020, when the sludge samples were found to be positive for SARS-CoV-2 in Milan. These sorts of findings demonstrate the sensitivity of environmental surveillance in detecting continuing pandemics in the general population. Viral identification in sewage, despite of the low prevalence of reported human illnesses, may be linked to sewage surveillance's capacity to estimate mild or asymptomatic cases following rigorous epidemiological modelling. Here lies the success of Wastewater-based epidemiology (WBE). The current pandemic can pose a long-term influence on the event of antimicrobial resistance (AMR), causing an additional burden to the environment. Antibiotics are given to up to 70% of COVID-19 patients in either an outpatient or hospital setting (Langford et al., 2021). Overuse of antimicrobial agents boosts the emergence of AMR in co-infecting pathogens (Tedijanto et al., 2018). Thus, this pandemic may place an even greater focus on the importance of wastewater monitoring.

Various crucial information about human health-related aspects such as pathogen distribution, potential disease threat, occurrence of antibiotic resistance genes can be well characterized through Metagenomics analysis of wastewater samples as a part of diagnostic metagenomics (Yoo et al., 2020; Breitwieser & Salzberg, 2020). Diagnostic metagenomics has the capacity to optimally trace all the pathogenic microbes as well as the diseases induced by corresponding microbes in a variety of samples such as faeces, urine, blood and even wastewater (Greninger, 2018). For instances, several tick-borne disease pathogens like *Anaplasma, Ehrlichia, Borrelia* species has been well identified using targeted metagenomic sequencing (Thoendel, 2020). In general microbes inhabit almost every niche on the earth, demonstrating the immense diversity of the microbial world. Many of them have been documented to be significant members of their own ecosystems playing key roles in a variety of environmental and host-related biological processes. Majority of these microbes are still unexplored and their impacts in the respective habitats are unknown due to their unculturability. The properties like poor growth rates, transitions into dormancy prevent the unculturable bacteria to grow in typical laboratory media (Bodor et al., 2020). Metagenomics, a modern culture-independent method for studying microbial community in various habitats, has opened up new paths for resolving common microbiological issues such as documentation of total microbial species composition in a given environment, assessment of their ecological impacts and many more (Datta et al., 2020). Metagenomics study of environmental DNA that is genetic material extracted from environmental samples can be used to improve public health. For example, environmental DNA can be used to assess the diversity of microbes and to track probable contamination sources in water bodies (Staley et al., 2018). DNA or RNA sequencing using next-generation sequencers, sequence assembly, gene prediction, functional and metabolic analysis, taxonomic binning and comparative analysis of the sequence data using specialized bioinformatics tools are the basic steps involved in a typical metagenomic project to estimate the number of species and the functional repertoire of an environment. Whereas, inherent difficulties with the metagenomic data, such as incomplete coverage, huge volumes of raw sequence data produced by next-generation sequencers, and short lengths of the sequence reads make each stage of the analysis more challenging (Zou et al., 2018). Metagenomics is an excellent approach to monitor AMR in sewage system. For thorough and accurate profiling of antimicrobial resistant genes, metagenomic data relies on high throughput sequencing of all community DNA and sequence depth. Global monitoring of untreated urban sewage through metagenomic study reported a significant abundance of ARGs conferring resistance to beta-lactams, tetracycline, macrolides, and aminoglycosides (Hendriksen et al., 2019). Kutilova et al., 2021 has conducted a metagenomic study of wastewater community DNA obtained from hospital and municipal waste effluent. High titer of pathogenic members of bacterial genera like *Pseudomonas, Escherichia, Klebsiella, Aeromonas, Enterobacter and Arcobacter* were noted in the samples. Hendriksen et al., 2019 has recommended metagenomic analysis of municipal sewage as a morally acceptable and cost-effective method for infectious disease surveillance and AMR prediction. Hence during the event of a public health emergency use of metagenomics databases allows prediction of potential disease threats in advance.

SUGGESTED PROTOCOL FOR WASTEWATER MONITORING

A comparative protocol has been formulated which utilizes both metagenomics guided data acquisition along with validation by standard culture dependent mechanisms.

From the metagenomics-based analysis, generation of a predicted pathogenic load network graph is possible which takes into account taxon enrichment and also enables to correlate the data with genus abundance profiles. The culture dependent segment can be used to validate and quantify the extent of antibiotic resistance in the samples under study. The detailed protocols are listed in Figure 1.

SAMPLE COLLECTION

Sludge samples need to be collected from a suitable collection site that accumulates community wastes. Followed by collection, samples should be kept in sterile containers at -20°C until they are ready to be processed. Subsequent culture based and culture independent steps need to be initiated within 10 hours of sample collection to obtain proper reproducible results. Stock samples can be stored at 4°C for future study.

Figure 1. Proposed pipeline for wastewater monitoring

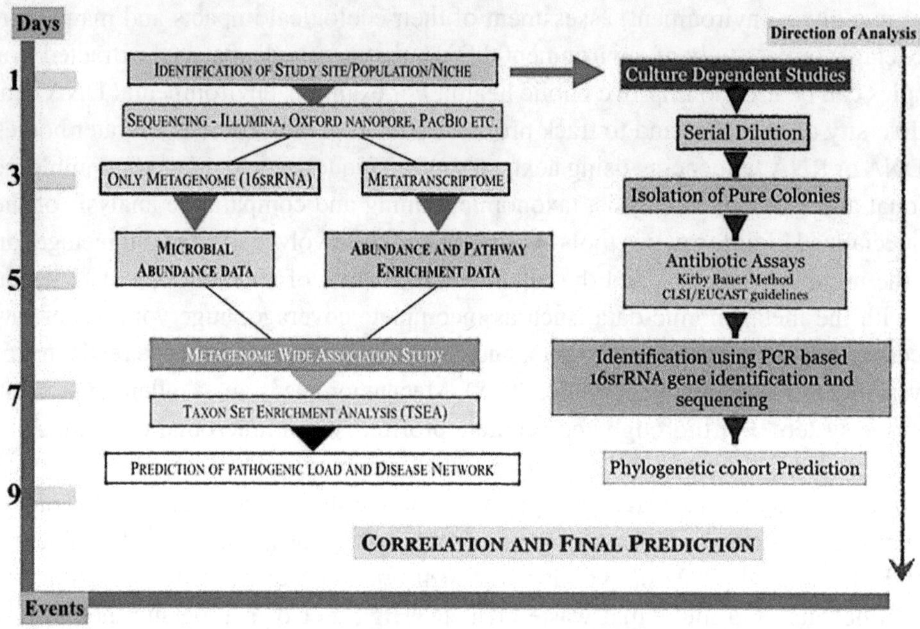

Protocol for Metagenomic Sequencing

Further processing of the raw samples should be initiated from DNA isolation following the protocol by Bonet et al., 2012. Followed by isolation, DNA quantification should be performed using Qubit dsDNA HS Assay kit (Life Tech). Then, Nextera XT Idex Kit (Illumina, Inc.) need to be employed in the 16S Metagenomic Sequencing Library preparation of amplicons. In the upcoming step, V3 and V4 hypervariable regions of 16S rDNA gene of bacteria and Archaea need to be targeted for primer preparation. Next task should be the amplification and purification of amplicons using Illumina adaptors and 1 × AMpureXP beads respectively. The purified amplicons ought to be checked and quantified in the successive steps. In the final step, sequencing of the libraries following Illumina sequencing chemistry must be performed (Singh et al., 2019).

Computational Analysis of Metagenomic Data

Thorough profiling of microbial community should be initiated from taxon calling and OTU clustering of the processed paired ended reads using suitable tool that can assign properties to the individual sequence reads. After obtaining the abundance data, set of unique genera across the test samples need to be identified using an appropriate interactive tool. Followed by identification, abundance data of common bacterial genera should be easily visualized through heatmap. Heatmap implementation function of RStudio user interface called pheatmap() are able to make the step easier (IDE; RStudio Team, 2020; Kolde, 2019). Metagenome-wide association studies, linking the microbes of sample to host health and diseases ought to the next mandatory step. Hence, pathogenic load and related disease networks of the microbial members must be predicted in the final step through Taxon Set Enrichment Analysis (TSEA) with respect to a pre-defined taxon set, sharing similar traits (Dhariwal et al., 2017).

Culture Dependent Studies

Microbiological Characterization

Microbiological culturing of the collected effluent samples should be initiated from serial dilution and plating onto nutrient agar plates (pH 6.6) followed by overnight incubation at 37°C. Subculturing of each plates containing isolated colonies in increasing dilution provides pure colonies. Finally, under 100x magnification of the objective lens of a Bright Field Light Microscope, morphological inspection of the subcultured colonies can be performed, followed by Gram staining. Gram positive and Gram-negative bacterial isolates can be presumably identified using a series of traditional morphological, cultural, and biochemical assays, as outlined in Bergey's Manual of Determinative Bacteriology.

Biochemical Characterization

Biochemical assays for Catalase, Oxidase, Nitrate reduction, Indole, Methyl red, Voges- Proskauer, Citrate utilization, Urease, Starch hydrolysis, and Carbohydrate fermentation are all useful for characterization of the isolates (Cappuccino & Sherman, 2005).

Antibiotic Assay

The Kirby-Bauer disc diffusion assay can assess the susceptibility of isolates to various antibiotics. The choice of antibiotics for bacterial susceptibility testing should be based on CLSI 2019 recommendations. Antibiotic discs from HiMedia Laboratories are recommended to use in the disc diffusion experiments (Mumbai, India). Sensitivity to antibiotics (intermediate or susceptible) can be calculated on the basis of the information provided in manufacturer's manual followed by result interpretation using CLSI guidelines. Isolate showing resistance to three or more types of antibiotics ought to be classified as Multi Drug Resistant (MDR).

Bacterial Identification Using PCR Based 16s rRNA Gene Identification and Sequencing

The very first step of bacterial identification involves genomic DNA isolation from bacterial isolates. Second step is amplification and sequencing of 16S rDNA fragment using universal 16S primers 27f(5′AGAGTTTGATCCTGGCTCAG3′) and 1492r (5′TACGGTTACCTTGTTACGACTT3′) (Gerhardt et al. 1994). In the third step, raw sequences are put together using Cap3 Contig Assembly –in silico programme. In the last step, all the obtained sequences can be utilized to identify the bacterium through BLASTn tool (https://www.ncbi.nlm.nih.gov/).

Phylogenetic Cohort Prediction

Construction of the phylogenetic tree is aided by Clustal omega (Sievers et al., 2011) and Muscle (Edgar, 2004) which perform multiple sequence alignment for a fast validation of sequence scores. The phylogenetic tree module of the CLC Genomics workbench can determine the phylogenetic tree based on the consensus multiple sequence alignment (Felsenstein, 1985) with 100 bootstrap repetitions. Both neighbour joining-based approaches (Saitou & Nei, 1987) and maximum likelihood-based methods (Felsenstein, 1981) must be used to evaluate the tree formed. Ultimately the best consensus tree should be chosen for analysis and interpretation. Discrimination of distinct clades using colour codes is possible by visualising the tree file in FigTree.

The efficiency of the protocol lies in the minimum time frame that is required to generate a plausible prediction in regard to water quality of collected samples. From our estimate this time frame is of 9 working days from the time of collection to the time of prediction of results (Figure 1).

SAMPLE DATA

Contamination of natural water with untreated sewage containing antibiotic residues, heavy metals, organic and inorganic chemicals exert a selective pressure for the rise of antibiotic resistant microorganisms posing a detrimental effect on human health. This in turn creates a socioeconomic burden on the community by directly reducing the therapeutic potency of widely used antibiotics and ultimately increasing community health care expenses (Wilson et al., 2020). Therefore, efficient management of clinical and household wastes has become a worldwide priority in recent years. Furthermore, as part of wastewater-based epidemiology (WBE), wastewaters and medical wastes are being analyzed on a regular basis to foretell the circulation pattern of pathogenic microbes as well as to predict possible disease network.

Keeping the success of WBE in mind, this particular study aims to compare the microbial community profile of wastewaters collected from Rural, Urban and Delta region in and around West Bengal, India. Rural sample represented the wastewater effluent collected from a medical facility located in Purulia. Whereas solid sludge and associated wastewaters obtained from a large medical facility in Central Kolkata was nominated as Urban sample. Effluent collected from the sewer system of a health care unit near the Deltaic region of Sundarbans was the Delta sample of this study. All the 3 samples were monitored using our proposed protocol. Metagenomics guided assessment provided a clear insight on abundances and pathogenic load of bacteria present in the collected sludge samples. An effort was made to predict possible disease network based on pathogenic load of the wastewater microbial community.

Comparison of wastewater microbial profile revealed that 19 bacterial genera were common throughout the Rural, Urban, and Delta samples under study. Noticeable common pathogenic genera were found to be *Pseudomonas, Escherichia, Staphylococcus, Lactobacillus, Prevotella, Acinetobacter* etc. Abundances of the respective bacterial genera have been represented through Table 1. Table 1 shows that *Cloacibacterium, Geobacter* and *Pseudomonas* are the highest abundant genera in Urban, Rural and Delta sample respectively. Common abundance data visualization through heatmap discloses that among the all genera, *Pseudomonas* exhibits highest abundance in Delta sample only. On the other hand, Urban sample shows much more diversity of different genera in terms of abundances compared to Rural and Delta sample. Surprisingly, all 19 common bacterial genera had lower titer of abundances in Rural sludge (Figure 2). Pathogenic load determination using taxon set enrichment analysis (TSEA) revealed higher probability of occurrence of multiple diseases in a network (Figure 3). Based on the disease network, Table 2 summarizes major diseases caused by drinking water contamination as a result of seepage of wastewater, such as diarrhoea, irritable bowel syndrome (IBS), liver cirrhosis, and ulcerative colitis, as well as their causative pathogens that are found in the samples. *Lactobacillus* is found to cause 3 major diseases such as Diarrhoea, Liver Cirrhosis and IBS. Liver Cirrhosis is also caused by bacterial genera *Prevotella* and *Streptococcus. Escherichia* and Streptococcus shows the potential to cause Ulcerative Colitis. Diseases

Table 1. Abundances of common microbial genera across the wastewater samples

Genus	Urban (%)	Rural (%)	Delta (%)
Cloacibacterium	5.22	0.06	0.17
Prevotella	2.85	0.36	0.13
Novosphingobium	2.02	0.11	0.10
Clostridium	1.44	0.24	0.01
Burkholderia	1.38	0.09	0.01
Acinetobacter	1.27	0.05	0.10
Escherichia	0.60	0.19	0.05
Streptococcus	0.30	0.23	0.02
Lactobacillus	0.30	0.17	0.09
Pseudomonas	0.29	0.07	2.40
Rhizobium	0.17	0.03	0.16
Sphingomonas	0.13	0.06	0.22
Acidovorax	0.11	0.13	0.01
Corynebacterium	0.10	0.05	0.01
Propionibacterium	0.10	0.18	0.12
Thauera	0.07	0.15	0.02
Staphylococcus	0.02	0.02	0.03
Rothia	0.02	0.02	0.17
Geobacter	0.02	0.38	0.06

Table 2. Noticeable wastewater related diseases and the causative pathogens obtained from predicted disease network of the common microbial genera across the wastewater samples

Disease	Associated Organisms	References
Diarrhea	*Lactobacillus*	(Agamennone et al., 2018)
Irritable Bowel Syndrome (IBS)	*Lactobacillus*	(Sadrin et al., 2020)
Liver Cirrhosis	*Lactobacillus, Prevotella, Streptococcus*	(Zhong et al., 2021)
Cirrhosis	*Streptococcus, Prevotella, Lactobacillus*	(Wang et al., 2019)
Ulcerative Colitis	*Escherichia, Streptococcus*	(Yang et al., 2020)
Chronic Obstructive Pulmonary Disease (COPD)	*Prevotella, Pseudomonas, Burkholderia, Lactobacillus*	(Deshpande et al., 2020)
Type I Diabetes	*Escherichia, Lactobacillus, Streptococcus*	(Etjahed et al., 2020)
Multiple Sclerosis	*Prevotella*	(Mirza et al., 2020)

like Chronic Obstructive Pulmonary Disease (COPD), Type 1 Diabetes, Multiple Sclerosis arises due to the secondary infections caused by *Prevotella, Pseudomonas, Burkholderia, Lactobacillus, Escherichia, Streptococcus* (Agamennone et al., 2018; Sadrin et al., 2020; Zhong et al., 2021; Wang et al., 2019; Yang et al., 2020; Deshpande et al., 2020; Etjahed et al., 2020; Mirza et al., 2020)

Figure 2. Heatmap representation of abundance data of the common genera

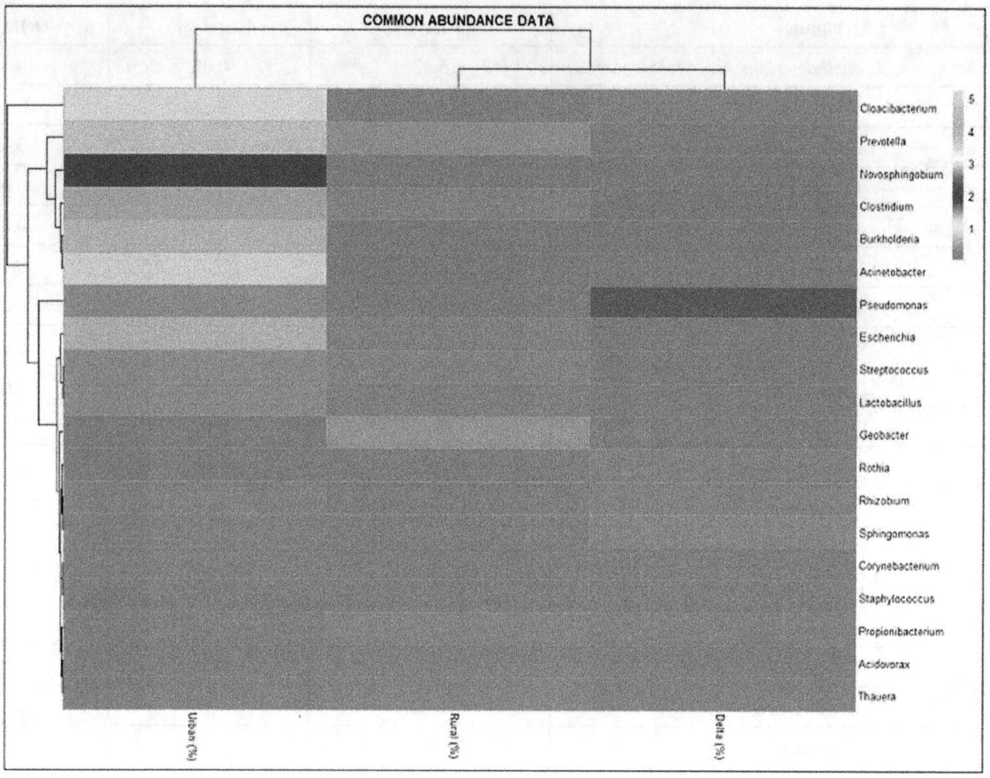

Figure 3. Predicted Disease Network from the common microbial genera across the wastewater samples

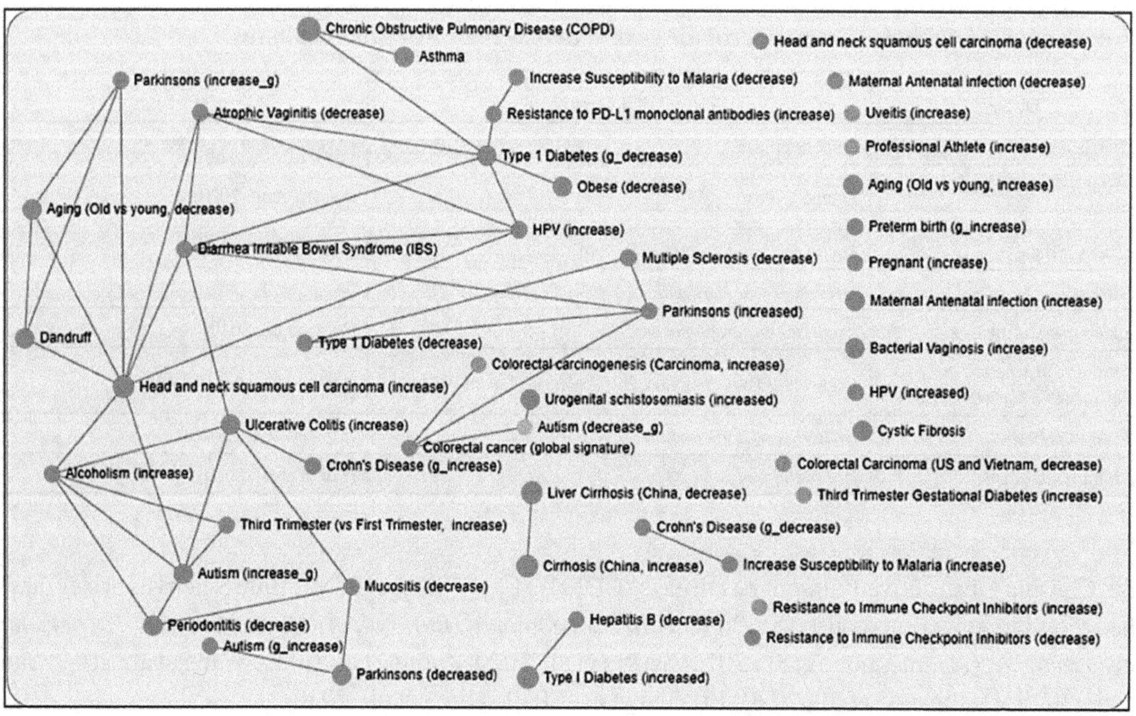

These findings led the authors to conclude that untreated medical and anthropogenic waste disposal can result in the spread of various pathogenic bacterial strains in wastewater environments of all the urban, rural and delta set ups.

Along with the computational analysis following metagenomic sequencing, this particular case study opted for a culture dependent analysis to correlate the computational data during real time lab-based experiments. Results are shown in figure 4. The figure provides the glimpses of the results of various biochemical tests (Figure 4A) followed by antibiotic sensitivity assays, the former providing initial information regarding the type of microorganism to support the Gram's staining results. These results would then be validated using PCR based amplification followed by sequencing of the 16srRNA gene from the pure culture of the individual bacterial members. In case of antibiotic sensitivity assays (Fig-

Figure 4. Glimpses of results obtained from the Biochemical and Antibiotic sensitivity assays in the wet lab-based segment of the proposed protocol. A) Representative images of results of Biochemical analysis; B) Representative images of Antibiotic sensitivity assays performed; C) List of antibiotics and their corresponding concentrations used for the analysis

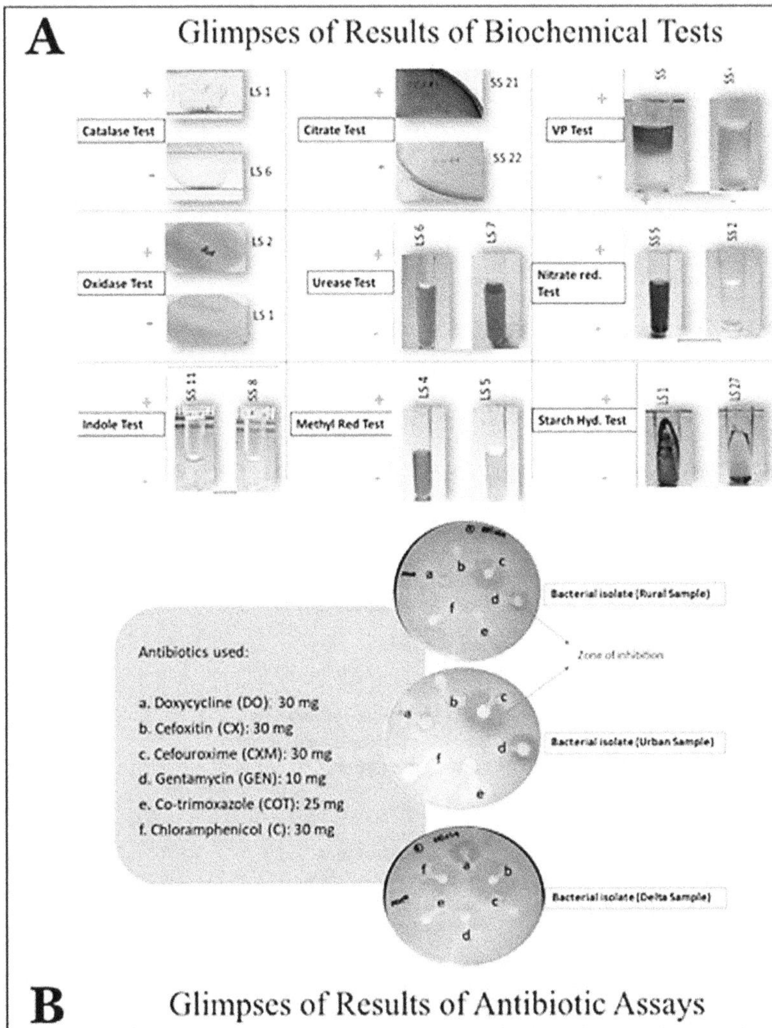

Antibiotic	Abb.	Conc. (µg)
Ampicillin	AMP	10
Methicillin	MET	30
Amoxyclav	AMC	30
Cephalothin	Cf	30
Ciprofloxacin	CIP	10
Norfloxacin	NOR	10
Ofloxacin	OFX	5
Nalidixic acid	NAL	30
Levofloxacin	LE	5
Amikacin	AK	30
Streptomycin	S	10
Gentamicin	GEN	10
Vancomycin	VAN	30
Erythromycin	ERY	15
Clindamycin	Cd	2
Tetracycline	TET	30
Doxycycline	DO	30
Ceftriaxone	CTR	30
Cefoxitin	CX	30
Cefotaxime	CTX	30
Cefouroxime	CXM	30
Chloramphenicol	C	30
Imipenem	I	10
Co-trimoxazole	COT	25

C Antibiotics used and their corresponding concentrations for performing the Antibiotic sensitivity assays

ure 4B), the presence and degree of inhibition zone (susceptibility) or the absence of inhibitory zone (resistance) enabled the authors to categorize the microorganisms into resistant and susceptible ones which were tested using various classes of antibiotics as indicated in Figure 4C. These were helpful for the authors to draw a conclusive evidence regarding the sample profiles. The culture-based method has the potential to reveal maximum levels of tolerance/minimum inhibitory concentrations, as well as the underlying antibiotic resistant mechanisms, in order to develop a comprehensive picture of antibiotic resistance in effluents and wastewaters.

APPLICATION OF THE PROTOCOL IN PUBLIC HEALTH MONITORING

The fast growth of megacities, as well as their increasing global connectivity, is becoming a major driver of new disease outbreaks. As part of risk-based surveillance, such cities should be included in early detection of anomalous illness emergence and dissemination (Nieuwenhuijse et al., 2020). The fundamentals of any outbreak teach us how to collect, document, and analyze descriptive data about people, places under specific time frames in order to generate a source hypothesis. Therefore, detection is going to be the most key facet highlighted in figure 5 as the node of interest from which the outcome of our protocol seems impactful. Local level public health response can be the primary control measures in terms of state level laboratory reference. This in turn leads to a state level public health response reporting followed by national level response. Afterwards, formal assessment as well as containment

Figure 5. Application of the protocol in outbreak

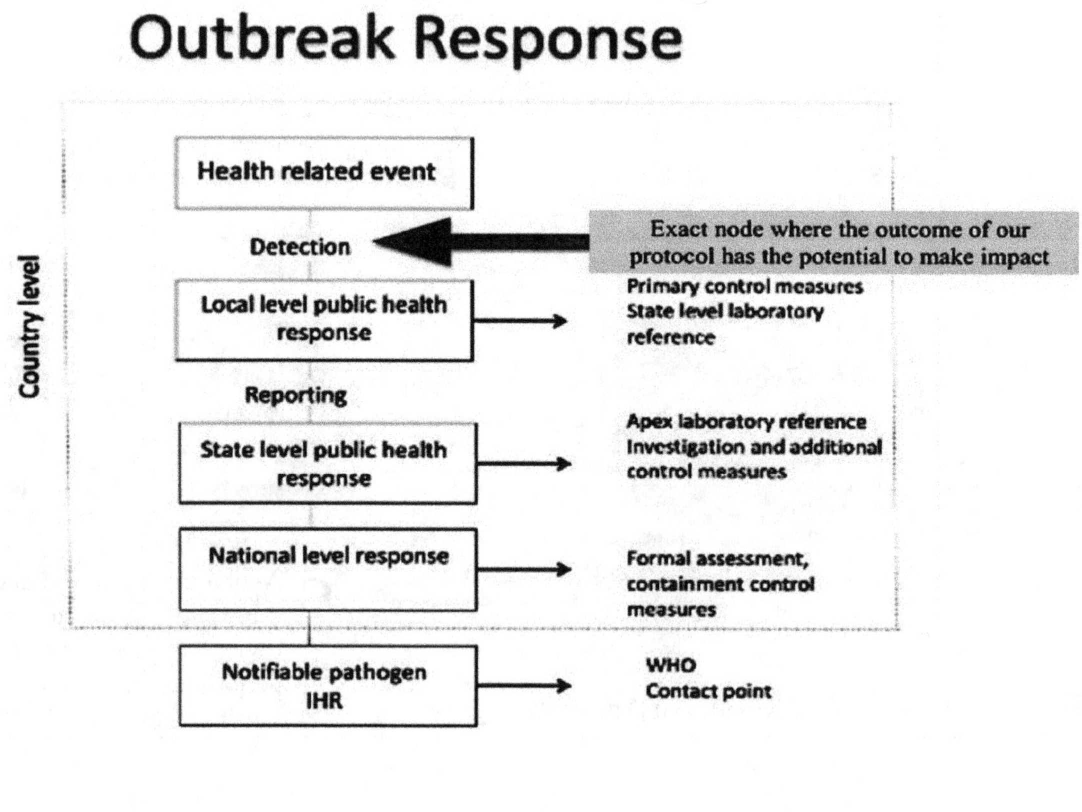

control measures can be subjected towards the application of the protocol. Finally, notifiable pathogens are subjected to international health regulations and emergency committees (IHR) in order to prevent, control, and respond to disease outbreaks around the world (Figure 5).

CONCLUSION

This research should help to understand how pathogenic and antibiotic resistant isolates propagate in wastewaters, mandating public awareness of sanitation measures. It should also allow to formulate a monitoring programme for evaluating wastewater health using a combination of culture independent and dependent molecular approaches to minimize the spread of these new pollutants in tropical environments. It can be believed that metagenomics guided environment impact assessment will be the next important topic of research in the near future, with the potential to transform policymaker's perspectives on climate change and linked healthcare, once all technical and ethical constraints are solved.

ACKNOWLEDGMENT

The authors acknowledge the contributions of Mr. Souradip Basu in the preparation of manuscript. They are also indebted to the authorities of the respective institutions for providing the necessary infrastructure for successful completion of the work.

REFERENCES

Abe, K., Nomura, N., & Suzuki, S. (2020). Biofilms: hot spots of horizontal gene transfer (HGT) in aquatic environments, with a focus on a new HGT mechanism. *FEMS Microbiology Ecology, 96*(5), fiaa031. doi:10.1093/femsec/fiaa031

Agamennone, V., Krul, C. A. M., Rijkers, G., & Kort, R. (2018). A practical guide for probiotics applied to the case of antibiotic-associated diarrhea in The Netherlands. *BMC Gastroenterology, 18*(1), 103. doi:10.118612876-018-0831-x PMID:30078376

Aghalari, Z., Dahms, H. U., Sillanpää, M., Sosa-Hernandez, J. E., & Parra-Saldívar, R. (2020). Effectiveness of wastewater treatment systems in removing microbial agents: A systematic review. *Globalization and Health, 16*(1), 13. doi:10.118612992-020-0546-y PMID:32013988

Ahmed, W., Angel, N., Edson, J., Bibby, K., Bivins, A., O'Brien, J. W., Choi, P. M., Kitajima, M., Simpson, S. L., Li, J., Tscharke, B., Verhagen, R., Smith, W., Zaugg, J., Dierens, L., Hugenholtz, P., Thomas, K. V., & Mueller, J. F. (2020). First confirmed detection of SARS-CoV-2 in untreated wastewater in Australia: A proof of concept for the wastewater surveillance of COVID-19 in the community. *The Science of the Total Environment, 728*, 138764. doi:10.1016/j.scitotenv.2020.138764 PMID:32387778

Ajiboye, T. O., Oyewo, O. A., & Onwudiwe, D. C. (2021). Simultaneous removal of organics and heavy metals from industrial wastewater: A review. *Chemosphere, 262*, 128379. doi:10.1016/j.chemosphere.2020.128379 PMID:33182079

Barcelo, D. (2020). An environmental and health perspective for COVID-19 outbreak: Meteorology and air quality influence, sewage epidemiology indicator, hospitals disinfection, drug therapies and recommendations. *Journal of Environmental Chemical Engineering*, *8*(4), 104006. doi:10.1016/j.jece.2020.104006 PMID:32373461

Bisseux, M., Colombet, J., Mirand, A., Roque-Afonso, A. M., Abravanel, F., Izopet, J., Archimbaud, C., Peigue-Lafeuille, H., Debroas, D., Bailly, J. L., & Henquell, C. (2018). Monitoring human enteric viruses in wastewater and relevance to infections encountered in the clinical setting: a one-year experiment in central France, 2014 to 2015. *Euro surveillance: bulletin Europeen sur les maladies transmissibles = European communicable disease bulletin, 23*(7), 17-00237. doi:10.2807/1560-7917.ES.2018.23.7.17-00237

Bodor, A., Bounedjoum, N., Vincze, G. E., Erdeiné Kis, Á., Laczi, K., Bende, G., Szilágyi, Á., Kovács, T., Perei, K., & Rákhely, G. (2020). Challenges of unculturable bacteria: Environmental perspectives. *Reviews in Environmental Science and Biotechnology*, *19*(1), 1–22. doi:10.100711157-020-09522-4

Breitwieser, F. P., & Salzberg, S. L. (2020). Pavian: Interactive analysis of metagenomics data for microbiome studies and pathogen identification. *Bioinformatics (Oxford, England)*, *36*(4), 1303–1304. doi:10.1093/bioinformatics/btz715 PMID:31553437

Bürgmann, H., Frigon, D., H Gaze, W., M Manaia, C., Pruden, A., Singer, A. C., F Smets, B., & Zhang, T. (2018). Water and sanitation: an essential battlefront in the war on antimicrobial resistance. *FEMS Microbiology Ecology, 94*(9), . doi:10.1093/femsec/fiy101

Cappuccino, J. G., & Sherman, N. (2005). Microbiology: A laboratory manual. *Benjamin Cummings: New York*, *2005*, 125–179.

Chen, J., Li, J., Zhang, H., Shi, W., & Liu, Y. (2019). Bacterial Heavy-Metal and Antibiotic Resistance Genes in a Copper Tailing Dam Area in Northern China. *Frontiers in Microbiology*, *10*, 1916. doi:10.3389/fmicb.2019.01916 PMID:31481945

Chin, A., Chu, J., Perera, M., Hui, K., Yen, H. L., Chan, M., Peiris, M., & Poon, L. (2020). Stability of SARS-CoV-2 in different environmental conditions. *The Lancet Microbe*, *1*(1), e10. doi:10.1016/S2666-5247(20)30003-3 PMID:32835322

Datta, S., Rajnish, K. N., Samuel, M. S., Pugazhendhi, A., & Ethiraj, S. (2020). Metagenomic applications in microbial diversity, bioremediation, pollution monitoring, enzyme and drug discovery. A review. *Environmental Chemistry Letters*, *18*(4), 1229–1241. doi:10.100710311-020-01010-z

Daughton, C. G. (2018). Monitoring wastewater for assessing community health: Sewage Chemical-Information Mining (SCIM). *The Science of the Total Environment*, *619-620*, 748–764. doi:10.1016/j.scitotenv.2017.11.102 PMID:29161600

Deshpande, R., & Zou, C. (2020). Pseudomonas Aeruginosa Induced Cell Death in Acute Lung Injury and Acute Respiratory Distress Syndrome. *International Journal of Molecular Sciences*, *21*(15), 5356. doi:10.3390/ijms21155356 PMID:32731491

Dhariwal, A., Chong, J., Habib, S., King, I. L., Agellon, L. B., & Xia, J. (2017). MicrobiomeAnalyst: A web-based tool for comprehensive statistical, visual and meta-analysis of microbiome data. *Nucleic Acids Research*, *45*(W1), W180–W188. doi:10.1093/nar/gkx295 PMID:28449106

Edgar, R. C. (2004). MUSCLE: Multiple sequence alignment with high accuracy and high throughput. *Nucleic Acids Research, 32*(5), 1792–1797. doi:10.1093/nar/gkh340 PMID:15034147

Edokpayi, J. N., Odiyo, J. O., & Durowoju, O. S. (2017). *Impact of Wastewater on Surface Water Quality in Developing Countries: A Case Study of South Africa, Water Quality, Hlanganani Tutu*. IntechOpen. Available from: https://www.intechopen.com/chapters/53194 doi:10.5772/66561

Ejtahed, H. S., Hoseini-Tavassol, Z., Khatami, S., Zangeneh, M., Behrouzi, A., Ahmadi Badi, S., Moshiri, A., Hasani-Ranjbar, S., Soroush, A. R., Vaziri, F., Fateh, A., Ghanei, M., Bouzari, S., Najar-Peerayeh, S., Siadat, S. D., & Larijani, B. (2020). Main gut bacterial composition differs between patients with type 1 and type 2 diabetes and non-diabetic adults. *Journal of Diabetes and Metabolic Disorders, 19*(1), 265–271. doi:10.100740200-020-00502-7 PMID:32550175

European Medicines Agency and the European Centre for Disease Prevention and Control. (2009). *The bacterial challenge – time to react. A call to narrow the gap between multidrug-resistant bacteria in the EU and development of new antibacterial agents*. European Medicines Agency. Available from: https://ecdc.europa.eu/sites/portal/files/media/en/publications/Publications/0909_TER_The_Bacterial_Challenge_Time_to_React.pdf

Felsenstein, J. (1981). Evolutionary trees from DNA sequences: A maximum likelihood approach. *Journal of Molecular Evolution, 17*(6), 368–376. doi:10.1007/BF01734359 PMID:7288891

Felsenstein, J. (1985). Phylogenies and the Comparative Method. *American Naturalist, 125*(1), 1–15. doi:10.1086/284325

Feng, L., Zhang, W., & Li, X. (2018). Monitoring of regional drug abuse through wastewater-based epidemiology—A critical review. *Science China. Earth Sciences, 61*(3), 239–255. doi:10.100711430-017-9129-x

Garcias-Bonet, N., Arrieta, J. M., de Santana, C. N., Duarte, C. M., & Marbà, N. (2012). Endophytic bacterial community of a Mediterranean marine angiosperm (Posidonia oceanica). *Frontiers in Microbiology, 3*, 342. doi:10.3389/fmicb.2012.00342 PMID:23049528

Gerhardt, P., Murray, R. G. E., Wood, W. A., & Krieg, N. R. (1994). *Methods for general and molecular bacteriology.* American Society for Microbiology. https://www.ncbi.nlm.nih.gov/

Greninger, A. L. (2018). The challenge of diagnostic metagenomics. *Expert Review of Molecular Diagnostics, 18*(7), 605–615. doi:10.1080/14737159.2018.1487292 PMID:29898605

Guerrero-Latorre, L., Ballesteros, I., Villacrés-Granda, I., Granda, M. G., Freire-Paspuel, B., & Ríos-Touma, B. (2020). SARS-CoV-2 in river water: Implications in low sanitation countries. *The Science of the Total Environment, 743*, 140832. doi:10.1016/j.scitotenv.2020.140832 PMID:32679506

Hassan, B., Yatoo, A. M., Khan, I. S., Ali, M. N., & Qadri, H. (2020). Determination of some commonly used antibiotic residues in hospital effluents of kashmir valley. *Uttar Pradesh Journal of Zoology, 41*(17), 1–7. http://www.mbimph.com/index.php/UPJOZ/article/view/1703

He, Y., Yuan, Q., Mathieu, J., Stadler, L., Senehi, N., Sun, R., & Alvarez, P. J. J. (2020). Antibiotic resistance genes from livestock waste: occurrence, dissemination, and treatment. *NPJ Clean Water, 3*(4). doi:10.1038/s41545-020-0051-0

Hendriksen, R. S., Munk, P., Njage, P., van Bunnik, B., McNally, L., Lukjancenko, O., Röder, T., Nieuwenhuijse, D., Pedersen, S. K., Kjeldgaard, J., Kaas, R. S., Clausen, P., Vogt, J. K., Leekitcharoenphon, P., van de Schans, M., Zuidema, T., de Roda Husman, A. M., Rasmussen, S., Petersen, B., ... Aarestrup, F. M. (2019). Global monitoring of antimicrobial resistance based on metagenomics analyses of urban sewage. *Nature Communications*, *10*(1), 1124. doi:10.103841467-019-08853-3 PMID:30850636

Hendriksen, R. S., Munk, P., Njage, P., van Bunnik, B., McNally, L., Lukjancenko, O., Röder, T., Nieuwenhuijse, D., Pedersen, S. K., Kjeldgaard, J., Kaas, R. S., Clausen, P., Vogt, J. K., Leekitcharoenphon, P., van de Schans, M., Zuidema, T., de Roda Husman, A. M., Rasmussen, S., Petersen, B., ... Aarestrup, F. M. (2019). Global monitoring of antimicrobial resistance based on metagenomics analyses of urban sewage. *Nature Communications*, *10*(1), 1124. doi:10.103841467-019-08853-3 PMID:30850636

Holmberg, T. (2019). Animal waste work. The case of urban sewage management in Sweden. *Contemporary Social Science*, *16*(1), 1–15. doi:10.1080/21582041.2019.1630669

Hossain, M. S., Rahman, N. N., Balakrishnan, V., Puvanesuaran, V. R., Sarker, M. Z., & Kadir, M. O. (2013). Infectious risk assessment of unsafe handling practices and management of clinical solid waste. *International Journal of Environmental Research and Public Health*, *10*(2), 556–567. doi:10.3390/ijerph10020556 PMID:23435587

Karkman, A., Do, T. T., Walsh, F., & Virta, M. (2018). Antibiotic-Resistance Genes in Waste Water. *Trends in Microbiology*, *26*(3), 220–228. doi:10.1016/j.tim.2017.09.005 PMID:29033338

Khan, F. A., Söderquist, B., & Jass, J. (2019). Prevalence and Diversity of Antibiotic Resistance Genes in Swedish Aquatic Environments Impacted by Household and Hospital Wastewater. *Frontiers in Microbiology*, *10*, 688. doi:10.3389/fmicb.2019.00688 PMID:31019498

Khan, F. A., Söderquist, B., & Jass, J. (2019). Prevalence and Diversity of Antibiotic Resistance Genes in Swedish Aquatic Environments Impacted by Household and Hospital Wastewater. *Frontiers in Microbiology*, *10*, 688. doi:10.3389/fmicb.2019.00688 PMID:31019498

Kocamemi, B. A., Kurt, H., Sait, A., Sarac, F., Saatci, A. M., & Pakdemirli, B. (2020). *SARS- CoV-2 detection in Istanbul wastewater treatment plant sludges.* doi:10.1101/2020.05.12.20099358

Kolde, R. (2019). *pheatmap: Pretty Heatmaps*. R package version 1.0.12, URL https://CRAN. R-project.org/package=pheatmap

Kraemer, S. A., Ramachandran, A., & Perron, G. G. (2019). Antibiotic Pollution in the Environment: From Microbial Ecology to Public Policy. *Microorganisms*, *7*(6), 180. doi:10.3390/microorganisms7060180 PMID:31234491

Kutilova, I., Medvecky, M., Leekitcharoenphon, P., Munk, P., Masarikova, M., Davidova-Gerzova, L., Jamborova, I., Bortolaia, V., Pamp, S. J., & Dolejska, M. (2021). Extended- spectrum beta-lactamase-producing Escherichia coli and antimicrobial resistance in municipal and hospital wastewaters in Czech Republic: Culture-based and metagenomic approaches. *Environmental Research*, *193*(110487). Advance online publication. doi:10.1016/j.envres.2020.110487 PMID:33232750

La Rosa, G., Iaconelli, M., Mancini, P., Bonanno Ferraro, G., Veneri, C., Bonadonna, L., Lucentini, L., & Suffredini, E. (2020). First detection of SARS-CoV-2 in untreated wastewaters in Italy. *The Science of the Total Environment, 736*, 139652. doi:10.1016/j.scitotenv.2020.139652 PMID:32464333

Lamastra, L., Suciu, N. A., & Trevisan, M. (2018). Sewage sludge for sustainable agriculture: Contaminants' contents and potential use as fertilizer. *Chemical and Biological Technologies in Agriculture, 5*(10), 10. Advance online publication. doi:10.118640538-018-0122-3

Langford, B. J., So, M., Raybardhan, S., Leung, V., Soucy, J. R., Westwood, D., Daneman, N., & MacFadden, D. R. (2021). Antibiotic prescribing in patients with COVID-19: rapid review and meta-analysis. *Clinical Microbiology and Infection, 27*(4), 520–531. doi:10.1016/j.cmi.2020.12.018

Li, X., Rensing, C., Vestergaard, G., Arumugam, M., Nesme, J., Gupta, S., Brejnrod, A. D., & Sørensen, S. J. (2021). Metagenomic evidence for co-occurrence of antibiotic, biocide and metal resistance genes in pigs. *Environment International, 158*, 106899. Advance online publication. doi:10.1016/j.envint.2021.106899 PMID:34598063

Mansfeldt, C., Deiner, K., Mächler, E., Fenner, K., Eggen, R., Stamm, C., Schönenberger, U., Walser, J. C., & Altermatt, F. (2020). Microbial community shifts in streams receiving treated wastewater effluent. *The Science of the Total Environment, 709*, 135727. doi:10.1016/j.scitotenv.2019.135727 PMID:31887504

Mirza, A., Forbes, J. D., Zhu, F., Bernstein, C. N., Van Domselaar, G., Graham, M., Waubant, E., & Tremlett, H. (2020). The multiple sclerosis gut microbiota: A systematic review. *Multiple Sclerosis and Related Disorders, 37*, 101427. doi:10.1016/j.msard.2019.101427 PMID:32172998

Mondal, S., & … . (2021). A Pipeline for Assessment of Pathogenic Load in the Environment Using Microbiome Analysis. In M. Nath, D. Bhatt, P. Bhargava, & D. K. Choudhary (Eds.), *Microbial Metatranscriptomics Belowground*. Springer. doi:10.1007/978-981-15-9758-9_23

Nascimento, A. L., Souza, A. J., Andrade, P., Andreote, F. D., Coscione, A. R., Oliveira, F. C., & Regitano, J. B. (2018). Sewage Sludge Microbial Structures and Relations to Their Sources, Treatments, and Chemical Attributes. *Frontiers in Microbiology, 9*, 1462. doi:10.3389/fmicb.2018.01462 PMID:30018612

Ngigi, A. N., Magu, M. M., & Muendo, B. M. (2019). Occurrence of antibiotics residues in hospital wastewater, wastewater treatment plant, and in surface water in Nairobi County, Kenya. *Environmental Monitoring and Assessment, 192*(1), 18. doi:10.100710661-019-7952-8 PMID:31820117

Nieuwenhuijse, D. F., Oude Munnink, B. B., Phan, M., Hendriksen, R. S., Bego, A., Rees, C., Neilson, E. H., Coventry, K., Collignon, P., Allerberger, F., Rahube, T. O., Oliveira, G., Ivanov, I., Sopheak, T., Vuthy, Y., Yost, C. K., Tabo, D., Cuadros-Orellana, S., Ke, C., … Koopmans, M. P. G. (2020). Setting a baseline for global urban virome surveillance in sewage. *Scientific Reports, 10*(1), 13748. doi:10.103841598-020-69869-0 PMID:32792677

Quintela-Baluja, M., Abouelnaga, M., Romalde, J., Su, J. Q., Yu, Y., Gomez-Lopez, M., Smets, B., Zhu, Y. G., & Graham, D. W. (2019). Spatial ecology of a wastewater network defines the antibiotic resistance genes in downstream receiving waters. *Water Research, 162*, 347–357. doi:10.1016/j.watres.2019.06.075 PMID:31295654

Rimoldi, S. G., Stefani, F., Gigantiello, A., Polesello, S., Comandatore, F., Mileto, D., Maresca, M., Longobardi, C., Mancon, A., Romeri, F., Pagani, C., Cappelli, F., Roscioli, C., Moja, L., Gismondo, M. R., & Salerno, F. (2020). Presence and infectivity of SARS-CoV-2 virus in wastewaters and rivers. *The Science of the Total Environment*, *744*, 140911. doi:10.1016/j.scitotenv.2020.140911 PMID:32693284

RStudio Team. (2020). *RStudio: Integrated Development for R*. RStudio. https://www.rstudio.com/

Sadrin, S., Sennoune, S., Gout, B., Marque, S., Moreau, J., Zinoune, K., Grillasca, J. P., Pons, O., & Maixent, J. M. (2020). A 2-strain mixture of Lactobacillus acidophilus in the treatment of irritable bowel syndrome: A placebo-controlled randomized clinical trial. *Digestive and Liver Disease, 52*(5), 534–540. doi:10.1016/j.dld.2019.12.009

Saitou, N., & Nei, M. (1987). The neighbor-joining method: A new method for reconstructing phylogenetic trees. *Molecular Biology and Evolution*, *4*(4), 406–425. doi:10.1093/oxfordjournals.molbev.a040454 PMID:3447015

Schneider, D., Aßmann, N., Wicke, D., Poehlein, A., & Daniel, R. (2020). Metagenomes of Wastewater at Different Treatment Stages in Central Germany. *Microbiology Resource Announcements*, *9*(15), e00201–e00220. doi:10.1128/MRA.00201-20 PMID:32273362

Serwecińska, L. (2020). Antimicrobials and Antibiotic-Resistant Bacteria: A Risk to the Environment and to Public Health. *Water (Basel)*, *12*(12), 3313. doi:10.3390/w12123313

Sievers, F., Wilm, A., Dineen, D., Gibson, T. J., Karplus, K., Li, W., Lopez, R., McWilliam, H., Remmert, M., Söding, J., Thompson, J. D., & Higgins, D. G. (2011). Fast, scalable generation of high-quality protein multiple sequence alignments using Clustal Omega. *Molecular Systems Biology*, *7*(1), 539. doi:10.1038/msb.2011.75 PMID:21988835

Singh, M., Ganguli, S., & Ghosh, M. M. (2019). Comparative metagenomic dataset of hospital effluent microbiome from rural and urban hospitals in West Bengal. *Data in Brief*, *25*, 104264. doi:10.1016/j.dib.2019.104264 PMID:31440549

Staley, Z. R., Chuong, J. D., Hill, S. J., Grabuski, J., Shokralla, S., Hajibabaei, M., & Edge, T. A. (2018). Fecal source tracking and eDNA profiling in an urban creek following an extreme rain event. *Scientific Reports*, *8*(1), 14390. doi:10.103841598-018-32680-z PMID:30258068

Sun, F., Xu, Z., & Fan, L. (2021). Response of heavy metal and antibiotic resistance genes and related microorganisms to different heavy metals in activated sludge. *Journal of Environmental Management*, *300*, 113754. Advance online publication. doi:10.1016/j.jenvman.2021.113754 PMID:34543965

Tedijanto, C., Olesen, S. W., Grad, Y. H., & Lipsitch, M. (2018). Estimating the proportion of bystander selection for antibiotic resistance among potentially pathogenic bacterial flora. *Proceedings of the National Academy of Sciences of the United States of America*, *115*(51), E11988–E11995. doi:10.1073/pnas.1810840115 PMID:30559213

Thai, P. K., Ky, L. X., Binh, V. N., Nhung, P. H., Nhan, P. T., Hieu, N. Q., Dang, N., Tam, N., & Anh, N. (2018). Occurrence of antibiotic residues and antibiotic-resistant bacteria in effluents of pharmaceutical manufacturers and other sources around Hanoi, Vietnam. *The Science of the Total Environment*, *645*, 393–400. doi:10.1016/j.scitotenv.2018.07.126 PMID:30029118

The Center for Disease. (2020). *Dynamics Economics & Policy. Resistance Map: Antibiotic resistance.* https://resistancemap.cddep.org/Antibiotic Resistance. php

Thoendel, M. (2020). Targeted Metagenomics Offers Insights into Potential Tick-Borne Pathogens. *Journal of Clinical Microbiology, 58*(11), e01893–e20. doi:10.1128/JCM.01893-20 PMID:32878948

UN WWAP (United Nations World Water Assessment Programme). (2017). *The United Nations World Water Development Report (2017). Wastewater: The Untapped Resource.* UNESCO.

Voigt, A. M., Zacharias, N., Timm, C., Wasser, F., Sib, E., Skutlarek, D., Parcina, M., Schmithausen, R. M., Schwartz, T., Hembach, N., Tiehm, A., Stange, C., Engelhart, S., Bierbaum, G., Kistemann, T., Exner, M., Faerber, H. A., & Schreiber, C. (2020). Association between antibiotic residues, antibiotic resistant bacteria and antibiotic resistance genes in anthropogenic wastewater - An evaluation of clinical influences. *Chemosphere, 241*, 125032. doi:10.1016/j.chemosphere.2019.125032 PMID:31622887

Wang, J., Chu, L., Wojnárovits, L., & Takács, E. (2020). Occurrence and fate of antibiotics, antibiotic resistant genes (ARGs) and antibiotic resistant bacteria (ARB) in municipal wastewater treatment plant: An overview. *The Science of the Total Environment, 744*, 140997. doi:10.1016/j.scitotenv.2020.140997 PMID:32755790

Wang, X., Lan, B., Fei, H., Wang, S., & Zhu, G. (2021). Heavy metal could drive co-selection of antibiotic resistance in terrestrial subsurface soils. *Journal of Hazardous Materials, 411*, 124848. doi:10.1016/j.jhazmat.2020.124848 PMID:33858075

Wang, Y., Pan, C. Q., & Xing, H. (2019). Advances in Gut Microbiota of Viral Hepatitis Cirrhosis. *BioMed Research International, 9726786*, 1–9. Advance online publication. doi:10.1155/2019/9726786 PMID:31886272

WHO Media Centre. (2017). *News Release. WHO publishes list of bacteria for which new antibiotics are urgently needed.* Available from: https://www.who.int/mediacentre/news/releases/2017/bacteria-antibiotics-needed/en/

Wilson, D. N., Hauryliuk, V., Atkinson, G. C., & O'Neill, A. J. (2020). Target protection as a key antibiotic resistance mechanism. *Nature Reviews. Microbiology, 18*(11), 637–648. doi:10.103841579-020-0386-z PMID:32587401

Xagoraraki, I., & O'Brien, E. (2020). Wastewater-Based Epidemiology for Early Detection of Viral Outbreaks. In Women in Water Quality (pp. 75-97). Springer. doi:10.1007/978-3-030-17819-2_5

Yang, H., Mirsepasi-Lauridsen, H. C., Struve, C., Allaire, J. M., Sivignon, A., Vogl, W., Bosman, E. S., Ma, C., Fotovati, A., Reid, G. S., Li, X., Petersen, A. M., Gouin, S. G., Barnich, N., Jacobson, K., Yu, H. B., Krogfelt, K. A., & Vallance, B. A. (2020). Ulcerative Colitis-associated E. coli pathobionts potentiate colitis in susceptible hosts. *Gut Microbes, 12*(1), 1847976. doi:10.1080/19490976.2020.1847976 PMID:33258388

Yang, Y., Li, B., Zou, S., Fang, H. H., & Zhang, T. (2014). Fate of antibiotic resistance genes in sewage treatment plant revealed by metagenomic approach. *Water Research, 62*, 97–106. doi:10.1016/j.watres.2014.05.019 PMID:24937359

Yasir, M. (2020). Analysis of Microbial Communities and Pathogen Detection in Domestic Sewage Using Metagenomic Sequencing. *Diversity (Basel)*, *13*(1), 6. doi:10.3390/d13010006

Yoo, K., Yoo, H., Lee, J., Choi, E. J., & Park, J. (2020). Exploring the antibiotic resistome in activated sludge and anaerobic digestion sludge in an urban wastewater treatment plant via metagenomic analysis. *Journal of Microbiology (Seoul, Korea)*, *58*(2), 123–130. doi:10.100712275-020-9309-y PMID:31875929

Yuan, Q. B., Zhai, Y. F., Mao, B. Y., & Hu, N. (2018). Antibiotic resistance genes and intI1 prevalence in a swine wastewater treatment plant and correlation with metal resistance, bacterial community and wastewater parameters. *Ecotoxicology and Environmental Safety*, *161*, 251–259. doi:10.1016/j.ecoenv.2018.05.049 PMID:29886312

Zhang, D., Ling, H., Huang, X., Li, J., Li, W., Yi, C., Zhang, T., Jiang, Y., He, Y., Deng, S., Zhang, X., Wang, X., Liu, Y., Li, G., & Qu, J. (2020). Potential spreading risks and disinfection challenges of medical wastewater by the presence of Severe Acute Respiratory Syndrome Coronavirus 2 (SARS-CoV-2) viral RNA in septic tanks of Fangcang Hospital. *The Science of the Total Environment*, *741*, 140445. doi:10.1016/j.scitotenv.2020.140445 PMID:32599407

Zhong, X., Cui, P., Jiang, J., Ning, C., Liang, B., Zhou, J., Tian, L., Zhang, Y., Lei, T., Zuo, T., Ye, L., Huang, J., & Chen, H. (2021). *Streptococcus*, the Predominant Bacterium to Predict the Severity of Liver Injury in Alcoholic Liver Disease. *Frontiers in Cellular and Infection Microbiology*, *11*, 649060. doi:10.3389/fcimb.2021.649060 PMID:33816353

Zhou, Y., Wylie, K. M., El Feghaly, R. E., Mihindukulasuriya, K. A., Elward, A., Haslam, D. B., Storch, G. A., & Weinstock, G. M. (2016). Metagenomic Approach for Identification of the Pathogens Associated with Diarrhea in Stool Specimens. *Journal of Clinical Microbiology*, *54*(2), 368–375. doi:10.1128/JCM.01965-15 PMID:26637379

Zou, Q., Lin, G., Jiang, X., Liu, X., & Zeng, X. (2018, September 18). (2018). Sequence clustering in bioinformatics: An empirical study. *Briefings in Bioinformatics*. Advance online publication. doi:10.1093/bib/bby090 PMID:30239587

KEY TERMS AND DEFINITIONS

Antibiotic Resistant Bacteria (ARB): Bacteria that have changed their properties from being antibiotic-sensitive to antibiotic-resistant.

Hospital Effluent: Solids and associated wastewaters containing antibiotic residues, disinfectants, anaesthetics, organic matters, heavy metals, excreta from patients, discharged into the sewage system from clinical facilities.

Illumina Sequencing: A sequencing platform widely used to generate DNA sequences which are generated in small clusters using sequencing by synthesis approach in designated flow cells.

Metagenomics Analysis: The study of environmental samples using genetic techniques aiding microbial profiling from a mixed sample.

Pathogenic Load: A measure of the amounts of microbes in an environmental sample that can cause infections. It has a direct correlation to the severity of infectious diseases.

Outbreak Response: Set of actions such as proper hygiene, contact tracing, mapping of disease clusters, physical distancing taken in an emergency basis to cease the spread of a disease.

Wastewater Monitoring: Surveillance of municipal and clinical sludges in order to assess wastewater health, which is linked to community wellbeing.

Chapter 6
Impact of Climate Change on Upper Ganga Ramsar Site of UP, India:
Sustainable Restoration of Wetland Ecosystem

Vinaya Satyawan Tari
https://orcid.org/0000-0002-8352-252X
University of Mumbai, India

Rashmi Gupta
Jindal Global Law School, Sonipat, India

Nabeela Siddiqui
https://orcid.org/0000-0001-7496-8127
CHRIST University (Deemed), India

ABSTRACT

According to the IPCC Second Assessment Report, climate change will lead to an alteration of the hydrological cycle and could have major impacts on regional water resources. India features a diverse range of wetlands, including high-altitude alpine lakes, littoral swamps in the form of mangroves and corals, and inland wetlands of various sorts. The Upper Ganga Ramsar Site is Uttar Pradesh's only Ramsar Site geographical distribution and may fluctuate as a result of climate change. Wetland reactions to climate change are frequently left out of global climate change models. The climate change adaptations must be incorporated into the economic development, planning, and implementation process.

DOI: 10.4018/978-1-7998-9498-8.ch006

INTRODUCTION

Wetlands are shallow water bodies, rich with natural resources, and immense environmental, economic, and social benefits, contributing naturally to a healthy environment and which hold great significance in terms of maintaining the ecological balance in the environment. They are also known as the kidneys of the landscape, as they help maintain a healthy and resilient wetlands ecosystem and are a source of various benefits. An immense variety of species of microbes, plants, insects, amphibians, reptiles, birds, fish, and mammals make part of a wetland ecosystem (EPA, n.d., Wetlands). Wetlands not only improve water quality but also provide flood protection, controls shoreline erosion, provides recreation opportunities and aesthetic appreciation as well. Inland wetlands, such as floodplains, rivers, lakes, and swamps, function like sponges, absorbing and storing excess rainfall and reducing flood surges. During dry seasons in arid regions, wetlands provide refuges for wildlife and supply water for communities and stock (Salimi et al., 2021). Wetlands also buffer coastlines from extreme weather. Yet, the value of wetlands remains largely unrecognized by policy and decision-makers.

Despite their natural resilience and the natural benefits, wetlands are extremely vulnerable to changes in our landscapes and weather patterns. We are losing wetlands three times faster than natural forests (Ramsar, n.d., The Global Wetland Outlook). Over time, changes in the environment affect wetlands and can lead to shifts in species distributions, species communities, loss of native species as well as biogeochemical changes in the soil. Warming temperatures, increased levels of carbon dioxide in the atmosphere, and changes in frequency and intensity of rainfall alter these ecological processes.

There are various ecosystem goods and services provided by a wetland that benefits the communities who depend on them for their survival, social and economic security. Wetlands primarily have the natural ability to retain water during dry seasons, which keeps the water table stable, smooth, and running. They also supply water for irrigation as well as for domestic consumption and provide swarms of freshwater fisheries for human consumption and water for recreation (Figure 2). The wetland types and their distribution in India were studied in detail by Patel et al. in 2009, by using 1:2,50,000 scale with the help of Resources at-1 (Indian Remote Sensing Satellite- P6) Advanced Wide Field Sensor (AWiFS) data of 2004-05. In this study, wetlands were categorized in 25 different classes under two headings i.e. coastal wetland and inland wetlands. Whereas, the area estimated under wetland was 11.69 Mha in various bio-geographical zones of India which roughly corresponds to about 3.66 percent of the geographic area of the country (Patel et al., 2009).

Climate change has started showing its impact on this important fragile ecosystem as well. Wetlands are likely to be affected by increased temperatures and changes to precipitation and more frequent or intense droughts, storms, and floods. Wetlands that are highly modified or degraded may be even more sensitive and less resilient to climate change (Australian Government, n.d., Wetlands and Climate Change).

As the rising temperature changes the biochemistry and function of a wetland, these wetlands will then no longer be able to provide the necessary water purification service and may adversely start to decompose and release nutrients to the surface water (Salimi et al., 2021). Climate change is predicted to increase flooding, drought, the number of high heat days, and the frequency of severe storms, all of which will affect wetlands. Wetlands are particularly vulnerable to changes in hydrology as they exist between fully aquatic and fully terrestrial ecosystems (Australian Government, n.d., Wetlands and Climate Change). Wetlands might cover only about 5–8% of the world's land surface but comprise 20–30% of the world's carbon pool and steps must be taken to protect it (Mitsch et al., 2013). One of the numerous services and benefits wetlands provide is their ability to sequester (or store) carbon dioxide from the atmosphere through photosynthesis. This ecosystem service keeps the rate of ongoing climate change slower than it would be otherwise. (Desai, 2018). Wetlands are estimated to store more than one-third of the world's terrestrial carbon. (Australian Government, n.d., Wetlands and Climate Change). Their destruction often results in major releases of greenhouse gases into the atmosphere. If wetlands are drained, burned, or cleared, they become a carbon source, releasing into the atmosphere centuries of stored carbon. Carbon dioxide emissions from drained and burned peatlands equate to about 10 percent of global annual fossil fuel emissions. (Australian Government, n.d., Wetlands and Climate Change).

Invasive species, pests, and diseases will take advantage of these changes and will increase. Severe storm events may further cause structural ecological changes from which our natural communities may not rebound easily. The biggest risk to this wetland service is the long-term effect of continued wetland drainage and a warming climate. (Gillian Davies, 2018) Climate change acts with other stressors, such as urbanization, pollution, invasive species, and land-use changes. Along with climate change, these stressors may disassemble existing ecosystems and lead to the emergence of new ones, further altering the benefits wetlands provide to human communities. (Gillian Davies, 2018)

It is difficult to project the exact implications of climate change on wetlands and to predict its precise impact on the various aspects of our surroundings. However, it may be understood from the studies over the decades that the future is gloomy, and the impact is harsh and irreversible. A thorough study of these models that can analyze the relationship between climate change and wetlands is required to define human actions that can mitigate those effects on wetlands. There is a need for more studies to understand these relationships and to help government and communities draw strategy that protects wetlands systems.

DEFINING THE WETLANDS

Wetlands may consist of a wide range of ecosystems in rivers, streams, and marshes. A comprehensive definition of wetlands was provided as, "wetlands as lowlands covered with shallow and sometimes temporary or intermittent waters referred to by such names as marshes, swamps, bogs, wet meadows, potholes, sloughs, and river-overflow lands" (Martin et al., 1953; Shaw & Fredine, 1956 and Tiner, 2005). The definition of wetlands as provided by the Ramsar Convention, which is also the most widely acclaimed one is as follows: "Areas of marsh, fen, peatland or water, whether natural or artificial, permanent or temporary, with water that is static or flowing, fresh, brackish or salt, including areas of marine water (Figure 1) the depth of which at low tide does not exceed six meters" (Finlayson and Moser, 1991). Ramsar Convention is an intergovernmental treaty that provides the framework for national action and international cooperation for the conservation and wise use of wetlands and their resources. Further, the Convention additionally explains that wetlands:" may incorporate riparian and coastal zones adjacent to the wetlands, and islands or bodies of marine water deeper than six meters at low tide lying within the wetlands." One of the well-accepted definitions of wetlands is: The wetlands are lands where saturation with water is the dominant factor determining the nature of soil development and the types of plant and animal communities living in the soil and on its surface (Cowardin et. al, 1979). Wetlands have also been popularly defined as 'lands transitional between terrestrial and aquatic eco-systems where the water table is usually at or near the surface of the land is covered by shallow water (Mitsch and Gosselink, 1986). The National Environment Policy (2006) identifies wetlands as important components of 'freshwater resources' and recommends integration in developmental planning, management based on 'wise use' strategies, promotion of ecotourism, and implementation of a regulatory framework. Clause 2 (1) (g)

Figure 1. Wetland ecosystems

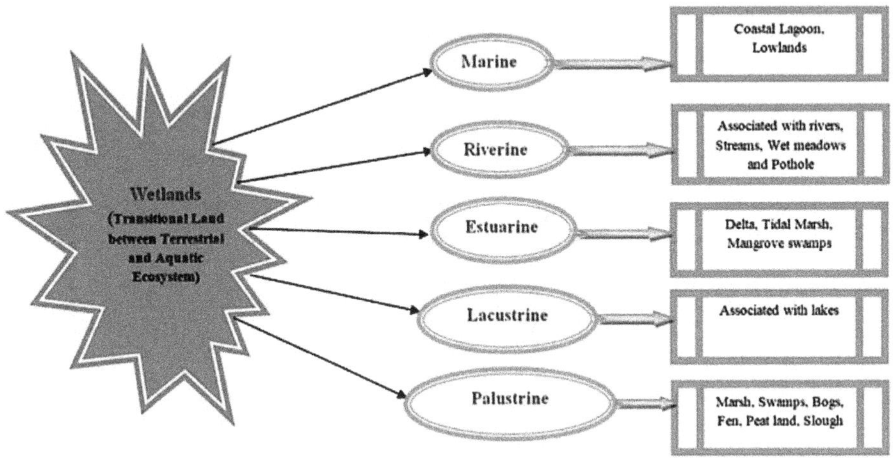

of the Wetlands (Conservation and Management) Rules, 2017, defines a wetland as 'an area of marsh, fen, peatland or water; whether natural or artificial, permanent or temporary, with water that is static or flowing, fresh, brackish or salt, including areas of marine water the depth of which at low tide does not exceed six meters, but does not include river channels, paddy fields, human-made water bodies/tanks specifically constructed for drinking water purposes and structures specifically constructed for aquaculture, salt production, recreation, and irrigation purposes'. Wetlands are a unique type of ecosystem as they have the characteristics of both terrestrial and aquatic ecosystems.

ECOLOGICAL FUNCTIONS OF WETLANDS

Major functions of wetlands are water storage in the surface water reservoir, stormwater retention, maintenance of shoreline, and control on flood, erosion, nutrients, sediments, carbon, pollutants, etc (Patel et al., 2009). These wetlands are extremely useful for groundwater recharge (Figure 2). They also help in flood control as they can be a potential alternative to dams, dikes, and embankments without really damaging the natural ecosystem (Boyd and Banzhaf, 2007). The wetland ecosystem causes an anaerobic process and which subsequently, makes the plants rooted in the ground hold it firmly, and hence help in flood control (Keddy, 2010). Wetlands also help in controlling water pollution by filtering the excessive nutrients in the area (Bystrom et al., 2000).

The majority of the inland wetlands in India are directly or indirectly dependent on the major rivers like Ganga, Brahmaputra, Godavari, Narmada, Krishna, Kaveri, Tapti. They occur in the hot arid regions of Rajasthan and Gujarat, the deltaic regions of the east and west coasts, highlands of central India, wet humid zones of south peninsular India, and the Andaman and Nicobar and Lakshwadeep islands (Meena, 2019). Wetlands in India exhibit huge diversity according to their geographic location, genesis, water regime and chemistry, dominant species, and soil and sediment characteristics (Meena, 2019). The southern states of Andhra Pradesh, Karnataka, and Tamil Nadu have the largest concentration of irrigation tanks, numbering 0.12 million (Palanisami *et al.*, 2010), and account for nearly 60% of India's tank-irrigated area. Similarly, there are traditional tank systems in the States of Bihar, Orissa,

Figure 2. Functions of wetlands

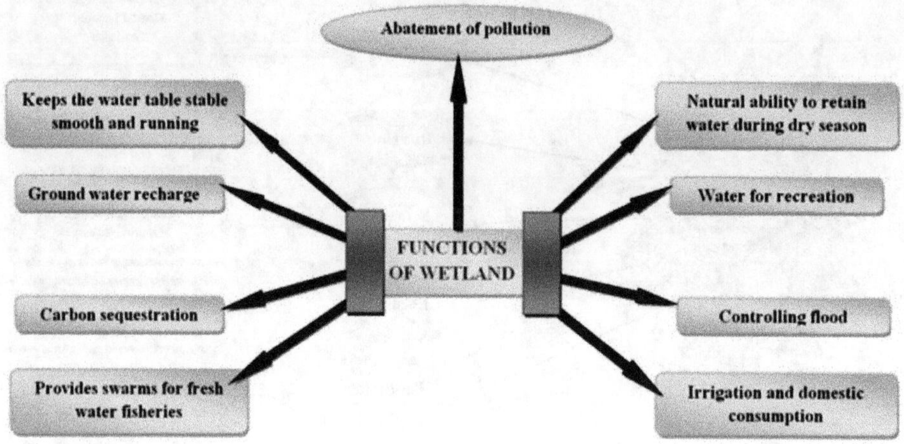

Uttar Pradesh, and West Bengal, accounting for nearly 25% of net tank irrigated areas (Pant and Verma, 2010). Tanks play a fundamental role in harvesting surface runoff during monsoon and then allowing it to be used later. Apart from irrigation, these tanks are also used for fisheries, as a source of water for domestic needs and nutrient-rich soils, fodder grass collection, and brick making. These uses have high worth in terms of nutrition, health, and household income for the poorest of the poor (Meena, 2019). Floodplains in river basins in India have been very popular for agriculture because of their natural fertility. Wetlands also make for some of the best regions to cultivate rice and wild rice. The freshwater ecosystems of Western Ghats, in southern India, is a biodiversity hotspot and home to hundreds of species of fish, molluscs, amphibians, and plants (Meena, 2019).

In the IPCC 2nd assessment report, it is reported that climate change may lead to a shift in the geographical distribution of wetlands also it can alter the hydrological cycle and could have negative impacts on the regional water reservoirs (Lakeman, 1999, Patel et al., 2009). Change in temperature, rainfall, evapotranspiration, runoff, etc. due to climate change is directly affecting the hydrological regime. Thus, the predictable changes in climate may affect wetlands and the composition and concentration of their flora and fauna.

CLIMATIC VARIABILITY AND ITS IMPACT ON WETLANDS

Wetlands will be affected by future climate change in two ways: the number of functioning wetlands (and their functional capacity) will diminish in most ecoregions, and the geographic position of certain types of wetlands will shift. Estimates suggest that there shall be a huge cost of climatic effects on Indian wetlands, whereby almost 84% of the total coastal wetlands and about 13% of saline wetlands will be lost due to a rise in the sea level (Blankespoor et al., 2012). This will further cause a variation of the hydrological cycle and will cause considerable and consequential impacts on wetlands.

However, these inland and coastal wetlands are very sensitive and extremely prone to climatic disruptions such as rises in environmental temperatures, uneven distribution of precipitation and rainfall, and sea-level rise, and are on the verge of fast degradation and extinction. Temperature changes in the atmosphere can adversely affect the species composition of the wetlands and can have serious implications on the overall ecological balance of wetlands (IPCC, 2002). These natural habitats are now extremely prone to degradation and will be further exacerbated by climate change. The Second Assessment Report the IPCC, in categorical terms states that anthropogenic effects on the environment will have significant implications on it and it will lead to climate change. Climate change has caused the temperatures to rise and has altered the precipitation patterns and distribution, which may cause adverse effects on wetlands along the Gangetic plains (Sinha, 2011). Moreover, Climate change will also affect the natural shifting mechanism and geographical distribution of wetlands.

The IPCC report also states that these anticipated changes are more or less very certain to happen and alter the wetlands climate permanently and are most likely to have long-term effects on their natural distribution and function. The communities depending on these wetlands will be most affected by these climatic variations. As a result, the communities which are dependent on wetlands ecosystems goods and services are the most widely affected as they will not be able to cope with the sudden and fast changes in climatic conditions. Patel et al. in 2009 analysis showed that coastal wetlands, high altitude wetlands, mangroves, and corals are the most sensitive classes that can be affected due to climate change.

IMPACT OF CLIMATE CHANGE ON UPPER GANGA RAMSAR SITE

The Upper Ganga Ramsar Site in India has a spacious band of diverse wetlands with a wide range of spatial distribution, which may include lakes, swamps, mangroves, corals, and various other types of wetlands in the region (MoEF, 1990, WWF, 1993). These wetlands, owing to their geographical distribution, are very rich in terms of biodiversity and provide various kinds of ecosystem services across the region (Wetlands Rules, 2010). According to a report by the Millennium Ecosystem Assessment, seven percent of the earth's surface area is covered by a wide range of diverse and rich wetlands ecosystems and provides ecosystems services worth $20 trillion a year (Source: www.MAweb.org). The coastal wetlands across the globe (Boorman, 1990) and high altitude wetlands in India (Patel et al., 2009) are the most vulnerable ecosystems prone to climate change.

We need to devise a mechanism involving the key diverse stakeholder groups highlighting the key issues facing wetlands in the Upper Ganga Ramsar Site basin, their status, and importance. Mainstreaming of the adaptation through strategic planning and effective implementation procedures can be the key to the ecological as well economic development with the engagement of the local community and civil society groups. Adaptation and mitigation strategies would enhance the benefits to communities that are the most vulnerable to climate change. For the conservation of wetlands in the Upper Ganga Ramsar, which requires an integrated approach for Ganga rejuvenation, the need is to strengthen the integration of wetlands into basin management programs across the country. The Ramsar Convention which India ratified in 1982 has categorized rivers as wetlands and the conservation of these wetlands may be streamlined and achieved through the Ganga River Basin Management Plan launched in 2015 under the National Mission for Clean Ganga (NMCG).

It is imperative to understand that wetlands need proper care and appropriate conservation and protection techniques with both short-term plans and long-term strategies, and the legal protection may be accorded to them under the Wetlands Act of 2017. The state of Uttar Pradesh consists of the most oxbows and floodplain wetlands, which fall under the Forest Department's jurisdiction (Wildlife Sanctuaries) or Irrigation Department. There are around 1,800 wetlands, which have been identified for restoration and management keeping in mind the effects of climate change on them. This will require prioritization of wetlands through rapid assessment of wetlands which will identify those wetlands that need to be prioritized and prepared for conservation on an urgent basis.

MANAGEMENT OF NATURALLY SENSITIVE WETLAND ECOSYSTEMS

The Constitution of India stipulates the importance of water resources (Jain, 2005) and the need for their conservation. Article 48A of the Constitution of India clearly states that "The State shall endeavor to protect and improve the environment and to safeguard the forests and wildlife of the country". Similarly, Article 51A of the constitution of India says it is the fundamental duty of each citizen to protect and improve the natural environment. "It shall be the duty of every citizen of India, to protect and improve the natural environment including forests, lakes, rivers, and wildlife, and to have compassion for living creatures." The first National Water Policy was formed in 1987 but it was poorly drafted. Conservation of water bodies, on the other hand, was only addressed at the national level in 2002, when the 1987 National Water Policy was revised. A rebirth of ancient systems was also emphasized in the amended policy. The National Environment Policy (NEP) 2006 policy framework was implemented to improve current environmental policies and correct flaws by offering a complete set of solutions. The policies were put in place to help India solidify its beneficial environmental contributions. The policy (NEP, 2006.) suggests methods for the sustainable maintenance of the wetlands that included:

1. Creating a legislative framework for the detection and prevention of wetlands.
2. Involve governmental agencies and multi-stakeholders in the implementation of new conservation plans and projects.
3. Traditional approaches should be promoted to avoid the high economic costs associated with large infrastructure projects.

For the first time in 2010, the Central Government issued legally enforceable wetlands management laws. They decentralized the wetland management process by expanding the scope of authorities vested in states. The powers were not limited to identifying endangered wetland locations; they also included monitoring forbidden activity to ensure their "wise usage." The Centre was given the authority to oversee the States'/UTs' implementation of the laws, control the management of transboundary wetlands, and ensure compliance with the Ramsar Convention. Due to inefficiency in the existing rules, lack of implementation, insufficient actions by the government, and clear effects of declining wetlands in India several public interest petitions were filed before the Supreme Court of India and various High Courts all across the country. The Supreme Court (Balakrishnan, 2001.) in February 2017 directed the Central Government to frame a policy to protect wetlands within 6 months. In this order, the SC directed the central and state governments to notify all 2,01,503 wetlands that were identified with the help of satellite imagery from ISRO and to draw up a phased plan of action to conserve the water bodies. The SC emphasized that the poor management of wetlands was very apparent as the state governments and UTs were unable to satisfy the judges as to how the wetlands management funds dedicated for the Ramsar wetland sites were being spent. The SC also directed various High Courts to monitor the management of 26 sites identified in the

Ramsar Convention of 1971 till there was some visible improvement as these were international heritage sites. The same year government reluctantly introduced the Wetland (Conservation and Management) Rules, 2017.

The Wetland Rules, 2017 provide for the establishment of a state-level authority to oversee the management of wetlands under its jurisdiction. The authority is led by the State's Environment Minister and consists of government officials and experts in the domains of ecology, hydrology, and biodiversity. The body guarantees that the principle of "smart usage" is followed by advising solutions, funding research studies, and implementing the results. Anthropogenic activities such as improper trash disposal, company, and industry expansions, effluents, and pollutants discharged into water bodies that threaten the existence of wetland habitats are monitored by the State authorities. The activities should be carried out following the regulations' notion of "smart use." The 2017 Rules require state officials to compile a list of wetland locations within their authority that must be disclosed to the federal government within six months. Based on the list provided by the authorities, the Centre generates a computerized inventory of all such places. The 2017 Rules also call for the establishment of a nationalized Wetlands Committee, chaired by the Secretary of the Ministry of Environment, Forestry, and Climate Change, to oversee the state's operations. The committee has been given the ability to advise the Centre on schemes, techniques, and ideas for promoting the appropriate use of wetlands, as well as to assist the Government in the designation of wetlands as wetlands of international importance under the Ramsar Convention. The Rules of 2017 however, were criticized for doing away with strong wetland monitoring systems and omitting important wetland types. Instead of delegating its powers to the state government, the center abdicated its power in favor of the state governments to evade liability. In the 2010 Rules, some related criteria were made explicit, such as natural beauty, ecological sensitivity, genetic diversity, historical value, etc. These have been omitted in the 2017 Rules. Both the Wetland Rules and the Coastal Regulation Zone Act were modified to make it easier to divert natural areas away which not only affects the water bodies but also thousands of people as well. Even today the governments have disregarded multiple judicial directions to demarcate the wetlands. Given the granularity of the issue, the federal complications, and the number of stakeholders involved (in addition to the governmental authorities, residents, fisherpersons, building developers, construction workers, and grass-cutters are just some of the different groups of affected communities), there are limitations to the court's ability to effect change on such a widespread and complex issue (Iyengar, 2019).

The National Water Mission, as part of the National Action Plan on Climate Change, also recommended lake and other water body conservation, including the expeditious implementation of a program for water body repair, renovation, and restoration in climate-sensitive areas /situations, developing an inventory of water bodies, particularly those with unique features, and mapping catchments, creating awareness among people on the importance of ecosystems of the water bodies (Kumar & Dinesh Kumar, 2020). The National Green Tribunal (NGT), several High Courts, and even the Supreme Court have been extremely active on the wetlands front, although this has had very little impact on wetlands and their governance in India, according to the overview of wetlands in India in 2020. This is mostly because, as the second main point of this overview points out, the federal and state governments have shown little interest, understanding, or willingness to conserve the wetlands. This is even though the Ramsar convention will be expanded to include many new Indian wetlands in 2020, as experience and this summary demonstrate that the Ramsar convention does not appear to be very beneficial to the wetlands' fate. The third point to note in the overview is that there is a lot of civil society effort in India to protect wetlands, both in terms of lobbying and on the ground. In reality, it is primarily owing to their efforts that we are seeing legal action in the NGT and courts. The few positive advances we've seen so far are the result of community and civil society activities.

In 2021, we hope to see a significant change in this situation. Let us hope that the legal action becomes more effective and results in a considerable improvement in wetlands governance. There is clear disregard and non-compliance with both the legislations and the court orders related to wetlands in India, nor is there a willingness or enthusiasm on part of the authorities. Wetlands are just too vital to be destroyed the way we are, and they will become increasingly more so as the climate changes. However, it appears that society does not recognize their importance, except for populations that are directly dependent on the wetlands, but who have very little, if any, say in wetlands decisions. It's a depressing scene, but let's hope things improve soon. Key government actions, Legal developments, other events, and studies & reports are the four broad areas of this overview.

RESTORATION OF WETLANDS

Although wetlands cover an estimated 4.7% of India's surface and perform a variety of critical ecosystem services (Bassi et al. 2014). India's protection for these critical habitats has been weak and inadequate. India's wetlands are at potential risk of either total loss in their composition or may face the challenge of degradation from various externalities including climate change, rapid urbanization, industrialization, and agricultural intensification. Agricultural stresses and other ancillary developmental activities caused potential damage to the wetland ecosystems (Ramsar Convention on Wetlands, Global Wetland Outlook, 2018). These potential threats result in a reduction of the natural functions and ecosystem services they provide (Bassi et al., 2014). Despite being recognized as the most ecologically sensitive zone, prone to climate and other anthropogenic exigencies, wetlands have been largely neglected, and not much has been done in terms of policy initiatives and measures for their restoration. It is only now that the wetlands are increasingly receiving due attention for restoration. Rivers and coastal wetlands ecosystems have enormous potential and value in terms of ecosystem goods and services and in providing buffer zones. These rivers and wetlands area is known as the lifelines of human sustenance as they provide clean potable water, as well food security amongst its other usefulness (Clarkson et al., 2013). Wetlands today have become the most pressing and critical issue, as a lot of discussions are going on regarding restoration and rejuvenation. The health of wetlands depends upon the existing water table around the wetlands and hence development activities around the wetlands, which may have an adverse effect, should be planned carefully beforehand (McLaughlin et al., 2019). Restoration strategies would require scientific understanding and geographical knowledge about the spatial distribution, ecology, and ground-level socio-economic conditions for the conservation and management of a wetland. It is the most pressing need of the hour that various gaps that remain in the conservation methods be identified and worked upon based on sound scientific research so that the restoration process can be started. One of the key issues facing the conservation of wetlands initiatives is the lack of a comprehensive and accurate scientific database which is a big impediment in conservation initiatives across the region.

Under the aegis of the United Nations Development Program (UNDP), a proposal was brought forth to identify and categorize conservation strategies for 200 inland wetlands as potential Ramsar sites by Salim Ali Centre for Ornithology and Natural History (SACON) (Vijayan et al., 2004). After this, a more comprehensive and elaborate list of wetlands was prepared by the Bombay Natural History Society

(BNHS) (Islam & Rahmani, 2008). Sustainable management of wetlands must be prioritized keeping in view that their natural ecosystem and biodiversity are preserved for a long time. In view that the wetlands should be competent to maintain their natural character and become resilient to any external pressures from any change that may be climatic or otherwise (Davis et al., 2010). The management needs to be holistic, multidisciplinary and based on scientific knowledge. Conservation measures have to be streamlined with the overall ecosystem development of coastal zones, keeping in view the socioeconomic development of wetland-dependent communities.

Specific adaptation strategies and techniques for specific kinds of wetlands could be used through ecosystem-based management while maintaining hydrological, morphological, and ecological integrity. Some of them could be the cost-effective mitigation and improvement procedures to make them climate change-resistant and resilient. Recognizing the relevance of an integrated approach towards managing, the IPCC (2000) has also addressed the significance of wetland and called for its restoration on an urgent basis. As with terrestrial farming, there may be opportunities for aquaculture operators to restore wetland areas and make a commitment not to convert existing wetlands for further aquaculture development. Conservation, restoration, and wise use of wetlands can be a cost-effective strategy for climate adaptation with strong benefits for poverty reduction and biodiversity conservation. Conversely, it is also believed that strategies for climate adaptation and development that do not address the continuing crisis in wetland loss and degradation will have strong limitations. Local-level planning and management strategies also must be evolved for comparing various alternative options suitable at different conditions. There is an urgent need to develop a strategic roadmap for integrating wetland conservation in the Upper Ganga Ramsar Site Management. Some important measures that can be taken on an urgent basis are:

1. Identification and documentation of the wetlands on the Upper Ganga Ramsar Site
2. Biophysical characterization of wetlands in the floodplains
3. Identification of the crisis at hand and their root causes
4. Integrated management and planning
5. Possible mitigation measures must be chalked out
6. Prioritizing and valuing conservation, restoration, and mitigation
7. Creating a list of priority wetlands
8. Continuous observation, monitoring, and review of wetlands after restoration
9. National laws based on international conventions and multi-lateral agreements
10. Strengthening all the stakeholders through capacity building
11. Devising effective institutional mechanism for integrated wetland management
12. Involving the local community and empowering them through education and active participation

The Ministry of Environment, Forest, and Climate Change in the year 2020 released detailed guidelines to support the state governments in implementing the Wetlands (Conservation and Management) Rules 2017. These guidelines helped states identify and prepare a list of wetlands. The guidelines mandate that the states must follow them under the Wetlands Rules, 2017 before identifying wetlands for notification. These guidelines also make good for delineating wetlands, wetlands complexes, and zone of influence. The states have to prepare a draft rule identifying a list of activities to be regulated and permitted around the wetland's zones. The most important point is developing an integrated management plan for wetlands, which are rich reservoirs of biodiversity. The guidelines say that all wetlands, irrespective of their location, ownership, biodiversity, size, or ecosystem services values, can be notified under the Wetlands

Rules 2017, except river channels, paddy fields, human-made water bodies specifically constructed for drinking water, aquaculture, salt production, recreation, irrigation, wetlands falling within areas covered under the Indian Forest Act, 1927, Forest (Conservation) Act, 1980, Wildlife (Protection) Act, 1972 & the CRZ Notification, 2011.

SOLUTION AND RECOMMENDATIONS

Ministry of Jal Shakti, Department of Water Resources, River Development and Ganga Rejuvenation through National Mission for Clean Ganga (NMCG), National River Conservation Directorate (NRCD) are responsible for coordinating the ecological monitoring of the wetlands; and it would be the duty of the Ministry of Environment, Forest and Climate Change to manage and conserve wetlands and river stretches within any Protected Area, such as National Park, Tiger Reserve, Sanctuary, Ramsar Site, and any other area protected under the provisions of the Wild Life (Protection) Act, 1972 (Hussain, S.A., et. al. 2021). Restoration of wetlands through mitigation and ecologically sound adaptation strategies and urgent measures is the key to the preservation of wetlands ecosystem and maintaining the ecological balance, which can also keep at bay the adverse effects of climate change on wetlands. The government of India is under an obligation to assess the existing condition of wetlands and accordingly formulate policies, legislation, and regulatory regimes for the conservation of wetlands. The assessment would include key aspects of ecosystem integrity, resilience capacity in the face of anthropogenic pressure and climate change, and threats (Rapport, 1995). One of the ways for restoration and mitigation could be to sensitize a sense of duty among the general masses towards environment protection and include adaptation measures as an agenda in the key policy initiatives at the local, national, regional level (Kaul, S., Kumar, R. 2019). It is with this spirit, Article 51–A (g) of the Indian Constitution internalizes and summarizes that "it shall be the duty of every citizen of India to protect and improve the natural environment including forests, lakes, rivers, and wildlife and to have compassion for living creatures." At the local level, however, community-level stakeholders can play a substantive role in the conservation and mitigation process. In so doing, the communities can also engage in participatory management which could also help improve the capacity development techniques at the beginning or baseline of or from the lowest strata of the community which may include the learning process among all the action groups involved. This can have better results and this technique has been appreciated by the Ramsar Convention as an inclusive approach towards conservation and management. The Convention further elucidates the concept of involving and motivating the local community through local planning and management so that they learn to appreciate the ecological values of the wetlands and in turn become the ground army in advocating for the conservation of wetlands.

FUTURE RESEARCH DIRECTIONS

There is much scope for research in the field of the present status of wetlands. The Physico-chemical characteristics, water quality index, carbon sequestration capacity, net productivity, various biodiversity indices, plastic pollution, oil pollution, nutrient level, etc. The repeated sampling, seasonal variations, long-time assessment, and observations concerning changing patterns of climate can give a better idea regarding key issues facing wetland resources. It will give baseline data for a better understanding of the status of wetlands. The reaction of wetlands to changing patterns of climate is poorly understood and therefore, there is scope for research. It is in this context that through the conservation, maintenance, and rehabilitation strategy, wetland ecosystems could be made more resilient towards the dangers of climate change. We need to create an institutional mechanism for mitigation.

CONCLUSION

The most daunting of facts remains that despite providing such varied and significant ecosystems goods and services, the key stakeholders and policymakers have yet not appreciated the real value of wetlands and sadly there has been little or no attention shown towards conservation and intelligent utilization of wetlands. Over time, climate changes affect wetlands and can lead to shifts in species distributions, species communities, loss of native species as well as biogeochemical changes in the soil. Warming temperatures, increased levels of carbon dioxide in the atmosphere, and changes in frequency and intensity of rainfall alter these ecological processes. Due to climate change, the number of functioning wetlands (and their functional capacity) will diminish in most eco-regions, and the geographic position of certain types of wetlands will shift. Well-planned and executed strategies through proper research activities are the well-felt needs towards protection, conservation, and restoration of wetlands. It is a prioritized need of interventions of government and policymakers for a bright future of a healthy environment.

REFERENCES

Australian Government, Department of Agriculture, Water and the Environment. (n.d.). *Wetlands and the Climate Change*. https://www.environment.gov.au/water/wetlands/publications/wetlands-climate-change

Balakrishnan, M. K. vs. Union of India. Civil (W.P.) No. 230 of 2001. Supreme Court of India. https://main.sci.gov.in/jonew/courtnic/rop/2000/17346/rop_886913.pdf

Bassi, N., Kumar, M. D., Sharma, A., & Pardha-Saradhi, P. (2014). Status of wetlands in India: A review of extent, ecosystem benefits, threats and management strategies. *Journal of Hydrology. Regional Studies, 2*, 1–19. doi:10.1016/j.ejrh.2014.07.001

Blankespoor, B., Dasgupta, S., & Laplante, B. (2012). *Sea-level rise and coastal wetlands: Impacts and costs* [Policy Research Working Paper 6277]. The World Bank. doi:10.1596/1813-9450-6277

Boorman, L. A. (1990). Impact of sea level changes on coastal areas. In M. M. Boer & R. S. de Groot (Eds.), *Landscape ecological impacts of climate change* (pp. 379–391). IOS Press.

Boyd, J., & Banzhaf, S. (2007). What are ecosystem services? The need for standardized environmental accounting units. *Ecological Economics, 63*(2–3), 616–626. doi:10.1016/j.ecolecon.2007.01.002

Bystrom, O., Andersson, H., & Gren, I. (2000). Economic criteria for using wetlands as nitrogen sinks under uncertainty. *Ecological Economics, 35*(1), 35–45. doi:10.1016/S0921-8009(00)00166-X

Clarkson, B. R., Ausseil, A. E., & Gerbeaux, P. (2013). Wetland ecosystem services. In J. R. Dymond (Ed.), *Ecosystem services in New Zealand - conditions and trends*. Manaaki Whenua Press.

Cowardin, L. M., & Carter, V. (1979). Classification of wetlands and deepwater habitats in the United States. U.S. Dept. Interior, Fish & Wildlife Service, FWS/OBS- 79/31.

Davies, G. (2018, May 24). *How will wetlands be affected by climate change?* Wisconsin Wetlands Association. https://www.wisconsinwetlands.org/updates/how-will-wetlands-be-affected-by-climate-change/

Davis, J., Sim, L., & Chambers, J. (2010). Multiple stressors and regime shifts in shallow aquatic ecosystems in antipodean landscapes. *Freshwater Biology*, *55*(s1), 5–18. doi:10.1111/j.1365-2427.2009.02376.x

Desai, A. (2018, May 24). *How can wetlands provide resilience against climate change?* Wisconsin Wetlands Association. https://www.wisconsinwetlands.org/updates/how-can-wetlands-provide-resilience-against-climate-change/

Finlayson, M., & Moser, M. (Eds.). (1991). *Wetlands*. International Waterfowl and Wetlands Research Bureau.

Hussain, S. A., Badola, R., Gopi, G. V., Johnson, J. A., Shivakumar, K., Dasgupta, N., Barthwal, S., & Tuboi, C. (2021). Framework for Ecological Monitoring of Ramsar Sites and other Wetlands in India. Ganga Aqualife Conservation Monitoring Centre (GACMC). Wildlife Institute of India.

IPCC. (2002). Climate change and biodiversity. IPCC Technical Paper V.

Islam, Z., & Rahmani, A. R. (2008). Potential and Existing Ramsar Sites in India. Bombay Natural History Society, Oxford University Press.

Iyengar, S., Dolšak, S., & Prakash, A. (2019, December). Selectively Assertive: Interventions of India's Supreme Court to Enforce Environmental Laws. *Sustainability*, *11*(24), 7234. Advance online publication. doi:10.3390u11247234

Jain, S. K. (2005). Water Resources of India. Water Encyclopedia. doi:10.1002/047147844X.wr243

Kaul, S., & Kumar, R. (2019). Wetland Conservation Ethos. New Delhi: Wetlands International. *South Asia*, 64.

Kumar, M. D., & Dinesh Kumar, M. (2020). Water management in India: the multiplicity of views and solutions. In Politics and Policies for Water Resources Management in India (pp. 1–15). doi:10.1201/9780429316234-1

Lakeman, J. A. (1999). Climate Change 1995 The Science of Climate Change. In Contribution of WGI to the Second Assessment Report of the Intergovernmental Panel on Climate Change (Vol. 30, Issue 3). Cambridge University Press. doi:10.1007/978-3-662-03925-0_1

Meena, A. K., & Sharma, C. (2019). *Wetlands in India: Ecosystem Benefits*. Threats and Management Strategies.

Ministry of Environment, Forest, and Climate Change. (2020). *Guidelines for implementing Wetlands (Conservation and Management) Rules, 2017*. Government of India. http://www.indiaenvironmentportal.org.in/files/file/Wetlands%20(Conservation%20and%20Management)%20Rules,%202017.pdf

Ministry of Environment & Forest. (2006). *National Environment Policy 2006*. Government of India. http://www.indiaenvironmentportal.org.in/files/nep2006e.pdf

Mitsch, W. J., Bernal, B., Nahlik, A. M., Mander, Ü., Zhang, L., Anderson, C. J., Jørgensen, S. E., & Brix, H. (2013). Wetlands, carbon, and climate change. *Landscape Ecology*, *28*(4), 583–597. doi:10.100710980-012-9758-8

Palanisami, K., Meinzen-Dick, R., & Giordano, M. (2010). Climate change and water supplies: Options for sustaining tank irrigation potential in India. *Economic and Political Weekly*, *45*(26-27), 183–190.

Pant, N., & Verma, R.K. (2010). *Tanks in Eastern India: A Study in Exploration International Water Management Institute*. TATA Water Policy Research Program and Centre for Development Studies.

Patel, J. G., Murthy, T. V. R., Singh, T. S., & Panigrahy, S. (2009). Analysis of the Distribution Pattern of Wetlands in India in Relation to Climate Change. *Archives XXXVIII8W3 Workshop Proceedings Impact of Climate Change on Agriculture*, 282–287. www.MAweb.org

Ramsar Convention on Wetlands. (n.d.). *The Global Wetland Outlook, About*. https://www.global-wetland-outlook.ramsar.org/

Salimi, S., Almuktar, S. A. A. N., & Scholz, M. (2021). Impact of climate change on wetland ecosystems: A critical review of experimental wetlands. *Journal of Environmental Management*, *286*, 112160. doi:10.1016/j.jenvman.2021.112160 PMID:33611067

United States Environment Protection Agency. (n.d.). *Why are Wetlands Important?* EPA. https://www.epa.gov/wetlands/why-are-wetlands-important

Vijayan, V. S., Prasad, S. N., Vijayan, L., & Muralidharan, S. (2004). Inland Wetlands of India: Conservation Priorities. Sàlim Ali Centre for Ornithology and Natural History.

Wetlands (Conservation and Management) Rules. (2010). *Ministry of Environment and Forests, Gazette of India*. https://thc.nic.in/Central%20Governmental%20Rules/Wetlands%20(Conservation%20And%20Management)%20Rules,%202010.pdf

Wetlands (Conservation and Management) Rules. (2017). *Ministry of Environment and Forests, and Climate Change. Gazette of India*. http://www.indiaenvironmentportal.org.in/content/447822/wetlands-conservation-and-management-rules-2017/

WWF. (1993). *Directory of Indian Wetlands*. World Wide Fund for Nature.

Chapter 7
Status and Threats of Wetland Change in Land Use Pattern and Planning:
Impact of Land Use Patterns and Urbanization

Dipanwita Sarkar (Paria)
Chandernagore College, India

Nibedita Maji
Chandernagore College, India

ABSTRACT

Wetland-related studies documented the loss of native species diversity and promotion of the biotic homogenization due to wetland loss. Excessive withdrawals of water from wetlands for residential, agricultural, or industrial use are responsible for wetland degradation. Constructions of dams impedes water flow and replenishment of wetlands, and it also creates a hazard to aquatic living organisms. Climate change causing some wetlands to disappear under rising sea levels, while others are severely impacted by changing climatic conditions, including drought. So necessary steps such as increase wetlands and prevention of the illegal swamping of wetlands, etc. should be taken for conserving the wetland biodiversity from the threatening of unplanned urbanization, purifying the environment and mainlining the sustainable development. Though the Ramsar Convention policy exists to persevere wetlands and achieve sustainable development throughout the world, mass consciousness, greater participation of local people, use of indigenous knowledge in the management strategies are needed to protect wetlands.

INTRODUCTION

Wetlands are characterized by the interaction between terrestrial and aquatic ecosystem. Therefore, they could be considered as the world's most productive ecosystem (Ghermandi et al., 2008). Wetlands provides many functions and services within local watersheds along with economic values in communi-

DOI: 10.4018/978-1-7998-9498-8.ch007

ties. Not only that, wetlands also supply habitat for birds & wildlife and create recreational opportunities (Keddy 2010). From the very beginning, wetlands have a significant role in human civilization. But in course of time the wetland based human civilization have been transformed into industry based urbanized civilization. For the last few decades, change in land use pattern due to the unplanned urbanization leads to the destruction of natural wetlands and related native species. Rapidly growing constructions and anthropogenic activities are the important cause of wetland loss at local, regional and global scales. Urbanization exerts significant influences on the structure and function of wetlands, mainly through modifying the hydro logical and sedimentation regimes and the dynamics of nutrients and chemical pollutants. Numerous wetland related studies documented the loss of native species diversity and promotion of the biotic homogenization due to wetland loss (Marty, 2005). So necessary steps such as increase wetlands & prevent the illegal swamping of wetlands etc. should be taken for conserving the wetland biodiversity from the threatening of unplanned urbanization, purifying the environment and maintaining the sustainable development.

BACKGROUND

A wetland is a land area that is saturated with water, either permanently or seasonally and are one of the most productive ecosystems, comparable to tropical evergreen forests. It plays a vital role in the ecological sustainability of a region. Wetlands are very essential part of our biosphere and they fulfill many crucial needs for life on earth such as drinking water, water purification, flood storage, transport, biodiversity conservation, climate stabilizers, flood storage, transport etc. Wetlands cover only a marginal area of the earth's surface 6.2 to 7.6 percent.

Our each and every organs are vital to our health for playing a different significant role. For instance, kidneys remove waste particles of our body and filter our blood. Similarly wetlands are often referred to as "Earth's kidneys" because they provide the same functions, absorbing wastes such as nitrogen and phosphorous and purifying the water that pass through them and distill out harmful chemicals, poisonous sediments and pollutants. When excess amounts of these substances and nutrients flow into waterways, it can create harmful algal blooms, hypoxia, and summer fish kills. Wetlands purify waste matter through a natural process of oxidation, radiation, bio-degradation and fish cultivation. Wetlands also regulate and maintain the planet's air & water cycle including the level of oxygen, nitrogen, sulphur and carbon-dioxide etc.

CHARACTERISTICS OF WETLANDS

There are three characteristics that describe a wetland:

- **Hydrology**: Wetlands must contain water at or near the surface of the land for a designated amount of time.
- **Soil**: Soil must be hydraulic or saturated with water to create oxygen free or anaerobic environment. It's mainly of three types- slit, sandy and clay soils.
    - ***Sandy Soil***: It has good drainage and aeration capacity. So it does not store water well and not suitable for most plants. It contains mineral grains ranging from 0.05 - 2 mm in diameter.

    - ***Slit Soil***: This type of soil made from minerals. It also known as rock flour or stone dust when produced by glaciers. It contains mineral grains ranging from 0.002 - 0.05 mm in diameter.
    - ***Clay soil***: This type of soil can hold water very well and water does not drain through it easily so it is also not suitable for most plants. It contains mineral grains ranging from 0.002 mm in diameter.
- **Plants**: Plants must be wetland plants or hydrophytes type that require lots of water and the anaerobic conditions that the hydraulic soil creates. Seasonally dry wetlands or wetlands with slow-moving water can often support trees and other sturdy vegetation. More frequently flooded wetlands have mosses or grasses.
    - Submergent - Plants that grow entirely under water. Most are rooted at the bottom and some may have flowers that extend above the water surface.
    - Floating leaves - Plants rooted to the bottom with leaves that float on the water surface. Flowers are normally above water.
    - Free Floating - Plants not rooted to the bottom and float on the surface.
    - Emergent - herbaceous or woody plants that have the majority of their vegetative parts above the surface of the water.

TYPES OF WETLANDS

Besides tidal and man made wetlands, it can be classified into various forms (Figure 1).

- **Marsh:** Standing or slow moving water bodies with plant cover greater than 25%. Often seasonally flooded and dominated by non woody plants with nutrient rich water for most of growing season neutral pH and rich species diversity.
- **Fen:** Fens also feature a build up of peat but instead of a lot of moss, sedges are the dominant vegetation. They are less acidic because of slow water movement.
- **Bog:** Bogs are acidic wetlands with a build up of slowly decaying vegetation called peat. They are dominated by sphagnum mosses and get their water only from rain or snow. Their nutrient level are poor.
- **Swamp:** Swamps are like wooded marshes. They are soggy landscapes with waterlogged trees and other plants. Their pH levels are neutral. They are mineral and oxygen rich and often flooded in rainy season or dry out in summer.
- **Vernal Pool:** Found only seasonally in shallow depressions in the land. i.e. temporary pool ecosystems. They can be covered in shallow water, but in the summer and fall, they can be completely dry. They may forms in open grasslands, forested landscapes, pine savannas and Flat woods. Many amphibian species depend upon vernal pools for spring breeding; these ponds provide habitat free from fish which eat the eggs and young of amphibians. An example is the endangered gopher frog (*Rana sevosa*).
- **Wet Meadow:** Wet meadows occur in Areas such as shallow lake basins, lowlying depressions and the land between shallow marshes and upland areas. They also occur on the edges of large lakes and rivers. Wet meadows often have very high Plant diversity and high densities of buried seeds. They are regularly flooded but are often dry in the summer.

Figure 1. Diagram of various types of wetlands
(Source: researchgate.net)

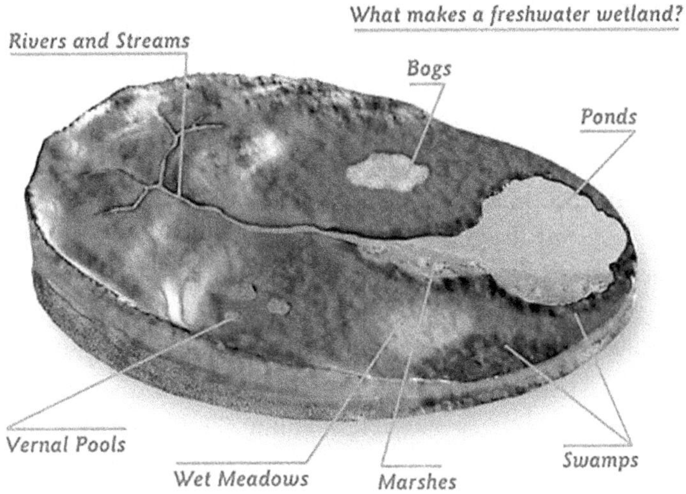

IMPORTANCE OF WETLANDS

Wetlands provides many functions and services within local watersheds along with economic values in communities (Figure 2). Not only that, wetlands also supply habitat for various species and create recreational opportunities. Benefits of wetlands include:

- ***Groundwater replenishment***: Periodically flooded wetlands are very effective in storing rainwater and also have innate capacity to recharge the ground waters. Ground water recharge may occurs through the mineral soils, found primarily around the edges of wetlands. Ground water recharge extension depends on the type of soil and its permeability, vegetation, sediment accumulation in the lake bed, surface area to volume ratio and water table gradient.
- ***Water storage and flood control***: In their natural condition, most wetlands store floodwaters temporarily, protecting downstream areas from flooding. Apart from this, wetlands maintain a constant flow regime downstream, preserve water quality and increase the biological productivity of the aquatic communities. This function is highly important in urban areas, where developmental activities such as breaching of wetlands for residential, commercial and industrial activities have gradually increased. The protection of wetlands are highly needed for minimizing flood damages in the future.
- ***Shoreline protection and stabilization***: Wetland vegetation can reduce shoreline erosion in several ways, including increase in durability of the sediment through binding, dampening waves through friction and reducing current velocity through friction, improving water quality. Coastal wetlands particularly mangroves help in shoreline stabilization and storm protection by dissipating the force by reducing the damage of wind and wave action.
- ***Toxic sediment and nutrient retention***: Wetlands, transition zones between land and water are efficient in filtering sediments. They can obstruct run-off from land before it reaches the water and help in filtering nutrients, wastes and sediments from flood waters. In some wetlands, plants are so

efficient in removing wastes that artificial waste water treatment systems use aquatic plants for the removal of pollutants from water. Wetlands remove nutrients, specially nitrogen and phosphorus, particulates and total biological oxygen demand from flooding waters for plant growth and help prevent eutrophication or over-enrichment of other forms of natural waters. Overloading a wetland with nutrients, beyond its threshold, prevents its ability to perform basic functions.

- ***Water purification*** : Wetlands play a vital role in improving the water quality by filtering sediments and nutrients from surface water. Aquatic vegetation helps to remove nearly 90% of the dissolved nutrients and also in adsorption of heavy metals. Dissolved materials may be retained in wetlands so the water quality may vary seasonally. Sediment load removal is also important because it often carry absorbed nutrients, pesticides, heavy metals & toxic materials that pollute water.
- ***Habitats for diverse aquatic plants and animals***: Wetlands are one of the most biologically productive natural ecosystems which are vital for the survival of diverse flora and fauna, including many threatened and endangered species by providing shelter and food. It is estimated that about 20% of the known species depend directly or indirectly on wetlands for their survival, as wetlands provide them permanent or seasonal habitats. Wetlands contain extraordinary wildlife diversity including various types of fishes, crustaceans, waterfowl and some mammal species such as foxes, minks, deer, even bears etc. Wetlands are also important habitat for migratory birds including ducks, egrets, geese etc (Pittock et al. 2015). Wetlands also serve as spawning and nursery grounds for many fish, turtles, frogs, snakes, and other reptiles and amphibians. Most of the wetlands species ensure food webs remain intact both near and far from wetlands.
- ***Recreation and tourism***: Wetlands play an important role in human civilization and are of significant religious, historical or archeological value to many cultures around the world. Often wetlands are attracted tourists for recreational purposes such as hiking, bird watching, wildlife photography, and hunting. These activities help to increase the local businesses and rise the economy.
- ***Wetlands provide sustainable livelihoods***: Nearly 62,000,000 people depend directly on aquaculture and fisheries for their living beings in India. Medicinal plants, timber, vegetable oil, stems and leaves for weaving all can originate from sustainably managed wetlands.
- ***Wetlands have high nature conservation values***: Wetlands are the most bio-diverse ecosystem types. They are habitat of a very diverse range of animal and plant species which live permanently in wetlands or rely on wetlands for at least part of their life cycle. They play a vital role for migratory species, especially migratory waterbirds. Because of the threats for anthropogenic activities in wetlands, high conservation priority is needed.

STATUS OF WETLANDS IN INDIA

Wetlands are among the most important but most threatened ecosystems in India and it covers 4.7% of the total geographical area of the country. They are a precious part of our cultural and natural heritage, providing and extremely important resource for many human interest and activities, as well as habitats that support a rich diversity of animal and plant life. But there is no proper regulatory framework for conservation of wetlands in India. One third of Indian Wetlands has already been wiped out has been

Figure 2. Diagram of various functions of wetlands
(Source: sciencedirect.com)

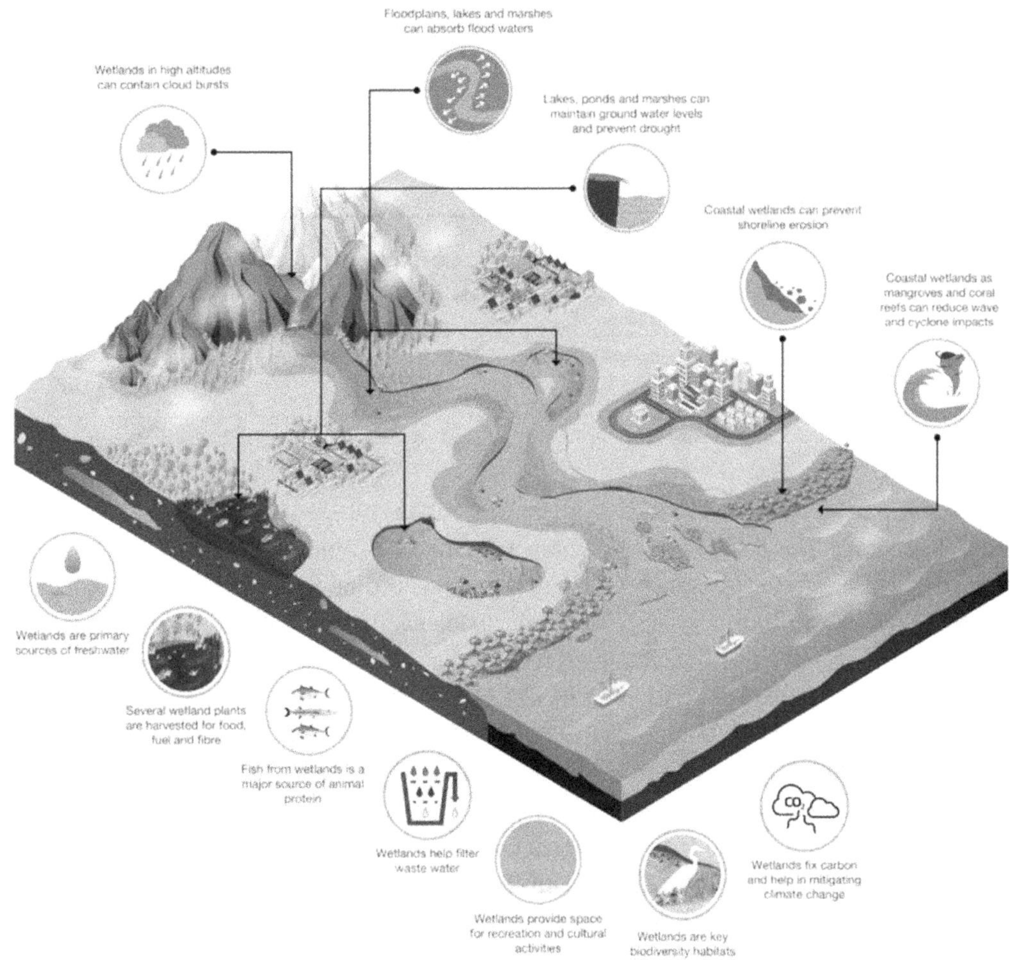

severely degraded. 70 to 80% individual fresh water marshes and lakes in the Gangetic flood plains have been lost in the last five decades. Indian mangrove have been halved almost from 7 lakh hectares in 1987 to 4.53 lakh hectares in 1995.

Distribution of Wetlands in India

There are around 67,429 wetlands in India, covering an area of about 4.1 million hectares (MoEF, 1990). Out of these wetlands, 2,175 are natural and 65,254 are man made. Excluding rivers, wetlands in India cover total 18.4% of the country's geographic area, of which 70% is under paddy cultivation.

A survey conducted by the Ministry of Environment and Forests (MoEF) in 1990 estimated that wetlands occupied around 4.1 million hectares of which 1.5 million hectares were natural and 2.6 million hectares man made excluding paddy fields, rivers and streams and mangroves occupying around 0.45 million hectares. Out of these nearly 80% of the mangroves were distributed in the Sunderbans of West

Table 1. Distribution of mangroves in Indian states and union territories (MoEF, 1990)

Sl. No	States / Union Territories	Area in sq. km
1	Andaman and Nicobar Islands	1190
2	West Bengal	4200
3	Orissa	150
4	Andhra Pradesh	200
5	Tamil Nadu	150
6	Karnataka	60
7	Goa	200
8	Gujarat	260
9	Maharashtra	330
	TOTAL	6740

Bengal and Andaman and Nicobar Islands, with the rest in the coastal states of Orissa, Andhra Pradesh, Tamil Nadu, Karnataka, Kerala, Goa, Maharashtra and Gujarat (Table 1). Wetlands occupy 58.2 million hectares or 18.4% of the country's area (excluding rivers) according to the Directory of Asian Wetlands (1989), of which 40.90 million hectares (70%) are under paddy cultivation. A preliminary inventory by the Department of Science and Technology, recorded a total of 1,193 wetlands, covering an area of about 3,904,543 ha, of which 572 were natural. The Directory of Indian Wetlands published by WWF and Asian Wetland Bureau in 1995 records total 147 wetland sites of which 68 are protected under the National Protected Area Network by the Wildlife Protection Act of 1972. State and Union Territory wise distribution of wetlands in India (Chatrath, 1992) is given in table 2 and figure 4. India has 16% of the world's population in 2.42% of the earth's surface. Out of these about 74% of human population is rural and subjected wetlands to stress from various anthropocentric activities.

From the beginning of civilization, human communities in India are closely associated with wetlands. The wetlands and their resources have been an integral part of the social and cultural ethos of human lives. Recently near about 170 million people that means 17% of India's total population in more than 3,800 coastal villages are scattered along the 7,500 km coastline. Communities living close to wetlands

Figure 3. Percentage of category wise distribution of various kinds of wetlands wetlands in India
(Source: Ministry of Environment and Forests, 1990)

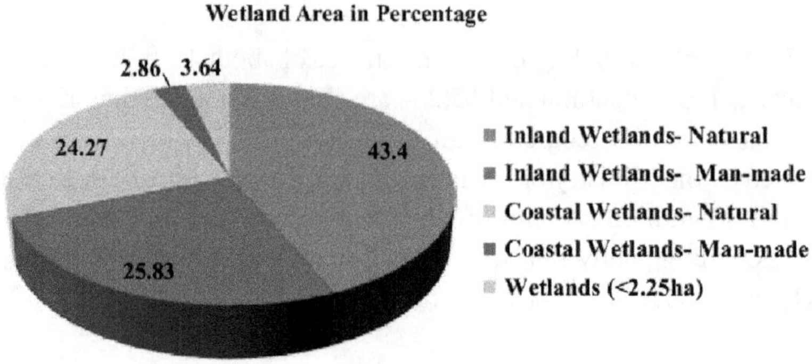

Table 2. Number of important wetlands in Indian states and union territories (Chatrath, 1992)

Sl. No	State and Union Territories	Natural		Artificial	
		Nos	Area (ha)	Nos	Area (ha)
1	Andhra Pradesh	219	100457	19020	425892
2	Arunachal Pradesh	2	20200	NA	NA
3	Assam	1394	86355	NA	NA
4	Bihar	62	224788	33	48607
5	Goa	3	12360	NA	NA
6	Gujarat	22	394627	57	129660
7	Haryana	14	2691	4	1079
8	Himation Pradesh	5	702	3	19165
9	Jammu & Kashmir	18	7227	NA	21880
10	Karnataka	10	3320	22758	539195
11	Kerala	32	24329	2121	210579
12	Madhya Pradesh	8	324	53	187818
13	Maharashtra	49	21675	1004	279025
14	Manipur	5	26600	NA	NA
15	Meghalaya	2	NA	NA	NA
16	Mizoram	3	36	1	1
17	Nagaland	2	210	NA	NA
18	Orissa	20	137022	36	148454
19	Punjab	33	17085	6	5391
20	Rajasthan	9	14027	85	100217
21	Sikkim	42	1107	2	3
22	Tamil Nadu	31	58068	20030	201132
23	Tripura	3	575	1	4833
24	Uttar Pradesh	125	12832	28	212470
25	West Bengal	54	291963	9	52564
26	Chandigarh	NA	NA	1	170
27	Pondicherry	3	1533	2	1131
	TOTAL	**2170**	**1460113**	**65254**	**2589266**

hampered the natural cycle of wetlands. Due to poor resource base and livelihood insecurity, the coastal villagers are compelled to lid an unsustainable dependence on coastal wetlands and change their characteristics leading to their destruction. Marine fisheries depends on the Arabian Sea and the Bay of Bengal, export approximately 307,337 tons of fish annually and about two thirds of this export is made up of shrimp (Ministry of Fisheries, 1999). Besides various ecological and economic values, wetlands also provide cultural value too. As a result most of the Indian villages are settled around dependable water sources for drinking etc.

Figure 4. Map showing the state wise total number of wetlands in India
(Source: researchgate.net)

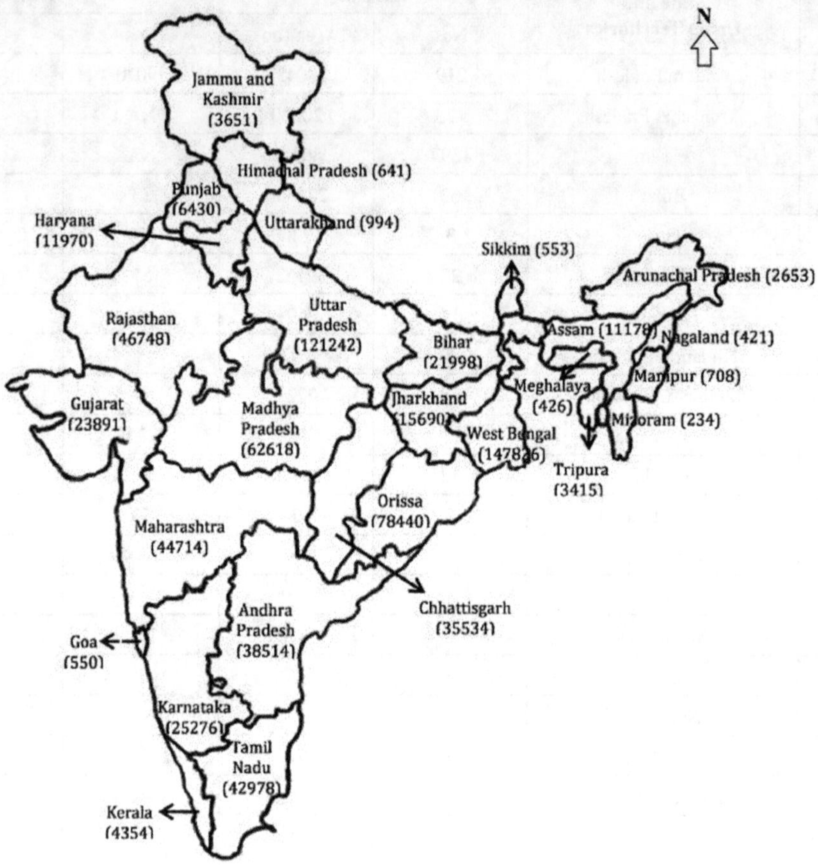

Threats

- Filling, ditching or draining of wetlands for highways, commercial and residential development, ports, waste disposal sites or agriculture.
- Invasive species due to urbanization displace local aquatic flora and fauna, alter water flow, quantity, quality and disrupt overall wetland ecosystem functionality.
- Excessive withdrawals of water from wetlands for residential, agricultural or industrial use (Fig 5).
- Pollution from sediment, agricultural wastes like fertilizers and pesticides, human sewage, animal waste, industrial wastes like heavy metals and toxic particles can contaminate or degrade wetlands and the ecosystems they support.
- Climate change causing some wetlands to disappear under rising sea levels, while others are severely impacted by changing climatic conditions, including draught.
- Constructions of dams impedes water flow and replenishment of wetlands and it also creates a hazard to aquatic living organisms.

Consequences of Losing or Degrading Wetlands

Land use pattern change due to urbanization causes loss of wetland, more than 60%. In fact, the urbanization is an inevitable trend of human society, and necessary for modernization. At the same time, it has severe impact on wetland ecosystem. surface, out of which about 4.8 million sq km are found in the tropic and sub-tropic regions (Fig 6). In 19th century, it was found that approximately 50% of the world's wetlands are lost. Current loss rates for Asian wetlands are estimated at about 5000 km2 per year (Zedler and Kercher 2005, Gopal 2013). The major anthropogenic activities responsible for wetlands loss are urbanization, drainage for agriculture, and water system regulation (O'Neill et al. 2016). The construction and development activities due to urbanization like excavation, filling, draining, and so forth, are the major destructive methods resulting in a significant loss of wetland spatial extent. The wetlands are in tremendous pressure and needs effective measures for its rejuvenation.

Land Use and Land Cover

There are two technical terms used in geography as land use and land cover. Both the terms are inter-related.

- **Land Use**: It is commonly described as a series of anthropogenic activities on land with the intention to obtain products & benefits through using land resources.
- **Land Cover**: It is generally defined as the vegetation or man-made constructions which occur on the earth surface. Water, bare rock, ice, sand and similar surfaces also count as land cover.

Land use and land cover have some fundamental differences. Land use refers to the purpose the land serves, for example, recreation, wildlife habitat or agriculture; it does not describe the surface cover on the ground. For example, a recreational land use could occur in a forest, shrub land, grasslands or on manicured lawns. Land cover refers to the surface cover on the ground, whether vegetation, urban infrastructure, water, bare soil or other; it does not describe the use of land, and the use of land may be different for lands with the same cover type. For instance, a land cover type of forest may be used for timber production, wildlife management or recreation; it might be private land, a protected watershed or a popular state park. In short, land use indicates how people are using the land, whereas land cover indicates the physical land type. Both types of data are most often obtained from analysis of either satellite or aerial images. Understanding both the land use and land cover of a track of land provides a comprehensive picture of a particular area (Fig 7). This data is a fundamental component of the planning and decision-making processes for many communities because it helps them to understand better where to plan for different types of growth and where to preserve; it also helps them to understand the connectivity or fragmentation of various features in their community.

Causes of Change in Land Use Pattern

Land cover change can be defined as the loss of natural vegetative areas along with water bodies due to urban expansion. Rapidly growing conversion of rural to urban areas for so called development have created a marked effect on the natural functioning of ecosystems including wetlands. wetlands provide numerous ecological goods and services but are under tremendous stress for land use changes due to rapid urbanization, industrialization and agricultural intensification, manifested by the shrinkage in their areal extent, and decline in the hydro-logical, economic and ecological functions they perform. The easy availability and low cost of wetlands tempted promoters and they convince the wetland owners to sell their wetland which are finally filled and transformed into various constructions (Gibbs 2011). Sometimes unplanned urban development are also the reason of wetland degradation (Fig 8).

- ***land use for residential purpose*:** Overpopulation become a great challenge to ensure residential security in the era of urbanization. For limitation of lands, in many cases wetlands are filled and used for flats, hotels etc. Constructions.
- ***land use for transportation purpose*:** As a part of transportation developmental system, the number of rods, railways, ports and air ports are gradually increasing and many of the wetlands are degraded for these reason.
- ***land use for commercial purpose*:** Urbanization leads to raise the number of shops, markets, shopping malls and many other commercial constructions. For limitation of lands, in many cases wetlands are filled and used for these purpose.
- ***land use for industrial purpose*:** Urbanization leads to Industrialization that has expanded the employment opportunities. For limitation of lands, some times wetlands are filled and used for construction of various kinds of industry.
- ***land use for agricultural purpose*:** It has become a great challenge to ensure food security in the era of urbanization because with the raise of population, the need for food also increasing. For limitation of agricultural lands, in many cases wetlands are filled and used for farm purpose.
- ***land use for institutional purpose*:** Urbanization leads to advanced education system like more schools, colleges and universities with modern technologies. For limitation of lands, in many cases wetlands are filled and used for construction of institutions.
- ***land use for medical purpose*:** Urbanization leads to convert wetland into hospitals and clinics to utilize better medical facilities.
- ***land use for other purpose*:** In urban life people can enjoy the recreational facilities such as art galleries, cinema halls, theatre, clubs, parks etc. that may be constructed by filling the wetlands. Wetlands also be lost for mining purpose.

Figure 5. Diagram of threats of wetlands
(Source: sciencedirect.com)

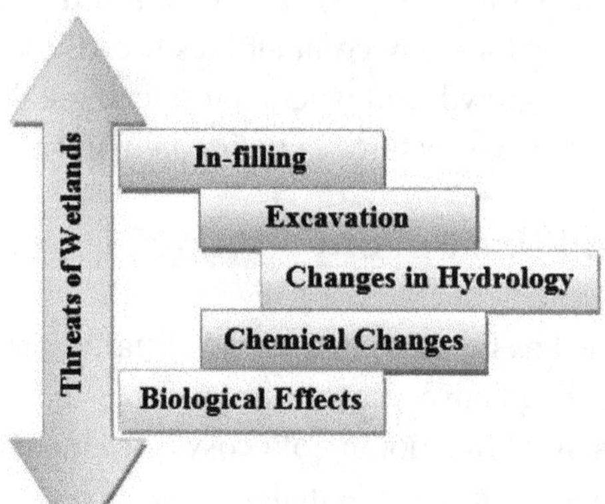

Figure 6. Continent wise distribution of wetlands
(Source: sciencedirect.com)

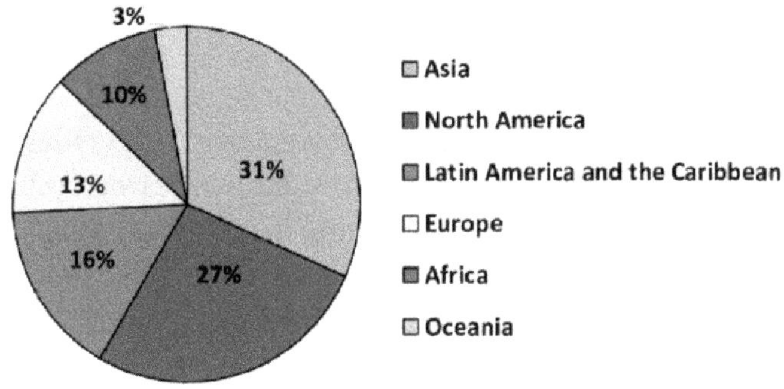

Region	Wetland area (%)
Asia	32
North America	27
Latin America and the Caribbean	16
Europe	13
Africa	10
Oceania	3

Figure 7. Methodological steps of study of land use and land patten change
(Source: semanticscholar.org)

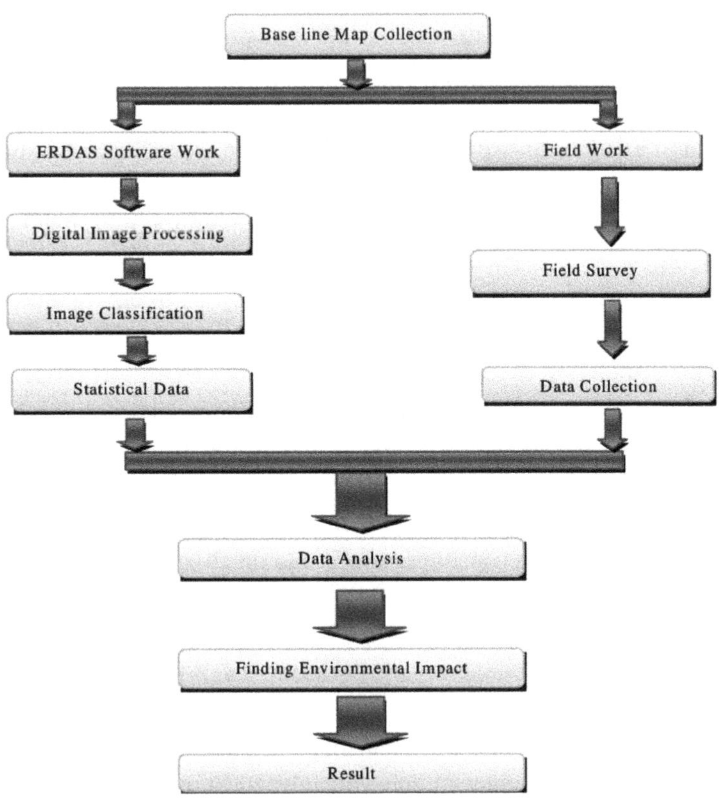

Process of Degrading Wetlands

- ***Ground Water Depletion***: Demand of water for irrigation purpose has increased dramatically in the last 20 years. In the light of the population's needs for water, the water budgets & desired surface flows of water to sustain wetlands must be put in perspective. In between the year 1980 & 1983, the rapid depletion of ground-water left over nearly 6000 villages in rural India with no source of drinking water (Chopra 1985). Simple hand withdrawal systems of water like hand-dug wells did not deplete ground-water significantly in the past and wetlands continued to exists. With the introduction of advent water withdrawal technologies like tube wells and large-capacity pumps for crop irrigation, water tables have been reduced quickly over entire regions, drying wetlands, emptying hand-dug wells, and ultimately pumping themselves dry (Chopra 1985). Water pumping continuously reduces the ground-water levels by 1.5 to 6 m per year in some areas. As a result, the river beds that once flowed continuously are now dry for most of the year. The excessive water extraction is becoming a serious problem through world-wide (Anon. 1991). The southern coastal plains of Saurashtra, over an area exceeding 100000 ha in Gujarat is affected by this phenomenon(Chopra 1985).
- ***Effects of Global Climate Change*** : Climate change contributes a vital role in wetland loss. Wetlands are the single largest global emission source for methane, a gas which is a major contributor to the atmospheric trapping of heat that leads to atmospheric warming. Atmospheric warming leads to drying, desertification and loss of wetlands at local and global scales. Apart from this, the predicted increase in global sea levels are of 10-21 cm by the year 2025 (Hoffman *et al.* 1986). Based on some of the earliest archaeological descriptions of the extensive forests that covered areas now occupied by plains and deserts for such climate change, occurs over decades or centuries. India was a much wetter region 2700 years ago (Tharpar 1963). Entire regions of India's land surface were altered through long-term human use and manipulation such as intensive cropping, deforestation, intensive grazing, and alteration of water flows. The population of India has been high enough, unlike most countries of the world to cause such land use changes. Relatively large populations influence to accumulate the incremental effects on wetlands over millennia. Warming and drying cycle of global change leads to the continued degradation of water and wetland resources.
- ***Degradation of water quality***: Diminution of water quality is directly related to human population and industry. Two thirds of the 98 metropolitan areas in the world including Bombay, Calcutta and Madras with a population in excess of two million, began as port cities or are within 100 km of the port area. These ports modify natural coastal wetlands and pollute the waters of enclosed bays and coastal lagoons to the extent that the potential for fisheries and recreational uses is significantly degraded or eliminated (Sorensen 1993). Coastal lagoons which have conjunction with urban runoff, exchange polluted water with the open ocean (Packer 1984). More than 50000 small and large Indian lakes are polluted (Chopra 1985) and the primary sources of pollution are human sewage, industrial pollution and agricultural runoff which may contain pesticides, fertilizers and herbicides. Gopal (1994) observed that 'about 70% of the 3100 towns in India have neither sewerage nor treatment facilities.' Extreme organic pollution levels from sewage can be put to constructive use in some situations. The wetland sewage system in Calcutta is an excellent example of a beneficial use of wetlands. Fish like carp and tilapia are introduced and raised after a brief algal bloom. This sewage fishery produces approximately 7000 ton of fish annually for the Calcutta

markets or 2.81 per ha per year (Edwards 1985). Solids in the sewage outfall and pond bottom sediments are removed and used as fertilizer for nearby vegetable farms. Highly nutrient rich soils allow farmers to produce nearly 30% of the vegetables supplied to Calcutta. The unrestricted dumping of industrial wastes into the sewage stream leads to toxic chemicals accumulation, causing a loss of fish production and native biodiversity loss. Sometimes pollution from mariculture operations in coastal area affects the mangroves and coral reefs. The world's other leading grain producing countries generally use ten times more fertilizer than their Indian counterparts (Brown *et al.* 1991).

- ***Water demands and diversions***: Anthropogenic disturbance and alteration to the hydrological regime is the most important and serious cause of tropical wetland damage (Gopal 1992). Canals and diversions have been constructed to transport water from upper to lower arid regions where crops are grown. Several of the largest water diversion projects in the world are constructed in India. The Indira Gandhi Canal Project removes water from the Sutlej River and diverts it via a canal system to the desert regions of Gujarat and Rajasthan to provide irrigation for cash crops. Some estimation claim that up to 22 million ha m of water could be made available countrywide by ditching and transferring water from waterrich basins into the adjacent low-flow, or fully utilized basins (Chopra 1985). Inter-basin water transfers offer great potential for increases in irrigation, yet associated with them are other ecological problems, such as soil water logging, soil chemical changes, exotic species introductions, more dam construction to provide increased head to divert water, the interruption of migration routes of people and animals, regional desertification and elimination of culturally sustainable lifestyles. Improper uses of water in water diversion projects for irrigation in areas of high evaporation potential and are usually plagued by soil salinization. This incidents are particularly acute where the water table is artificially raised through irrigation, land clearing removes deep-rooted native vegetation, and dams are constructed (Abrol et al. 1988). Continuous surface evaporation draws saline water to the soil surface, where the water evaporates into water vapour and the salt remains as crystals in the soil (Pessarakli 1991). Rivers gradually being isolated from their floodplains leading to lowered water tables, loss of ground-water recharge, human development in flood prone areas and increased flooding potential downstream due to faster flood-water drainage (Gopal & Sah 1993).
- ***Agricultural Conversion*:** As a result of rice cultivation, the gross area of wetlands on the Indian subcontinent are gradually decreased (Verma, 2001). Rice farming, a wetland-dependent activity, has been developed in virtually all suitable floodplains, riparian zones, river deltas and savannah areas in India. Notable rice farm conversion areas include Hokarsar Lake (Kashmir), Depar bil (Assam) and the Pyagpur and Sitadwar Jheels (Uttar Pradesh). Koleru Lake in Andhyra Pradesh has lost 34000 ha of natural wetlands for agricultural purpose (Anon. 1991). Apart from this, as a result of captured precipitation and the diversion of stream flows into catchment areas, aquaculture and rice farming also occupy areas that were not previously wetlands, thereby robbing water from downstream natural wetlands. Approximately 1.6 million ha area are occupied by freshwater fishponds in India (Anon. 1993). A greater portion of rice fields and shallow fish ponds in India are technically wetlands, even though they rarely function like natural wetlands. Though rice fields along with the Ganges plain can produce great quantities of human food but such agricultural wetlands lack many of the other wetland functions that provide benefits to society (Fernando 1977). Specially functions of wetlands such as flood storage capacities, water-quality enhancement, sediment trapping, sewage filtration, wildlife habitat and small-scale landscape diversity are reduced

due to these agricultural systems (Abell et al. 2016). In the dry areas of neighbouring country Sri Lanka, linked complexes of rain catchments, rice paddy fields and buffalo wallows support fish populations that traditional villagers depend upon (Ulluwishewa 1995). Coastal wetlands with mangroves and salt flats have been converted to salt producing ponds in the western coastal zone of Gujarat as well as in Pulicat lake of Tamil Nadu. Salt is extracted from the marine water when sea-water is trapped at high tide, then it is stored in expansive shallow ponds to evaporate the water and concentrate brine to form salt crystals. These systems greatly alter the local resident biodiversity. Although brine ponds are used by some migratory birds like flamingos, ecosystem losses appear in the form of reduced tidal energy and export of decaying plant and animal materials to ocean (de la Cruz 1978; Montague *et al.* 1987). The flow of detritus and soluble soil nutrients initiates a food-chain in natural estuaries where microscopic organisms thrive, shrimps and crabs feed on detritus and microorganisms and the birds, wild mammals & humans feed on the fish. Obstacles of such exchanges clearly impacts the ecological dynamics of the estuary (Savenije & Pages 1992).

- ***Direct deforestation in wetlands***: Generally mangroves are flood and salt tolerant shrubs and trees that grow along the coasts in the tropics. In India, mangrove forests are valued for production of fish and shellfish, livestock fodder, fuel wood, building materials, local medicine, honey and beeswax, and for extracting chemicals used in tanning leather (Costanza et al. 1997). Transformed land-use pattern of farming and fisheries production have replaced many mangrove areas and continue to pose threats to the forests. Eighty percent of India's mangrove forests with area of 4240 km^2 occur in the Sundarbans and the Andaman and Nicobar Islands (Government of India 1991) where 59 species of mangroves occur (Singh 1994). In many coastal areas, mangroves are under severe pressure. The demand for shrimp has provided an economic incentive to clear coastal mangroves and develop shallow, open water shrimp ponds. In the estuaries of Bhitarkanika National Park, the illegal clearing of mangroves for prawn ponds in interior tidal areas constitutes a management and enforcement problem for resource managers. Mangrove regeneration can be relatively rapid because mangrove species sort themselves along hydrologic and salinity gradients according to competitive interactions (Smith 1987). The replacement of an original mangrove species may take a long time if the hydrology has been changed. One estimate is that 85% of India's original mangrove cover has been lost since pre-agricultural times and that only 1894 km^2 remain (Anon. 1991). Whatever the rate, India's mangrove loss is likely to follow the pattern documented by Primavera (1991), who showed that 70% of the Philippines' mangroves have been cleared and replaced by shrimp ponds during the previous 15 years. The production of cultivated shrimp in 1990 was almost seven times the amount measured in 1982 in south-east Asia (Conner & Skladany 1991). Crucial wetlands functions such as buffer zones against storm surges, nursery grounds and escape cover for commercially important fish, sustainable levels of minor forest product extraction, and sediment trapping ability were lost. It besides, shrimp ponds are also required excessive freshwater withdrawal and produced pollution (increased lime, organic wastes, pesticides, chemicals and disease organisms) that was flushed into the remaining mangroves adjoining the ponds. Chilka Lake (Orissa) is the largest brackish lake in the country. The lake is being filled with sediments and the total lake area has been reduced from 860 km2 to 605 km2 in the last 60 years due to the growth of shoreline mud flats and an invasion of wetland and upland plants (Malini *et al.* 1993). The tidal exchange channels are clogged by sediments which formerly connected the lake to the Bay of Bengal. The downstream causes of soil erosion in mountainous areas are silt deposition, severe flooding during the wet seasons and lower than normal water levels in the dry season (Hawting & Mateo 1987). Along with these changes has come a reduction in the fish output from Chilka from 1970 to the present and the impoverishment of the 12250 fishermen who fish there (Das *et al.* 1990). The ecological linkages between land alterations, such as deforestation, and the downstream effects on wetlands, lakes, fisheries and people are apparent. There is a striking set of Indian deforestation estimates: India sustained a 4.1% annual loss of forested area which translates into approximately 1.5 million ha of woodlands per year for the 1975-82 period of study (Anon. 1990). Except for the relatively small Costa Rican forests, India sustained the highest rate of tropical deforestation in the world, primarily through agricultural conversion, fuel wood harvesting and land-uses (Chatterjee, 1995).

- ***Hydro-logic alteration***: Hydro-logic alteration can change in the characters, functions, values and appearance of wetlands. The most extreme changes in hydrology involve changes which either remove water from wetlands or raise the land-surface elevation to the point where it no longer floods. The growth of Salt Lake City in the eastern outskirts of Calcutta, is an example of wetland filling that has converted over four thousand hector of productive wetlands into a filled area that is ready for development. This wetland replacement leads to the loss of 25000 metric tonnes of commercial fish production from Calcutta's markets (Chopra 1985). Erosion of cleared hillsides enhance to eliminate wetlands by filling ponds and river courses and building up river out wash areas as described in Issue 8 in relation to siltation.
- ***Defoliation (overgrazing, biomass removal)***: India has relatively more grazing animals than any other countries in the world. Worldwide, India supports twenty percent of all cattle, fifty percent of all water buffaloes and fifteen percent of all goats (Hulkarni 1970). The foraging activities of these animals change the vegetation and soils of the country, especially since over ten percent of the cattle graze in forests. Intensive stocking of livestock on wetland soils can directly reduce herbage on the site through grazing but it can also reduce the ability of the land to produce vegetation through soil composition and long-term changes in soil structure (Adams & Akhtar 1994), as well as the selective removal of the most palatable plant species, leaving a preponderance of poor forage for wildlife and future grazers.
- ***Inundation by dammed reservoirs***: Recently more than 1550 large reservoirs covering area around 1450000 ha and more than 100000 small and medium reservoirs covering area around 110000 ha are present in India (Gopal 1994). Impediment of water for controlled release is usually carried out in areas that formerly supported river corridor and backwater wetlands. By storing water, the hydrology of an area is significantly changed, providing an opportunity to use moving water as a source of energy. Indian government plans to harness that energy with dams and reservoirs for the production of electrical power in the mountainous regions. Indian civilizations have a long history of damming waterways. Irrigation and the storage of waters behind human-constructed dams have been influential landscape features since the era of Chandragupta. Dam near Ginar in western India was dated by stone inscriptions as having been in constant use for over 800 years (Tharpar 1963). According to the time, impoundments of water influences the change of climate, vegetation, and animal species within the affected region. Though in some points of view, the human-altered landscape has existed with greater constancy than some natural ecological communities. It is a frame of reference that is usually under-appreciated in industrialized countries. Small earthen dams generally used for water harvesting and these are considered ecologically sound with few negative environmental im-

Figure 8. Causes of degradation of wetlands
(Source: link.springer.com)

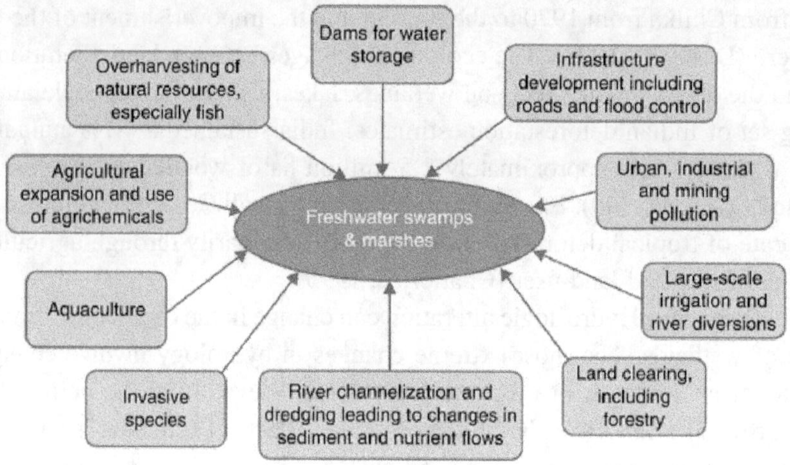

pacts. The potential for shoreline wetlands to develop and mitigate the losses of river bottom and riparian zones is minimal because of the large and erratic alterations of impounded water levels. Variable dam releases for power generation are responsible for the wetted area that does not follow a predictable seasonal pattern and prevention of stable wetland flora and wildlife community development. There are few alleviating beneficial effects from reservoir construction for particular species. Pandey (1993) noticed an increased bird species diversity resulting from the construction of the Pong Dam Reservoir in Himachal Pradesh. There are also some economic benefits like fishing from reservoirs and the availability of standing water in large reservoirs provides the potential for nearby wetland development. Apart from this, zones of deep water will be converted to shallow wetlands because of high sediment loads fill portions of the reservoir. Ultimately the reservoir efficiency will be reduced to marginal benefits and there may be some opportunities for restoration of the flooded areas to emergent wetlands.

NECESSITY OF LAND USE PLANNING FOR WETLAND CONSERVATION

Land use planning is a public policy that encircle various disciplines which seek to order and regulate land use in an efficient and ethical way. It is necessary to overcome the following problems of

- Haphazard, uncontrolled development and destruction of important wetlands.
- Deteriorating environmental quality
- Extinction of local hydrophyte medicinal plants
- Loss of fish and habitat of wetland species
- Gradual reduction of wetland related migratory birds
- Reduction of ground water recharge potential
- Flood and erosion of shoreline

Though land use change and wetland loss occurs at the local scale, but its impacts spread across regional as well as global scales (Foley et al. 2005). Despite the human demands for arable lands, living space, irrigation water, hydroelectric power and wetland materials, India still contains many crucially important wetlands deserving protection and management. There are 170 Indian wetlands covering 580000 km^2 described in the Directory of Indian Wetlands (Anon. 1993). In another listing by Scott and Pole (1989), of 93 key Indian wetlands only 27% of them were protected at 28 sites and 45% of all Indian wetlands are considered moderately to highly threatened. Through passage of laws, fencing, privatization of land for mechanized farming, water developments and commercial forestry, public ac-cess to land and wetlands has been reduced. Jodha (1992) calculated that in semi-arid regions spanning eight Indian states, the area of the commons declined by 30 to 50% between 1950 and 1980, and taking population growth over that span into account, the area per ca-pita declined on average by two thirds. In Indian, gradually increasing population, diminished quality off wetlands and decreasing access to wetland functions is an example of 'Tragedy of the commons'. The Indian Ministry of Environment and Forests (MOEF) provides financial assistance to the states for carrying out activities related to conservation and management of wetlands. The MOEF has also identified key research institutions to provide scientific

Figure 9. Distribution of Ramsar sites in India
(Source: researchgate.net)

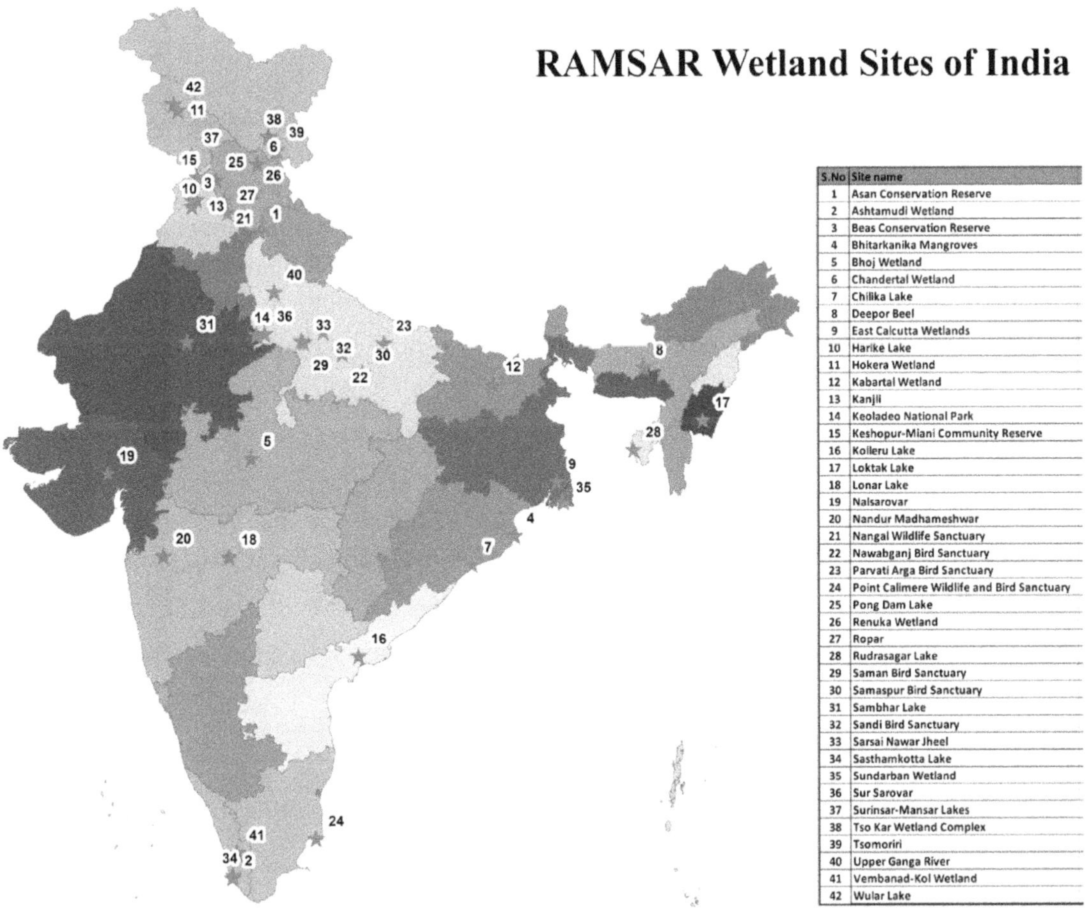

S.No	Site name
1	Asan Conservation Reserve
2	Ashtamudi Wetland
3	Beas Conservation Reserve
4	Bhitarkanika Mangroves
5	Bhoj Wetland
6	Chandertal Wetland
7	Chilika Lake
8	Deepor Beel
9	East Calcutta Wetlands
10	Harike Lake
11	Hokera Wetland
12	Kabartal Wetland
13	Kanjli
14	Keoladeo National Park
15	Keshopur-Miani Community Reserve
16	Kolleru Lake
17	Loktak Lake
18	Lonar Lake
19	Nalsarovar
20	Nandur Madhameshwar
21	Nangal Wildlife Sanctuary
22	Nawabganj Bird Sanctuary
23	Parvati Arga Bird Sanctuary
24	Point Calimere Wildlife and Bird Sanctuary
25	Pong Dam Lake
26	Renuka Wetland
27	Ropar
28	Rudrasagar Lake
29	Saman Bird Sanctuary
30	Samaspur Bird Sanctuary
31	Sambhar Lake
32	Sandi Bird Sanctuary
33	Sarsai Nawar Jheel
34	Sasthamkotta Lake
35	Sundarban Wetland
36	Sur Sarovar
37	Surinsar-Mansar Lakes
38	Tso Kar Wetland Complex
39	Tsomoriri
40	Upper Ganga River
41	Vembanad-Kol Wetland
42	Wular Lake

and technical inputs. The need for research and policy specialists is apparent in wetland issues. Enhanced mass awareness of various environmental issues, specially the growing concern for the consequences of uncontrolled land use change has motivated the establishment of sustainable land use policies in Asian countries like Philippines, China, Mongolia and Japan (Boyer 2000). In 1998, Chinese government announced a wetland restoration program right after the catastrophic flood that grabbed about 9770 km^2 of land. For example, Central Yangtze River project was successfully implemented through growing hydrophytes with high economic value, raising livestock on the seasonal grasslands, planting commercial seasonal vegetables, developing aquaculture in the paddy field and encouraging ecotourism (Liu et al. 2004). Such type of innovative sustainable land use strategies are crucial not only for the environmental soundness, economic viability and social acceptability of the countries in Asia, but also for global sustainable development. We end this discussion with three important points. Firstly, we follow Singh (1995) who points out 'The profound crisis of population growth in India is being trivialized by the bulk of society. Ecologists have to tell all concerned people about this, before they recommend doing research on sustainability of natural resource use.' Our second message is that wetlands provide basic, sustainable benefits to people throughout India and they deserve protection and restoration. Last but not the least, mistakes are inevitable when conservation attempts are undertaken because our knowledge of wetland ecosystems is so incomplete. The awareness of wetland values and losses puts the consequences of failed conservation efforts in perspective and calls for renewed attempts with more effective methods.

CONCLUSION

From the point of local as well as global environmental conservation, Wetlands are very important ecosystems. In most cases, human encroachment and unplanned developmental activities are accelerating the degradation process of wetlands. Conservation of wetlands should be given a prime importance among all the environmental conservation policies. Mass consciousness, greater participation of local people, use of indigenous knowledge in the management strategies and application of scientific techniques are needed to protect wetlands. Most importantly we must be sincere enough to increase awareness among public regarding the importance of wetlands and its biodiversity conservation.

REFERENCES

Abell, R., Lehner, B., Thieme, M., & Linke, S. (2016). Looking beyond the fenceline: Assessing protection gaps for the world's rivers. *Conservation Letters*. https://onlinelibrary.wiley.com/doi/10.1111/conl.12312/full

Abrol, A. P., Yadav, J. S. D., & Massoud, F. I. (1988). Salt-affected soils and their management. FAO Soils Bulletin, 39.

Adams, W. A., & Akhtar, N. (1994). The possible consequences for herbage growth of waterlogging compacted pasture soils. *Plant and Soil*, *162*(1), 1–17. doi:10.1007/BF01416085

Anon. (1990). *World Resources 1990-91*. Oxford University Press.

Anon. (1991). *World Resources 1991-92*. Oxford University Press.

Anon. (1993). *Directory of Indian Wetlands*. World Wildlife Federation.

Boyer, B. (2000). Institutional mechanisms for sustainable development: A look at national councils for sustainable development in Asia. *Glob Environment Change-Hum Policy Dimens, 10*(2), 157–160. doi:10.1016/S0959-3780(00)00019-4

Brown, L., Flavinc, Q., & Pootel, S. (1991). *State of the World. A World Watch Institute report on progress towards a sustainable society*. World Watch Institute India.

Chatrath, K. J. S. (1992). *Wetlands of India – Major threats faced by wetlands of India and their conservative measures*. Ashish Publication House.

Chatterjee, N. (1995). Social forestry in environmentally degraded regions of India: Case-study of the Mayurakshi Basin. *Environmental Conservation, 22*, 20–30.

Chopra, R. (1985). *The State of India's Environment*. Ambassador Press.

Conner, B., & Skladany, M. (1991). Aquacultural development in tropical Asia: A re-evaluation. *Natural Resources Forum, 15*(1), 66–73. doi:10.1111/j.1477-8947.1991.tb00111.x

Costanza, R., d'Arge, R., de Groot, R., Farber, S., Grasso, M., Hannon, B., Limburg, K., Naeem, S., O'Neill, R. V., Paruelo, J., Raskin, R. G., Sutton, P., & van den Belt, M. (1997). The value of the world's ecosystem services and natural capital. *Nature, 387*(6630), 253–260. doi:10.1038/387253a0

Das, H. S., Mishra, R. K., & Sinha, B. N. (1990). Some observations on Chilka wetland ecosystem. In B. N. Sinha (Ed.), *Ecosystem degradation in India* (pp. 85–93). Ashis Publishing House.

de la Cruz, A. A. (1978). Production and transport of detritus in wetlands. In *Wetland Functions and Values: The State of our Understanding*. American Water Resources.

Edwards, P. (1985). *Aquaculture: A Component of Low-Cost Sanitation Technology*. The World Bank.

Fernando, C. H. (1977). Investigations on the aquatic fauna of tropical ricefields with special reference to southeast Asia. *Geo-Eco-Trop, 1*, 166–188.

Foley, J. A., DeFries, R., Asner, G. P., Barford, C., Bonan, G., Carpenter, S. R., Chapin, F. S., Coe, M. T., Daily, G. C., Gibbs, H. K., Helkowski, J. H., Holloway, T., Howard, E. A., Kucharik, C. J., Monfreda, C., Patz, J. A., Prentice, I. C., Ramankutty, N., & Snyder, P. K. (2005). Global consequences of land use. *Science, 309*(5734), 570–574. doi:10.1126cience.1111772 PMID:16040698

Ghermandi, A., van den Bergh, J. C. J. M., Brander, L. M., & Nunes, P. A. L. D. (2008). *The Economic Value of Wetland Conservation and Creation: A Meta-Analysis* [Working Paper 79]. Fondazione Eni Enrico Mattei.

Gibbs, J. P. (2011). Wetland loss and biodiversity conservation. *Conservation Biology, 14*(1), 314–317. doi:10.1046/j.1523-1739.2000.98608.x

Gopal, B. (1992) Tropical wetlands: degradation and need for rehabilitation. In Ecosystem Rehabilitation Vol. 2: Ecosystem analysis and synthesis. Academic Publishing.

Gopal, B. (1994) Conservation of inland waters in India: an overview. *Verhandlungen der Internationalen Vereinigung fiir Theoretische und Angewandte Limndlogie, 25*, 2494-7.

Gopal, B. (2013). Future of wetlands in tropical and subtropical Asia, especially in the face of climate change. *Aquatic Sciences, 75*(1), 39–61. doi:10.100700027-011-0247-y

Gopal, B., & Sah, M. (1993). Conservation and management of rivers in India: Case-study of the River Yamuna. *Environmental Conservation, 20*(3), 243–254. doi:10.1017/S0376892900023031

Government of India. (1991). *India 1990: A Reference Annual*. Research and Reference Division, Ministry of Information and Broadcasting, Government of India.

Hawting, G., & Mateo, N. (1987). Farming on high International Development Research Center, Markham, Ontario. *Report, 16*(1), 4–5.

Hoffman, J. S., Wells, J. B., & Titus, J. G. (1986). Future global warming and sea level rise. In J. G. Titus (Ed.), Greenhouse effect, sea level rise and coastal wetlands. US Government Printing Office, EPA-230-05-86-0I3.

Hulkarni, D. H. (1970). *Forestry in India*. Oregon State University Printing Dept.

Jodha, N. S. (1992). *Common Property Resources ~ A Missing Dimension of Development Strategies*. Report No. 169. The World Bank.

Keddy, P. A. (2010). *Wetland Ecology: Principles and Conservation*. Cambridge University Press. doi:10.1017/CBO9780511778179

Liu, X. P., Kelin, W., & Geli, Z. (2004). Perspectives and policies: Ecological industry substitutes in wetland restoration of the Middle Yangtze. *Wetlands, 24*, 633–641.

Malini, B. H., Rao, K. S., & Rao, K. N. (1993) Evolution and dynamics of the Chilka Lake. In *Proceedings of the eighth symposium on Coastal and Ocean Management*. American Society of Civil Engineers.

Marty, J. T. (2005). Effects of cattle grazing on diversity in ephemeral wetlands. *Conservation Biology, 19*(5), 1626–1632.

MOEF. (1990). Wetlands of India - A directory. Ministry of Environment and Forests, Govt. of India.

Montague, C. L., Zale, A. V., & Percival, H. F. (1987). Ecological effects of coastal marsh impoundments: A review. *Environmental Management, 11*(6), 743–756.

O'Neill, B. J., Rogers, D. C., & Thorp, J. H. (2016). Flexibility of ephemeral wetland crustaceans: Environmental constraints and anthropogenic impacts. *Wetlands Ecology and Management, 24*(3), 279–291.

Pandey, S. (1993). Changes in waterbird diversity due to the construction of Pong Dam Reservoir, Himachal Pradesh, India. *Biological Conservation, 66*, 125–130.

Pessarakli, M. (1991). Formation of saline and sodic soils and their reclamation. *Journal of Environmental Science and Health. Part A, Environmental Science and Engineering & Toxic and Hazardous Substance Control, 26*(7), 1303–1307.

Pittock Jet al. (2015). Managing freshwater, river, wetland and estuarine protected areas. 569–608. In G. L. Worboys, M. Lockwood, A. Kothari, S. Feary, & I. Pulsford (Eds.), *Protected Area Governance and Management*. ANU Press.

Primavera, J. H. (1991). Intensive prawn farming in the Philippines: Ecological, social, and economic implications. *Ambio*, *20*, 28–33.

Ramsar Secretariat. (2013). *The List of Wetlands of International Importance*. The Secretariat of the Convention on Wetlands.

Savenije, H. H. G., & Pages, J. (1992). Hypersalinity: A dramatic change in the hydrology of Sahelian estuaries. *Journal of Hydrology (Amsterdam)*, *135*, 157–174.

Scott, D. A., & Pole, C. M. (1989). *A Status Overview of Asian Wetlands*. Asian Wetlands Bureau.

Singh, B. P. (1994). Biodiversity conservation in mangrove environment. *Indian Forester*, *120*, 827–836.

Singh, S. P. (1995). India's population problem, a global issue of environment and sustainability: An appeal for cooperation. *Bulletin of the Ecological Society of America*, *76*, 55–59.

Smith, T.J., HI. (1987). Effects of light and intertidal position on seedling survival and growth in tropical tidal forests. *Journal of Experimental Marine Biology and Ecology*, *110*, 133–146.

Sorensen, J. (1993). The management of enclosed coastal waterbodies: the need for a framework for international information exchange. In *Proceedings of the Eighth Symposium on Coastal and Ocean Management*. American Society of Civil Engineers.

Tharpar, R. (1963). *A History of India* (Vol. 1). Penguin Books.

Ulluwishewa, R. (1995). Traditional practices of inland fishery resources management in the dry zone of Sri Lanka: Implications for sustainability. *Environmental Conservation*, *22*, 127–3.

Verma, M. (2001). *Economic Valuation of Bhoj Wetlands for Sustainable Use* [EERC Working Paper Series: WB-9]. Indian Institute of Forest Management.

Zedler, J. B., & Kercher, S. (2005). Wetland resources: Status, trends, ecosystem services, and restorability. *Annual Review of Environment and Resources*, *30*, 39–74.

Chapter 8
Ecological Succession of Wetlands:
A Review of the Current Scenario

Dwaipayan Sinha
https://orcid.org/0000-0001-7870-8998
Government General Degree College, Mohanpur, India

Soumi Datta
https://orcid.org/0000-0003-0360-4910
Jamia Hamdard, India

Vidhi Chaudhary
Daulat Ram College, University of Delhi, India

Debolina Banerjee
Kidderpore College, India

Rituparna Kundu Chaudhuri
Krishnagar Government College, India

ABSTRACT

The change in community structure induced by environmental perturbances over time is called ecological succession and results in modification of community structure in a particular ecosystem. However, with climatic changes and pollution, wetlands are most likely to have modified their successional trails, especially as these ecosystems are exposed to various fluctuating water rhythms enhanced carbon dioxide, salinity invasion, and climatic temperatures. The chapter is an attempt to review the present condition of wetlands with respect to pollution and its impact on successional patterns.

DOI: 10.4018/978-1-7998-9498-8.ch008

BACKGROUND

Wetlands are regions in which land either remains covered by water, or hydrated throughout the year or for varying periods. They are characterized by varied life forms. Wetlands are intricately related to human civilization down the ages with major ancient settlements growing near a sustainable water source in the form of river or adjoining wetlands. The Mesopotamian civilization is recorded as the oldest civilization established 4000 years ago in the fertile region between Tigris and Euphrates (Morrisson et al., 2021). In Nile Valley, another ancestral civilization emerged in 3200 BC indicating the importance of water. The inhabitants of these regions are evidence of significance for biotic diversity, and thus prove that the wetlands are a very crucial ecosystem and are the most fertile habitats.The sunderbans form a distinctive wetland in the gangetic delta situated in the present day India and Bangladesh. They form the largest contiguous forested wetland in he world and covers about 10,000 square kilometres. Aside from the multilayered forest, the Sundarban has freshwater, brackish, and saline marshes, rice fields, and shrimp farms. The Sundarbans, which are extraordinarily rich in biodiversity, are the Royal Bengal tiger's sole known mangrove habitat (Gopal and Chauhan, 2018). A continual supply of sediment from the Mekong River keeps the Mekong Delta, one of South East Asia's Cenozoic river deltas, afloat. Natural inland wetlands constitute roughly 7% of the total area of the Vietnamese Mekong Delta (VMD), and are mostly restricted to regions of closed floodplains and back swamps due to broad-scale land use change (Quan et al., 2018). The Vietnamese Mekong Delta has a wide variety of wetlands, ranging from coastal systems like tidal flats, sandy ridges, and tidal back swamps to river floodplain areas like large depressions, peat swamps, and alluvial levees (Dinh, 2016). The Mahanadi delta also forms a wetland in Odhisha. Wetlands largely in the form of small fresh waterbodies evolved as depressions on both flanks of the Mahanadi and Brahmani rivers in the upper delta region, which has a fanning distributary system with intervening alluvial plains. Wetlands typically exist in the form of floodplains linked to active river channels as well as paleo-channels in the central delta, which has primarily clayey soil. Transient wetlands in the alluvial floodplain zones are also found in the central delta. Water and sediment distribution patterns within the delta have been considerably affected by extensive hydrological fragmentation and landuse changes, as well as a loss of connectivity between river channels and floodplains. Between 1975 and 2010, the wetland area decreased by at least 32% due to a variety of stresses (Kumar and Patnaik, 2016).The wetlands are different from other terrain or marine forms concerning their typical foliage of water flora (Butler, 2010) and their adaptation to distinctive hydric soil. They are formed when the water is in proximity to the land surface or where the land is covered by water (Ramsar Convention, 2018) and these are an exceptional ecosystem whose conservation is a global concern; and thus formed the Ramsar International Convention in 1971.

Wetlands imitate diverse behavioral responses to the surroundings since they are widely dispersed from polar to arid zones, peatlands, rivers coral reefs, deltas, lakes, rivers, mangroves, marshes, overflowing forests, and are intermediary regions between marine and terrain environments. Thus, when wetlands are classified, differences occur. Depending upon the water quantity, they are categorized into permanent or seasonally flooded; while construction style divides wetland into natural and man-made. However, factors like physical, chemical, habitat differences, topography, abiotic and biotic surroundings, and quality of sediment decide the role and classification of the wetlands (Finlayson and Spiers, 1999). Wetlands play a pivotal role in hydrology, distillation of water, water retention, dispensation of carbon and other nutrients, shoreline stabilization, and sustain an ecosystem. The performance of the function depends on specific features of that particular wetland, terrain, and aquatic bodies close to it (Dorney et al., 2018). Wetland loss occurs for a variety of reasons, namely urbanization, pollution, reduced biotic diversification, invasive species, terrain deprivation, and changes in climate. Multidisciplinary perception in basic and social sciences has proved effective for the evaluation of wetlands. The management reforms and its implementations of their capacity to absorb the majority of atmospheric carbon are considered to be approximately more than five times as compared to tropical forests; thus wetlands pave their way in betterment of future climate (Davidson, 2014; Sizo et al., 2015; Barros and Albernaz, 2014; Meng et al., 2016).

SUCCESSIONAL PATTERNS IN WETLANDS-CONCEPTS

The eventual alteration of foliage coerced by turbulence and the senescence of flora is referred to as *ecological succession*. The changes of structure and patterns of wetland or any ecosystem are also known as "*Succession*". These changes occur due to natural or anthropogenic disturbances, the appearance of new species, population changes, or sometimes the death or extinction of species gradually with time. Plant ecologists Frederic E. Clements and Henry A. Gleason of the early twentieth century provide a classic case of different approaches to explaining vegetation, with Clements allegedly committed to a law of vegetation despite numerous exceptions, and Gleason denying the law in favour of less organised phenomena (Eliot, 2007). The ''Clementsian'' school of thought implies that the change in vegetation through several stages represents linear progression towards a climatic state, and each stage is encompassed of distinct interlocked species (Middleton, 2016). However, the ''Gleasonian'' theory states that species within the communities are more autonomous of each other, and are not a result of the successive replacement of related communities. A seminal example of Gleasonian theory was reported by van der Valk (2005). in the Prairie wetlands (North America), where the succession advances cyclically, alternating with flood and drought conditions, and thereby alters particular species, to characteristically respond to behaviour and characteristic of water and other instabilities force transformation in environmental sieve model. The assumption that lakes and bogs reflect the initial periods of succession, followed by terrestrialization, is redundant. As ecosystems are open to diverse climatic variations like inundation, famine, salinity infringement, and elevated CO_2 as a result of change in climate, the wetlands possibly have modified successional trails.

If we consider the pattern of succession, then there are two types- primary succession and secondary succession (Figure 1).

Figure 1. Primary and secondary succession

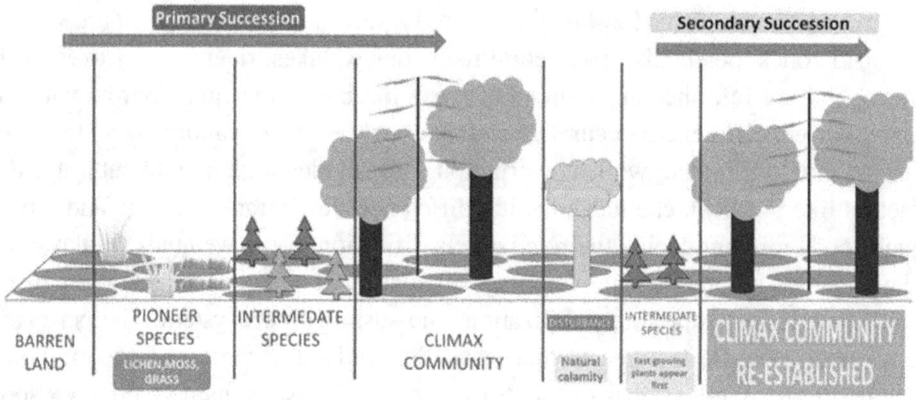

When a simple bare rocky place or a pond or a hardened lava floor turns slowly over 1000 years into useful land then it is termed as primary succession. The factors of primary succession are volcanos, earthquakes, landslides, etc. The species which first started occupying the bare land or clear watery pond is called the pioneer species. By the species decomposition, it creates nutrients in the soil which helps to grow or to colonize other species gradually.

When the changes happen on a prevailing ecosystem due to some disturbances (storms, fires, logging, anthropogenic factors), it is called secondary succession. The origin of disturbances may be biological or non-biological. Sometimes flood carries some species in the ecosystem to develop a stable community (Moseley, 2018). Allogenic (external) influences or autogenic (intrinsic, mostly biotic) features of ecosystem components can induce successional shifts (Burton et al., 2020). Direct or indirect competition for habitat area, as well as changes in the environment as a result of colonisation, and stochastic factors like ecological drift, dispersal etc all play a role in autogenic succession (Jurburg et al., 2017). In contrast, allogenic succession occurs by external environmental abiotic factors such as erosion, leaching, climate, etc (Hastings et al., 2014).

FACTORS AFFECTING SUCCESSION IN A WETLAND

The plant succession is the gradual replacement of one plant community with the other community. The successional pattern is a complicated process and is governed by environmental and biological parameters. The changes in vegetation can be caused by the effects of autogenic and allogenic succession. In the Autogenic succession, the soil and vegetation structure is changed by the effects of organisms that grow on that place and allogenic succession occurs by external environmental factors like erosion, leaching, climate, etc. At present, pollution is rapidly engulfing the wetlands thereby causing detrimental effects on the aquatic species (Xu et al., 2019; Smalling et al., 2019; Tornabene et al., 2020). Thus, pollution also has an effect on the succession of plant communities in a wetland (Figure 2). In this section selected parameters that govern the succession and plant communities (including microbes) in wetlands would be discussed.

1. **Heavy metals and nutrient enrichment:** Heavy metals(HMs) pose a significant harm to the environment and have become a matter of global concern due to their continuous emission from industries (Meena et al., 2018).Plants behave in a varied ways in response to HMs. Thy have come up with a number of mechanisms to deal with the toxicity. The most straightforward technique is to inhibit metal uptake from the soil or to exclude it entirely, therefore preventing metal transfer into

Figure 2. Factors affecting succession in wetlands

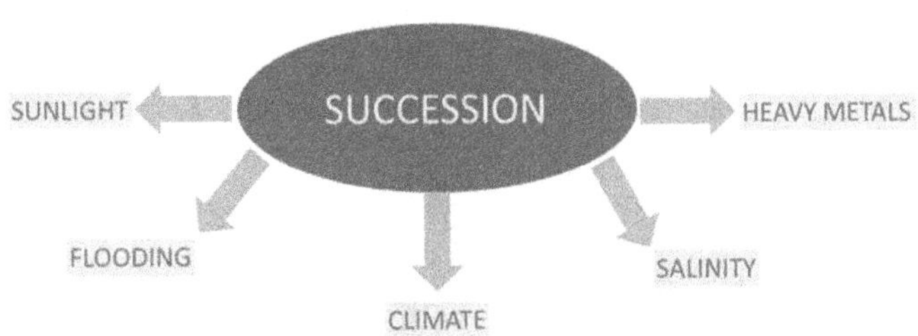

Figure 3. Strategies to combat heavy metal stress by a plant

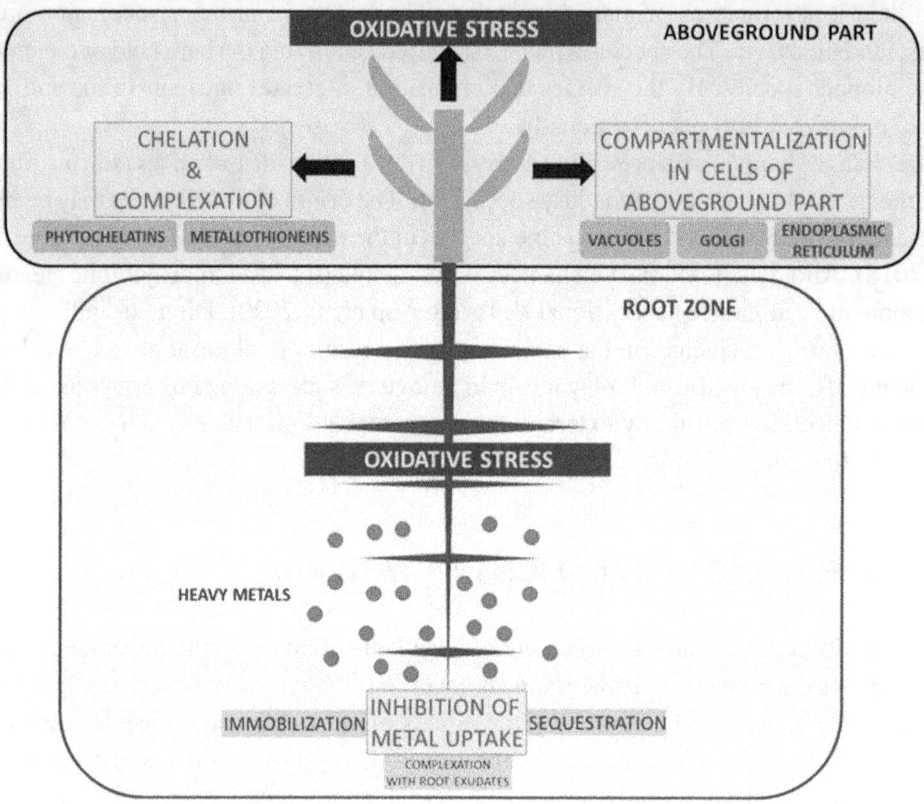

the shoots (Viehweger,2014) . This can be accomplished through a variety of processes, including metal immobilisation through mycorrhizal association, metal sequestration, or complexation by the exudation of organic molecules from the root (Emamverdian et al., 2015). If these measures fail and HMs manage to reach plant tissues, tolerance detoxifying mechanisms such as metal sequestration and compartmentalization in distinct intracellular compartments are activated (Sharma et al.,2016) . Additionally, trafficking pf heavy metals, binding of heavy metals to cell wall, biosynthesis or buildup of osmolytes and osmoprotectants, such as proline, intracellular complexation or chelation of metal ions by organic acids, polysaccharides, phytochelatins, and metallothioneins also helps to counteract the HM toxicity in plants. Finally oxidative stress builds up in case the HM concentration is too high and other defence mechanisms turn out to be ineffective (Emamverdian et al., 2015) (Figure 3).

The fresh water ecosystem bears the brunt of anthropogenic pollution to a large extent (Häder et al., 2020). The water bodies are often clogged with pollutants emitted from households and industries which results in serious consequences (Hasan et al., 2019; Ferronato and Torretta, 2019). The contamination of HMs in waterbodies has long-lasting impacts on the biotic components of the ecosystem. The HMs have prolonged persistence, tend to bioaccumulate in the organisms, and ultimately magnify in the food chain (Bonanno et al., 2017; Wang et al., 2019; Nkwunonwo et al., 2020). Furthermore, HMs are non-biodegradable and continue to accumulate in waterbodies., and ultimately accumulate in the sediments

(Kumar et al., 2019). This results in toxicity amongst the microbial, floral and faunal members (Zeng and Wu, 2013; Farkhondeh et al., 2020). Heavy metals have a high impact on the microbial communities that live in the sediments of water bodies.. A study looked at the impact of HMs on microbial communities in Huangjinxia Reservoir sediments.. It was observed that HMs, namely chromium and manganese negatively affected the alpha diversity of microbial communities. In addition, chromium, lead, and zinc exhibited uniform negative associations concerning the abundance of *Nitrospirae*, *Bacteroidetes,* and *Verrucomicrobia* (Li et al., 2020). A recent study reports the variation of microbial communities in Qingliu river sediments flowing through the urbanized location of Chuzhou city in China. It was found that the microbial communities showed remarkable variations and were related to the enrichment of a particular zone. It was also observed that family Rhodocyclaceae and genus *Dechloromonas* exhibited high relative abundance whereas family Sphingomonadaceae and *Sphingomonas* exhibited a low relative abundance. In addition, correlation with environmental factors and composition of bacterial communities revealed a positive correlation between pH, cadmium with Actinobacteria, and Acidobacteria. On the other hand, *Proteobacteria* and *Chloroflexi* were negatively connected with Copper and Zinc (Zhang et al., 2020). The heavy metal pollution accompanied by the over-enrichment of waterbodies also results in eutrophication (Guo et al., 2020; Vasiliu et al., 2020). Studies indicate that in lakes that are characterized by eutrophication and reduced nitrogen to phosphorus ratio, cyanobacteria form the dominant community (Bravo-Inclan et al.,2008). Similar reports were also obtained from earlier studies where prolonged nutrient enrichment resulted in increased thickness of anoxic hypolimnion in lake Atezca of Mexico and chlorophycean members were replaced by cyanophycean members, especially *Microcystis aeruginosa* (Díaz-Pardo et al., 1998). The dominance of *Anabaena flos-aquae* have also been reported in lake Chapala, Mexico in response to nutrient enrichment. The vast amounts of anthropogenic waste which reach the lake through the Lerma River, enhanced phytoplankton population (deAnda and Shear, 2001). Cyanobacterial blooms with dominance of *Aphanizomenon gracile*, *Microcystis aeruginosa, Microcystis pulvera* and *Anabaena affinis* were also observed in Lake Pátzcuaro, Michoacán State, Mexico (Tomasini-Ortiz et al., 2012). The high phosphate levels accompanied by low nitrogen to phosphorus ratio resulted in cyanobacterial bloom in Lake Atitlan of Guatemala. Various species of *Lyngbya* constituted the members in cyanobacterial blooms (Rejmánková et al., 2011). The effect of arsenic on diatoms of the fresh water ecosystem is also explored in lakes of Yunnan, China. It was observed that populations of arsenic resistant species of diatom namely *Achnanthidium minutissimum* increased while the sensitive representatives namely *Fragilaria construens* and *Fragilaria crotonensis* were reduced (Chen et al., 2015).

2. **Salinity:** The salinity of the waterbodies influences the structure of the plant community.. Anthropogenic activities account for an upsurge of salinity in the waterbodies. Inorganic salts and other pollutants are commonly found in saline and/or alkaline wastewater, which may also have a high pH. Agriculture, aquaculture, and other industrial sections are the main sources of this type of effluent (Cheng et al., 2020). The direct discharge of saline and/or alkaline wastewater has had a deleterious influence on receiving water bodies due to the presence of both salt and other contaminants, which destroys aquatic ecosystems (Cheng et al., 2020). Increased salinity results in osmotic stress in the submerged aquatic plants. They also affect growth and reproductive patterns (Tootoonchi and Gettys, 2019). The salinity of inland waterbodies may also arise naturally. In arid regions where there is no option of outflow of water, evaporation results in a concentration of salts in the waterbodies and ultimately results in the formation of salt lakes. Classic examples of such conditions are the Great salt lakes, Salton Sea and Mono Lake in USA, and the Dead Sea

in Israel. In a study on Australian wetlands, it was observed that *Nitella* spp. exhibited a reduction in germination at salt concentrations between 300 to 1000 mg per liter. A similar response was also observed in the case of *Lythrumhys sopifolia* while *Juncus* spp. exhibited tolerance towards salinity (Nielsen et al., 2003). In another study on Ejina Desert Oasis of China, it was found that *Populus euphratica, Tamarix chinensis,* and *Phragmites communis* were main plant representatves. It was also observed that the distribution of plant communities is governed by the salinity to a large extent in addition to distance from river and ground water parameters (Xi et al., 2016). In another study, it was observed that high salt concentration increased in phytoplankton. It was further observed that the biomass of microalgae *Nitella* decreased with an increase in salt concentration (Lind et al., 2018). In another report, it was observed that salinity was a strong determining factor in influencing species richness of phytoplanktons of Great Salt Lake in USA. The results indicate that *Dunaliella* spp. was greatly influenced by salinity and nutrient enrichment and its abundance was directly proportional to the mentioned parameters (Larson et al., 2013). Another study found that *Arthrospira fusiformis* and *Picocystis salinarum* coexist in the salty lake of Dziani Dzaha on the Indian Ocean's Mayotte Island. (Bernard et al., 2019).Salinity also have a direct effect on the microbial communities that reside in the soil of the wetlands. Soil microorganisms are considered sensitive environmental indicators because of their huge surface area to volume ratio, porous cellular membrane, and rapid turnover rates, which allow them to respond quickly to changes in nutrient availability (Chambers et al., 2016).A study on coastal estuarine wetland ecosystem showed an increase in abundance *Gammaproteobacteria, Bacteroidetes* and *Firmicutes* with increase in salinity while *Acidobacteria, Chloroflexi* and *Cyanobacteria* were more prevalent in regions having low salinity. Moreover, Salinity had a negative correlation with the abundance of most functional genes involved in carbon decomposition and nitrogen cycle (Zhang et al., 2021).

3. **Sunlight:** The species makeup of a region is heavily influenced by the amount of available sunshine. (Gossner et al., 2009; Dormann et al., 2020). Sunlight directly affects the photosynthesis of the plant and affects primary productivity (Tanabe et al., 2019). Photosynthesis rates are influenced by the intensity and spectrum quality of light, which are associated to the water's optical properties and depth in the aquatic environment, as well as the rate of circulation of phytoplanktonin in the water column (Pierson, 2012). The phytoplanktons in the waterbodies are directly linked to primary productivity (Dokulil and Qian., 2021; Li et al., 2020). It is evident from certain reports that illumination and shading affect the species diversity in a water body. One of the reports states that increase in illumination results in an increase in biomass of filamentous algae which resulted in an overall increase in the productivity of the aquatic body. This in turn shifted the invertebrate community away from mosquitoes which are filter feeders and toward anuran tadpoles which graze on algae (Mokany et al., 2008). In another study, five species of *Carex* and four types of grasses namely *Agrostis canina, Anthoxanthum odoratum, Holcus lanatus, Molinia caerulea* were subjected to artificial shading in a field experiment spanning three years. It was observed that there is a dominance of grass species in the treatment plots whereas *Carex* dominated in the control plots. With time, shading reduced the biomass of *Carex* species and also reduced the species diversity. (Edelkraut and Güsewell, 2006). The influence of *Juncus effusus* on plant species richness of wetlands was studied. It was reported that plots that contained larger and heavily shaded tussocks resulted in lesser species richness in the surrounding plant communities as compared to less shaded plots. The species diversity and richness increased by artificial reduction of shading by *Juncus* sp. which also correlated with an improvement in aerial cover (Ervin & Wetzel, 2002).

4. **Flooding:** Wetland plant community structures and functions are affected by variations in flooding regimes caused by natural and human activity. Flooding is a key factor in shaping the plant community composition across many spatial scales namely at the landscape scale, inside habitat spots, across larger assortments, and across floodplains (Richards et al., 2020). The local performance of different species at the patch scale depends on tolerances to the environment and the capability to contest against one another (Richards et al., 2020). The performance of a species is influenced by both competitions with other species and environmental tolerances because the balance of competition between species is influenced by environmental factors (Tootoonchi and Gettys, 2019). It is also observed that the species which are competitive under drought condition may be adversely affected by flooding as they are unable to tolerate root aeration stress (Richards et al., 2020). Thus during flooding, plant species that are more tolerant to flood can thrive and those who cannot withstand water logging may be temporarily eliminated. Several studies have been done to ascertain the effects of flooding on plant communities. A study was conducted to assess the responses of 3 macrophytes namely *Carex brevicuspis, Miscanthus sacchariflorus*, and *Polygonum hydropiper* growing in Dongting Lake wetlands in China during flooding. According to the findings, flooding had a greater impact on *Polygonum* and *Carex* community qualities than it did on Miscanthus community traits. It was further observed that the biomass of the *Miscanthus* community decreased due to shorter flooding periods, while the *Carex* and *Polygonum* communities' biomass rose.. During the earlier flood recession, the species richness and Shannon index of a diversity of the *Polygonum* and *Miscanthus* communities improved., but those of the *Carex* community decreased (Huang et al., 2021). In Seoul National University in South Korea, a mesocosm experiment was carried out to ascertain the growth of wetland plants and species dominance as a function of the flooding regime. The results indicated enhanced water level and flooding frequency increased the number and biomass of stems of *Zizania latifolia*, but not those of the other plants. This indicated that *Scirpustabernaemontani* and *Typhaorientalis* were outcompeted by *Zizania latifolia* irrespective of water regimes (Byun et al., 2017). Another study was undertaken to evaluate the response of height of plant, richness of species, and aerial biomass in response to inundation patterns in the wetlands of north eastern China. For two *Carex* populations (*Carex lasiocarpa*, and *Carex appendiculata*), only the depth of the water table had a substantial impact on species richness. while no impact on *Calamagrostis angustifolia*. The findings of the study are indicative of the fact that on continual drying of the wetlands, the Carex communities are likely to be replaced by *Calamagrostis angustifolia* (Lou et al., 2016).
5. **Climatic condition:** Climatic conditions are instrumental in determining the species composition of a wetland. Wetland ecosystems are more prone to being affected by global climate change than many other habitats restricted to the terrestrial or marine environment. These plant communities face anthropogenic and climatic stresses in developed places often compounding the biotic and abiotic pressure (Short et al., 2016). An interesting study done on Badesø/ Kangerlussuaq Tasia in Greenland revealed that *Callitriche hamulata* growing in the waters of the lakes is responsive to warm temperature regime, particularly above 10 degrees Celsius. It was thus predicted that future warming in the arctic region will result in a deeper extension of the thermocline and can stimulate growth under varied light conditions. Moreover, it was further predicted that along with increased growth there is a distribution to the macrophytes further northwards (Lauridsen et al., 2020). In another experiment, the effect of rising temperature on the growth and stoichiometry of three aquatic plants namely *Potamo getonlucens, Vallisneria spiralis*, and *Elodea nuttallii* was studied.

It was observed that an increase in temperature increased the growth of all three plants while there was a decrease in nitrogen and phosphorus content (Zhang et al., 2019). A shallow eutrophic lake simulator was created for studying the effect of temperature on the growth and competition of two cyanobacteria namely *Microcystis aeruginosa* and *Oscillatoria mougeotii*. It was observed that *Microcystis* was repressed by *Oscillatoria*, which was the greater competitor at a temperature less than 60 degrees Celcius (Chu et al., 2007). A phytoplankton community structure analysis was performed in Lake Taihu in China between 1992 to 2012. It was observed that *Cryptomonas* was dominant during the spring of the early 1990s which later on shifted to *Ulothrix* and also by *Microcystis* in a later stage (2007-2012). The results of the study indicate that the composition of the phytoplankton community in the lake is governed by nutrient enrichment as well as temperature (Deng et al., 2014).

THE SUCCESSIONAL PATTERN IN WETLANDS-WITH EXAMPLES

Majority of transitory wetland plants has a clear trend of zonation, through a succession between open water to dry ground, which is substantially associated with the time and periodicity of flooding (Johnson and Rogers, 2003). Different aquatic plant species have differences in the degree of adaptation which is

Table 1. Study on successional pattern in selected location

Location	Type of Study	Important Findings	Reference
Olentangy River Wetland Research Park, Ohio State University, United States of America	Study on successional pattern, development of soil, changes in quality of water and nutrition dynamics- A study done for 15 years in a created wetland	1.Plant species richness increased from 13 to 116 in the study period. 2. Planted wetland had higher species diversity than the unplanted one. 3. Increase in organic carbon content by a factor of 3 in 15 years. 4. Conversion of soil into hydric type.	Mitsch et al., 2012
Durham, North Carolina, United States of America	Investigation of floral succession in the context of natural processes post restoration of wetlands.	1.Increase in species richness from 24 to 38 in low marshes. 2.Decrease in upland species communities from 69 to 29 in the region of hgh marsh which were subjected to flooding. 3. Slow turnover of species succession in low marsh as compared to high marsh. 4. Decimation of *Ligustrum sinense* and dominance of *Microstegium vimineum.*	Ho and Richardson, 2013
Great Lakes, United States of America	The effects of hydrologic variability on plant succession	1. Human habitation were mostly responsible for vegetation changes. 2. Climate-induced hydrologic changes had a stronger impact on changes in plant community than self-generated processes.	Wilcox, 2004
Portsmouth, New Hampshire	Freshwater mitigation wetland's successional dynamics (Documentations in year 2002 and 2020)	1. In the 2020 floral census, 129 plant species were discovered, divided into 54 families and 85 genera. 2. Since the last study in 2002, the community has grown by 19 species, 14 genera, and 6 families. 3. Presence of *Cardamine bulbosa* and *Potamogeton foliosus*, and, *Sparganium eurycarpum* (Threatened). 4. Shifting of plant communities and presence of exotic species observed.	McKown et al., 2021

reflected in the hydrosere. During succession, the plant's species can be easily distinguished based on their survival according to modified conditions or significant changes in environmental factors (; Walker et al., 2010). It shows the clear trends of community-level plant traits along the gradients of succession (Holdaway et al., 2011). Changes in surroundings, as well as interactions amongst biotic communities cause functional variety at the time of succession (Craven et al., 2018). Few examples of study of successional pattern in wetland is illustrated briefly (table 1).

WETLAND SUCCESSION IN A PERMAFROST DEGRADATION

Perturbations such as fire, flooding, permafrost degradation, and drainage influence the spatial patterns of vegetation in Alaska's low-lying flood plains. Permafrost melting can affect ecosystem structure by increasing carbon buildup and methane production in peats, as warming soils elevate water tables and release nutrients. (Camill et al., 2001; Turetsky et al., 2008). The thawing of permafrost can also cause water bodies to drain, (Riordan et al., 2006) allowing aerobic decomposition and carbon emission in wetlands. A study related to fire-scarred black spruce forest succession, and the relationships between fire and thermokarst in *Sphagnum*-dominated permafrost breakdown was undertaken. According to the findings, drought circumstances could limit the growth of both black spruce and *Sphagnum*, as well as the long-term carbon storage capacity of the ecosystem (Myers- Smith et al., 2008).

EFFECT OF SUCCESSIONAL PATTERN OF WETLAND ON THE ECONOMY

Wetlands are critically important for the maintenance of vast biodiversity and offer protection from floods, erosion of shorelines or coastal areas as well as creates values, unlike other ecosystems. Climate change and its mitigation, loss of biodiversity can be restored through wetland conservation. Wetlands are also important for providing clean water which gets filtered through swamps and marshes and removes pollutants. They also are important to generate recreational and aesthetic values and potent producers of natural products. These hold an important place for the global economy, nearly one billion people depend on it for earning their livelihoods and at present world economical essential services amounts a USD47 trillion that is provided by wetlands (United States Environmental Protection Agency Webpage) An earlier estimation during 1997, the worth of wetlands globally was estimated to be $14.9 trillion in one study. (Costanza et al.,1997). This chapter holds a brief understanding of the economic impact of wetland succession and its present global scenario (Figure 4).

1. Economic Valuation of Wetland

Wetlands being an individual ecosystem with high ecological value are a major source of primary productivity and support numerous species of plants and animals. This harbors a great and enriched biodiversity along with diverse flora and microbes. However, this highly complex and interdependent community structure can vary according to the size, shape, climate, and location of the wetland. Water and nutrient sources are also crucial in the diversity, dominance, and distribution of species. Wetland can be well described as a 'biological supermarket' in terms of numerous biological and natural products

Figure 4. Schematic representation of wetland succession on economy

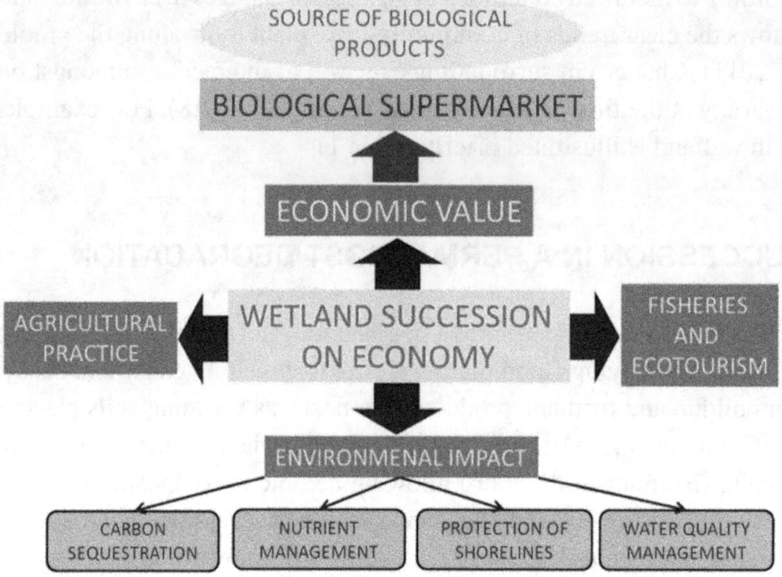

through which they well serve for human welfare and can impart economic impact by alleviating poverty level. Local communities are particularly benefitted from wetland ecosystems in terms of goods and services, however, productivity is highly hindered by poor actualization of economic values (Gimenez and Mas, 2020).

The estimated total economic values of a wetland can be categorized as direct, indirect, and no use values (Baral et al., 2016). The major aspects that contribute towards direct values are goods consumption and tourism that requires market-based techniques for analysis. This involves holding interviews with house residents, visitors, and stakeholders of the wetland community in a district. Non-use values are those encompassing desire to pay for the retention of goods, which is independent towards its potential use in the present or future. They often include existence value, altruistic value, and bequest value. Besides Total Economic valuation, recreational economic value analysis can be performed based on two methods. These are Travel Cost Method and the Contingent Valuation Method (Rietbergen-McCracken and Abaza, 2000; Mendelsohn and Olmstead, 2009; Ling et al., 2018).

As the wetlands are susceptible to various alterations in climatic conditions, namely temperature, flooding, drought, and salinity augmentation, and elevated CO_2 alteration, all these factors could affect the economy of the local community. Other than natural forces anthropogenic activities are also concerned to bring about successional changes.

2. Fisheries and Ecotourism

Wetland offers direct revenue generation through fisheries and ecotourism worldwide. Shrimp-fishery was established in the Gulf of Mexico (Engle, 2011; Bhowmik, 2020). India harbors wetland-dependent fisheries in Kerala, Goa, Odisha, Jammu, and Kashmir, Assam, Bihar, Andhra Pradesh, Gujarat (Bassi et al., 2014).

Further ecotourism-based famous places are Chilka lake of Odisha, Rann of Kutch, backwaters of Kerala which help to earn foreign currency. Coastal wetlands and the mangroves present at the shoreline of Bhitarkanika protect the locality from the impact of repeated cyclones and high-rising tidal waves as well as protects from soil erosion (Hashimet and Catherine, 2013). Thus, direct benefits procured from wetlands contribute towards local community development and economical growth.

3. Carbon Sequestration, Nutrient and Water Quality Management

Carbon sequestration in wetland reservoirs is primarily affected by wetland drainage. Rapid oxidation and peat mining also intensify the process (Bhowmik, 2020). An augmentation in sequestration is found through run-off waste deposition and increased CO_2 concentration. Alongi et al., 2015 reported that mangroves and marshes of Indonesian maritime wetland can trap almost 17% of the world's total blue carbon. Bassi et al., 2014, reported 200 times greater sequestration of carbon in the soil in contrast to vegetation. Increased salinity triggered less carbon accumulation in mangrove vegetation (Mitsch et al., 2016).

Another important function of the wetland is to remove nitrogen using various biogeochemical processes, that might vary depending upon several factors like hydrology, the area concerned, soil permeability, etc. which are also responsible for regulating the whole process of removal of nutrients (Ardón et. al., 2010).

Wetland systems of various forms, such as natural and constructed wetlands, have the ability to sequester carbon. Each wetland can act as a carbon sink for the atmosphere. Wetlands can sequester carbon in a magnitude of 830Tg per year (Rosli et al., 2016). This phenomena of sequestration lead to the concept and coining of the term blue carbon(BC) which is fundamentally organic carbon collected and retained by the oceans and coastal ecosystems, particularly vegetated coastal ecosystems such as seagrass meadows, tidal marshes, and mangrove forests. The potential for BC to mitigate climate change while providing co-benefits such as coastal protection and fisheries enhancement has piqued international interest (Macreadie et al., 2019). The blue carbon forms an intrinsic part of blue economy which is use of ocean for benefit in terms of economy (Steven et al., 2019). Blue carbon ecosystems give jobs and money to local economies, enhance water quality, maintain healthy fisheries, and provide coastal protection, in addition to carbon storage benefits. Mangroves operate as natural barriers, stabilizing shorelines and reducing wave energy, reducing the risk of flooding in coastal areas due to storm surge and rising sea levels (Reef Resilience Network).

4. Water Quality Management and Agricultural Practices

Removal of contaminants in form of HMs, harmful sewage microbes, other sedimentary residues is an important function of wetland and this ensures the availability of usable water (Bhowmik, 2020). Health hazards are reduced while the diversity remains unaffected. The water obtained in this way can be utilized for irrigation purposes. The stored water is channelized to supply the nearby field for agriculture during the dry season. However due to silting and sedimentation problem that arises due to agricultural practices wetland can be diminished that poses a threat to species richness of biodiversity. In a recent study, Western Himalaya harboring Pong Dam wetland in Himachal Pradesh shows a land cover change due to anthropogenic activities. The remote sensing imageries recorded from 2001 to 2016 exhibited destruction within catchment areas along with agricultural land increased by 159 percent, while barren land dropped by 53 percent. Such agricultural activity not only reduced the aquatic plankton diversity, also waterfowl diversity was affected. Hence agroforestry is suggested herein and planting of forest trees is recommended to ensure the preservation of wetlands (Malik and Rai, 2019).

It can be noted that in the last three centuries, 87 percent of the world's wetlands have vanished due to pollution, drainage, and conversion of lands either due to natural causes or anthropogenic interference. Considering its importance 170 countries have pledged to preserve this ecological resource in Ramsar Convention and 2000 wetlands worldwide have been identified that spreads over more than 250 million hectares which are within 13 to 18% of total wetlands present worldwide (United States Environmental Protection Agency Webpage). The preserved wetlands have proved to be of immense benefit to the humans. Wetlands produce essential natural resources, which are vital to the rural economy. They bring numerous significant benefits not only to the local community, but also to those who live far away. They are well-known around the world for their crucial role in supporting a diverse range of biodiversity and supplying goods and services, as well as important natural resource sources on which rural economies rely (Kakuru et al., 2013). A large number of freshwater aquatic and wetland systems are integrated in agricultural landscapes, with commodity production accounting for the majority of water demand. As a result, agricultural decisions are inextricably tied to establishing a sustainable wetland future, which requires well functioning hydrology. As a result, it can be proposed that holistic policy directives that recognise the relevance and interconnectivity of agricultural economics and wetland ecosystem services can compete with technical improvements in providing a viable, long-term solution to water sustainability. To uncover conservation inefficiencies and opportunities, such directions require a fundamental understanding of water law, farm economies, and water consumption (King et al., 2021).

CONCLUSION

Wetlands are characterized by a moist environment and growth of diverse and unique flora which in turn houses wide varieties of fauna and are considered to be highly productive. Since they are enriched with biodiversity, they experience an unique successional pattern and this is often most prominent with respect to floral succession. The ideal condition generally happens from the pioneer successors and ends in climax vegetation. It is thus observed that the occurrence of climax vegetation is often associated with drying up of the wetlands. However, this ideal condition often varies in reality as each stage in wetland succession or more appropriately, the 'seral stage' is often associated with intricate and unique community structure which often affects growth of one another. However this successional pattern is often governed by a number of natural factors such as flooding, light intensity, precipitation, drought, pH. In addition to it, environmental pollution caused by humans which leads to enrichment of water bodies in terms of ions and nutirents also affects the successional pattern. High enrichment results in eutrophication which not only generates toxic compounds in the waterbody but also inhibits the entry of sunlight to the lower layers thereby limiting the growth of benthic flora. Thus eutrophication leads to an alteration of community structure and hence the successional patterns. Flooding, drought and climatic changes similarly affect the successional pattern by promoting growth of species which are tolerant to the stressed condition. The eutrophication and consequent generation of toxins in the waterbodies affects the humans. This leads to occurrence of several diseases. Eutrophication also affects the economy of the region as wetlands are often associated with recreational and commercial activities. Eutrophication renders these wetlands unsuitable for any use. Agricultural activites are also affected by polluted inland water bodies as irrigation is often associated to them. This often leads to biomagnification of toxic compounds from the agricultural products and also changes microbial successional pattern. Thus the primary motive would be to continuously monitor the pollution levels of the wetlands. On the other hand, minute observation and documentation of the successional pattern in the wetland should also be done and any variation should be further investigated with respect to the pollution pattern. Variation in successional pattern should also be documented and model system requires to be constructed to predict the probable change in vegetation with respect to change in climatic condition and pollution levels. This will enable to understand the successional dynamics in more efficient way. Monitoring of the pollution levels as well as successional pattern is also important for economic activity. Wetlands which are frequented by tourists should also be monitored with top prority in order to check the pollution level and protect the visitors from any harmful effect. This will not only keep the tourists free from any pollution- related diseases but also help a region to develop a sustainable multipurpose system centered on the wetland which can then serve a dual purpose of reacreation and economy. The majority of countries have insufficient management planning and reporting on the ecological nature of their Ramsar Sites, necessitating a greater focus on a vision and objectives, as well as regular reporting of important indicators to guide management. Using modern techniques and available indicators, there are more options to thoroughly track ecological character. It is vital that the world's wetlands be protected, with the Ramsar Convention or the Convention on Wetlands at the forefront (Kingsford et al., 2021). Strict law enforcement is also required from the policy makers to ensure the harmony in biodiversity and successional pattern in these waterbodies. This will bring totality to the protective approach and help in restoring the balance of the ecosystem.

REFERENCES

Alongi, D. M., Murdiyarso, D., Fourqurean, J. W., Kauffman, J. B., Hutahaean, A., Crooks, S., & Wagey, T. (2015). Indonesia's blue carbon: A globally significant and vulnerable sink for seagrass and mangrove carbon. *Springer. Wetlands Ecology and Management*, *24*(1), 3–13. doi:10.100711273-015-9446-y

Ardón, M., Montanari, S., Morse, J. L., Doyle, M. W., & Bernhardt, E. S. (2010). Phosphorus export from a restored wetland ecosystem in response to natural and experimental hydrologic fluctuations. *Journal of Geophysical Research*, *115*(G4), G04031. doi:10.1029/2009JG001169

Baral, S., Basnyat, B., Khanal, R., & Gauli, K. (2016). A total economic valuation of wetland ecosystem services: An evidence from Jagadishpur Ramsar site, Nepal. *TheScientificWorldJournal*, *2016*, 2016. doi:10.1155/2016/2605609 PMID:27830175

Barros, D. F., & Albernaz, A. L. M. (2014). Possible impacts of climate cane on wetlands and its biota in the Brazilian Amazon. *Brazilian Journal of Biology*, *74*(4), 810–820. doi:10.1590/1519-6984.04013 PMID:25627590

Bassi, N., Kumar, M. D., Sharma, A., & Pardha-Saradhi, P. (2014). Status of wetlands 2014 in India: A review of extent, ecosystem benefits, threats and management strategies. *Elsevier. Journal of Hydrology. Regional Studies*, *2*, 1–19. doi:10.1016/j.ejrh.2014.07.001

Bernard, C., Escalas, A., Villeriot, N., Agogué, H., Hugoni, M., Duval, C., Carré, C., Got, P., Sarazin, G., Jézéquel, D., Leboulanger, C., Grossi, V., Ader, M., & Troussellier, M. (2019). Very Low Phytoplankton Diversity in a Tropical Saline-Alkaline Lake, with Co-dominance of *Arthrospira fusiformis* (Cyanobacteria) and *Picocystissalinarum* (Chlorophyta). *Microbial Ecology*, *78*(3), 603–617. doi:10.100700248-019-01332-8 PMID:30729265

Bhowmik, S. (2020). *Ecological and Economic Importance of Wetlands and Their Vulnerability: A Review, in Current State and Future Impacts of Climate Change on Biodiversity*. Retrieved October 9, 2021, from: https://reefresilience.org/blue-carbon/blue-carbon-introduction/blue-carbon-benefits/

Bonanno, G., Borg, J. A., & Di Martino, V. (2017). Levels of heavy metals in wetland and marine vascular plants and their biomonitoring potential: A comparative assessment. *The Science of the Total Environment*, *576*, 796–806. doi:10.1016/j.scitotenv.2016.10.171 PMID:27810764

Bravo-Inclan, L. A., Saldana-Fabela, M. P., & Sanchez-Chavez, J. J. (2008). Long-term eutrophication diagnosis of a high altitude body of water, Zimapan Reservoir, Mexico. *Water Science and Technology*, *57*(11), 1843–1849. doi:10.2166/wst.2008.279 PMID:18547939

Burton, P. J., Jentsch, A., & Walker, L. R. (2020). The ecology of disturbance interactions. *Bioscience*, *70*(10), 854–870. doi:10.1093/biosci/biaa088

Butler, S. (Ed.). (2010). *Macquarie Concise Dictionary* (5th ed.). Macquarie Dictionary Publishers.

Byun, C., Nam, J. M., & Kim, J. G. (2017). Effects of flooding regime on wetland plant growth and species dominance in a mesocosm experiment. *Plant Ecology*, *218*(5), 517–527. doi:10.100711258-017-0707-0

Camill, P., Lynch, J. A., Clark, J. S., Adams, J. B., & Jordan, B. (2001). Changes in biomass, above ground net primary production, and peat accumulation following permafrost thaw in the boreal peatlands of Manitoba, Canada. *Ecosystems (New York, N.Y.)*, *4*(5), 461–478. doi:10.100710021-001-0022-3

Chambers, L. G., Guevara, R., Boyer, J. N., Troxler, T. G., & Davis, S. E. (2016). Effects of salinity and inundation on microbial community structure and function in a mangrove peat soil. *Wetlands*, *36*(2), 361–371. doi:10.100713157-016-0745-8

Chen, G., Shi, H., Tao, J., Chen, L., Liu, Y., Lei, G., Liu, X., & Smol, J. P. (2015). Industrial arsenic contamination causes catastrophic changes in freshwater ecosystems. *Scientific Reports*, *5*(1), 17419. doi:10.1038rep17419 PMID:26615891

Cheng, R., Zhu, H., Cheng, X., Shutes, B., & Yan, B. (2020). Saline and Alkaline tolerance of wetland plants—What are the most representative evaluation indicators? *Sustainability*, *12*(5), 1913. doi:10.3390u12051913

Chu, Z., Jin, X., Iwami, N., & Inamori, Y. (2007). The effect of temperature on growth characteristics and competitions of *Microcystis aeruginosa* and *Oscillatoria mougeotii* in a shallow, eutrophic lake simulator system. In B. Qin, Z. Liu, & K. Havens (Eds.), *Eutrophication of Shallow Lakes with Special Reference to Lake Taihu, China. Developments in Hydrobiology* (Vol. 194). Springer. doi:10.1007/978-1-4020-6158-5_24

Costanza, R. R., d'Arge, R., de Groot, R., Farber, S., Grasso, M., Hannon, B., Limburg, K., Naeem, S., O'Neill, R. V., Paruelo, J., Raskin, R. G., Sutton, P., & van den Belt, M. (1997). The value of the world's ecosystem services and natural capital. *Nature*, *387*(6630), 253–256. doi:10.1038/387253a0

Craven, D., Hall, J. S., Berlyn, G. P., Ashton, M. S., & van Breugel, M. (2018). Environmental filtering limits functional diversity during succession in a seasonally wet tropical secondary forest. *Journal of Vegetation Science*, *29*(3), 511–520. doi:10.1111/jvs.12632

Davidson, N. C. (2014). How much wetland has the world lost? Long-term and recent trends in the global wetlands area. *Marine and Freshwater Research*, *65*(10), 936–941. doi:10.1071/MF14173

deAnda, J., & Shear, H. (2001). Nutrients and Eutrophication in Lake Chapala. In A. M. Hansen & M. van Afferden (Eds.), *The Lerma-Chapala Watershed*. Springer. doi:10.1007/978-1-4615-0545-7_8

Deng, J., Qin, B., Paerl, H. W., Zhang, Y., Wu, P., Ma, J., & Chen, Y. (2014). Effects of nutrients, temperature and their interactions on spring phytoplankton community succession in Lake Taihu, China. *PLoS One*, *9*(12), e113960. doi:10.1371/journal.pone.0113960 PMID:25464517

Díaz-Pardo, E., Vazquez, G., & López-López, E. (1998). The phytoplankton community as a bioindicator of health conditions of Atezca Lake, Mexico. *Aquatic Ecosystem Health & Management*, *1*(3-4), 257–266. doi:10.1080/14634989808656922

Dinh, Q. T. (2016). Vietnam-Mekong Delta Integrated Climate Resilience and Sustainable Livelihoods (MD-ICRSL) Project: Environmental assessment (No. SFG1760, pp. 1-209). The World Bank.

Dokulil, M. T., & Qian, K. (2021). Photosynthesis, carbon acquisition and primary productivity of phytoplankton: A review dedicated to *Colin Reynolds. Hydrobiologia*, *848*(1), 77–94. doi:10.100710750-020-04321-y

Dormann, C. F., Bagnara, M., Boch, S., Hinderling, J., Janeiro-Otero, A., Schäfer, D., & Hartig, F. (2020). Plant species richness increases with light availability, but not variability, in temperate forests under storey. *BMC Ecology*, *20*(1), 1–9. doi:10.118612898-020-00311-9 PMID:31941480

Dorney, J., Savage, R., Adamus, P., & Tiner, R. (Eds.). (2018). *Wetland and Stream Rapid Assessments: Development, Validation, and Application*. Academic Press.

Edelkraut, K. A., & Güsewell, S. (2006). Progressive effects of shading on experimental wetland communities over three years. *Plant Ecology*, *183*(2), 315–327. doi:10.100711258-005-9042-y

Eliot, C. (2007). Method and metaphysics in Clements's and Gleason's ecological explanations. *Studies in History and Philosophy of Science Part C Studies in History and Philosophy of Biological and Biomedical Sciences*, *38*(1), 85–109. doi:10.1016/j.shpsc.2006.12.006 PMID:17324810

Emamverdian, A., Ding, Y., Mokhberdoran, F., & Xie, Y. (2015). Heavy metal stress and some mechanisms of plant defense response. *TheScientificWorldJournal*, *2015*, 756120. doi:10.1155/2015/756120 PMID:25688377

Engle, V. D. (2011). Estimating the Provision of Ecosystem Services by Gulf of Mexico Coastal Wetlands.*Springer. Wetlands*, *31*(1), 179–193. doi:10.100713157-010-0132-9

Ervin, G. N., & Wetzel, R. G. (2002). Influence of a dominant macrophyte, Juncuseffusus, on wetland plant species richness, diversity, and community composition. *Oecologia, 130*(4), 626–636. doi:10.100700442-001-0844-x PMID:28547266

Farkhondeh, T., Naseri, K., Esform, A., Aramjoo, H., & Naghizadeh, A. (2020). Drinking water heavy metal toxicity and chronic kidney diseases: A systematic review. *Reviews on Environmental Health, 36*(3), 359–366. doi:10.1515/reveh-2020-0110 PMID:33128529

Ferronato, N., & Torretta, V. (2019). Waste Mismanagement in Developing Countries: A Review of Global Issues. *International Journal of Environmental Research and Public Health, 16*(6), 1060. doi:10.3390/ijerph16061060 PMID:30909625

Finlayson, C. M., & Spiers, A. G. (1999). *A compendium of information for managing and monitoring wetlands in tropical Australia. Supervising Scientist Report, 148*. Supervising Scientist.

Gimenez, F. V., & Mas, C. R. (2020). The Valuation of Recreational Use of Wetlands and the Impact of the Economic Crisis. *International Journal of Environmental Research and Public Health, 17*(9), 3228–3242. doi:10.3390/ijerph17093228 PMID:32384709

Gopal, B., & Chauhan, M. (2018). The Transboundary Sundarbans Mangroves (India and Bangladesh). In C. Finlayson, G. Milton, R. Prentice, & N. Davidson (Eds.), *The Wetland Book*. Springer.

Gossner, M. M. (2009). Light intensity affects spatial distribution of Heteroptera in deciduous forests. *European Journal of Entomology, 106*(2), 241–252. doi:10.14411/eje.2009.032

Guo, C., Chen, Y., Xia, W., Qu, X., Yuan, H., Xie, S., & Lin, L. S. (2020). Eutrophication and heavy metal pollution patterns in the water suppling lakes of China's south-to-north water diversion project. *The Science of the Total Environment, 711*, 134543. doi:10.1016/j.scitotenv.2019.134543 PMID:31812425

Häder, D. P., Banaszak, A. T., Villafañe, V. E., Narvarte, M. A., González, R. A., & Helbling, E. W. (2020). Anthropogenic pollution of aquatic ecosystems: Emerging problems with global implications. *The Science of the Total Environment, 713*, 136586. doi:10.1016/j.scitotenv.2020.136586 PMID:31955090

Hasan, M. K., Shahriar, A., & Jim, K. U. (2019). Water pollution in Bangladesh and its impact on public health. *Heliyon, 5*(8), e02145. doi:10.1016/j.heliyon.2019.e02145 PMID:31406938

Hashimet, A. M., & Catherine, S. M. P. (2013). A Laboratory Study on Wave Reduction by Mangrove Forests. *Elsevier. APCBEE Procedia, 5*, 27–32. doi:10.1016/j.apcbee.2013.05.006

Hastings, K. L., Smith, L. E., Lindsey, M. L., Blotsky, L. C., Downing, G. R., Zellars, D. Q., Downing, J. K., & Corena-McLeod, M. (2014). Effect of microalgae application on soil algal species diversity, cation exchange capacity and organic matter after herbicide treatments. *F1000 Research, 3*(281), 281. doi:10.12688/f1000research.4016.1

Ho, M., & Richardson, C. J. (2013). A five-year study of floristic succession in a restored urban wetland. *Ecological Engineering, 61*, 511–518. doi:10.1016/j.ecoleng.2013.05.001

Holdaway, R. J., Richardson, S. J., Dickie, I. A., Peltzer, D. A., & Coomes, D. A. (2011). Species- and community-level patterns in fine root traits along a 120000-year soil chronosequence in temperate rain forest. *Journal of Ecology, 99*(4), 954–963. doi:10.1111/j.1365-2745.2011.01821.x

Huang, Y., Chen, X. S., Li, F., Hou, Z. Y., Li, X., Zeng, J., & Xie, Y. H. (2021). Community Trait Responses of Three Dominant Macrophytes to Variations in Flooding During 2011–2019 in a Yangtze River-Connected Floodplain Wetland (Dongting Lake, China). *Frontiers in Plant Science, 12*, 957. doi:10.3389/fpls.2021.604677 PMID:34122461

Johnson, P., & Rogers, G. (2003). *Ephemeral wetlands and their turfs in New Zealand*. Department of Conservation.

Jurburg, S. D., Nunes, I., Stegen, J. C., Le Roux, X., Priemé, A., Sørensen, S. J., & Salles, J. F. (2017). Autogenic succession and deterministic recovery following disturbance in soil bacterial communities. *Scientific Reports, 7*(1), 1–11. doi:10.1038rep45691 PMID:28383027

Kakuru, W., Turyahabwe, N., & Mugisha, J. (2013). Total economic value of wetlands products and services in Uganda. *The Scientific World Journal*. doi:10.1155/2013/192656 PMID:24163614

King, S. L., Laubhan, M. K., Tashjian, P., Vradenburg, J., & Fredrickson, L. (2021). Wetland Conservation: Challenges Related to Water Law and Farm Policy. *Wetlands, 41*(5), 1–17. doi:10.100713157-021-01449-y

Kingsford, R. T., Bino, G., Finlayson, C. M., Falster, D., Fitzsimons, J. A., Gawlik, D. E., Murray, N. J., Grillas, P., Gardner, R. C., Regan, T. J., Roux, D. J., & Thomas, R. F. (2021). Ramsar wetlands of international importance–improving conservation outcomes. *Frontiers in Environmental Science, 9*, 53. doi:10.3389/fenvs.2021.643367

Kumar, R., & Patnaik, P. (2016). Wetlands of Mahanadi Delta (India). In C. Finlayson, G. Milton, R. Prentice, & N. Davidson (Eds.), *The Wetland Book*. Springer. doi:10.1007/978-94-007-6173-5_28-4

Kumar, V., Parihar, R. D., Sharma, A., Bakshi, P., Singh Sidhu, G. P., Bali, A. S., Karaouzas, I., Bhardwaj, R., Thukral, A. K., Gyasi-Agyei, Y., & Rodrigo-Comino, J. (2019). Global evaluation of heavy metal content in surface water bodies: A meta-analysis using heavy metal pollution indices and multivariate statistical analyses. *Chemosphere, 236*, 124364. doi:10.1016/j.chemosphere.2019.124364 PMID:31326755

Larson, C. A., & Belovsky, G. E. (2013). Salinity and nutrients influence species richness and evenness of phytoplankton communities in microcosm experiments from Great Salt Lake, Utah, USA. *Journal of Plankton Research, 35*(5), 1154–1166. doi:10.1093/plankt/fbt053

Lauridsen, T. L., Mønster, T., Raundrup, K., Nymand, J., & Olesen, B. (2020). Macrophyte performance in a low arctic lake: Effects of temperature, light and nutrients on growth and depth distribution. *Aquatic Sciences, 82*(1), 18. doi:10.100700027-019-0692-6

Li, C., Quan, Q., Gan, Y., Dong, J., Fang, J., Wang, L., & Liu, J. (2020). Effects of heavy metals on microbial communities in sediments and establishment of bioindicators based on microbial taxa and function for environmental monitoring and management. *The Science of the Total Environment, 749*, 141555. doi:10.1016/j.scitotenv.2020.141555 PMID:32841857

Li, Y., Meng, J., Zhang, C., Ji, S., Kong, Q., Wang, R., & Liu, J. (2020). Bottom-up and top-down effects on phytoplankton communities in two freshwater lakes. *PLoS One, 15*(4), e0231357. doi:10.1371/journal.pone.0231357 PMID:32271852

Lind, L., Schuler, M. S., Hintz, W. D., Stoler, A. B., Jones, D. K., Mattes, B. M., & Relyea, R. A. (2018). Salty fertile lakes: How salinization and eutrophication alter the structure of freshwater communities. *Ecosphere*, *9*(9), e02383. doi:10.1002/ecs2.2383

Ling, M. A., King, S., Mapendembe, A., & Brown, C. A. (2018). Review of Ecosystem Service Valuation Progress and Approaches by the Member States of the European Union. UNEP-WCMC.

Lou, Y., Pan, Y., Gao, C., Jiang, M., Lu, X., & Xu, Y. J. (2016). Response of plant height, species richness and aboveground biomass to flooding gradient along vegetation zones in floodplain wetlands, Northeast China. *PLoS One*, *11*(4), e0153972. doi:10.1371/journal.pone.0153972 PMID:27097325

Macreadie, P. I., Anton, A., Raven, J. A., Beaumont, N., Connolly, R. M., Friess, D. A., Kelleway, J. J., Kennedy, H., Kuwae, T., Lavery, P. S., Lovelock, C. E., Smale, D. A., Apostolaki, E. T., Atwood, T. B., Baldock, J., Bianchi, T. S., Chmura, G. L., Eyre, B. D., Fourqurean, J. W., ... Duarte, C. M. (2019). The future of Blue Carbon science. *Nature Communications*, *10*(1), 3998. doi:10.103841467-019-11693-w PMID:31488846

Malik, M., & Rai, S. C. (2019). Drivers of land use/cover change and its impact on Pong Dam Wetland. *Environmental Monitoring and Assessment*, *191*(4), 203–217. doi:10.100710661-019-7347-x PMID:30834470

McKown, I. D., Gregg, E., Moore, I. D., Payne, A. R., Natalie, A. W., & Gibson, J. L. (2021). Successional dynamics of a 35 year old freshwater mitigation wetland in southeastern New Hampshire. *PLoS One*, *16*(5), e0251748. doi:10.1371/journal.pone.0251748 PMID:33999932

Meena, R., Sathishkumar, P., Ameen, F., Yusoff, A., & Gu, F. L. (2018). Heavy metal pollution in immobile and mobile components of lentic ecosystems-a review. *Environmental Science and Pollution Research International*, *25*(5), 4134–4148. doi:10.100711356-017-0966-2 PMID:29247419

Mendelsohn, R., & Olmstead, S. (2009). The economic valuation of environmental amenities and disamenities: Methods and applications. *Annual Review of Environment and Resources*, *34*(1), 325–347. doi:10.1146/annurev-environ-011509-135201

Meng, L., Roulet, N., Zhuang, Q., Christensen, T. R., & Frolking, S. (2016). Focus on impact of climate cane on wetland ecosystems and carbon dynamics. *Environmental Research Letters*, *11*(10), 100201. doi:10.1088/1748-9326/11/10/100201

Middleton, B. (2016). Succession in Wetlands. In *The Wetland Book*. Springer. doi:10.1007/978-94-007-6172-8_1-2

Mitsch, W. J., Marchio, D. A. Jr, Savarese, M., & Bovard, B. (2016). Carbon Sequestration and Sedimentation in Mangrove Swamps Influenced by Hydrogeomorphic Conditions and Urbanization in Southwest Florida. *Forests*, *7*(12), 116. doi:10.3390/f7060116

Mitsch, W. J., Zhang, L., Stefanik, K. C., Nahlik, A. M., Anderson, C. J., Bernal, B., Hernandez, M., & Song, K. (2012). Creating Wetlands: Primary Succession, Water Quality Changes, and Self-Design over 15 Years. *Bioscience*, *62*(3), 237–250. doi:10.1525/bio.2012.62.3.5

Mokany, A., Wood, J. T., & Cunningham, S. A. (2008). Effect of shade and shading history on species abundances and ecosystem processes in temporary ponds. *Freshwater Biology, 53*(10), 1917–1928. doi:10.1111/j.1365-2427.2008.02076.x

Morrison, K. D., Hammer, E., Boles, O., Madella, M., Whitehouse, N., Gaillard, M. J., Bates, J., Vander Linden, M., Merlo, S., Yao, A., Popova, L., Hill, A. C., Antolin, F., Bauer, A., Biagetti, S., Bishop, R. R., Buckland, P., Cruz, P., Dreslerová, D., ... Zanon, M. (2021). Mapping past human land use using archaeological data: A new classification for global land use synthesis and data harmonization. *PLoS One, 16*(4), e0246662. doi:10.1371/journal.pone.0246662 PMID:33852578

Moseley, K. (2018). *Wetland Ecology- Basic Principles. Regional ESS.* Soil Science Division.

Myers-Smith, I. H., Harden, J. W., Wilmking, M., Fuller, C. C., McGuire, A. D., & Chapin Iii, F. S. (2008). Wetland succession in a permafrost collapse: Interactions between fire and thermokarst. *Biogeosciences, 5*(5), 1273–1286. doi:10.5194/bg-5-1273-2008

Nielsen, D. L., Brock, M. A., Crosslé, K., Harris, K., Healey, M., & Jarosinski, I. (2003). The effects of salinity on aquatic plant germination and zooplankton hatching from two wetland sediments. *Freshwater Biology, 48*(12), 2214–2223. doi:10.1046/j.1365-2427.2003.01146.x

Nkwunonwo, U. C., Odika, P. O., & Onyia, N. I. (2020). A Review of the Health Implications of Heavy Metals in Food Chain in Nigeria. *The Scientific World Journal.* doi:10.1155/2020/6594109 PMID:32351345

Pierson, D. C. (2012). Light and Primary Production in Lakes. In L. Bengtsson, R. W. Herschy, & R. W. Fairbridge (Eds.), *Encyclopedia of Lakes and Reservoirs. Encyclopedia of Earth Sciences Series.* Springer. doi:10.1007/978-1-4020-4410-6_261

Quan, N. H., Toan, T. Q., Dang, P. D., Phuong, N. L., Anh, T. T. H., Quang, N. X., Quoc, D. P., Quoi, L. P., Hanington, P., & Sea, W. B. (2018). Conservation of the Mekong Delta wetlands through hydrological management. *Ecological Research, 33*(1), 87–103. doi:10.100711284-017-1545-1

Ramsar Convention on Wetlands. (2018). *Global Wetlands Outlook: State of the World's Wetlands and their services to people.* Ramsar Convention Secretariat.

Rejmánková, E., Komárek, J., Dix, M., Komárková, J., & Girón, N. (2011). Cyanobacterial blooms in lake Atitlan, Guatemala. *Limnologica, 41*(4), 296–302. doi:10.1016/j.limno.2010.12.003

Richards, D. R., Moggridge, H. L., Warren, P. H., & Maltby, L. (2020). Impacts of hydrological restoration on lowland river floodplain plant communities. *Wetlands Ecology and Management, 28*(3), 403–417. doi:10.100711273-020-09717-0

Rietbergen-McCracken, J., & Abaza, H. (2000). *Environmental Valuation: A Worldwide Compendium of Case Studies.* Earthscan.

Riordan, B., Verbyla, D., & Mc Guire, A. D. (2006). Shrinking ponds in subarctic Alaska based on 1950 2002 remotely sensed images. *Journal of Geophysical Research, III*(G4), G04002. doi:10.1029/2005JG000150

Rosli, F. A., Lee, K. E., Goh, C. T., Mokhtar, M., Latif, M. T., Goh, T. L., & Simon, N. (2017). The Use of Constructed Wetlands in Sequestrating Carbon: An Overview. *Nature Environment & Pollution Technology, 16*(3), 813–819.

Sharma, S. S., Dietz, K. J., & Mimura, T. (2016). Vacuolar compartmentalization as indispensable component of heavy metal detoxification in plants. *Plant, Cell & Environment*, *39*(5), 1112–1126. doi:10.1111/pce.12706 PMID:26729300

Short, F. T., Kosten, S., Morgan, P. A., Malone, S., & Moore, G. E. (2016). Impacts of climate change on submerged and emergent wetland plants. *Aquatic Botany*, *135*, 3–17. doi:10.1016/j.aquabot.2016.06.006

Sizo, A., Noble, B., & Bell, S. (2015). Futures analysis of urban land use and wetland cane in Saskatoon, Canada. An application in strategic environmental assessment. *Sustainability*, *7*(1), 811–830. doi:10.3390u7010811

Smalling, K.L., Anderson, C.W., Honeycutt, R.K., Cozzarelli, I.M., Preston, T., & Hossack, B.R. (2019). Associations between environmental pollutants and larval amphibians in wetlands contaminated by energy-related brines are potentially mediated by feeding traits. *Environmental Pollution, 248*, 260–268.

Steven, A. D. L., Vanderklift, M. A., & Bohler-Muller, N. (2019). A new narrative for the Blue Economy and Blue Carbon. *Journal of the Indian Ocean Region*, *15*(2), 123–128. doi:10.1080/19480881.2019.1625215

Tanabe, Y., Hori, M., Mizuno, A. N., Osono, T., Uchida, M., Kudoh, S., & Yamamuro, M. (2019). Light quality determines primary production in nutrient-poor small lakes. *Scientific Reports*, *9*(1), 1–8. doi:10.103841598-019-41003-9 PMID:30874599

Tomasini-Ortiz, C., Moeller Chávez, G., Sánchez Chávez, J., & Bravo Inclán, L. (2012). Cianobacterias y cianotoxinas en el lago de Pátzcuaro, Michoacán, México. *Revista AIDIS de Ingeniería y Ciencias Ambientales: Inestigación, desarrollo y práctica, 5*(2), 93-101.

Tootoonchi, M., & Gettys, L. A. (2019). Testing salt stress on aquatic plants: Effect of salt source and substrate. *Aquatic Ecology*, *53*(3), 325–334. doi:10.100710452-019-09692-6

Tornabene, B. J., Breuner, C. W., & Hossack, B. R. (2020). Relative Toxicity and Sublethal Effects of NaCl and Energy-Related Saline Wastewaters on Prairie Amphibians. *Aquatic Toxicology (Amsterdam, Netherlands)*, *228*, 105626. doi:10.1016/j.aquatox.2020.105626 PMID:32992088

Turetsky, M. R., Wieder, R. K., Vitt, D. H., Evans, R. J., & Scott, K. D. (2008). The disappearance of relict permafrost in boreal North America: Effects On peatland carbon storage and fluxes. *Global Change Biology*, *13*(9), 1922–1934.

United States Environmental Protection Agency. (n.d.). *Wetlands Protection and Restoration*. Retrived June 24, 2021, from: https://www.epa.gov/wetlands

van der Valk, A. G. (2005). The prairie potholes of North America. *Hydrobiologia*, *539*, 171–188.

Vasiliu, D., Bucse, A., Lupascu, N., Ispas, B., Gheablau, C., & Stanescu, I. (2020). Assessment of the metal pollution in surface sediments of coastal Tasaul Lake (Romania). *Environmental Monitoring and Assessment*, *192*(12), 749. doi:10.100710661-020-08698-0 PMID:33155147

Viehweger, K. (2014). How plants cope with heavy metals. *Botanical Studies (Taipei, Taiwan)*, *55*(1), 35. doi:10.1186/1999-3110-55-35 PMID:28510963

Walker, L. R., Wardle, D. A., Bardgett, R. D., & Clarkson, B. D. (2010). The use of chronosequences in studies of ecological succession and soil development. *Journal of Ecology*, *98*(4), 725–736. doi:10.1111/j.1365-2745.2010.01664.x

Wang, X., Sun, Y., Li, S., & Wang, H. (2019). Spatial distribution and ecological risk assessment of heavy metals in soil from the Raoyanghe Wetland, China. *PLoS One*, *14*(8), e0220409. doi:10.1371/journal.pone.0220409 PMID:31398209

Wilcox, D. A. (2004). Implications of hydrologic variability on the succession of plants in Great Lakes wetlands. *Aquatic Ecosystem Health & Management*, *7*(2), 223–231. doi:10.1080/14634980490461579

Xi, H., Feng, Q., Zhang, L., Si, J., Chang, Z., Yu, T., & Guo, R. (2016). Effects of water and salinity on plant species composition and community succession in Ejina Desert Oasis, northwest China. *Environmental Earth Sciences*, *75*(2), 138. doi:10.100712665-015-4823-7

Xu, T., Weng, B., Yan, D., Wang, K., Li, X., Bi, W., Li, M., Cheng, X., & Liu, Y. (2019). Wetlands of International Importance: Status, Threats, and Future Protection. *International Journal of Environmental Research and Public Health*, *16*(10), 1818. doi:10.3390/ijerph16101818 PMID:31121932

Zeng, H., & Wu, J. (2013). Heavy metal pollution of lakes along the mid-lower reaches of the Yangtze River in China: Intensity, sources and spatial patterns. *International Journal of Environmental Research and Public Health*, *10*(3), 793–807. doi:10.3390/ijerph10030793 PMID:23442559

Zhang, G., Bai, J., Tebbe, C. C., Zhao, Q., Jia, J., Wang, W., Wang, X., & Yu, L. (2021). Salinity controls soil microbial community structure and function in coastal estuarine wetlands. *Environmental Microbiology*, *23*(2), 1020–1037. doi:10.1111/1462-2920.15281 PMID:33073448

Zhang, L., Tu, D., Li, X., Lu, W., & Li, J. (2020). Impact of long-term industrial contamination on the bacterial communities in urban river sediments. *BMC Microbiology*, *20*(1), 1–16. doi:10.118612866-020-01937-x PMID:32795344

Zhang, P., Grutters, B., van Leeuwen, C., Xu, J., Petruzzella, A., van den Berg, R. F., & Bakker, E. S. (2019). Effects of Rising Temperature on the Growth, Stoichiometry, and Palatability of Aquatic Plants. *Frontiers in Plant Science*, *9*, 1947. doi:10.3389/fpls.2018.01947 PMID:30671079

Chapter 9
An Overview of the Determination of Biodiversity of Wetlands

Moumit Roy Goswami
Netaji Nagar College for Women, Kolkata, India

Aniruddha Mukhopadhyay
University of Calcutta, India

ABSTRACT

Wetland ecosystems support rich and unique biodiversity. Biodiversity of a given ecosystem in general and wetlands in particular provide important insights to the ecological health of an area. The Ramsar Convention 1971 identified nine criteria for identifying wetlands of international importance. Out of the nine criteria, eight are linked to biodiversity of which three are based on sites of international importance for conserving biological diversity, two are specific for water birds, two are specific for fish, and one criterion for other taxa. Hence, determination of biodiversity of wetlands is of utmost importance. In order to understand that birds, fishes, amphibians, odonates, mammals, and aquatic plants were particularly selected as indicators of wetland biodiversity, the chapter discusses the different methodologies about determination of each of these taxa under different criteria as mentioned above. These methodologies will help various stakeholders in appropriate determination of biodiversity of wetlands of a particular area.

INTRODUCTION

Wetlands are considered as one of the significant ecosystems of the globe. They are termed as "biological supermarkets." for their role in supporting extensive food webs and rich biodiversity. Wetland ecosystems encompasses both aquatic and terrestrial conditions which makes them among the most complex ecosystems of the world. As ecotones, wetlands are embraces various habitats and numerous microhabitats, each one holding atypical, complex, and rich communities. Wetlands are exposed to ecological and topographical gradients, which effect the distribution patterns and zonation of fauna and vegetation. In

DOI: 10.4018/978-1-7998-9498-8.ch009

both inland and coastal wetlands, a gradient exists between almost truly terrestrial sites to permanently flooded plateaus. As a result soil infauna, microbenthic fauna with limited mobility, and vegetation are distributed according to the topographical features determining the gradient itself. Also the environmental features of wetlands are determined largely by hydrologic processes that may fluctuate daily or seasonally in relation to regional climate and geographic conditions. All these factors led to great variety of wetland types globally. Wetlands are broadly classified into five categories i.e., estuarine, marine, riverine, palustrine and lacustrine (Cowardin *et.al.* 1979, 1992) with recent classification including man-made wetlands. Due to wide range of variety amongst wetland types, the variety of living organisms inhabiting and adapted to the different wetlands tends to be high comprising major taxonomic groups of plants and animals. Wetland vegetation includes representatives of all taxonomic groups - from unicellular algae through bryophytes, mosses and ferns to woody angiosperms. Physiological and morphological adaptations have been reported in plant species of wetlands for long-term survival of flooding and even occasionally with remarkable high plant productivity (Mitsch and Gosselink 2000a; Larcher 2003). The faunal diversity includes representatives of almost all taxonomic categories, from protozoa to mammals. Wetland fauna comprises of two kinds of animals namely those depending entirely on wetlands for their whole life, and those utilizing wetlands for a part of their life cycle or for particular needs. Apart from plant and animal species inhabiting in wetlands they are also utilized by visitor species as well, which in turnintensify the overall biodiversity of wetlands. Wetlands are considered as favoured feeding and resting grounds along migratory flyways for ducks, shorebirds, and waders benefiting from diversity of food organisms. Thus a notable concentration of avian diversity can be seenin different wetlands of the globe. The high avian diversity of wetlands attract large numbers of raptors, making them real reservoirs of biodiversity.Six categories of wetland faunaincludingresidents of the wetland, regular migrants from deep water habitats,migrants from terrestrial uplands, regular migrants from other wetlands, fauna visiting the wetland occasionally and fauna indirectly dependent on wetland biota has been documented by Gopal and Junk (2000).Many regional case studies have documented the value of wetlands as habitat for migratory birds. Two million shorebirds are estimated to visit both the Banc d'Aguin National Park, Mauritania, and the Wadden Sea, Northern Europe; while the St. Lucia estuarine system (a Ramsar site in South Africa) supports 350 species of birds. The Productivity levels of many wetland ecosystems are among the highest in the world. The high productivity levels are directly associated with reservoirs of biodiversity supported by wetlands. These is related to both invertebrate and vertebrate species spending only a particular stage of their lifecycle within the wetlands and vertebrate species taking advantage of seasonal variations in water levelto feed on the wetland's resources. According to Van den Wyngaert and Bobbink (2006) the structure and functioning of wetlands can be strongly influenced by vertebrate herbivory. Wetlands provide myriad of ecosystem services which include food in the form of fish, prawn, rice and many other plants, wastewater purification, hydrological regulation of floods and droughts, regulation of climate etc. The ecosystem services provided by wetlands are critical to the entire life sustaining process in Earth. These ecosystem services are related directly or indirectly to the interactions between different biodiversity components of wetlands and abiotic factors. Finlayson *et al.* (2005) discussed the ecosystem services provided by different kinds of wetlands. The relationshipsexisting between the processes,ecosystem components, characteristicsand functions of wetlands with goods and services provided by wetlands has been discussed by Maltby (2009). Oertli *et al.* (2002) reported that many small wetlands together support more biological diversity than one large wetland though some species may occur only in large wetland. Wetlands improve aesthetics and support a widerange of livelihoods apart from various cultural/recreational activities. Overall wetlands contribute to human wellbeing, cultural identity and economic growth. All these factors related to wetland and its biodiversity is possibly the major reason which lead to the fact that conservation and wise use of wetlands gained high priority worldwide. A series of conferences and technical meetings culminated in the 'Convention on Wetlands of International Importance especially as Waterfowl Habitat' (well known as the Ramsar Convention) which came into force in 1975. Therefore, wetlands became the only single group of ecosystems to have their own international convention.Also multilateral environmental agreements related to implementation of biodiversity conservation included wetland conservation policies. In this regard Ramsar convention, Convention on Biological Diversity (CBD) and United Nations Framework Convention on Climate Change (UNFCCC) set up mission that are inter-linked and overlapping. As per as wetland biodiversity is concerned The 5th International Wetlands Conference of INTECOL, held in 1996 in Perth, Australia, was devoted specifically to an analysis of the biodiversity of wetlands. The outcome of this symposium was the publication of two volumes (Gopal *et al.*, 2000; 2001) which became the first major global impressions on wetland biodiversity. Also it has been reported that four overarching goals and nineteen specific targets that directly support achievement of both the Sustainable Development Goals (SDGs) as well as the Aichi Targets set up by the Convention on Biological Diversity that have been identified by The Ramsar Convention's fourth Strategic Plan (2016-2024).Biodiversity of a given ecosystem in general and wetlands in particular provide important insights to ecological health of the said area.

The Ramsar Convention 1971 identified nine criteria for identifying wetlands of international importance. Out of the nine criteria, eight are linked to biodiversity of which three are based on sites of international importance for conserving biological diversity, two are specific for water birds, two specific for fish and one criteria for other taxa.In this chapter an endeavour has been made to discuss the different methodologies about determination of each of these taxa under different criteria as mentioned above. These methodologies will help various stakeholders in appropriate determination of biodiversity of wetland of a particular area. This will also provide immense help to the stakeholders in relating wetland biodiversity with ecological characterization, economic valuation, identifying threats and evolving measures for ecological restoration.

BIO-DIVERSITY ASSESSMENT

1. Requirement of Ecological Data

The baseline ecological data primarily required for the assessment of biodiversity are represented in Table 1 & Table 2 below. It includes habitat particulars, class, brief methods, frequency and duration/time of sampling. The sample size may be decided at a later stage once the interventions are finalised.

METHODOLOGY

Information Gathering

Information shall be gathered through following sources:

1. Desktop Study:
    a. Searching of available scientific literature for biodiversity & ecological studies of a region from internet databases.

Table 1. Floral study details

Habitat	Particulars	Methods	Frequency	Duration / Time Period of Sampling
Terrestrial	Trees	1. Line Transect Method for Linear Plantation (Transect of 100 m length) 2. Quadrat Method for non-linear Plantation (Quadrat of 10m x 10m)	Once during the study period	Daily (actual time to complete may vary from 7-15 days on the sample size)
	Shrubs	Quadrat Method (Quadrat of Size 5m x 5m)	Once during the study period	Daily (actual time to complete may vary from 7-15 days on the sample size)
	Herbs	Quadrat Method (Quadrat of Size 3m x 3m)	Thrice during the study period	Once in a month
	Grasses and Climbers	Quadrat Method (Quadrat of Size 1m x 1m)	Thrice during the study period	Once in a month
	Lichens	Quadrat Method	Thrice during the study period	Once in a month
Aquatic	Phyto-Planktons Zoo-Planktons	Collection of samples from water body (Upstream, Downstream and inside Reservoir) Sub Task 1.1	Morning, Afternoon and Evening (Once during the study period) Sub Task 1.2	Once in a month Sub Task 1.3
	Benthos	Grab Sampling method (Upstream, Downstream and inside Reservoir)	Morning and Evening (Once during the study period)	Once in a month
	Macro-Invertebrates	Dip Net Method using stratified random sampling	Morning and Evening (Once during the study period)	Once a week
Riparian	Macrophytes	Quadrat or Intercept Method (500 m at one Location)	Once during the study period	Once in a month

   b. Collection of secondary information from Forest & Wildlife department on rare, vulnerable, migratory endangered species, critical natural habitat areas & protected areas.
   c. Collection of secondary information from Fishery department about fishes in in an area / reservoir
   d. Collecting necessary data from WII-ENVIS, IBAT & IUCN data repository for endangered species & protected areas.
2. Field Study:
   a. In absence of availability of secondary data for the specified area of study, primary ecological & biodiversity surveys/ studies shall be conducted to collect up-to-date baseline data.
3. Discussions and meetings with following stakeholders:
   a. NGOs
   b. Community
   c. Forest& Wild life Officials
   d. Fishermen near the area / Reservoir

Table 2. Faunal study details

Class	Particulars	Methods	Frequency[1]	Duration / Time Period of Sampling
Mammals	Wild and Domestic	Visual Encounter Survey through Transect Walk	Once during the study period	Morning and Evening (Once a week)
Birds	Terrestrial and Aquatic	Point Count Method	Once during the study period	Morning and Evening (Once a week)
Reptiles	Terrestrial and Aquatic	Visual Encounter Survey through Transect Walk	Once during the study period	Morning and Evening (Once a week)
Amphibians	-	Torch Count Method or Pitfall Method	Once during the study period	Morning and Evening (Once a week)
Insects	Butterfly and Moths	Belt Transect Method	Once during the study period	Morning and Evening (Once a week)
	Dragonfly and Damselfly	Belt Transect Method	Once during the study period	Morning and Evening (Once a week)
Invertebrates	Molluscs etc	Pit Fall Method	Once during the study period	Morning and Evening (Once a week)
Fishes	-	Random Fishing	Morning Four times during the study period	Morning (Once a week)

Offsite Analysis

An offsite analysis of gatheredinformation shall be made to establish the biodiversity status of an area.

Prediction of Impacts

The collected data shall be analysed and potential impacts on the biodiversity and natural habitats shall be predicted and its significance level shall be established. A careful analysis of the functions served by the affected habitats and ecosystems shall be made and the geographical distribution along with benefits of development at state & national levels shall be made.

Mitigation Measures and Management Plan

Options for eliminating, reducing to acceptable levels or mitigating impacts on biodiversity shall be analysed and suitable measures for conservation and mitigation to be designed and incorporated in Environmental Management Plan (EMP).

FIELD STUDY METHODOLOGY

Riparian Ecology

The wetland ecosystem includes the water body and the embankment. Generally, it is observed that physically and financially, it is not possible to make a 100 percent inventory in the wetland environment.

Hence, it is necessary to devise a sampling system to provide as accurate a measure of the attribute as possible with acceptable cost and effort.

1. Riparian Flora / Macrophytes

The survey team shall conduct an extensive biodiversity assessment on the riparian ecology. The forests around the wetland too shall be classified based on scientific literature collected in earlier steps. The sampling method to be used for the study involves Quadrat Method or Intercept Method.

a. Quadrat Method

This method is used for measuring abundances, density, frequency, cover or biomass of macrophytes. They are used to define sample areas within the study area and are made of usually four strips of wood or plastic frame which are fixed together to form a square that can float and can be used to sample floating or emergent vegetation on the water surface. Quadrat size of 5m x 5m can be used for the sampling of aquatic macrophytes.

b. Intercept Method

This method can be used to study the aquatic fauna along the embankment area. A walk through of 500 m at x locations will be undertaken to identify and prepare the inventory of the species. In case of hilly area, it can be reduced to 100 m. and locations can be decided based on area of reservoir. It can be selected for upstream, downstream, reservoir and at edges of lentic and lotic system. Density, frequency and abundance shall also be estimated with each longitudinal traverse measuring a distance of 500m. at x locations.

Aquatic Ecology

1. Fish Survey

The population of fishes in an aquatic system are a result of the physical, biological, and chemical factors surrounding them and especially the biotic factors in the trophic levels below them. The sampling of these trophic level organisms is not only expensive but also often impossible; however, because fishes are dependent on these lower organisms, the understanding of ecosystem functioning can be gained from the fishes themselves. The size, structure, and growth rates of the fish population helps in determining the aquatic habitat conditions that existed in the past few years. Sampling of fish populations must be done accurately because freshwater fishes are notorious for wide fluctuations in year-class strengths.

Use of electrofishing, explosive prima cord, spot explosive concussion, toxicants, nets, scuba or snorkel, redd counts, Seine Netting, Random fishing, Capture, Mark and Release method are some common field techniques to obtain data to estimate fish population and biomass, fish species composition, and fish health and survival. The Seine Netting (Heavily leaded seine with mesh) method is described below.

a. Seine Netting (Heavily Leaded Seine With Mesh) Method

A seine net is the most common type of net used on fish farms to harvest fish. It is a long net with ropes at each end and is pulled to collect the fish and then drawn into a circle to trap them. Random experimental fishing using seine net is to be done in the lotic water of by local made fishing net of size 152 x 900 cm. at upstream and downstream of the site once a week. The sizes of the aperture of the local nets that can be used are 1.0 x 1.0 cm, 1½ x 1½ cm and 3.0 x 3.0 cm respectively. Altogether, six stations shall be made for random sampling once a week. The percentage; average weight and size of each species will be determined from the total number of collected specimens. In addition physio-chemical characteristics of the water samples of study area along with average depth & deepest depth of the waterbodyis to be recorded. The surveys shall also find spawning grounds, if any within the study area.The fish sampling data is to be analysed and categorised as abundant, rare, threatened or endangered.

At each site 500 m reach of stream sampling is to be done with a heavily leaded seine with mesh during each visit. The technique consists of slow downstream seining, as close to the shore and cover as possible.

- During each site visit – 20 seine hauls of adequate length, covering the same amount of surface area & the same segment of the shoreline will be taken. In addition, a mesh, non-filament gill net will be placed across the site.
- Dissolved oxygen (DO) to be measured during each visit & so will be the other water qualities like temperature, turbidity etc.
- Information on fish, fish habitat, and environmental factors along with weather information is to be recorded. Several photographs will also be taken to record the conditions at that visit.
- The specimen samples of fishes caught will be preserved in 10% formalin (Formaldehyde). In the laboratory, fish will be sorted & identified. Max & min. length & weight is to be measured. Standard length, total biomass, densities (fish/m^2), biomass densities (g/m^2), species diversity (based on no. of individual/species & biomass) to be calculated.
- Spies diversity is to be calculated from the numerical & biomass data with the Shannon-Weaver equation
- Diversity is to be calculated for each collection from each site. The individual diversities will be used to calculate the mean annual numerical/biomass diversity for a specific site in a given year to the mean numerical diversity (MND) for a specific site over the sampling period.

The percentage, average weight and size of each species shall be determined from the total number of collected specimens using sampling data sheet represented in Table 3 below

Table 3. Fish sampling data sheet

Sl. No.	Name of Species	Total Count	Average Length (in cm)	Average Weight (in gm)	Percentage From Grand Total Count
1.					
2.					
3.					

Note: The total length of a fish is measured from the tip of the snout to the end of the tail fin.

Macroinvertebrates

Macroinvertebrates are animals without backbones and are large enough to be seen without magnification. These includes annelids, crustaceans, flatworms, molluscs, and insects (usually predominant). These are the important intermediaries in the utilization of plant material, such as algae, vascular hydrophytes, leaves, and wood, and the recycling of nutrients in aquatic environments. These animals are a major food source for fishes and serve to determine the well-being of those populations. Besides, these animals possess several characteristics that make them useful for detecting environmental perturbations:

1. Most members of this community possess limited mobility; hence, their status reflects conditions in the immediate vicinity of the collection site.
2. Most of the organisms (mussels are the main exception) have life spans of several months to a few years. Thus, their characteristics are a function of conditions during the relatively recent past, including sporadic influences that would be difficult to detect by periodic microbial or chemical analysis.

Hence, sampling of these animals is important to study any aquatic system. The sampling location should be selected among the reservoir area, upstream and downstream area. The stratified random sampling method using Dip Net should be used as the distribution of Macroinvertebrates is heterogeneous largely as a result of the non-random distribution of micro environmental features, especially the substratum and the current. In stratified sampling, the first unit in the sample is selected at random and the next units are established at fixed intervals from the first. The sampling method selected for the study is Dip Net method.

a. Dip Net Method

This method is designed to sample major habitats in proportional representation within a sampling reach. Benthic macro invertebrates will be collected systematically from all available stream habitats by kicking the substrate or jabbing with a D-frame dip net. A total of 5 jabs (or kicks) will be taken from all major habitat types in the reach resulting in sampling of approximately 80% of habitat. For example, if the habitat in the sampling reach is 50% snags, then 50% or 10 jabs should be taken in that habitat.

Standard Sampling Gears:

- D-frame dip net: Dimensions of frame are 0.3 m width and 0.3 m height and shaped as a "D" where frame attaches to long pole. Net is cone or bag-shaped for capture of organisms. It can be used in a variety of habitat types and used as a kick net, or for "jabbing", "dipping", or "sweeping".
- Rectangular dip net: Dimensions of frame are 0.5 m width and 0.3 m height and attached to a long pole. Net is cone or bag-shaped. Sampling is conducted similarly to the D-frame.

Sampling Procedure:

A 100 m reach that is representative of the characteristics of the wetland should be selected.

- Before sampling, complete the physical/chemical field sheet to document site description, weather conditions, and land use. After sampling, review this information for accuracy and completeness.
- Draw a map of the sampling reach. This map should include attributes (e.g., riffles, falls, fallen trees, pools, bends, etc.) and important structures, plants, and attributes of the embankment. Indicate the areas that were sampled for macro invertebrates on the map. If available, use hand-held GPS for latitude and longitude determination taken at the furthest point of the sampling reach.
- Sampling begins at the downstream end of the reach and proceeds upstream. A total of 5 jabs or kicks will be taken over the length of the reach; a single jab consists of forcefully thrusting the net into a productive habitat for a linear distance of 0.5 m. A kick is a stationary sampling accomplished by positioning the net and disturbing the substrate for a distance of 0.5 m upstream of the net.
- The jabs or kicks collected from the multiple habitats will be composited to obtain a single homogeneous sample. Every 1 jab, more often if necessary, wash the collected material by running clean stream water through the net two to three times. If clogging does occur that may hinder obtaining an appropriate sample, discard the material in the net and redo that portion of the sample in the same habitat type but in a different location. Remove large debris after rinsing and inspecting it for organisms; place any organisms found into the sample container. Do not spend time inspecting small debris in the field.
- Transfer the sample from the net to sample container(s) and preserve in enough 95% ethanol to cover the sample. Forceps may be needed to remove organisms from the dip net. Place a label indicating the sample identification code or lot number, date, stream name, sampling location, and collector name into the sample container. The outside of the container should include the same information and the words "preservative: 95% ethanol". If more than one container is needed for a sample, each container label should contain all the information for the sample and should be numbered (e.g., 1 of 2, 2 of 2, etc.). This information will be recorded in the laboratory.
- Document observations of aquatic flora and fauna. Make qualitative estimates of macro-invertebrate composition and relative abundance as a cursory estimate of ecosystem health and to check adequacy of sampling.
- Return samples to laboratory.

Quality Control in Field:

- Sample labels must be properly completed, including the sample identification code, date, stream name, sampling location, and collector's name and placed into the sample container. The outside of the container should be labelled with the same information.
- After sampling has been completed at a given site, all nets, pans, etc. that have come in contact with the sample should be rinsed thoroughly, examined carefully, and picked free of organisms or debris. Any additional organisms found should be placed into the sample containers. The equipment should be examined again prior to use at the next sampling site.
- Replicate (1 duplicate sample) 10% of the sites to evaluate precision or repeatability of sampling technique or collection team.

b. Benthos

Benthos are the bottom dwelling organisms in an aquatic ecosystem. They feed upon algae and other organic matters in bed of water body. It includes Diatoms, Sponges, Bivalves, and Polychaetes etc. They are found adhered to substratum like pebbles, gravels etc. These organisms are food source to fishes, snails, crustaceans etc. Hence, their population affects the population of other organism in larger trophic levels. For study of these organisms, Grab Sampling Method will be followed. Grabs at different locations in water body should be undertaken to identify these species. Grab sampler will be used for grabbing the sample. The grabs shall be taken in both banks and in middle of the reservoir.

c. Plankton

Planktons are those organisms which are unable to propel themselves against a water current. They act as a source of food to many small and large aquatic animals like fishes, bivalves etc. The presence or absence of these organisms represent the quality of aquatic ecosystem hence their analysis plays an important role in understanding the changes in aquatic environment. These organisms are not randomly distributed across a water body, due to variations in flow and nutrients. So, samples should be taken at various locations. The sampling of planktons is to be done using plankton net.

In the study area, the water body will be sampled using a plankton net to cover a larger area and to concentrate the plankton sample. A plankton net consists of a fine-mesh net (e.g. 30 μm) ending with a bottle to hold the plankton. The net can be fixed either to a pole or to towing lines. For the latter the net can be towed either by person or by a boat. The sample collected will be preserved using Lugol's Solution immediately after collection. The sample will be then transferred to Laboratory. Species Identification and Density estimation can be determined through a microscope.

TERRESTRIAL ECOLOGY

Mammals

Visual Encounter Survey Method Through Transect Walk

The survey to understand the faunal presence and their density in the study area will be carried out using Visual Encounter Survey Method through standard line transect walk. A sampling unit in the form of a long continuous strip in which only those individuals which are observed on the transect line are assumed to be completely counted. In this method the transect line were chosen randomly perpendicular with the road / distribution line axis. The length of the transect lines are dependent on the habitat and the terrain. This transect lines has been drawn during day time. Randomly selected alternate transects has been made on both the sides of the road. Thus, for every kilometre there are two transects, one on either side of the road / distribution line.

Nocturnal random surveys shall be carried out in every alternate day to identify the presence of nocturnal species in the study area in general and also to study the nocturnal movements of various faunal species in particular. The entire study stretch shall be monitored during different time intervals starting from 7 PM in the evening till 5 AM. A general checklist shall also be prepared for any and all encounter that happened in the study area.

Birds

Point Count Method

A sampling unit in the form of a predefined route within a predefined survey unit is undertaken. The avifauna is counted at regular intervals along the route and for a given duration at each point. It is a preferred survey method in many situations. It is highly adaptable and can be used in terrestrial and aquatic systems. It can be used to survey individual species, or groups of species. This method can be used to examine bird–habitat relationships and can be used to derive relative and absolute measures of bird abundance.

Breeding pairs, individuals, can be detected by walking along the shores of the rivers. This method has essentially the same assumptions as the line transect method (Emlen, 1977). Birds are relatively easily detectable on smaller systems because of the observer, although they can fly past unnoticed if observers are not constantly watching the river. These birds are constantly moving up and down so observers need to note the direction and numbers of birds moving past.

Reptiles

Visual Encounter Survey Method Through Transect Walk

The survey for study of reptiles will be carried out using Visual Encounter Survey Method through transect walk. A sampling unit in the form of a long continuous strip in which only those individuals which are observed on the transect line are assumed to be completely counted. In this method the transect line were chosen randomly perpendicular with the road / distribution line axis. The length of the transect lines are dependent on the habitat and the terrain. This transect lines has been drawn during day time. Reptiles seen or encountered along transect will be noted (Garden *et al.*, 2007).

Amphibians

Torch Count Method and Pitfall Method

The survey for study of amphibians will be carried out using Torch Count Method or Pitfall Method. In Torch Count Method, the amphibians along water body will be searched using torch at night and the species seen will be noted. In pitfall method, pits around the water body will be made and the amphibians found in the pits will be noted. The pits will be made in all directions of the water body so that amphibians from all directions can be noticed (Garden *et al.*, 2007).

Butterflies, Dragonflies and Other Insects

Belt Transect Method

The survey for study of Insects will be carried out using Belt Transect Method. A belt transect of 5 m width and 500 m length will be used for survey of insects. The survey will be carried out at morning and evening (Sutherland, 2006).

Invertebrates

Surface-living invertebrates such as larger beetles, spiders, moths, night flying insects and ants etc. will also be surveyed. The Methods used for invertebrate study are mentioned below.

a. Pit Fall Trap Method

This method is used for trapping larger beetles, spiders, ants in lower vegetation and bare ground cover. Pitfall traps consist of straight-sided containers sunk level with the surface of the ground into which invertebrates inadvertently fall. The rim must be flush with the soil – even a small ridge or gap reduces the trap's effectiveness. Any size or type of container with smooth sides can be used, but tough polythene tubs about 8 cm in diameter and 10 cm deep with fitting lids have been found to work well and have the advantages that they are less easily crushed by stock, do not crumple when pushed into stony ground, and can be sealed quickly and replaced with a new tub without the bother of emptying the sample into a polythene bag. Traps with a larger circumference will catch more invertebrates, but this should be balanced against the need to set enough traps in different micro-habitats to ensure that the catches are representative of the site as a whole. Catches can be increased by erecting barriers that direct invertebrates to the traps (McCarvy, 2018; Sutherland, 2006).

Preservative such as Propylene Glycol solutions are usually put in the trap to arrest decay and prevent invertebrates from eating each other. It is recommended that the traps are emptied at fortnightly intervals in warm weather and checked to ensure that they have not become blocked up with leaves, have not overflowed with rain water, have not caught small vertebrates, which will act as carrion bait, and have not been upset by large vertebrates. Use of a cover can reduce these problems. Pitfall traps may be used without a preservative for short periods (a few days at most), but large animals will eat the small ones. Small vertebrates often fall in, but this problem can be reduced using some sort of raised lid or grid over the top, leaving a small gap (about 10 mm). Pitfall traps can also be baited with raw meat, fish, cheese or fermenting fruit to attract beetles. Preservatives should not be used in baited traps, since these may mask the smell of the bait. Such traps therefore have to be checked daily.

b. Light Trap Method

Light traps are used to catch moths, flies and other night-flying insects. Many night-flying insects, particularly moths, are attracted towards light, particularly at the UV end of the spectrum. Insects attracted to a light can be caught using a net or within a collecting container. Catches can be increased by positioning the light beside a white wall or next to or above a white sheet.

A light trap (most types are usually referred to as moth traps) can be used. These traps consist of a container with an opening and a light inside it. Insects attracted towards the light spiral around it and hit the baffles which cause them to fall down into the trap. Catches will be maximised if the trap is positioned so that its light shines into as much habitat as possible. The response to light traps varies greatly among species. Some species tend to settle near the trap, rather than entering it, so it is worth searching vegetation in the vicinity of the trap to reveal additional species (McCarvy, 2018; Sutherland, 2006).

c. Aerial Attractants and Traps Method

These methods are used for flies, butterflies, moths, dung beetles and wasps. In this method, baits are placed in traps that contain a funnel or narrow opening that guides the insects into a collecting chamber where they can be held alive or killed. Examples of baits are rotting fruit for fruit flies and butterflies (over-ripe bananas are particularly good); dung for dung flies and dung beetles (McCarvy, 2018; Sutherland, 2006).

Terrestrial Flora Survey

Sessile plants like trees, shrubs, herbs, grasses etc. are usually arranged over a substrate (soil, sediment, etc.) and can be found, identified and examined easily. This characteristic makes easier to census plants than it is to census other organisms and estimates of, e.g., density, species number and composition and distribution of a species are more accurate for plants. The methods used for sample of plants are described below (Gilbert *et al.*, 1998; Higgins Pearce *et al.*, 2008).

a. Line Transect Method

Transects are commonly used to survey changes in vegetation along an environmental gradient or through differing habitats. The length of the transect can be several centimetres or hundreds of kilometres, depending on the vegetation and the aim of the study. This method is used to study the density of plant. This method will be used for survey of tree species and densities along linear plantation in the study area as less variation in number of species is observed along linear plantation. Hence, at such locations, transects are easier to implement than quadrats. In study area, normally transects of 100m will be undertaken at various locations and quantitative and qualitative data for tree species will be noted.

b. Quadrat Method

This method is used for measuring abundances of sessile species in any vegetation, including aquatic macrophytes. It can be used to measure density, frequency, cover or biomass. They are used to define sample areas within the study area and are usually four strips of wood, metal or rigid plastic, which are fixed together to form a square. Quadrats are the basic tool of plant ecologists. They are very easy to use and can be employed in a wide range of studies. Quadrat method, which is being adapted to study the local vegetation as well as the habitat structure is an internationally accepted standard method.

Quadrat of 10m x 10m is taken for trees survey, quadrat of 5m x 5m is taken for shrubs survey, quadrat of 3m x 3m is taken for herbs survey and quadrat of 1m x 1m is taken for grasses, climbers and lichens survey. At least 60-80% of area should be selected for survey to decrease biasness in the study.

As the study area is a terrestrial system and shows homogeneity in the vegetation structure hence representative quadrat of fixed size in identified terrains put on a random manner helps to ascertain the habitat structure.

The study of communities will also be undertaken by using qualitative characteristics and quantitative characteristics. Qualitative characteristic mainly involves presence or absence of the species, genera and family. This showed the community structures, composition and other characteristics which can be readily described by visual observation without actual measurements. The quantitative analysis involved the structure and composition of vegetation across vegetation types and compared in terms of frequency, density and abundance.

OFF SITE ANALYSIS

The following parameters is to be analysed from the data collected (Buckland *et al.*, 1993; Sutherland, 2006).

Density

Density is an expression of the numerical strength of a species where the total number of individuals of each species in all the quadrats is divided by the total number of quadrats studied.

Density is calculated by the equation:

$$Density = \frac{Total\,number\,of\,individuals\,of\,a\,species\,in\,all\,quadrats}{Total\,number\,of\,quadrats\,studied}$$

Frequency (%)

This term refers to the degree of dispersion of individual species in an area and usually expressed in terms of percentage occurrence. It was studied by sampling the study area at several places at random and recorded the name of the species that occurred in each sampling units. It is calculated by the equation:

$$Frequency(\%) = \frac{Number\,of\,quadrats\,in\,which\,the\,species\,occurred \times 100}{Total\,number\,of\,quadrats\,studied}$$

Abundance

It is the study of the number of individuals of different species in the community per unit area. By quadrats method, samplings are made at random at several places and the number of individuals of each species was summed up for all the quadrats divided by the total number of quadrats in which the species occurred. It is represented by the equation:

$$Abundance = \frac{Total\,number\,of\,individuals\,of\,a\,species\,in\,all\,quadrats}{Total\,number\,of\,quadrats\,in\,which\,the\,species\,occurred}$$

Importance Value Index

This index is used to determine the overall importance of each species in the community structure. In calculating this index, the percentage values of the relative frequency, relative density and relative dominance are summed up together and this value is designated as the Importance Value Index or IVI of the species (Curtis, 1959).

Relative Density

Relative density is the study of numerical strength of a species in relation to the total number of individuals of all the species and can be calculated as:

$$Relative\ density = \frac{Number\ of\ individuals\ of\ the\ species \times 100}{Number\ of\ individuals\ of\ all\ the\ species}$$

Relative Frequency

The degree of dispersion of individual species in an area in relation to the number of all the species occurred.

$$Relative\ frequency = \frac{Number\ of\ occurrences\ of\ the\ species \times 100}{Number\ of\ occurrences\ of\ all\ the\ species}$$

Relative Dominance

Dominance of a species is determined by the value of the basal cover. Relative dominance is the coverage value of a species with respect to the sum of coverage of the rest of the species in the area.

$$Relative\ dominance = \frac{Total\ basal\ area\ of\ the\ species \times 100}{Total\ basal\ area\ of\ all\ the\ species}$$

The total basal area was calculated from the sum of the total diameter of immerging stems. In trees, poles and saplings, the basal area was measured at breast height (1.37m) and by using the formula πr^2.

The floral communities were identified based on IVI values of trees. A single species representing >50% of the total IVI was designated as a single species dominated community, whereas two or more species contributing 50 or >50% of the total IVI were named as a mixed community.

Analysis of Community

The data obtained is to be used for community analysis by calculating the following indices.

Shannon-Weiner Diversity Index

Shannon-Weiner Diversity Index is widespread diversity index in use by ecologists (Danilov and Ekelund, 2001; Simboura and Reizopoulou, 2007). Shannon-Weiner Diversity Index is calculated using the formula

$(H) = -\sum PiInPi$

where, pi represents proportion of total number of species made up of the ith species.

Simpson's Index of Diversity

Simpson's index of diversity is calculated using the mathematical formula as given below:

$D=1-\Sigma (ni/N)^2$ (Southwood, 1978)

where,ni represents number of individuals of a particular species N represents total number of all individuals of all species

Margalef Species Richness Index

Margalef's index is calculated using the following mathematical formulae given below (Margalef, 1958).

$D_{Mg} = (S - 1) / In\ N$

where,S represents total number of species and N represents total number of individuals in the sample

Evenness Index

This index is calculated for analysing the degree of equality in species abundance from a particular the sample. The index is calculated using the mathematical formula as given below: (Pielou, 1966)

$E= (H)/Hmax$

Where, H represents Shannon-Weiner Index and Hmax = Maximum Species diversity

It is a measure of the uniformity of different species in a community.

BIODIVERSITY ASSESSMENT AND PEOPLE'S BIODIVERSITY REGISTER

To understand the change in a wetland ecosystem over a year there are two other methods which are proving to be important one is developing citizen science platform and also to use the People's Biodiversity Register (PBR) as per Convention on Biological Diversity. This gives the people's perspective to document the change and usage of a particular wetland (Gadgil *et al.*, 2006).

The Sample questionnaire like

- What type of birds visits the wetland?
- What are the fishes available?
- Which one are edible?
- What are the changes that have taken place in the wetland in the last 10 years?
- Is there any change in the submerged vegetation?

Developing the questionnaire depends on the objective and outlook which helps in formulating the questions and also helps in developing a better insight about the use of wetland among the local community and its temporal changes.

USE OF REMOTE SENSING AND GEOGRAPHIC INFORMATION SYSTEM (GIS) FOR ASSESSMENT OF BIODIVERSITY

Remote sensing and Geographic Information System (GIS) can be used as important tools separately or in combination for application in studies of forestry and biodiversity. Remote Sensing and GIS methodologies use the satellite imageries to detect the change in an area. This can be done by using satellite imageries available from Landsat or Sentinel at the same time using softwares like Arc-Gis or QGIS software to create the classes of which the change can be studied. Like creating Productivity value detecting changes in productivity or vegetation cover or the surface and shape change. Identifying the land use land cover change in the wetland and its adjoining areas. These remote sensing data they can be ground verified and a prediction system can be generated by using these data (Qisheng, 2018).

WETLAND BIODIVERSITY, ECOSYSTEM SERVICES AND CONSERVATION

The Ecosystems of the world provide a myriad of services that are vitalfor human well-being, health, livelihoods, and survival (Costanza *et al.,* 1997; Millennium Ecosystem Assessment (MEA), 2005; TEEB, 2010). The Millennium Ecosystem Assessment (2005) categorised the ecosystem services into four groups namely provisioning services, regulating services, supporting and cultural Services. Over the decades, different ecologists have discussed the relationship between biodiversity and ecosystem functions (Schulze and Mooney, 1993; Loreau, 2000). All ecosystem services directly or indirectly linked with biodiversity. A number of drivers governed has been recognized to govern biodiversity which affect processes and function and consequently the ecosystem services. The ecosystem services of wetlands has been reported to be diverse in nature (Barbier *et al.*,1997; Russi *et al.*, 2013). Although wetlands cover only 1.5% of the Earth's surface, but it has been reported that wetlands provide around 40 per cent of global ecosystem services disproportionately (Zedler and Kercher, 2005). Wetlands differ in terms of geology, climate and anthropogenic factors. Different wetland types therefore shows variation in drivers and wetland processes and accordingly the ecosystem services linked to it. Therefore all kinds of ecosystem services can be obtained from any particular wetland or a kind of wetland. Finlayson *et al.* (2005) discussed the ecosystem services provided by different kinds of wetlands. The ecosystem services of wetlands includes climate regulation that occurs as wetlands capture carbon dioxide, methane and nitrous oxide; nursery provision for fish population; flood protection to inland settlements; regulating groundwater recharge; filtering nutrients and pollutants from surface water; and providing opportunities for recreational and educational activities (Dise, 2009; Pramod *et al.,* 2011; Sarkar and Upadhyay, 2013).Wetland services are significant in the context of food security as various food products such as fish, rice etc. are available from wetlands and also various crops are found to be grown along the edges of wetlands (Kakuru *et al.*, 2013). The TEEB (2013) study explained the values of ecosystem services in terms of dollar. Wetlands throughout the world provide different valuable ecosystem services however they have been systematically utilized to support agriculture, urban expansion, and other developments leading to deterioration and loss of wetland area. It has been reported about 50% of the world's original wetland area has been lost (Clarkson *et al.*, 2014) ranging from relatively minor losses in boreal countries to extreme losses of >90% in parts of Europe (Mitsch and Gosselink 2000b). According to National Research Council (2001) and TEEB (2013) wetland area and condition continue to decline in many developing and developed countries in spite of strong regulatory protection. Urgent remediation measures are required for many wetlands if key functions and associated ecosystem services from the wetlands are to be maintained. Therefore restoration of degraded wetland has become an important priority in the current scenario. These restoration practices will help to recover much of the biodiversity and associated ecosystem services (Zedler, 2005). Major shift in policies related to land and water use are to be looked upon for conservation of wetlands in the context of protecting biodiversity and obtaining optimum benefits from them. The real causes of wetland loss and degradation need to be thoroughly addressed. The total biodiversity of all wetlands should be assessed properly and periodical monitoring and all ecosystem services of wetlands should be assessed and valued in economic terms. The economic valuation should address the benefits to the local community and their livelihoods and the action for conservation and wise use of wetlands must clearly be taken at local, national and international levels.

CONCLUSION

The wetland ecosystem is one of the most complex ecosystems of the world and provide essential ecosystem services. Wetland supports rich biodiversity. Wetland biodiversity is the crucial natural resource for maintaining sustainability. Many of these wetlands have already lost their biodiversity and ecosystem services because of natural and anthropogenic interventions. Various methodologies are involved in determination of different flora and fauna inhabiting in wetlands. The methodologies comprises both field study and offsite analysis to determine the overall biodiversity of wetland of a particular area. Monitoring, determination and assessment of wetland biodiversity is highly essential in the context of ecological characterization, associated ecosystem services and its economic valuation, identifying threats and evolving measures for ecological restoration. This will provide immense help to stakeholders for policy or decision-making process for conservation of wetland and their sustainable use.

REFERENCES

Barbier, E. B., Acreman, M., & Knowler, D. (1997). *Economic Valuation of Wetlands: A Guide for Policy Makers and Planners*. Ramsar Convention Bureau.

Buckland, S. T., Anderson, D. R., Burnham, K. P., & Laake, J. L. (1993). *Distance Sampling: Estimating Abundance of Biological Populations*. Chapman and Hall.

Clarkson, B. R., Ausseil, A. G. E., & Gerbeaux, P. (2014). Wetland ecosystem services. In J. R. Dymond (Ed.), *Ecosystem Services in New Zealand* (pp. 192–202). Manaaki Whenua Press.

Costanza, R., d'Arge, R., de Groot, R., Farber, S., Grasso, M., Hannon, B., Limburg, K., Naeem, S., O'Neill, R. V., Paruelo, J., Raskin, R. G., Sutton, P., & van den Belt, M. (1997). The value of the world's ecosystem services and natural capital. *Nature*, *387*(6630), 253–260. doi:10.1038/387253a0

Cowardin, L. M., Carter, V., Golet, F. C., & LaRoe, E. T. (1979). *Classification of Wetlands and Deep Water Habitats of the United States*. US Department of the Interior, Fish and Wildlife Service, Office of Biological Services.

Cowardin, L. M., Carter, V., Golet, F. C., & LaRoe, E. T. (1992). *Classification of Wetlands and Deepwater Habitats of the United States*. U.S. Department of the Interior, Fish and Wildlife Service.

Danilov, R. A., & Ekelund, N. G. A. (2001). Comparative studies on the usefulness of seven ecological indices for the marine coastal monitoring close to the shore on the Swedish East coast. *Environmental Monitoring and Assessment*, *66*(3), 265–279. doi:10.1023/A:1006364317956 PMID:11281424

Dise, N. B. (2009). Peatland response to global change. *Science*, *326*(5954), 810–811. doi:10.1126cience.1174268 PMID:19892972

Emlen, J. T. (1977). Estimating breeding season bird densities from transect counts. *The Auk*, *94*, 455–468.

Finlayson, C. M., D'Cruz, R., Aladin, N., Barker, D.R., Beltram, G., Brouwer, J., Davidson N., Duker, L., Junk, W.J., Kaplowitz, M.D., Ketelaars, H., Kreuzberg-Mukhina, E., Espino, G.L., Lévêque, C., Lopez, A., Milton, R.G., Mirabzadeh, P., Pritchard, D., Revenga, C., Rivera, M., … Steinkamp, M. (2005). Inland Water Systems. In Ecosystems and Human Well-being: Current State and Trends (pp. 551-583). Washington, DC: Island Press.

Garden Jenni, G., Mcalpine Clive, A., Possingham Hugh, P., & Jones Darryl, N. (2007). Using multiple survey methods to detect terrestrial reptiles and mammals: What are the most successful and cost-efficient combinations? *Wildlife Research*, *34*(3), 218–227. doi:10.1071/WR06111

Gilbert, G., Gibbons, D. W., & Evans, J. (1998). *Bird Monitoring Methods*. RSPB Sandy.

Gopal, B., & Junk, W. J. (2000). Biodiversity in wetlands: An introduction. In B. Gopal, W. J. Junk, & J. A. Davis (Eds.), *Biodiversity in Wetlands: Assessment, Function and Conservation* (Vol. 1, pp. 1–10). Island Press.

Gopal, B., & Junk, W. J. (2001). *Assessment, determinants, function and conservation of biodiversity in wetlands: Present status and future needs*. Academic Press.

Gopal, B., Junk, W. J., Kakuru, W., Turyahabwe, N., & Mugisha, J. (2013). Total economic value of wetlands products and Services in Uganda. *The Scientific World Journal*, 1–13.

Larcher, W. (2003). *Physiological plant ecology*. Springer. doi:10.1007/978-3-662-05214-3

Loreau, M. (2000). Biodiversity and ecosystem functioning: Recent theoretical advances. *Oikos*, *91*(1), 3–17. doi:10.1034/j.1600-0706.2000.910101.x

Maltby, E., & Barker, T. (2009). *The Wetlands Handbook*. Blackwell Science. doi:10.1002/9781444315813

Margalef, R. (1958) Temporal succession and spatial heterogeneity in phytoplankton. In Perspectives in Marine biology (pp. 323-347). University of California Press. doi:10.1525/9780520350281-024

McCravy, K. W. (2018). A Review of Sampling and Monitoring Methods for Beneficial Arthropods in Agroecosystems. *Insects*, *9*(4), 170. doi:10.3390/insects9040170 PMID:30477100

Millennium Ecosystem Assessment (MEA). (2005). *Ecosystems and human well-being: a framework for assessment. Wetlands and water*. World Resources Institute.

Mitsch, W. J., & Gosselink, J. G. (2000a). The Value of Wetlands: Importance of Scale and Landscape Setting. *Ecological Economics*, *35*(1), 25–33. doi:10.1016/S0921-8009(00)00165-8

Mitsch, W. J., & Gosselink, J. G. (2000b). *Wetlands*. John Wiley.

National Research Council. (2001). *Compensating for wetland losses under the Clean Water Act*. DC, National Academy Press.

Oertli, B. D. A., Joye, D. A., Castella, E., Juge, R., Cambin, D., & Lachavanne, J.-B. (2002). Does size matter? The relationship between pond area and biodiversity. *Biological Conservation*, *104*(1), 59–70. doi:10.1016/S0006-3207(01)00154-9

Pearce-Higgins, J. W., Stephen, L., Langston, R. H. A., & Bright, J. A. (2008). Assessing the cumulative impacts of wind farms on peatland birds: A case study of golden plover *Pluvialis apricaria* in Scotland. *Mires and Peat*, *4*, 1–13.

Pielou, E. C. (1966). The measurement of diversity in different types of biological collections. *Journal of Theoretical Biology*, *13*, 131–144. doi:10.1016/0022-5193(66)90013-0

Pramod, A., Kumara, V., & Gowda, R. (2011). A study on Physico-Chemical Characteristics of Water in Wetlands of Hebbe Range in Bhadra Wildlife Sanctuary, Mid-Western Ghat Region. *Indian Journal of Experimental Sciences*, *2*(10), 9–15.

Russi, D., ten Brink, P., Farmer, A., Badura, T., Coates, D., Förster, J., Kumar, R., & Davidson, N. (2013). *The Economics of Ecosystems and Biodiversity for Water and Wetlands*. IEEP.

Sarkar, A., & Upadhyay, B. (2013). Assessment of the Variations in Physico-Chemical Characteristics of Water Quality of the Wetlands in District Manipuri (UP), India. *International Journal of Geology, Earth and Environmental Sciences*, *3*(1), 95–103.

Schulze, E. D., & Mooney, H. A. (1993). *Biodiversity and ecosystem function*. Springer-Verlag. doi:10.1007/978-3-642-58001-7

Simboura, N., & Reizopoulou, S. (2007). A comparative approach of assessing ecological status in two coastal areas of Eastern Mediterranean. *Ecological Indicators*, *7*(2), 455–468. doi:10.1016/j.ecolind.2006.05.003

Southwood, T. R. E. (1978). *Ecological Methods*. Chapman and Hall.

Sutherland, W. J. (2006). *Ecological Census Techniques: A Handbook*. Cambridge University Press. doi:10.1017/CBO9780511790508

TEEB. (2010). *The Economics of Ecosystems and Biodiversity: Ecological and Economic Foundations* (P. Kumar, Ed.). Earthscan.

TEEB. (2013). The economics of ecosystems and biodiversity for water and wetlands. Institute for European Environmental Policy (IEEP) & Ramsar Secretariat.

Van den Wyngaert, I. J. J., & Bobbink, R. (2006). The influence of vertebrate herbivory on ecological dynamics in wetland ecosystems. In E. Maltby (Ed.), *The wetlands handbook*. Blackwell Scientific.

Wu, Q. (2018). GIS and Remote Sensing Applications in Wetland Mapping and Monitoring. In B. Huang (Ed.), *Comprehensive Geographic Information Systems* (Vol. 2, pp. 140–157). Elsevier. doi:10.1016/B978-0-12-409548-9.10460-9

Zedler, J. B. (2005). Progress in wetland restoration. *Trends in Ecology & Evolution*, *10*, 402–407. PMID:10998517

Zedler, J. B., & Kercher, S. (2005). Wetland resources: Status, trends, ecosystem services and restorability. *Annual Review of Environment and Resources*, *30*(1), 39–74. doi:10.1146/annurev.energy.30.050504.144248

Chapter 10
Solid Waste Disposal on Inland Water Wetlands:
A Sustainable Restoration of Wetland Ecosystems

Chandani Bhattacharjee
H.R. College of Commerce and Economics, India

ABSTRACT

Generation of solid waste precedes the surge of urbanization. The earliest waste dumping is recorded in Greece as early as 500 BCE, the conservancy workers in France, wastewater treatment in London, and aqueduct systems in oriental civilizations. The magnitude of waste has been compounding annually with the rise of global population, urbanization, and economic growth. Waste has been overtly and irresponsibly dumped in inland water bodies and the wetlands around it causing inherent damage to the fluvial, pond, or riverine ecosystems. The United Nations has declared this decade to be for ecosystem restoration, and hence, this chapter intends to ponder and establish the concerns of health, species modification, ecosystem endangering, pollution of the surface and subsurface water, impact on the vegetation along the water stretches, to name a few. The objective of this chapter is to evaluate the impact on the ecospheres while arriving at sustainable restoration options.

INTRODUCTION

Loosely described as discarded and used products, wastes have been one of the realities of modern day living. The urban solid waste varies in composition from country to country. The nature and composition is more organic in developing and under developed country, than the developed nations. In India, the State Pollutions Control Boards working in conjunction with the Central Pollution Control Boards, have defined solid waste as, *those undesirable useless and unwanted materials and substances that arise from*

DOI: 10.4018/978-1-7998-9498-8.ch010

animal and human activities (MPCB, 2016). Generically solid waste can be categorised into, depending on their source, into, Household waste (also called municipal waste), Industrial waste as hazardous waste, Biomedical waste or hospital waste as infectious waste, Hazardous Waste and electronic waste. Historically wastes has been generated in all countries and have been despised within and in the outskirts of the cities, towns or urban agglomerations. Over time with the rise in population and the compounding increase per capita of the waste generation, the waste volumes have also headed northwards. Dumpsites and open dumping have been the most preferred way to dispose the wastes effectively and economically. With the increasing pressure on land and their occupation for urban infrastructures, in many cities the dumping yards have been located on or next to inland rivers, creeks, *nalas* (a local name for storm water channel) or estuarine regions. India has several cities which have such dumpsites, by the inland water bodies, namely, the Dhapa Dumpyard (East Kolkata Wetlands), Gorai, Mahim, Creek and Thane Creek (Manori Creek, Mumbai), Mavallipura (Bangalore), Astamudi Lake (Kerala) Adyar River (Chennai), Sheeshambada (Dehradun), Yamuna and Gazipur (Delhi), to mention a few. In Africa, Agbogbloshie (Odaw River, Ghanna), Shey Wenz River, Ethiopia are some examples.

Globally the amount of waste produced is a whopping 11.2 billion, out of which solid waste is 2.01 billion (World Bank[1] 2016), which is expected to rise to 3.04 billion in 2050(World Bank, 2016). Different countries produce unequal amounts of Municipal Solid Waste depending on the level of development of the country, i.e. developed nations have more inorganic and recyclable wastes in their municipal solid waste while developing and under developed nations have more organic waste than the inorganic waste in their municipal solid waste bins and yards. United States, is the leading producer of municipal solid waste, at, 258 million metric tonnes, followed by China(220.4 million metric tonnes) and India(168.4 million metric tonnes) (Tiseo, 2020).Other countries in Africa, Nigeria(27.61 million metric tonnes) South Africa (18.46 million metric tonnes) while EU as per Eurostat has reduced its waste generation to 53 million tonnes in 2019 (Eurostat, 2021),from 121 million tonnes in 1995, owing to waste management, recycling, bioremediation, incineration among others. Waste disposal, be it MSW, Hazardous waste or even Plastic Waste has been creating waste gluts in the world. Countries with high volumes of waste have been dumping them into nations thereby creating waste flows. The waste flows generally have been dumped in East Asian nations and post 2016, the trends have been to dump the waste in the Latin American countries.

BACKGROUND

Waste once generated needs to be managed through various methods. One of the oldest has been dumping them on the outskirts of the city. Ancient Indian cities as with many older European cities ensured that carts travelling to the city were loaded with the refuse of the city to be disposed in the outskirts. These have evolved into landfills. India for example the Dhapa Dumping Yard, the Gazipur Dumping Ground or the Deonar Dumping Ground has been the same. Waste management can also be undertaken by segregating the waste into the dry and wet waste. The Dry waste is recycled to recover the resources while the wet waste is reduced through composting and bio remediation. The hazardous waste of the Municipal Solid Waste and also the biomedical waste is incinerated under controlled conditions. The other ways to manage the waste is to generate energy through the Waste to Energy generation units set up in the facilities of waste collection. These help harness the landfill gases. Finally some of the recyclables have been brought to alternative uses such as bricks, stationary, reuse of plastic, paper etc.

The generation of solid waste is a major point of concern in all developing and underdeveloped countries. The issues of dumping arise from the fact that, in many sites, the dumpyards have become exhausted and have unscientifically closed, paving way for more pollutants to enter the surface and the ground water. In limited cases, we have seen that the scientific closure and capping for landfill gases have also been attempted such as in the case of the Gorai Dumpsite in Mumbai, India. In several cases the dumping sites of the MSW have been the wetland and inland riverine, estuarine or mangroves. The outcome of dumping in these wetlands have resulted in the consistent reduction and endangering of the wetlands. Wetlands International South Asia (WISA) reports that from 1970 onwards up to 2014, cities have been encroaching into the wetlands for their needs and waste disposal falls into city's sustenance requirement.

As urban cities have grown in the last few centuries, the volume of waste has also risen. According to the United Nations at the turn of the century in 2000, there were 371 urban cities with one million inhabitants, which rose in 2018 to 548 and were expected to increase to 706 in 2030. (UN, 2018)[2] It is interesting to note that according to the World Cities 2018, the two continents of Asia and Africa have the fastest growing urban cities in the regional blocks. The percapita waste generated in Urban areas globally have risen from 0.64 kg per day per capita to 1.2 per day per capita(World Bank,2016) in 2016. This immense quantity of waste has been disposed and dumped in city dumpsites. In several occasions the dumpsites have been along the water bodies, inland rivers, lakes, causing damage to the riverine areas. Occasionally the dumpsites may be also the Ramsar sites of the country.

India is home to 42 Ramsar Sites, after Japan and China, and hence the conservation and protection should be a priority for the country. Climate change is a reality, location of a water body geographically protects and saves cities from disasters which are on the rise. Cities have fallen in BGI (Blue Green Index) as the water bodies which were the urban lungs have dried, been encroached upon or dumped with the city's refuse. Unsanitary dumping, open dumping and burning of wastes near water bodies have degraded the quality of water in the wetlands. The native flora and fauna have been impacted causing long term damage to the entire ecosystem.

The study highlights a comparative study of the countries of South Asia and Africa, the nature of dumping, the extent of damage to environment, and the socio-cultural conditions of the place which

Table 1. List of Ramsar sites in select Asian and African area

Select Countries in South Asia& Africa	Ramsar Sites
India	42
Pakistan	19
Nepal	10
Bhutan	3
Bangladesh	2
Nigeria	11
South Africa	17
Egypt	4
Congo	3
Algeria	50

Source:(Ramsar,2021)

determine the dumping patterns, the economic condition and the political and civic sense that predominates. Comparative discussion of the state of the soil and water impacted by the prolonged dumping of the solid waste in and around the water bodies has been undertaken for Tepi Town, Ethiopia, Lagos in Nigeria and Kolkata, Mumbai, India. Besufekad Mekonnen.*et al.,* (2019), has worked on the Tepi town in Ethiopia on the direct impact of waste dumping in the water and soils. The study has highlighted alkaline status of the water with a higher range of PH, TDS higher that the WHO standards, higher turbidity in open dumping areas, high dissolved inorganic substances in ionized form in and around the solid dump yard and the COD have been higher than the permissible limit, all indicating the change in the quality of the water around the dumping yard. The situation in Mumbai is also of major concern. The three yards of Deonar, Gorai and Kanjurmarg all have location along the water bodies, the Thane Creek, the Manori Creek or Marve Creek. The Water Quality Status Report of Maharashtra, 2018-2019, states that the COD and the BOD have been wavering between the various station in and around the city. The spots where the dumpyards are located and are over 20 years old are Mulund, Gorai, Deonar and Malad. Most of these areas have been reported to have poor water quality.

DEGRADATION AND RESTORATION

Wetlands and Mangroves

The wetlands inland riverine, estuarine and mangrove regions have a unique ecosystem. Wetlands are one of the most productive and diverse ecosystems in the world compared to any other. They range from saline, brackish, coastal mangroves, inland marshes, swamps, peat lands, riparian wetland, to name a few, though broadly there are three coastal and non-coastal or inland and manmade wetlands (Ramsar Convention). These ecosystems allow birds, mammals, reptiles, fish, invertebrate species and even humans to survive not to mention the varieties of plants and microorganism. They have been a support to the humans by providing food, genetic diversity, medicines, water supply, fisheries, agriculture, nutrient retention, water table management, forest cover to maintain the Carbon dioxide, protection of sea coasts and coastal and inland erosion. Globally 123 countries in the tropical and subtropical regions have mangroves. Asia has the largest cover of mangroves with South Asia having the highest percentage of 33.5%, while West Central Africa is 13.2%, East and South Africa is 5.2% and other East Asia is 0.1% (India State of Forest Report, 2017:56). These mangroves need to be conserved and preserved for the future sustenance of the coastal and riverine ecosystems. India for example, with over 30% of South Asian mangrove cover, has a National Mangrove Committee (1976) which advices the Government of India on the mangrove conservation.

Population and Urbanisation

The dumping of the solid wastes in the wetlands have resulted in a unique ecosystem, laden with leachates, fauna and flora impacts, encroachment of the area covered by the dumpyards and more predominantly, a shift in the overall essence of the wetland. South Asia includes countries of India, Bangladesh, Bhutan, Myanmar, Sri Lanka, Nepal, Afghanistan and Maldives. North Africa includes the countries of Algeria, Egypt, Libya, Morocco, Sudan Tunisia and western Sahara, and West Africa include Benin, Burkina Faso, Cape Verde, Côte D'Ivoire, Gambia, Ghana, Guinea, Guinea-Bissau, Liberia, Mali, Mauritania, Niger, Nigeria, Senegal, Sierra

Leone, Togo . One of the concerns in both the blocks is increasing population. This is an issue in both the study blocks of South Asia and Northern and Central Africa. The South Asia has countries like India which has a population of 136.65 crores, Pakistan 21.66 Crores, Nepal 2.86 crores, Sri Lank 2.18 Crores, Bangladesh 16.3 crores, Myanmar 5.4 crores, Afghanistan 3.8 crores and Maldives 5.31 Lakhs.

The population of North Africa is believed to be 3.14% of the global population, with Egypt with 102,334,404, Algeria 43,851,044, Sudan 43, 849,260, Morocco 36,910,560, Tunisia 11,818,619, Libya 6,871,292 and Western Sahara 597,339. However it needs to be mentioned here that the mere population size is not an indicator of the volume of MSW generated. It is an indication however of the nature of the waste collected. The total waste generated in 2012 in Central and North Africa as per the World Bank was 125 MT and the per capita waste generated was 0.6 kg per person. The waste collection is very low at 55%, leaving a lot of scope improvement in the waste collection.

Population rise remains as one of the important anthropogenic causes for the decline of the wetlands in the study areas of Asia and Africa. Population pressure and urbanisation have a twining effect on the pressure of the land. In megacities of India, there is a dearth of land, for example, the city of Mumbai has merely 476.24 sq. kilometres and a population of 1,24,42,373(Census, 2011) and a waste collection of 6900(MCGM, 2019), hence pushing the dumpyards to the otherwise unused river, creek sides and other inaccessible areas. The actual challenge is when cities have grown and the urban fringe areas have been included within the city limits. The dumpyard neighbourhoods thereafter become the hotbed of pollution, water borne illness and other ailments including respiratory.

Waste Management and Restoration

Waste management is one of the most necessary aspects that need to be managed for the restoration of the wetland ecosystems. The types of materials that get disposed include plastics, cardboards, garbage and domestic household waste, construction and demolition waste, bio medical waste materials, electronic waste, hazardous domestic waste, etc. There are two broad categories of waste that needs to be addressed, organic and perishable waste and inorganic and non-perishable waste. The inorganic, hazardous waste tends to hamper these ecosystems, more due to its non-decomposable nature. Management of the waste will have to be more planned and organised in the study area as these fall in the developing and underdeveloped regions of the world. In developed nations the focus of waste management has been waste minimisation. In countries such as India and the Nigeria, Ethiopia, there is a higher percentage of organic waste in the bins. This provides the scope to segregate and treat the organic as through the bio processes of composting and vermin composting. The dry waste is encouraged to be recycled and reused through the various informal sectors like the ragpickers.

Controversies in Select Dumpsites

To study the controversy, the problem and the solutions, given under are few case examples of India and Africa.

- **Agbogbloshie, Ghana**

Ghana, ranked 138 out of 198 in the HDI and a medium developed country, has Agbogbloshie, one of the largest e waste dumping site in the world, next to the Odaw River and in the capital city of Accra. It has a population of 3.04 crore (World Bank, 2019). It became the illegal dumping ground for electronic and automobile waste of the industrialised nations. Over time the implementation of the Basal Convention, the e waste has been generated within the country and selects adjoining nations. Covering roughly 20 acre, the waste yard has become a home bred for toxic wastes that have infiltrated the waters as it was originally a wetland area. With a huge potential for the circular economy and a steady resource for the recycling industry, this yard is a hotbed of pollutants and impacts the residents of the place.

- **Deonar, Mumbai, India**

Deonar is one of the oldest dumping sites of Mumbai, established in 1927 and is still open to dumping. It has 127 hectares under it and the amount of disposal is around 1400 MTPD. Located in the mangroves and in extreme fragile environment, this dumpsite has been in the process of relocating on the instructions of the higher authorities. However, Thane Creek which circumvents a large part of the metropolis of Mumbai is impacted as have the dense settlements around it due to air pollution due to recurrent fires, water pollution and land degradation.

- **Dhapa, Kolkata, India**

Located in the Eastern wetlands of Kolkata, this is a colonial wetland set up as the dumping ground for the city. It is around 23.3 hectares, over 50 feet in height and a mountain of waste. The location on the marshes, also a Ramsar Site, poses a threat to the ecosystems of the region. Unscientific dumping has been countered by an unique urban agricultural pattern introduced in 1941, wherein, the vegetables were cultivated on these dumping areas for the nearby city.

- **Olusosun, Lagos, Nigeria**

One of the largest dumpsites, receiving, 2.1 million tonnes, of waste annually. It covers 43 hectares and is 18 meters deep. It was established around 1992. The residual population living around it is about 5 million in a radius of 10 Kilometers.The nature of the dumping is municipal solid waste and electronic waste. Studied extensively, this waste yard has been known to pollute the surrounding with heavy metals particularly Lead and Zinc (Omotayo AI, 2019), expectedly from the battery and other e wastes, paints and other chemicals.

The above mentioned four sites are not the exclusive sites in the regions under study, but typify it in the uniqueness, size, impact and possibilities of environmental sustainability to manage the wetlands they degrade. The larger issues of the solid waste and other wastes dumped in the wetlands have been remarkably damaging to the biotic, i.e. the plant and the animal life surviving in them. The Mumbai regions have mangroves which have been threatened, in Nigeria, the world's largest e waste site in on the wetlands, in Delhi, one of world's largest dumpyards are in Gazipur has threated the urban residents, just to mention a few issues.

Problems

In the Indian context, the Central pollution Control Board has in its report mentioned that there are 280 contaminated sites in 21 states in India. It defines a contaminated site as one in which "*constituent char-*

acteristics of the toxic and hazardous substances, caused by humans, exists at all levels and conditions which pose a threat to human health or environment"(CPCB,2019). It is worthwhile to note that some of the contaminated sites mentioned in the list are the MSW. Some of the sites which are relevant to this study have been, taken from Central Pollution Control Board Report on Contaminates Sites, 2019:-

- Gazipur Landfill Delhi- Contamination in soil with Zinc, Lead Nickle
- Manikui-Chandil Subarnarekha River Polluted Site
- Cauvery River downstream Nanjagund, Mysore- MSW dumping
- Mavallipura Dumpsite Yelahanka- Huge accumulated waste
- Begur Lake Bangalore- very high concentrations of Zinc, Chromium, Copper, Nickle
- Kuzhi Kandom Thodu Creek, Kerala
- Mithi River Mumbai,

The select sites mentioned clearly indicate that the presence of the dumpyards may be a threat to the wetland if located on them. The Ramasar Wetlands have been identified to preserve and conserve the fragile ecosystems in the world. The instances of Solid waste dumping in the wetlands have become the biggest threat to the vegetation such as mangrove, shrubs, etc. The bio geophysical factors such as population, urbanisation, waste load, industrial disposal, existing rules, pollution, agricultural waste, etc. have also posed a threat to the existing wetland. Wetlands also need a study of the vegetation, the endemic, the invasive, and the altered species due to the pollution from the cities, industrial effluents and removal for construction purposes. Apathy, lack of implementation, lack of involvement of the local communities, uncontrolled population and urbanisation, lack of alternative disposal methods and lack of use of the sanitary landfills may be the biggest challenges hamper the proactive conservation of the wetlands.

SOLUTIONS AND RECOMMENDATIONS

The rise of environmentalism as a movement has been to tackle the excess of anthropological impacts. The concept of sustainable management of ecosystem has been implemented in nations through a series of systems of conventions, policy and community based approaches. Some of the innovations that can be devised for innovative ecosystem management include:-

1. Valuing the Ecosystem

The most crucial aspect of the ecosystem is the manner in which they are handled by the individuals, communities and corporate. There are three basic benefits that can be identified, direct use, indirect use, aesthetic benefits which may be derived from any given ecosystem According to this study, the wetland ecosystem needs to be mapped for direct use benefits, indirect use benefits and the aesthetic benefits

Direct use benefits are based on the conscious use of environmental assets in consumption and production. (Wills1997:147) .The Indirect use benefits are those natural resources that support human existence and involuntarily aid our existence. Dependence of humans on the trees for the oxygen required by the human body and the regulation of the carbon and the oxygen cycles is an example. The aesthetic benefits are innately built into the human existence in the form of socio-cultural ethics that have evolved with human life. The respect for nature, the religious taboo around nature and the practices are some examples. Most innovations would need to be cantered on the values that can be attached to the ecosystems. The idea to protect and preserve ecosystems would evolve from the concept of attaching value to the components of these niche systems.

2. Ethical Standards Prevailing in the Society

Environmental ethics define the manner in which a society perceives nature. The idea that nature needs to be exploited or conserved has been discussed at length by the environmentalist. From the Judeo Christian ideologies to the more modern biocentric and ecocentric ethics have all hovered around the different aspects of respect for nature. The amount of exploitation and degradation of the ecosystems have begun to show a distinct pattern in the use of technology. The greater the consumption, the greater the use to technology in production process, larger the finished products and more the degradation. Also there are two concepts shallow ecology and deep ecology. The shallow ecology movement centres on the idea that man needs to protect the environment from its pollution, degradation and respect the natural elements. The concept of deep ecology focuses on relating human ego with nature and hence tries to build on the basic ethos of the civilizations of preservation of nature and protection of natural elements through the ideological, theological and other non-tangible aspects. It laid more emphasis on the use of self to generate conservation mechanism.(Bhattacharjee,2012).

3. Pollution and Degradation Control

The inevitable in the economic process of conversion of the resources is the generation of waste which eventually creates pollution and degradation. Innovativeness in pollution control has resurfaced as clean technology rather than a clean-up technology. Thus the newer establishments in the production process have to acquire environmental clearances through the environmental management system. It includes the environmental impact assessment of the project, environmental audit of the inflow and outflows from the production process and calculation of the carbon footprint of the project. Clean development mechanisms are encouraged so that the carbon footprints are minimal and cause the least damage to the environment.

Pollution of the environment is dependent on the amount of waste accumulated in nature and the capability of the ecosystem to use its homeostatic mechanism to rectify the same. Very often natural systems exist in the ecosystem which enables a cleansing of the pollutant. The nature of the pollutant in these cases is important along with the available biotic or non-biotic organisms which can manage the change in the ecosystem. Bio magnification is a major consequence when the pollutant harbours the growth of specific plants destroying the balance in the ecosystems.

Zero carbon emission is the ultimate that the administration can seek to achieve with the increased pressures on the ecosystems to prolong the balance in them. The manner of achieving the least carbon emission has to emerge in every step of the utilization of carbon elements. Reduced consumptions of carbon based energy with a focus on the alternative generation of power is the most important step towards the carbon reduction process. This would secure the fuels and ensure lower environmental pollution.

4. Community, Economy and Polity

The economies of the different political regimes have been characteristic in their differing ideologies. Every polity has evolved its own environmental procedures and norms. The global community as a whole focused on the environmental protection much after the extensive degradation that had crept into the world with the industrial revolution, world wars and the rise in consumerism. With increased production capacities in the factories and the quick emergence of the new markets with colonization, rampant destruction of the ecology was seen. It was later when the oil started to dwindle in the western world and rivers got polluted that the need to protect the environment was felt. From the essays on the 'Limits to Growth' and the Club of Rome, efforts began and resulted in global consciousness in the developed world, seeping gradually into the developing and the underdeveloped world. Protection at the local, regional and national level was encouraged, through the multiple conventions, meets leading to a responsible world attempting to protect its natural ecosystems. The global commons i.e., the oceans, the atmosphere and the Antarctic were also protected through the Treaty and Conventions.

Innovativeness can be encouraged at the community level, largely because all environmental catastrophes are directly affecting the local communities living in the regions. There has been an increase in the pollution, natural disasters and manmade disasters. The nature of the environmental catastrophes has to be identified with the vulnerability of the ecosystems and communities involved in the process of managing the problems. Community based measures to conserve through routine monitoring and removal of the harmful pollutants along with the reduction in usage of the polluting agents through policy would go a long way in managing the ecosystems. With the differences in polity, ideologies need to equip themselves with knowledge based environmental management measures rather than emulate those of other polities.

Finally the larger management of the wetlands would need to ensure urban settlements keep the blue green index (BGI) balanced. Instead of regular unsanitary landfills the path forwards needs to be sanitary landfills where no alternatives are possible. Otherwise as the world has attempted, waste minimisation would be the best way to reduce the waste dump sites. In developing countries, uniform Handling Rules for Solid Waste Management should be encourages, with better civic sense, proper implementation of the policy and governmental efforts. The backwardness of the two regions have meant that there is a growth of an informal recycling and sorting segment of people like the ragpickers. This has been due to the larger concerns of poverty, illiteracy and unemployment. Segregation at source and recycling as an option needs more encouragement from the community and the governments. Inclusion of Community Role in waste handling and management, such as the Advanced Local Management Committees in Mumbai, may also be of help.

CONCLUSION

The economic way of thinking about the environment begins with the recognition that environmental problems are people problems." (Wills,1997). The study of the ecosystem management sustainably also is a people centric approach. It impacts humans who are the beneficiaries of the ecosystem. All innovations in the efficient management have to be people centric and involve all the components in the ecosystem. All ecospheres have to incorporate the changes after an assessment of the problems faced by the specific ecosystem. It needs to be added that every ecosystem is unique in its problem and hence the management method has to adopt the uniqueness while solving the environmental issues. There are no man made borders in the ecospheres, only natural limits. The deliberation and the solutions to the environmental imbalances thus would require more than national identities to tackle them. The management notions have to transcend borders and the line of countries and states have to be removed for effective management of the ecosystems. The delicateness of the ecospheres which harbour multiple micro niches has to be dealt with innovative techniques of modern man. Yet, the assessments and the uniqueness of each such ecosystem have to be preserved and protected for the future generations. The Sustainable Development Goals call for protection of the earth, its waters, land, species, air, land and humans. It's time the steps are taken in such a manner that the inland and estuarine environments are capable of restoring themselves organically without much human intervention in the future.

REFERENCES

Africa Solid Waste Management Data Book. (2019). Available online at https://africancleancities.org/assets/data/JICA_databook_EN_web_20191218.pdf https://www.ramsar.org/sites/default/files/documents/library/info2007-01-e.pdf https://vedas.sac.gov.in/vedas/downloads/atlas/Wetlands/NWIA_National_atlas.pdf

Akhilesh, S. (2013). Indian Society: Issues and Problems. Academic Press.

Bigio & Bharat. (2004). *Urban Environment and Infrastructure Towards Liveable Cities*. The World Bank.

Eurostat. (2021). *Municipal Waste Statistics*. Retrieved from https://ec.europa.eu/eurostat/statistics-explained/index.php?title=Municipal_waste_statistics#Municipal_waste_generation

Ganesh, K., & Kanhaiya, A. (Eds.). (2006). *Environmental Issues of Development*. The Associated Publishers.

Jain, A. K. (2005). *A Sustainable Vision for Urban India*. Kalpaz Publications.

Martin, H. (1997). From Care to Action making a Sustainable World, IUCN and Earthscan Publication Ltd.

Mekonnen, B. (2019). Assessment of the Effect of Solid Waste Dump Site on Surrounding Soil and River Water Quality in Tepi Town, Southwest Ethiopia. *Journal of Environmental and Public Health*. PMID:32587623

Melosi Martin, V. (2005). *Garbage in the cities, refuse, reform and the environment* (revised edition). University of Pittsburgh Press.

NWIA National Wetland Atlas. (2018). *Ministry of Environment and Forests, Government of India; As a part of the project on National Wetland Inventory and Assessment (NWIA)*. https://www.un.org/en/events/citiesday/assets/pdf/the_worlds_cities_in_2018_data_booklet.pdf)

Tchobanoglous, G. (1993). *Integrated Solid Waste Management: Engineering Principles and Management Issues*. McGraw-Hill.

Tiseo, I. (2020). *Generation of municipal solid waste worldwide in 2017, by select country (in million metric tons)*. Statista. Retrieved from https://www.statista.com/statistics/916749/global-generation-of-municipal-solid-waste-by-country/

Urvashi. (2006). *Sustainable Solid Waste Management: Issues Policies and Structures*. Academic Foundation.

ENDNOTES

1 World Bank Report, What a waste: 2.0A Global Snapshot of Solid Waste Management to 2050 p:3, https://www.un.org/en/events/citiesday/assets/pdf/the_worlds_cities_in_2018_data_booklet.pdf

Chapter 11
Global Climate Change by Wetland Greenhouse Gas Fluxes:
Mechanisms, Effects, and Control

Madhavi Konni
https://orcid.org/0000-0002-2022-5528
Department of Basic Sciences and Humanities, Dadi Institute of Engineering and Technology, India

Vara Saritha
https://orcid.org/0000-0003-0473-993X
Department of Environmental Science, GITAM Institute of Science, GITAM University (Deemed), India

Pulavarthi Madhuri
Department of Basic Sciences and Humanities, Dadi Institute of Engineering and Technology, India

K. Soma Sekhar
Department of Basic Sciences and Humanities, Dadi Institute of Engineering and Technology, India

Manoj Kumar Karnena
https://orcid.org/0000-0001-9241-2990
Department of Environmental Science, GITAM Institute of Science, GITAM University (Deemed), India

ABSTRACT

Wetlands (WLs) in the landscapes are important for the GHGs production, ingesting, and exchange with the atmosphere. In this chapter, the authors illustrated how the WLs influence climate change, even though it is typical for determining the climatic role of WLs in the broader perspective. The conclusions might be wary based on the radiative balance as the radiative forcing since the 1750s or climatic roles are continuously changing in the wetlands. Degradation of WLs leads to reducing their functioning, and GHG fluxes might

DOI: 10.4018/978-1-7998-9498-8.ch011

change and alter the climatic roles of the WLs. The chapter demonstrated that WL disturbances might cause global warming for a longer duration even though the WLs are restored or managed by replacing them with the mitigation WLs. Thus, activities that cause disturbance in the WLs leading to carbon oxidation in the soils should be avoided. Regulating the climate is an ecosystem service in the WLs; during the planning of the WLs, protection, restoration, and creation, environmental management should be considered.

INTRODUCTION

Wetlands (WLs) are defined as the soils that are periodically drenched with water by the presence of plants adapted to the waterlogged conditions (Lakitan et al., 2021; Wang et al., 2021). WLs are a hotspot for the biogeochemical cycles in the landscapes. Molecular oxygen is consumed faster in saturated soil, leading to the lack of oxygen for the soil. Natural replenishment of oxygen from the atmosphere to these soils is lesser than the average molecular oxygen consumption (Molina et al., 2021; Graversgaard et al., 2021). Nevertheless, microaerobic zones are created by soil microbes and roots of plants by which the availability of oxygen to the soils are compensated to some extent. The organic carbon from the adjacent WLs surface or biota can be added to the WLs to enhance these levels. Nitrogen and phosphorous availability to the soil can be delivered by the movement of water across the WLs (Silver et al., 2021). The opposition of oxic and anoxic groups with the reactive organic carbon or nutrients help to form biogeochemical cycles in the WLs. These compounds undergo a series of reactions to set up a dynamic environment. The biogeochemical cycle activity in the WLs is essential to the environment. They regulate the climate and helps in sequestering the carbon, nutrient cycles, soil formation, and harmful substances removal (Reddy and D'Angelo, 1997; (Belyea and Clymo, 2001).

The current chapter discusses the importance of WLs and their biogeochemical cycles that are influencing global climate change. According to the literate available, we found that the warming/cooling effects on the climate of WLs are they act as both source/sink of the greenhouse gases (GHGs), such as methane, carbon dioxide and oxides of nitrogen (White et al., 2021; Limpens et al., 2008). However, determining the complete role of the WLs is very challenging in climate change as the conclusions might vary based on WLs radiative balance for a specific time or radiative forcing for WLs for a lifetime. In this chapter, we used the dynamic model approach by considering how these WLs restorations and mitigations affect the greenhouse gas dynamics and explained the climate regulation related to the WLs. Along these lines, the effects of biogeochemical cycles of the WLs extended beyond the individual WLs that are influencing regional and global climate change.

WETLANDS AS THE SOURCE AND SINK FOR GREENHOUSE GAS

The nitrogen and carbon biogeochemical cycles in wetlands are essential for producing, consuming, and exchanging GHGs with the troposphere. Generally, wetlands remove the carbon dioxide from the surrounding via photosynthetic activity (Belyea and Malmer, 2004) and act as a sink for the organic carbon in the soils (Chmura et al., 2003). Methane was released into the atmosphere by natural WLs and paddy crops, contributing more than 35% to global methane emissions (Saunois et al., 2016). Nitrous oxides are released or used up by the soils of wetlands (Moseman-Valtierra et al., 2011). Carbon dioxide, methane and nitrous oxides are the major gases responsible for climate change (Ingole, 2021; Kang,

2021). Changes in the biogeochemical cycles in the wetlands are mediated greenhouse gas production/consumption even cause climate change. Later sections discussed the primary control rates and the greenhouse gas exchange in the WLs.

In addition to these gases, WLs also add water vapour to the atmosphere through evapotranspiration, a significant greenhouse gas (Helbig et al., 2021; Valach et al., 2021). The evapotranspiration denotes a heat flux from wetlands to the atmosphere. About 50% of the radiation by sunlight reaches the troposphere due to the specific heat of vaporization. Nevertheless, water vapour is not considered a role of climate change in the case of the WLs as it contributes to these reasons. Primarily, the lifetime of the water vapour is only a few days compared to the other gases, as they range from years (Ferrelli et al., 2021). Secondly, the capacity of the water vapour to hold in the air depends on the temperature of the air and doesn't rely on the rate of water vapour added to the atmosphere (Daou et al., 2006). However, without the longer-term mechanism that helps to enhance the temperature of the air, the water vapour added to the troposphere immediately precipitate out; it will not contribute anything to climate change. Even though WLs water vapour in the atmosphere influences the local climate, it will not contribute anything to climate change.

WLs act as the best sink for carbon dioxide when the ecosystem's primary production rate exceeds the combined respiration rates of autotrophs and heterotrophs. Moreover, WLs receive sediments consisting of carbon from the adjacent estuaries and rivers and are preserved longer in the WLs soils (Noe and Hupp, 2005). The carbon obtained is stored as organic carbon on the grounds of the WLs and woody biomass. In some WLs the soils contain inorganic carbon as carbonates (Hansen and Nestlerode, 2014). The WLs have more significant strengths to sink the carbon dioxide at the biome or landscape level as the WLs can exchange the carbon with the adjacent estuaries and the rivers (Chapin III et al., 2006). Partially, carbon sequestration at the higher rates is the function of the increased productivity, which happens only when there is stress linked with the growth in WLs environment. Pulsed systems measure hydrologically many WLs. The water level varies seasonally or is less predictable based on precipitation input (Day et al., 2007; Bragg, 2002). The pulse system subsides the energy to the WLs, supplies the nutrients, enhances the soil's redox conditions, and removes the pollutants from the soil; this feature enhances the WLs plant productivity and carbon inputs (Odum et al., 1995); as the WLs plant growth depends on the available nutrients than the newly added nutrients input, thus effective nutrient recycling system in the WLs help to increase the primary production at more significant levels.

The high carbon sequestration rates are due to the low decomposition rate; in a broader perspective, this might even relate to the anoxic conditions in the WLs (Reddy and DeLaune, 2008). In contrast, in some WLs, it also depends on the availability of nutrients, pH and temperature (Megonigal et al. 2004). In the absence of molecular oxygen, the microorganisms in the WLs utilize the alternate electron acceptors from other compounds, which might require lower thermodynamic energy and thus leading to slow decomposition. When the water level drops, the oxygen might penetrate the WLs soil, increasing the decomposition rate (Chimner and Cooper, 2003). Anoxia conditions in the WLs limit the oxidization of the phenols, which is an essential enzyme for the breaking down of the polymers of the plant and other composite organic substances; this will enhance the concentrations of the phenolic compounds in the WLs and limit the enzymatic activity required to break down the organic matter (Freeman et al., 2001). Further, the absence of oxygen reduces the fungi abundance, which is well than bacteria in breaking down the toxic substances (Thormann, 2006). These factors are linked with an oxygen abundance, leading to lesser decomposition of the organic matter and increasing the carbon preservation efficiencies in the WLs.

The methanogenesis (anaerobic condition) promotes the storage of the soil carbon, leading to methane gas production; drying up of the WLs leads to a decrease in methane production (Megonigal et al., 2004). An optimistic association is observed between methane and primary production (Vann and Megonigal, 2003). Methanogenesis rates are higher at the surface soils as these zones are abundant with fresh organic carbon, which might be fermented to acetate molecules required for the methanogens. Methanogenesis requires low energy compared to catabolic activity; the rate of methanogenesis is limited in the saline WLs owing to higher sulfate content (Poffenbarger et al., 2011). The plant aerenchyma tissues help vent the methane gases in the soil into the atmosphere efficiently by bypassing the aerobic methanotrophy active sites. More than 55 per cent of the methane produced by the soils of the WLs are oxidized into carbon dioxide before it enters the atmosphere; oxidation might occur under anaerobic or aerobic environments; however, in the current chapter, we only discuss the prominence of the anaerobic methane oxidation in the WLs (Segarra et al., 2015; Megonigal et al., 2004). The coalition of WLs methane fluxes signifies that most WLs are the primary methane sources to the atmosphere (Petrescu et al., 2015). Microbes in the WLs produce nitrous oxide by the process of nitrification/denitrification. The denitrification process is significant for the inanimate nitrogen removal from the WLs by reducing nitrate to nitrogen gas.

On the contrary, nitrification is an anaerobic process where the ammonia is oxidized to the nitrate. These are multistep reactions, requires multiple enzymes for the responses and executions. Several factors like an abundance of the electron acceptors and donors, moisture content, hydrogen sulphide concentration and pH might interfere by various processes results in the reactions be partial and produce nitrous oxide (Moseman-Valtierra, 2012; Hefting et al., 2003). The independent factors like the ratio of nitrogen to nitrous oxide and nitrate to nitrous oxide accelerate inorganic nitrogen availability. Thus, the rates of nitrous oxide emissions enhance with insemination (Syakila et al., 2010). On the other hand, if the availability of inorganic nitrogen is low, denitrifying bacteria might use nitrous oxide as an acceptor of the electron leading to the removal of the nitrous oxide from the atmosphere (Majumdar, 2013).

WETLAND'S IMPACTS ON GLOBAL CLIMATE CHANGE

As the WLs exchange the GHGs with the atmosphere, they can influence and alter the climate. The three significant ways to contemplate about these issues are primarily warming caused by the emissions of the WLs are lesser or greater than the cooling (removing of GHGs by the WLs)? And this phenomenon is called radiative balance and calculated over a specific time (generally for 100 years). Furthermore, it is questionable that if wetlands are contributing to the recent climate change. If the flux of the greenhouse gases between the atmosphere and WLs are different from before the 1750s than today, then the radiative balance of the WLs has affected climate change. Contrarily, suppose the radiative balance is the same as before the 1750s; then, WLs did not affect global climate change. Furthermore, it can consider the atmospheric influence at a specific point if we consider the lifetime of sequestration of GHGs and emissions. Transitions in WLs from warming affect the cooling due to their long period of carbon sequestration (Frolking and Roulet, 2006).

Radiative Balance

Defining the atmospheric impacts of the WLs fluxes is challenging as GHGs, and other radiative efficiencies are constant in the environment for various time intervals (Weston et al., 2014). For instance, methane's radiative efficiencies are twenty-six times higher than carbon dioxide; however, the average climatic lifetime of carbon dioxide is longer than methane. Since 1990, IPCC has reported the GWP compared to the atmospheric impact of GHGs. The WLs scientists used the GWP for comparing the cooling effect of WLs carbon dioxide sequestration with the warming of methane. As the GWP addresses the one-time emission of GHGs, it is not suitable for the environmental gas fluxes that typically occur and might be absorbed into or out of the environment. Addressing these issues, later the researchers suggested approaches like the GWP based on GHGs fluxes than the one-time emission. The SGWP is the radiative forcing linked with the time owing to the constant emissions relating to the carbon dioxide sequestration of equal mass. The SGCP is radiative forcing linked with the time due to the uptake of the emissions relating to the carbon dioxide sequestration of similar mass. The sustained flux (S) is identical to the step-change indices introduced by the climate science community but never adopted by the scientist of the WLs (Berntsen et al. 2005). The difference between these fluxes is significantly less and mainly used to compare the long live gases with short live gases to measure global warming potential.

Radiative Forcing

WLs radiative balance change from the 1750s implies that WLs are causative of worldwide radiative forcing. The WLs radiative proportions might vary due to the shifts in the biogeochemical cycles areal rates for a process commonly involved in the greenhouse gas dynamics of the WLs. For instance, the sulfate pollution from the peatlands reduces the production and emissions of methane gases, causing similar effects with salinized or inland systems (Gauci et al., 2004; Herbert et al., 2015). Seaside zones enhancement with nutrients around the globe enhances plant production rates, results in the formation of sediments and enhances the soil carbon sequestration; however, it also accelerates the emissions of nitrogen (Morris et al., 2002).

WLs biochemical cycles respond to the changes in the environment and the factors that drive global warming. For instance, methane production and decomposition rate enhance with higher temperatures, and global heating has increased by 1 ^{0}C since the industrial era. WLs hydrology variations result from variabilities in the temporal region that might alter the water table (Hartmann et al., 2013). WLs changes affect the production and consumption of methane and nitrous oxide (Moseman-Valtierra, 2012). The primary production rates of methane increase with the increase of carbon dioxide and rise to 50% since 1750 (Keeling et al. 2018; Rasse et al. 2005). According to the literature available, there are no global-scale studies conducted and available to assess whether the radiative forcing of WLs changed the biogeochemical cycles in the rates of GHGs production or consumption of WLs. The WLs radiative balance might even be altered as the WLs locations are different. It was estimated that since 1800 the worlds wetlands area a reduced up to 35% (Bridgham et al., 2006). Early 1985 in northern America and Europe, it was estimated that more than 65% of the coastal WLs and inland WLs are converted for agricultural activities.

Similarly, 27% of the WLs are drained for agriculture in the Aisa (Vedogbeton and Johnston, 2020). In the last 200 years, the global peatland area has declined by 15% worldwide (Frolking and Roulet, 2011). On the contrary, anthropogenic activities lead to the expansion of WLs around the world. For instance, deforestation on a larger scale and improper management of the soil in the northern parts of America enhances the sediments on the coastal zones; it forms a tidal wetland area (Khan and Brush 1994). There has been a recent expansion of WLs on the Tibetan plateau due to the retreat of glaciers and enhanced meltwater (Niu et al. 2012). Nevertheless, a significant reduction of the WLs is observed over the last 250 years. Thus, it is evident that in the absence of considerable changes in the WLs aerial rate of GHGs production/consumption, there is a possibility of losing the WLs area at a greater level for decades, shows that the GHGs balance of WLs is changed and contributing the radiative forcing to the environment.

CLIMATIC EFFECTS

As stated before, WLs acts as a sink for carbon dioxide, the source for methane generation and the source/sink for nitrous oxide (Imer et al., 2013). The WLs radiative balance can act as negative (cooling) or positive (warming) to the environment based on the ratios of the carbon dioxide uptake to methane emissions. The nitrous oxide fluxes are neglected in the preceding sections as their fluctuations are relatively minimal compared to the carbon dioxide and methane in the WLs (Johnson et al., 2021). Counter-intuitively, the WLs climatic effects over its lifetime might differ from its radiative balance over hundred years (Jiang et al., 2021). The dynamic modelling discusses the issues relevant to the climatic impacts of WLs and the reason for the lifetime effects changing with the Restoration because these models allow the GHGs fluxes to change over time. Atmospheric perturbation models are one among them developed by Frolking et al. (2006). The carbon dioxide, methane and nitrous oxide gases are replaced with the one-meter square areas of the WLs in these models. The surface oceans act as non-atmospheric reservoirs of the carbon dioxide equilibrium and are maintained by adding or removing the carbon dioxide by the WLs (Hall, 2021). The methane/nitrous oxide produced from the WLs can be removed from the environment by oxidation and photolysis. Indirectly, these models assume that the biogeochemical cycles in organic carbon and GHGs gases in the soil pore of the WLs are in a steady-state till the WLs are distributed. However, if the WLs are disturbed by the anthropogenic activities, the organic carbon was oxidized to the carbon dioxide results in zero emissions of methane. This phenomenon will stimulate the hydrology of the WLs and dries the soils of the WLs. Concerns about the GHGs are turned to radiation balance by using radiative competencies to estimate the gases' indirect climatic effects.

WARMING-COOLING SWITCH OF WETLANDS

WLs that acts as a carbon dioxide sink will have a cooling effect if their life is considered. The warming due to the WLs methane will reach the plateau only after 50 years; after this point, the methane emissions rates are balanced by the methane oxidations in the environment (Valach et al., 2021). On the contrary, the carbon dioxide equilibrium in the biotic/abiotic reservoirs does not research the balance thoroughly. Thus, warming caused by methane will attain a constant state. In contrast, cooling caused by the sequestration of carbon dioxide will be rising as the age of the WLs increases.

Dynamic Modelling for GHGS

These models allow evaluating the impacts of the climatic change on the WLs, and these models do not require radiative forcing/balance usage like other approaches (Harriss et al., 1982). For instance, reducing the water levels in the WLs leads to organic carbon oxidation in the soil, which reduces methane emissions (Harriss et al., 1982). Suppose the hydrology of the WLs is restored. The radiative balance might be theoretically like the year in the 1750s, so that radiative forcing of WLs might be zero and not affect the climate (Rodrigues et al., 2021). Similarly, the same conclusions can be drawn if WLs are converted into the plain for human activities and replacing with the same operational extenuation WLs. Nevertheless, disturbances to the WLs might have climatic effects for longer durations even the WLs are restored or replaced by the mitigation WLs. The perturbation models determine the role of climate that are affected the WLs biogeochemical cycles disturbances. These models assume that the WLs are formed in the year zero, with a drainages system that occurred in the 18th century and succeeding rebuilding of WLs to normal conditions in 1950 (Ivčević et al., 2021; Junk et al., 2021). The WLs methane flux was reduced to zero during the WLs disturbance; these disturbances reduced the productivity rate to zero resulting in organic carbon oxidation at 4.5% per year. Due to this, the soil organic pool ratios are reduced by one per cent value within 1000 years (Manna and Ghosh, 2021). Granek and Ruttenberg, 2008 reported that after clearing the mangrove forest, the soil organic oxidation per cent was found to be 4.5% per year, which is the highest to date (Harris et al., 2021). However, after following the Restoration, the sequestration rates of GHGs are the same as before the disturbance. The Restoration of the WLs might take a decade to have a similar rate of carbon cycling equivalent to the long-standing natural WLs (Cornell et al., 2007; Costanza et al., 2021); nevertheless, the accurate modelling of the WLs restoration process is beyond the scope of the present chapter.

WLs Restoration

The Restoration of the WLs around the globe has become a priority for the enhancement of the environment. Clarkson et al., 2014 stated that half of the WLs worldwide had been already lost, and the remaining are continuously deteriorating. Despite the disproportion of the WLs aerial extent, it has considerable biodiversity and human benefits. However, WLs failed to gain the attention of humans for their management and conservation. Meli et al., 2014 found that restored WLs offered advanced provisioning and regulating the environment than degraded WLs. To ensure the WLs health, restoring the WLs to a normal state is essential by considering integral units and the resources for a sustainable environment (Halls,1997). The efficiencies of approaches for attaining the predicted goals of the conservation differs depending on the WLs conditions, methods employed and ecosystem services. In addition, the application of the restoration technique for the WLs is considered one of the disturbances. However, by the restoration methods, lots of ecosystem services of the WLs are recovered. The methods that are used for the restoring of degraded WLs are categorized as follows:

- WLs hydrology management and reestablishment.
- Controlling of toxicants that are affecting the WLs.
- Native biota reestablishment.

The general design for the WLs restoration was shown in Figure 1 in detail.

Figure 1. WLs restoration designs

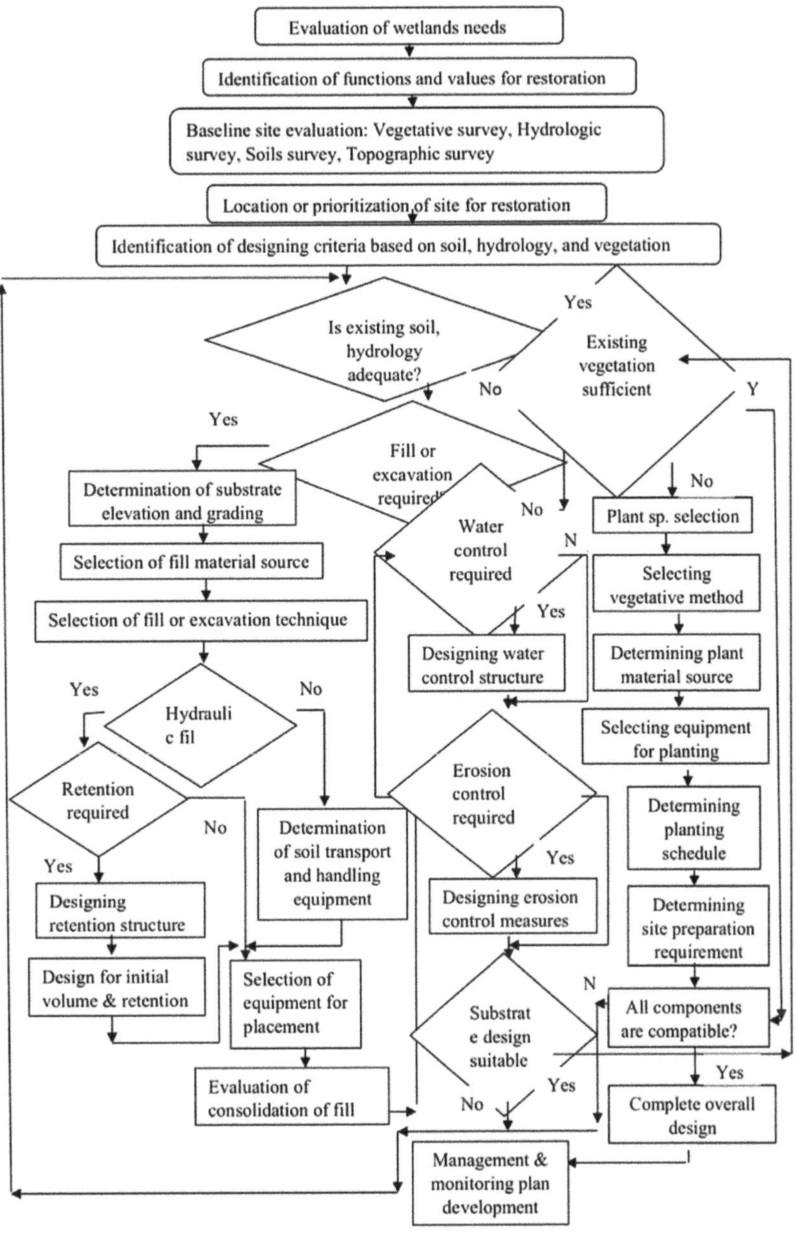

Suggestions for the Conservation and Management of WLs

WLs degradation might reduce its function efficiencies. GHGs flux might alter and lead to the soil oxidation to the carbon dioxide/methane, which changes the complete climatic role in the WLs. Restoration and mitigation activity restore the lost functions of the WLs. Our investigations show that anthropogenic activities disturbing WLs and leading to organic carbon oxidation must be avoided to conserve WLs (Billings et al., 2021; Xi et al., 2021). Whenever the WLs are disturbed, the soils' organic carbon is oxidized and reduces the cooling effects of the WLs, and even in some cases, these disturbances might lead to warming effects. The radiative forcing analysis over time for individual WLs are not available till now; however, the WLs types available in the literature are thousands of years old (Frolking and Roulet, 2007). Older WLs act as a reservoir for the soil organic carbon by assimilation and shows cooling effects to the environment for a lifetime. The disturbances of WLs should be managed to maintain their constancy for the organic carbon pool. Conservation practises required as the sites of the WLs are saturated with the water, which minimizes the aerobic decomposition of the carbon pool. Agriculture and development activities by humans reduced the areas of the WLs.

In the disturbed WLs the permitted anthropogenic activities should have mitigation ratios, which should be enough to compensate for the loss of the carbon sequestration (Su et al., 2021). These ratios are required for the mitigation and accumulation of organic carbon oxidation in WLs. Even though mitigation measures in the study do not become a part of the climate, several degraded WLs should be replaced with mitigation WLs to avoid adverse environmental effects (Sandi et al., 2021). The carbon cycles rates might be restored within the years if the WLs are undisturbed; however, in the case of the mitigation WLs, having a similar restoration might take hundreds of years (Biagi et al., 2021; Bartolucci et al., 2021). The mitigation WLs should be appropriately designed to sequester the methane and other emission releases from the WLs to the environment. For instance, Kroeger et al., 2017 stated that climatic effects of the WLs could be reduced by introducing the saline water tides to the freshwater WLs, as the methane gases emissions are suppressed by the introduction of saline water to the WLs (Poffenbarger et al., 2011). Zhang et al., 2016 affirmed that methane emissions in the paddy fields could be reduced by managing the water levels.

CONCLUSION

WLs have reverse effects on climate change as these acts as both sources and sinks for the GHGs. Conventional views of WLs radiative balance/forcing failed to explain the complexities involved in how the WLs is impacting the climate for their lifetime. It might take ten thousand years to have the cooling effects of the newly formed WLs compared to the old WLs. The ratios of carbon dioxide and methane depends on time, and radiative forcing switch overtime of the WLs is less when a higher percentage of GHGs is sequestrated. As the methane emissions from the WLs stabilize after 50 years, the organic carbon accumulation on the WLs predominates the climatic impacts; indicating that restoration and construction activities of WLs cause warming initially, except in the brackish and saline WLs the methane emissions are low. From the chapter evaluation, we recommend prioritizing the existing WLs conservation instead of constructing new WLs. WLs are considered the actual cool houses of the earth as they sequester large amounts of carbon dioxide and are sustained for longer durations as these are not drained. The peatland systems are converted into warming systems when the carbon is released after the drainage. Methane gases are released in higher amounts even due to the fluctuations of the water levels in the WLs. Even though we dint focused on nitrous oxide in the current chapter, many researchers are conducting studies on the emission of nitrous oxide from the WLs. Many researchers argue that the WLs are economically valuable in the ecosystem, and there is a need to develop plans for the WLs protection and Restoration. The existing WLs provide more cooling effects than the new WLs; new WLs require decades to perform similarly. These issues are not accepted or recognized in the WLs restoration programme to date. WLs services like pollutant removal, biodiversity enhancement, and flood control are much needed to restore and manage the WLs. These services are economical and worthy of enhancing the WLs.

REFERENCES

Bartolucci, N. N., Anderson, T. R., & Ballantine, K. A. (2021). Restoration of retired agricultural land to wetland mitigates greenhouse gas emissions. *Restoration Ecology*, *29*(3), e13314. doi:10.1111/rec.13314

Belyea, L. R., & Clymo, R. S. (2001). Feedback control of the rate of peat formation. *Proceedings of the Royal Society of London. Series B, Biological Sciences*, *268*(1473), 1315–1321. doi:10.1098/rspb.2001.1665 PMID:11410160

Belyea, L. R., & Malmer, N. (2004). Carbon sequestration in peatland: Patterns and mechanisms of response to climate change. *Global Change Biology*, *10*(7), 1043–1052. doi:10.1111/j.1529-8817.2003.00783.x

Berntsen, T. K., Fuglestvedt, J. S., Joshi, M. M., Shine, K. P., Stuber, N., Ponater, M., & Li, L. (2005). Response of climate to regional emissions of ozone precursors: Sensitivities and warming potentials. Tellus B. *Chemical and Physical Meteorology*, *57*(4), 283–304. doi:10.3402/tellusb.v57i4.16549

Biagi, K. M., Clark, M. G., & Carey, S. K. (2021). Hydrological functioning of a constructed peatland watershed in the Athabasca oil sands region: Potential trajectories and lessons learned. *Ecological Engineering*, *166*, 106236. doi:10.1016/j.ecoleng.2021.106236

Billings, S. A., Lajtha, K., Malhotra, A., Berhe, A. A., de Graaff, M. A., Earl, S., & Wieder, W. (2021). Soil organic carbon is not just for soil scientists: Measurement recommendations for diverse practitioners. *Ecological Applications*, *31*(3), e02290. doi:10.1002/eap.2290 PMID:33426701

Bragg, O. M. (2002). Hydrology of peat-forming wetlands in Scotland. *The Science of the Total Environment*, *294*(1-3), 111–129. doi:10.1016/S0048-9697(02)00059-1 PMID:12169001

Bridgham, S. D., Cadillo-Quiroz, H., Keller, J. K., & Zhuang, Q. (2013). Methane emissions from wetlands: Biogeochemical, microbial, and modeling perspectives from local to global scales. *Global Change Biology*, *19*(5), 1325–1346. doi:10.1111/gcb.12131 PMID:23505021

Chapin, F. S. III, Woodwell, G. M., Randerson, J. T., Rastetter, E. B., Lovett, G. M., Baldocchi, D. D., & Matson, P. A. (2006). McGuire, AD, Melillo, JM, Mooney, HA, Neff, JC. Houghton, RA, Pace, ML, Ryan, MG, Running, SW, Sala, OE, Schelesinger, WH, and Schulze, ED: Reconciling carbon-cycle concepts, terminology, and methods. *Ecosystems (New York, N.Y.)*, *9*(7), 1041–1050. doi:10.100710021-005-0105-7

Chimner, R. A., & Cooper, D. J. (2003). Influence of water table levels on CO_2 emissions in a Colorado subalpine fen: An in situ microcosm study. *Soil Biology & Biochemistry*, *35*(3), 345–351. doi:10.1016/S0038-0717(02)00284-5

Chmura, G. L., Anisfeld, S. C., Cahoon, D. R., & Lynch, J. C. (2003). Global carbon sequestration in tidal, saline wetland soils. *Global Biogeochemical Cycles*, *17*(4), n/a. doi:10.1029/2002GB001917

Clarkson, B. R., Ausseil, A. G. E., & Gerbeaux, P. (2013). *Wetland ecosystem services. Ecosystem services in New Zealand: conditions and trends*. Manaaki Whenua Press.

Cornell, J. A., Craft, C. B., & Megonigal, J. P. (2007). Ecosystem gas exchange across a created salt marsh chronosequence. *Wetlands*, *27*(2), 240–250. doi:10.1672/0277-5212(2007)27[240:EGEAAC]2.0.CO;2

Costanza, R., Anderson, S. J., Sutton, P., Mulder, K., Mulder, O., Kubiszewski, I., & Dee, G. (2021). The global value of coastal wetlands for storm protection. *Global Environmental Change*, *70*, 102328. doi:10.1016/j.gloenvcha.2021.102328

Daou, K., Wang, R. Z., & Xia, Z. Z. (2006). Desiccant cooling air conditioning: A review. *Renewable & Sustainable Energy Reviews*, *10*(2), 55–77. doi:10.1016/j.rser.2004.09.010

Day, R. H., Williams, T. M., & Swarzenski, C. M. (2007). Hydrology of tidal freshwater forested wetlands of the southeastern United States. In *Ecology of tidal freshwater forested wetlands of the Southeastern United States* (pp. 29–63). Springer. doi:10.1007/978-1-4020-5095-4_2

Ferrelli, F., Brendel, A. S., Perillo, G. M. E., & Piccolo, M. C. (2021). Warming signals emerging from the analysis of daily changes in extreme temperature events over Pampas (Argentina). *Environmental Earth Sciences*, *80*(12), 1–13. doi:10.100712665-021-09721-4

Freeman, C., Ostle, N., & Kang, H. (2001). An enzymic' latch' on a global carbon store. *Nature*, *409*(6817), 149–149. doi:10.1038/35051650 PMID:11196627

Frolking, S., Roulet, N., & Fuglestvedt, J. (2006). How northern peatlands influence the Earth's radiative budget: Sustained methane emission versus sustained carbon sequestration. *Journal of Geophysical Research. Biogeosciences*, *111*(G1).

Frolking, S., & Roulet, N. T. (2007). Holocene radiative forcing impact of northern peatland carbon accumulation and methane emissions. *Global Change Biology*, *13*(5), 1079–1088. doi:10.1111/j.1365-2486.2007.01339.x

Gauci, V., Matthews, E., Dise, N., Walter, B., Koch, D., Granberg, G., & Vile, M. (2004). Sulfur pollution suppression of the wetland methane source in the 20th and 21st centuries. *Proceedings of the National Academy of Sciences of the United States of America*, *101*(34), 12583–12587. doi:10.1073/pnas.0404412101 PMID:15297612

Granek, E., & Ruttenberg, B. I. (2008). Changes in biotic and abiotic processes following mangrove clearing. *Estuarine, Coastal and Shelf Science*, *80*(4), 555–562. doi:10.1016/j.ecss.2008.09.012

Graversgaard, M., Jacobsen, B. H., Hoffmann, C. C., Dalgaard, T., Odgaard, M. V., Kjaergaard, C., & Tonderski, K. (2021). Policies for wetlands implementation in Denmark and Sweden–historical lessons and emerging issues. *Land Use Policy*, *101*, 105206. doi:10.1016/j.landusepol.2020.105206

Hall, A. (2021). *The Influence of Anthropogenic Climate Change on the 2015-2017 Hydrological Drought in the South-Western Cape, South Africa* (Master's thesis).

Halls, A. J. E. (1997). *Wetlands, biodiversity and the Ramsar convention: the role of the convention on wetlands in the conservation and wise use of biodiversity* (Vol. 13). In Ramsar Convention Bureau.

Hansen, V. D., & Nestlerode, J. A. (2014). Carbon sequestration in wetland soils of the northern Gulf of Mexico coastal region. *Wetlands Ecology and Management*, *22*(3), 289–303. doi:10.100711273-013-9330-6

Harris, N. L., Gibbs, D. A., Baccini, A., Birdsey, R. A., De Bruin, S., Farina, M., & Tyukavina, A. (2021). Global maps of twenty-first century forest carbon fluxes. *Nature Climate Change*, *11*(3), 234–240. doi:10.103841558-020-00976-6

Harriss, R. C., Sebacher, D. I., & Day, F. P. Jr. (1982). Methane flux in the great dismal swamp. *Nature*, *297*(5868), 673–674. doi:10.1038/297673a0

Hartmann, D. L., Tank, A. M. K., Rusticucci, M., Alexander, L. V., Brönnimann, S., Charabi, Y. A. R., & Zhai, P. (2013). Observations: atmosphere and surface. In Climate change 2013 the physical science basis: Working group I contribution to the fifth assessment report of the intergovernmental panel on climate change (pp. 159-254). Cambridge University Press.

Hefting, M. M., Bobbink, R., & de Caluwe, H. (2003). Nitrous oxide emission and denitrification in chronically nitrate-loaded riparian buffer zones. *Journal of Environmental Quality*, *32*(4), 1194–1203. doi:10.2134/jeq2003.1194 PMID:12931872

Helbig, M., Gerken, T., Beamesderfer, E. R., Baldocchi, D. D., Banerjee, T., Biraud, S. C., & Richardson, A. D. (2021). Integrating continuous atmospheric boundary layer and tower-based flux measurements to advance understanding of land-atmosphere interactions. *Agricultural and Forest Meteorology*, *307*, 108509. doi:10.1016/j.agrformet.2021.108509

Herbert, E. R., Boon, P., Burgin, A. J., Neubauer, S. C., Franklin, R. B., Ardón, M., & Gell, P. (2015). A global perspective on wetland salinization: Ecological consequences of a growing threat to freshwater wetlands. *Ecosphere*, *6*(10), 1–43. doi:10.1890/ES14-00534.1

Imer, D., Merbold, L., Eugster, W., & Buchmann, N. (2013). Temporal and spatial variations of CO2, CH4 and N2O fluxes at three differently managed grasslands. *Biogeosciences Discussions*, *10*, 2635–2673.

Ingole, M. P. D. (2021). Environment and global warming in current scenario. *EPRA International Journal of Agriculture and Rural Economic Research*, *9*(6), 16–19.

Ivčević, A., Statzu, V., Satta, A., & Bertoldo, R. (2021). The future protection from the climate change-related hazards and the willingness to pay for home insurance in the coastal wetlands of West Sardinia, Italy. *International Journal of Disaster Risk Reduction*, *52*, 101956. doi:10.1016/j.ijdrr.2020.101956

Jiang, A. L., Lee, M. C., Zhou, G., Zhong, D., Hawaria, D., Kibret, S., & Hsu, K. (2021). Predicting distribution of malaria vector larval habitats in Ethiopia by integrating distributed hydrologic modeling with remotely sensed data. *Scientific Reports*, *11*(1), 1–14. doi:10.103841598-021-89576-8 PMID:33980945

Johnson, M. S., Matthews, E., Bastviken, D., Deemer, B., Du, J., & Genovese, V. (2021). Spatiotemporal methane emission from global reservoirs. *Journal of Geophysical Research: Biogeosciences*.

Junk, W. J., da Cunha, C. N., Thomaz, S. M., Agostinho, A. A., Ferreira, F. A., de Souza Filho, E. E., & Kawakita, K. (2021). Macrohabitat classification of wetlands as a powerful tool for management and protection: The example of the Paraná River floodplain, Brazil. *Ecohydrology & Hydrobiology*.

Kang, S. (2021). Microbes' Many Roles in Climate Change: Contribution, Consequence, Mitigation, and Model System. *Microbes: The Foundation Stone of the Biosphere*, 187-194.

Keeling, C. D., Piper, S., Bollenbacher, A., & Walker, S. (2018). *The keeling curve: A daily record of atmospheric carbon dioxide from Scripps Institution of Oceanography at UC San Diego*. https://scripps.ucsd. edu/programs/keelingcurve/, nd.

Khan, H., & Brush, G. S. (1994). Nutrient and metal accumulation in a freshwater tidal marsh. *Estuaries*, *17*(2), 345–360. doi:10.2307/1352668

Kroeger, K. D., Crooks, S., Moseman-Valtierra, S., & Tang, J. (2017). Restoring tides to reduce methane emissions in impounded wetlands: A new and potent Blue Carbon climate change intervention. *Scientific Reports*, *7*(1), 1–12. doi:10.103841598-017-12138-4 PMID:28931842

Lakitan, B., Kartika, K., Susilawati, S., & Wijaya, A. (2021). Acclimating leaf celery plant (Apium graveolens) via bottom wet culture for increasing its adaptability to tropical riparian wetland ecosystem. *Biodiversitas Journal of Biological Diversity, 22*(1).

Limpens, J., Berendse, F., Blodau, C., Canadell, J. G., Freeman, C., Holden, J., & Schaepman-Strub, G. (2008). Peatlands and the carbon cycle: From local processes to global implications–a synthesis. *Biogeosciences*, *5*(5), 1475–1491. doi:10.5194/bg-5-1475-2008

Majumdar, D. (2013). Biogeochemistry of N2O uptake and consumption in submerged soils and rice fields and implications in climate change. *Critical Reviews in Environmental Science and Technology*, *43*(24), 2653–2684. doi:10.1080/10643389.2012.694332

Manna, M. C., & Ghosh, A. (2021). Soil Organic Carbon Dynamics, Stabilization, and Environmental Implication. In *Soil Science: Fundamentals to Recent Advances* (pp. 13–33). Springer. doi:10.1007/978-981-16-0917-6_2

Megonigal, J. P., Hines, M. E., & Visscher, P. T. (2004). Anaerobic metabolism: Linkages to trace gases and aerobic processes. *Biogeochemistry*.

Meli, P., Rey Benayas, J. M., Balvanera, P., & Martínez Ramos, M. (2014). Restoration enhances wetland biodiversity and ecosystem service supply, but results are context-dependent: A meta-analysis. *PLoS One*, *9*(4), e93507. doi:10.1371/journal.pone.0093507 PMID:24743348

Molina, V., Eissler, Y., Fernandez, C., Cornejo-D'Ottone, M., Dorador, C., Bebout, B. M., & Hengst, M. (2021). Greenhouse gases and biogeochemical diel fluctuations in a high-altitude wetland. *The Science of the Total Environment*, *768*, 144370. doi:10.1016/j.scitotenv.2020.144370 PMID:33454466

Morris, J. T., Sundareshwar, P. V., Nietch, C. T., Kjerfve, B., & Cahoon, D. R. (2002). Responses of coastal wetlands to rising sea level. *Ecology*, *83*(10), 2869–2877. doi:10.1890/0012-9658(2002)083[2869:ROCWTR]2.0.CO;2

Moseman-Valtierra, S. (2012). Reconsidering the climatic role of marshes: Are they sinks or sources of greenhouse gases. In Marshes: Ecology, Management and Conservation. New York: Nova Science.

Moseman-Valtierra, S., Gonzalez, R., Kroeger, K. D., Tang, J., Chao, W. C., Crusius, J., & Shelton, J. (2011). Short-term nitrogen additions can shift a coastal wetland from a sink to a source of N2O. *Atmospheric Environment*, *45*(26), 4390–4397. doi:10.1016/j.atmosenv.2011.05.046

Niu, Z., Zhang, H., Wang, X., Yao, W., Zhou, D., Zhao, K., & Gong, P. (2012). Mapping wetland changes in China between 1978 and 2008. *Chinese Science Bulletin*, *57*(22), 2813–2823. doi:10.100711434-012-5093-3

Noe, G. B., & Hupp, C. R. (2005). Carbon, nitrogen, and phosphorus accumulation in floodplains of Atlantic Coastal Plain rivers, USA. *Ecological Applications*, *15*(4), 1178–1190. doi:10.1890/04-1677

Odum, W. E., Odum, E. P., & Odum, H. T. (1995). Nature's pulsing paradigm. *Estuaries*, *18*(4), 547–555. doi:10.2307/1352375

Petrescu, A. M. R., Lohila, A., Tuovinen, J. P., Baldocchi, D. D., Desai, A. R., Roulet, N. T., & Cescatti, A. (2015). The uncertain climate footprint of wetlands under human pressure. *Proceedings of the National Academy of Sciences of the United States of America*, *112*(15), 4594–4599. doi:10.1073/pnas.1416267112 PMID:25831506

Poffenbarger, H. J., Needelman, B. A., & Megonigal, J. P. (2011). Salinity influence on methane emissions from tidal marshes. *Wetlands*, *31*(5), 831–842. doi:10.100713157-011-0197-0

Rasse, D. P., Peresta, G., & Drake, B. G. (2005). Seventeen years of elevated CO2 exposure in a Chesapeake Bay Wetland: Sustained but contrasting responses of plant growth and CO2 uptake. *Global Change Biology*, *11*(3), 369–377. doi:10.1111/j.1365-2486.2005.00913.x

Reddy, K. R., & D'angelo, E. M. (1997). Biogeochemical indicators to evaluate pollutant removal efficiency in constructed wetlands. *Water Science and Technology*, *35*(5), 1–10. doi:10.2166/wst.1997.0152

Reddy, K. R., & DeLaune, R. D. (2008). *Biogeochemistry of wetlands: science and applications*. CRC Press. doi:10.1201/9780203491454

Rodrigues, A., Sardinha, R. A., & Pita, G. (2021). Fundamentals of Global Carbon Budgets and Climate Change. In *Fundamental Principles of Environmental Physics* (pp. 267–308). Springer. doi:10.1007/978-3-030-69025-0_8

Sandi, S. G., Rodriguez, J. F., Saco, P. M., Saintilan, N., & Riccardi, G. (2021). Accelerated Sea-Level Rise Limits Vegetation Capacity to Sequester Soil Carbon in Coastal Wetlands: A Study Case in South-Eastern Australia. *Earth's Future*.

Saunois, M., Bousquet, P., Poulter, B., Peregon, A., Ciais, P., Canadell, J. G., & Zhu, Q. (2016). The global methane budget 2000–2012. *Earth System Science Data*, *8*(2), 697–751. doi:10.5194/essd-8-697-2016

Segarra, K. E. A., Schubotz, F., Samarkin, V., Yoshinaga, M. Y., Hinrichs, K. U., & Joye, S. B. (2015). High rates of anaerobic methane oxidation in freshwater wetlands reduce potential atmospheric methane emissions. *Nature Communications*, *6*(1), 1–8. doi:10.1038/ncomms8477 PMID:26123199

Silver, W. L., Perez, T., Mayer, A., & Jones, A. R. (2021). The role of soil in the contribution of food and feed. *Philosophical Transactions of the Royal Society B, 376*(1834).

Su, J., Friess, D. A., & Gasparatos, A. (2021). A meta-analysis of the ecological and economic outcomes of mangrove restoration. *Nature Communications*, *12*(1), 1–13. doi:10.103841467-021-25349-1 PMID:34413296

Syakila, A., Kroeze, C., & Slomp, C. P. (2010). Neglecting sinks for N2O at the earth's surface: Does it matter? *Journal of Integrative Environmental Sciences*, *7*(sup1, S1), 79–87. doi:10.1080/1943815X.2010.497492

Thormann, M. N. (2006). Diversity and function of fungi in peatlands: A carbon cycling perspective. *Canadian Journal of Soil Science*, *86*(Special Issue), 281–293. doi:10.4141/S05-082

Valach, A. C., Kasak, K., Hemes, K. S., Anthony, T. L., Dronova, I., Taddeo, S., & Baldocchi, D. D. (2021). Productive wetlands restored for carbon sequestration quickly become net CO2 sinks with site-level factors driving uptake variability. *PLoS One*, *16*(3), e0248398. doi:10.1371/journal.pone.0248398 PMID:33765085

Vann, C. D., & Megonigal, J. P. (2003). Elevated CO 2 and water depth regulation of methane emissions: Comparison of woody and non-woody wetland plant species. *Biogeochemistry*, *63*(2), 117–134. doi:10.1023/A:1023397032331

Vedogbeton, H., & Johnston, R. J. (2020). Commodity consistent meta-analysis of wetland values: An illustration for coastal marsh habitat. *Environmental and Resource Economics*, *75*(4), 835–865. doi:10.100710640-020-00409-0

Wang, J., Chen, J., Wen, Y., Fan, W., Liu, Q., & Tarolli, P. (2021). Monitoring the coastal wetlands dynamics in Northeast Italy from 1984 to 2016. *Ecological Indicators*, *129*, 107906. doi:10.1016/j.ecolind.2021.107906

Weston, N. B., Neubauer, S. C., Velinsky, D. J., & Vile, M. A. (2014). Net ecosystem carbon exchange and the greenhouse gas balance of tidal marshes along an estuarine salinity gradient. *Biogeochemistry*, *120*(1), 163–189. doi:10.100710533-014-9989-7

White, E. E. Jr, Ury, E. A., Bernhardt, E. S., & Yang, X. (2021). Climate Change Driving Widespread Loss of Coastal Forested Wetlands Throughout the North American Coastal Plain. *Ecosystems (New York, N.Y.)*, 1–16. doi:10.100710021-021-00686-w

Xi, Y., Peng, S., Ciais, P., & Chen, Y. (2021). Future impacts of climate change on inland Ramsar wetlands. *Nature Climate Change*, *11*(1), 45–51. doi:10.103841558-020-00942-2

Chapter 12
Variation of the Descriptive Grain Size Distribution Along Cores From the Sebkha of Sidi El Hani, Eastern Tunisia

Elhoucine Essefi
University of Gabes, Tunisia

Soumaya Hajji
Faculty of Sciences of Sfax, Tunisia

ABSTRACT

In this chapter, cores were the object of descriptive classifications of the grain size distribution, which were meant to describe the grain size continuous variability within cores and to correlate between them. The statistical treatment of the crude data was done on the basis of two different methods (the method of moments statistics and the method of inclusive graphic statistics) to compute statistical parameters of the grain size distribution such as mean and median. The correlations between cores were done on the basis of sand/silt/clay percentages. Even though it has given special care to test different methods of studying the grain size distribution, this study has not deviated from its primary purpose of investigating the filling of the playa; correlations between different cores were meant to infer their sedimentary dynamics.

INTRODUCTION

Since the saline environment of Sidi El Hani is sedimentologically a clay pan (Essefi, 2009), sediments grain size in its cores is generally fine; it ranges mainly between $\Phi \geq 9$ (less than 2 µm) and $\Phi = -1$ (2mm). Even though traditional methods of classification (e.g., Cailleux & Tricart, 1962) give an idea about the origin (e.g., the cumulative curves of Cailleux & Tricart (1962) are divided into aeolian, deltaic, etc..), they are still far from being efficient to classify sediments within cores where slight modifications in the grain size distribution are still undetectable and the interplay between different depositional

DOI: 10.4018/978-1-7998-9498-8.ch012

environments risks to be interpreted as one same context. Thus, these classification schemes could not distinguish such a fine grain size distribution. The task, then, becomes more complicated when trying to correlate between cores. Consequently, to do an efficient study of the grain size distribution and a coherent correlation between cores, it is a must first to calculate weights of the fine grains by a statistical extrapolation and then to find the appropriate method of classification; which may be a genetic or a descriptive classification.

As for the descriptive classification, a good few of new or modified approaches has been proposed or investigated, e.g., a simplified classification of intertidal Wadden Sea sediments on the basis of sand/mud ratios (Reineck & Siefert, 1980), a hydrodynamically based classification of fine-grained estuarine sediments using sand/silt/clay ratios (Pejrup, 1988) and a classification for coarse-grained sediments based on gravel/sand/mud ratios (Blair and McPherson, 1999); but differences between the adopted descriptive systems of classification such as the difference between Schepard (1954) and Folk (1966, 1971, 1974), impose the question about their usefulness. Genetically speaking, the genetic approach of Cailleux and Tricart (1962) has always been trusted to infer the depositional environment of any sample; but none can deny that the comparison has always been based on a visual similarity between cumulative curves of the studied sample and these reference curves. Actually, an as yet open debate is still between geologists whether to use descriptive or genetic classification. On the one hand, Rodgers (1950) made it clear that a descriptive, objective and precise classifications and nomenclatures in sedimentology are ideally adequate rather than a genetic distinction. On the other hand, since the interpretation of sedimentary deposits looks for reconstructing depositional processes, environments and facies, it may be actually desirable for classifications to have genetic basics (Folk et al., 1970). In this study, the descriptive method was tested. In order to solve the problem of having fine grain size, this method followed two lines of attack: Inclusive Graphic Statistics (**IG**) and Moments Statistics (**MM**). Eventhough it has given special care to methods of classification, this study has not deviated from its primary purpose of investigating the sedimentary dynamics in Sidi El Hani clay pan. Accordingly, the use of these methods permitted correlations between Sidi El Hani cores, which in turn permitted the identification of the sedimentary dynamics in different areas from Sidi El Hani clay pan.

GEOGRAPHIC AND GEOMORPHOLOGIC FRAMEWORK

The modern sabkha system of MCSH consists of three interconnected sub-systems: Mechertate, Chrita, and Sidi El Hani (Fig. 1). Topographically, it is made of various hills that barely reach more than 150 m of elevation. This topographic contrast helps to create a moderate hydrological network within the 3,300 km^2 of its hydrological basin and enhances the erosion of the highlands. The sub-system of Mechertate covers approximately 600 km^2; it drains toward the sabkha of Mechertate (20 km^2) that is located at ~80 m above sea level. This sub-system collects water, and erodes and transports sediment from its watershed. In the middle of the system, the sub-system of Chrita covers 800 km^2 around an area of 70 km^2; the altitude of Chrita depression is ~60 m. It cumulatively collects the sum of water from Mechertate and Chrita basins. It accumulates water transported sediment from the Chrita sub-system. In spite of erosive along its sinuous edges, its areal quadratic form is quite¦ apparent¶. The Chrita saline lake drains toward the discharge playa of ¦Sidi¦ ¦El¦ ¦Hani¦ by means of the permant hydrographic network Oued Chrita¦. Going eastward, the sub-system of Sidi El Hani covers approximately 1500 km^2. The ssabkha of Sidi

El Hani is a NW-SE lengthened depression (Fig. 1) and its discharge playa (terminology according to Yechieli & Wood (2002)) is approximately 370 km^2, with an average water depth of 0.4 m that can locally reach 0.8 m in some locations. The principal highest point, which is located in the northeast corner of the depression, can reach 33 m elevation. Hydrologically and hydrogeologically, the depression of Sidi El Hani represents the basal part of the endorheic (closed basin) system (Ben Jmaa, 2008; Essefi, 2009), in which Kairouan, Souassi and Zarmedine aquifers converge after leaching older domes of salt and/or transporting salty paleo-water located in the subsurface. Hydrogeological studies on Kairouan aquifer (Besbes, 1975; Feuillette, 2001; Jarraya-Horriche, 2004; Essefi, 2009) proposed a multilayered structure of this aquifer. These studies also suggested that the deep layers feed the shallow levels; thus, groundwater upwelling and/or seepage is a hydrological mechanism to feed the surface of the system with water. Geomorphologically, the drainage basin of the system is characterized by the outcrop of aeolian and aqueous sediments and structures (Ben Jmaa, 2008; Essefi, 2009) (wet aeolian sediments). The deposition of these wet aeolian sediments is directly controlled by groundwater seepage and the fluctuation of the water table within the depressions of Mechertate, Chrita, Sidi El Hani, and their vicinities. Around these depressions, sediments are less threatened by erosion; because the water table is near the surface to preserve the deposits. Once this water evaporates, quantities of evaporate minerals cement and preserving the consolidated (indurated) sediment from erosion. In contrast, depressions are the local base level that collect sediment and water. In the basinward directions of the depressions, the water table falls and thus sediment is unconsolidated and more threatened by aeolian and water erosion. Due to evaporation, erosion is also enhanced by the suffusion process (Ben Jmaa, 2008). Suffusion process is related to the fluctuation of the water table. Once water evaporates, evaporites with a volume bigger than voids within sediment destroy cohesion and attack sediment mechanically.

MATERIALS AND METHODS

The manual coring of eight cores (Table 1) in Sidi El Hani saline environment was done during three field missions (coordinate was taken during field missions by using a Global Positioning System GPS and localized on Google Earth satellite image). The first mission spanned for one day; it focused on the northwest side of Sidi El Hani saline environment (*stricto sensu*), in the zone of Oued Elmaktaa. The

Table 1. Geological coordinates of Sidi El Hani saline environment cores

Core	Latitude	Longitude
Core 1	35°29'37.90"N	10°34'42.89"E
Core 3	35°33'38.31"N	10°28'47.94"E
Core 4	35°36'47.47"N	10°27'45.04"E
Core 5	35°35'45.23"N	10°29'00.18"E
Core 6	35°29'45.36"N	10°29'07.48"E
Core 7	35°29'45.32"N	10°29'15.60"E
Core 9	35°27'52.97"N	10°31'44.12"E
Core 10	35°35'45.49"N	10°21'39.91"E

Figure 1. Summary map of the endorheic system of Mchertate-Chrita-Sidi El Hani. Cores locations within the Sidi El Hani wetland
(Ben Jmaa, 2008; completed and modified; Essefi, 2013)

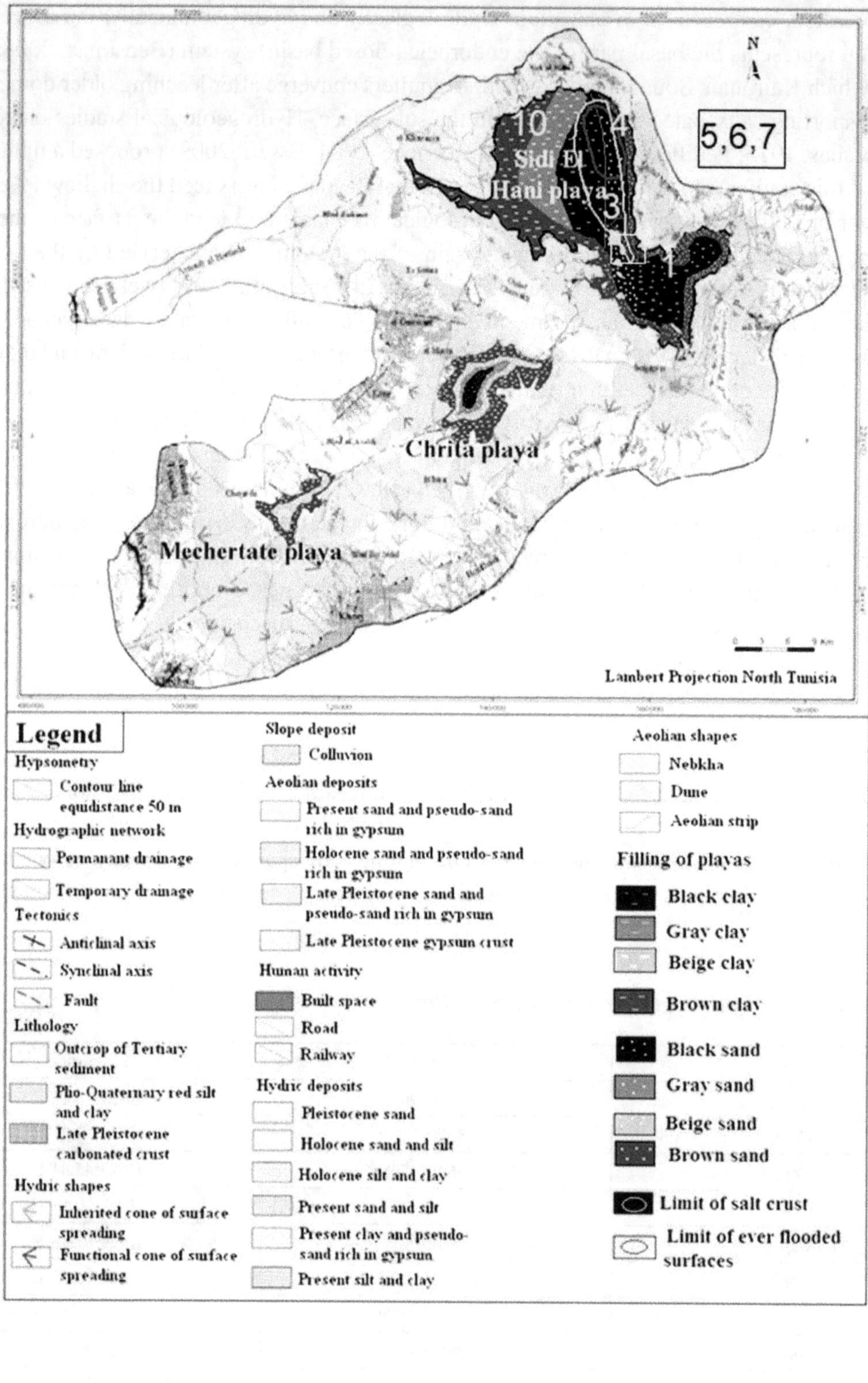

main purpose of this mission was to have an idea about the manual coring and the used materials. A 310cm core was done (core4). The second mission took place in the field of the margin of the Beta Island OuledMoussa-Aalalcha. Two other cores with length of 55cm (core5) and 72cm (core7) were done in the vicinity of the margin Beta Island-playa. Eventually, Author has done the third and longest mission, it spanned for 30 days (mission of July 2008). After taking the key concepts of the manual coring, it was judged to do cores with our hands.

In this context, five cores were done in different zones of the saline environment; their lengths are 310cm (core10), 210cm (core1), 193cm (core3), 155cm (core6) and 120cm (core9). It is worth to be noted that, for each core, coring was stopped because of serious technical problems resulting from the used materials and/or sediment types. Some cores were aborted at small depths. For an aborted core, another core was restarted approximately in the same location. For instance, core2 and core8 were aborted and replaced respectively by core3 and core9.

Materials

Seeing its specificity, the study of Sidi El Hani sediments needs special treatment to obtain all the fine grain size. Added to the AFNOR series, the sieves 50μm and 40μm are added to overcome the technical difficulty of having fine grain size. But it is revealed that much is still to be done to recuperate the entire fine fraction; hence, it is judged to do a statistical extrapolation.

Statistical Extrapolation

As for extrapolation, two methods (Poppe et al., 2004) may be used (Fig.2). On the one hand, the linear extrapolation tends to slightly overestimate the amount of clay present in a typical distribution and may be used by operators to account for material within the colloidal fraction. On the other hand, exponential extrapolation tends to slightly underestimate the amount of clay present in a typical distribution and may be used by operators who want data that represents the minimal amount of clay. Although an operator may select linear extrapolation, exponential extrapolation, or the mean of both extrapolations; the mean usually provides the most accurate estimation and is, therefore, the recommended solution for most applications (Poppe et al., 2004). In the case of Sidi El Hani cores, extrapolation is done by software in order to obtain the fine fraction, which dominates the grain size distribution. This software extrapolates even between the clayey and silty fractions; because, technically, it is difficult to manually separate grains that are smaller than 40μm. After the acquisition of untreated data, compilations are done to generate an output table for each core according to the method of Moments Statistics and the method of Inclusive Graphic Statistics.

Descriptive Classifications of the Grain Size Distribution

After obtaining crude data by manual separation and statistical extrapolation, statistical parameters, such as mean and median, are computed by two different methods: Moments Statistics (**MM**) (Fig.3) and Inclusive Graphic Statistics (**IG**) (Fig.4). The evolution of these parameters, then, is plotted along cores.

Figure 2. Linear, exponential and average extrapolations
(Poppe et al., 2004)

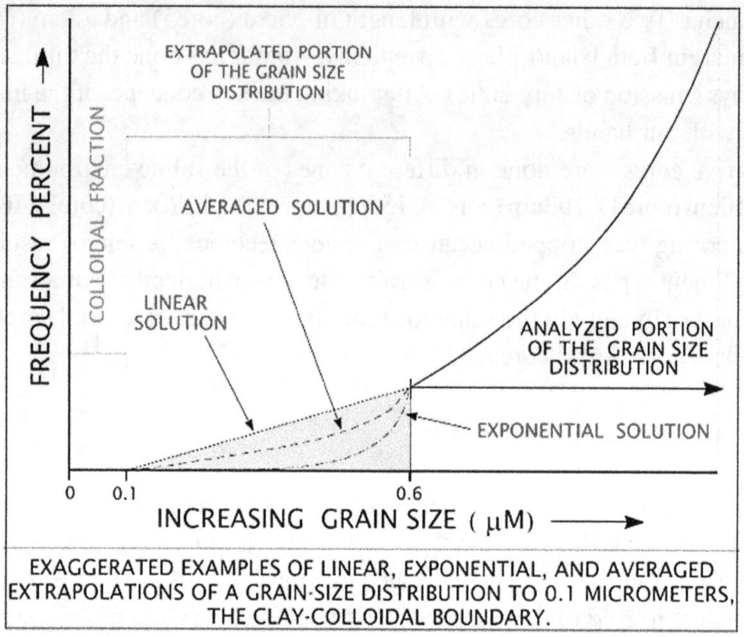

Method of Moments Statistics

The method of Moments Statistics (Fig.3) is based on equations of Krumbein and Pettijohn (1938) and Collias et al. (1963), and includes the Schepard's correction to the second and fourth moments (Kenny and Keeping, 1954). It allows computing of the statistical parameters such as mean, standard deviation, skewness and kurtosis (Fig.3).

Method of Inclusive Graphic Statistics

The inclusive Graphic statistics (Fig.4) is based on the equations of Folk (1974). To prevent overshoots found in classic spline methods, curves for the inclusive Graphic statistics were generated by constrained cubic spline interpolation (Kruger, 2003). The same parameters calculated with Moments Statistics are graphically measured from the cumulative curve (Fig.4).

The use of two methods was meant to unsure of their concordance. Since clay and even fine silt fraction was not detected during manual analysis. The extrapolation of grain-size distribution between 40μm and 2μm was done. The verbal description was based on works of Schepard (1954) and algorithms of Breslau (Schlee and Webster, 1966). Phi notation was based on that of Krumbein (1934) and Inman (1952). To compile these two methods and to give verbal description along cores, the output values were entered.

As for the software output, each output table contains thirty-one fields including: sample identifier, %gravel, %sand, %silt, %clay, sediment classification, median, mean, standard deviation, skewness, kurtosis, verbal description, and the frequency or cumulative frequency percentages of the whole-phi fractions from 13 phi (-(log, 2)) through -5 phi. If inclusive Graphic statistics is selected, the verbal-description field will be populated. If an extrapolation is selected, the two additional phi-fraction fields

Figure 3. Method of moments statistics
(Collias et al., 1963)

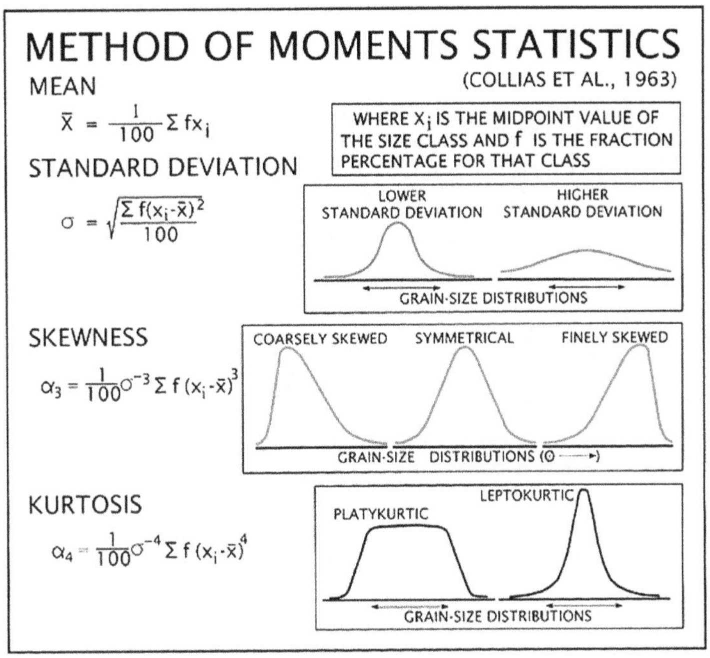

Figure 4. Method of Inclusive Graphic Statistics
(Folk, 1974)

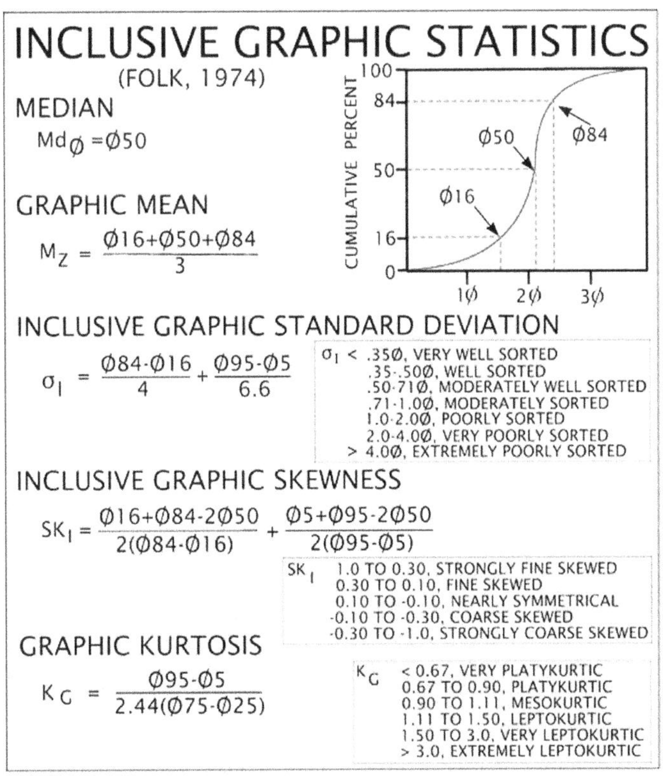

will be populated with data. The first line of the output table is a header that lists the field attributes. As for the calculated parameters, Mean size (continuous function) is one of the most useful parameters, by which bands were defined. It becomes skewed toward the coarser grades and unreliable, however, where the proportion of detrital plant tissues rise much above normal, typically for some centimetres above and below peats. The modal size, although discontinuously estimated, is more helpful where the mean fails in this way. The weight of clay is especially valuable, being insensitive to the present plant tissues. The weight of sand is also helpful, but has the same limitation as the mean. All the computed parameters are plotted along cores.

RESULTS

The statistical treatment of the crude data and the used classification must reach distinction within cores; because, the major characteristic of a successful classification is distinguishing the maximum of different bands within cores. Results will focus first on testing the efficiency of classification schemes. Then, they focus on investigating the different stages of sedimentation in Sidi El Hani clay pan.

Descriptive Grain Size Distribution in Cores of Sidi El Hani Clay Pan

For all cores, the percentage of gravel, sand, silt and clay computed by the methods of Moments Statistics and Inclusive Graphic Statistics are the same. Similarly, the mean, median and standard deviation remain similar by using the two methods. Thus, these parameters are used in studying the grain size distribution and correlating between cores. On the other hand, skewness and kurtosis differ according to the used method. Hence, it is not recommended to use them. The Schepard's classification, Folk classification and verbal description are to show, at the very same time, how important the combination of two classification schemes such as of Schepard (1954) and Folk (1974) is and how debatable the results of the descriptive grain size distribution are. The computed statistical parameters will refine distinction of the grain size distribution within cores.

In order to establish cores in Time, the correlation with sebkha Mhabeul previously done should be taken into account; the reference bands visually recognized in the non destructive analysis should be confirmed and used as references of time. The description of remaining parts of cores, in which no attempts of dating were done, tries not to fall into over or miss interpretations; because 1m of sediments may depose in one day as it may depose in thousands of years. Hence, it is better to be satisfied only with descriptive results.

Failure and Success of the Descriptive Grain Size Distribution in Core5 and Core6

It is not sure to reach the distinction between bands within cores on the basis of the descriptive grain size distribution. The success and failure of the descriptive grain distribution are related to the degree of distinctiveness between successive bands in the studied core. For example, the Figure 5 represents the success of the classifications of Schepard and Folk combined together to classify sediment within cores. By following the long of the core5, we do not find two successive bands having the same classifications according Schepard and Folk schemes.

Figure 5. The Schepard's classification, folk classification and verbal description of core5

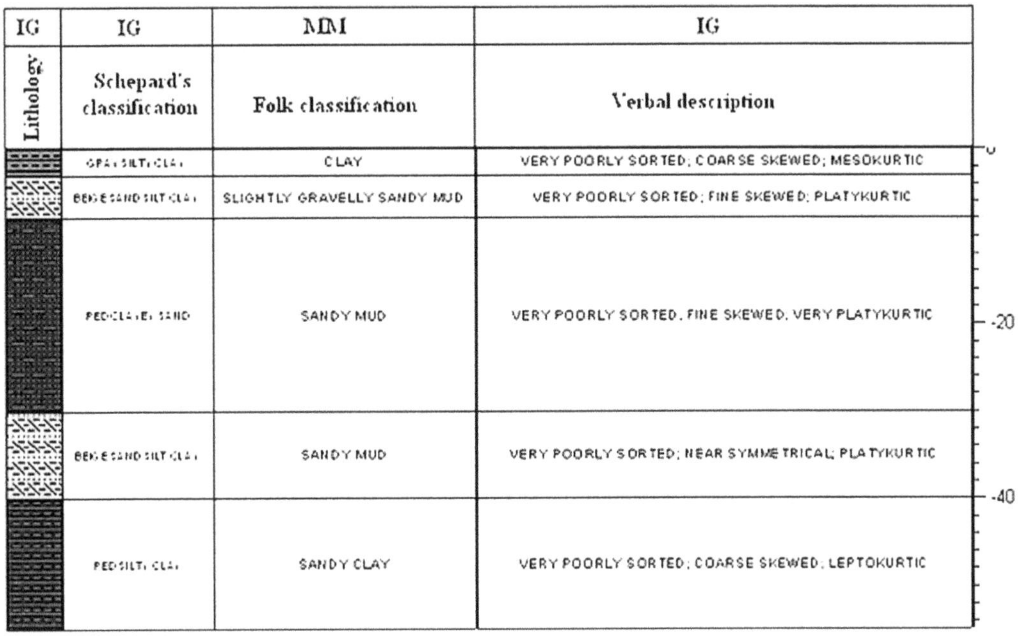

On the other hand, the Figure 6 represents the failure of the classifications of Schepard and Folk combined together to classify sediment within cores. For instance, all bands of core6 are classified as sand/silt/clay in the classification of Schepard and as sandy mud in the classification of Folk. Sometimes, the verbal description distinguishes between bands on the basis of the Kurtosis; such a success is noticed with core6 between band10 (PLATYKURTIC) and band11 (MESOKURTIC); other times, it fails to reach such a distinction, e.g., band1 and band2 in core6.

Descriptive Grain Size Distribution in Core1 on the Basis of the Computed Statistical Parameters

By plotting the percentages of sand, silt and clay of different samples along the cores (Fig.7), the quality of distinction of the grain size distribution increases and visualizes distinguishable bands. Hence, the task of following the grain size distribution along core1 becomes easier. Moreover, the computing of the statistical parameters (mean, median, standard deviation) along the core makes the visualisation of different bands more evident.

By following the major variations of percentages of sand, silt and clay of different samples along the core1, three episodes are distinguished. First, the dominance of the sandy fraction is noticed for the first 34cm. Second, the clayey fraction takes the super voice from 34cm to 52cm. Third, from 52cm to the end of core1, the sandy fraction dominates the grain size distribution. By following the evolution of the statistical parameters along the plot, this distinction may be refined to do an efficient correlation between cores.

Figure 6. Schepard's classification, folk classification and verbal description of core6

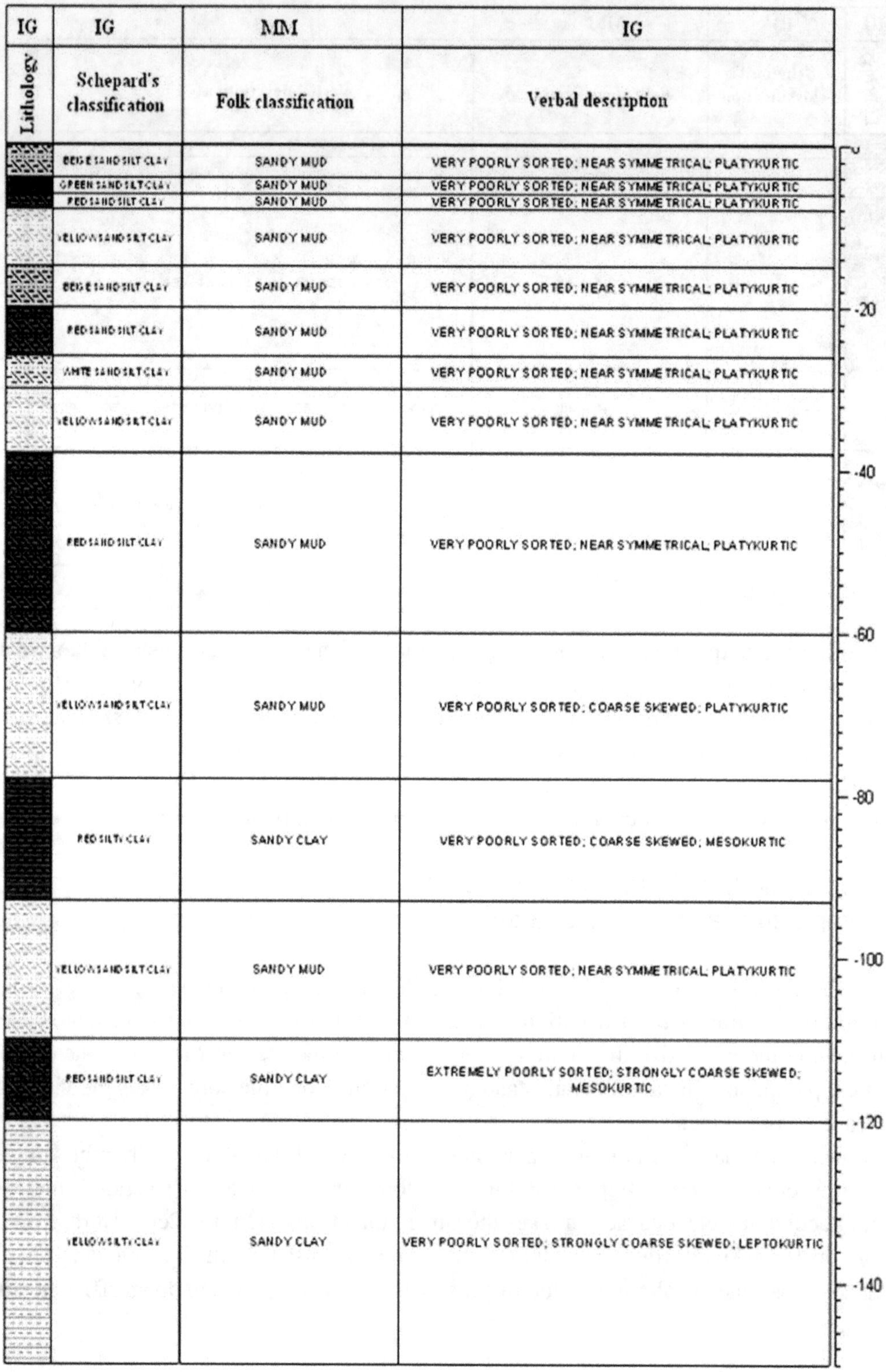

IG	IG	MM	IG
Lithology	Schepard's classification	Folk classification	Verbal description
	BEIGE SAND SILT CLAY	SANDY MUD	VERY POORLY SORTED; NEAR SYMMETRICAL; PLATYKURTIC
	GREEN SAND SILT CLAY	SANDY MUD	VERY POORLY SORTED; NEAR SYMMETRICAL; PLATYKURTIC
	RED SAND SILT CLAY	SANDY MUD	VERY POORLY SORTED; NEAR SYMMETRICAL; PLATYKURTIC
	YELLOW SAND SILT CLAY	SANDY MUD	VERY POORLY SORTED; NEAR SYMMETRICAL; PLATYKURTIC
	BEIGE SAND SILT CLAY	SANDY MUD	VERY POORLY SORTED; NEAR SYMMETRICAL; PLATYKURTIC
	RED SAND SILT CLAY	SANDY MUD	VERY POORLY SORTED; NEAR SYMMETRICAL; PLATYKURTIC
	WHITE SAND SILT CLAY	SANDY MUD	VERY POORLY SORTED; NEAR SYMMETRICAL; PLATYKURTIC
	YELLOW SAND SILT CLAY	SANDY MUD	VERY POORLY SORTED; NEAR SYMMETRICAL; PLATYKURTIC
	RED SAND SILT CLAY	SANDY MUD	VERY POORLY SORTED; NEAR SYMMETRICAL; PLATYKURTIC
	YELLOW SAND SILT CLAY	SANDY MUD	VERY POORLY SORTED; COARSE SKEWED; PLATYKURTIC
	RED SILTY CLAY	SANDY CLAY	VERY POORLY SORTED; COARSE SKEWED; MESOKURTIC
	YELLOW SAND SILT CLAY	SANDY MUD	VERY POORLY SORTED; NEAR SYMMETRICAL; PLATYKURTIC
	RED SAND SILT CLAY	SANDY CLAY	EXTREMELY POORLY SORTED; STRONGLY COARSE SKEWED; MESOKURTIC
	YELLOW SILTY CLAY	SANDY CLAY	VERY POORLY SORTED; STRONGLY COARSE SKEWED; LEPTOKURTIC

Figure 7. Computed grain size distribution along core1

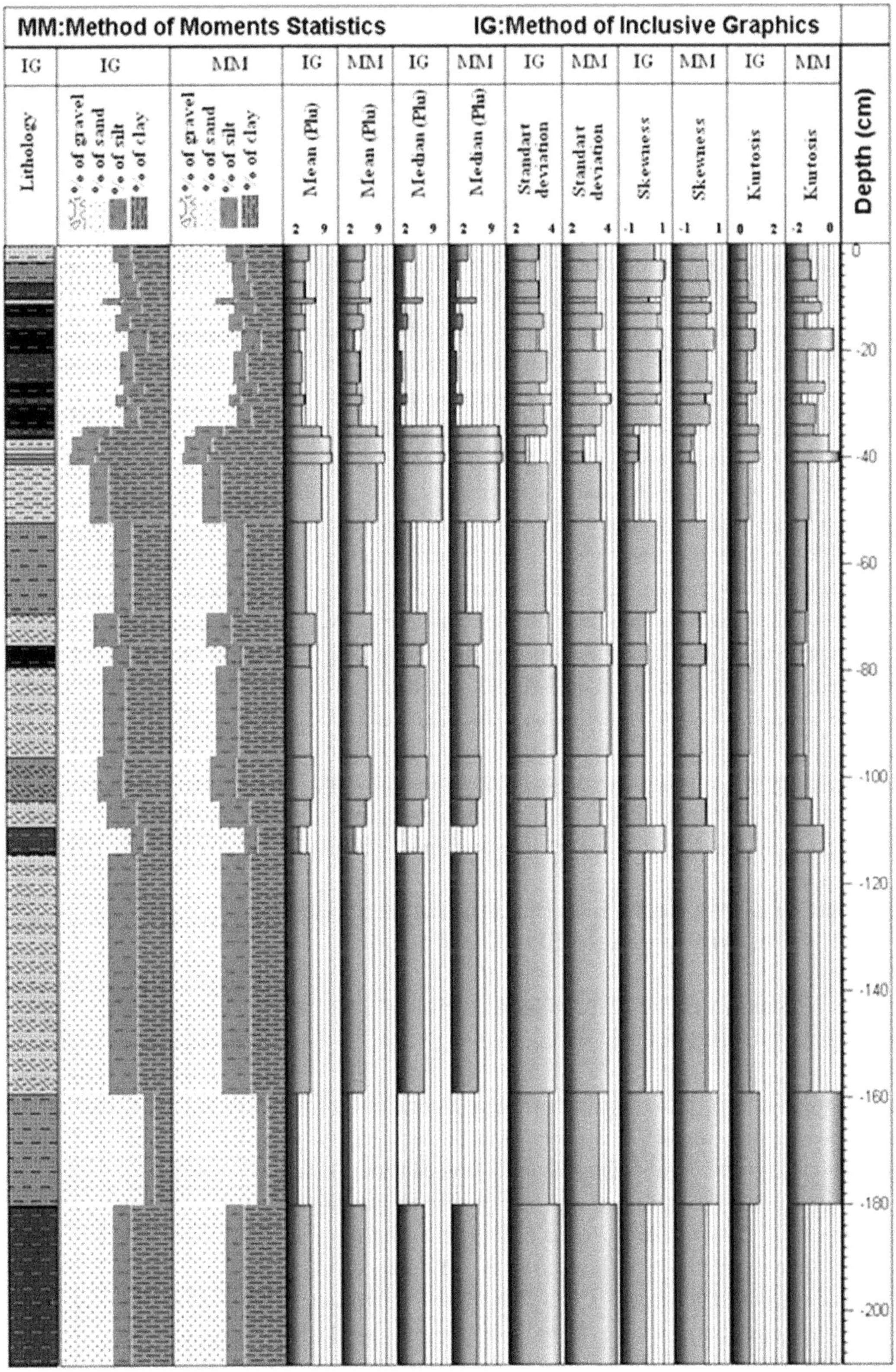

Combined Descriptive Grain Size Distribution in Core3

After testing the two results in the distinction of different bands in a core, the combination of the classification schemes and the computed statistical parameters will be applied on core3.

In the Figure 8, the classification of Schepard's and Folk combined together may classify sediment within cores; if two sediments escape from the classification of Schepard, they are more than likely to be distinguished by Folk classification. This distinction is itself in Sidi El Hani clay pan is an achievement. Because, the grain size distribution is dominated by the clayey and silty fraction. For instance, Folk classification fails to distinguish between the fifth and the sixth bands in core3 (Fig.8) by classifying them similarly as sandy clay. The distinction is given by using the classification of Schepard, which classifies the fifth band as silty clay and maintains the sixth band as sandy clay.

The computed statistical parameters of the grain size distribution in core3 (Fig.9) shows an evident variation of the grain size. In a general view, **three** parts are quite distinguishable. **First**, along the first 10cm, the sandy fraction tends to dominate. Second, the clayey fraction takes the super voice; it spans between 10cm and 82cm. **Third**, to the end of the core, the sandy fraction takes again the super voice; the percentage of sand becomes greater and the sandy fraction is more and more dominant. It is noticed that the sandy fraction is anti-correlated with the grey scale; the dark sediment is more likely to have a high clayey fraction and the clear sediment is more likely to have a high sandy fraction. The exception is made with the third reference band where a sandy sedimentation is combined with high grey scale. As it is quite evident, the descriptive grain size distribution is still far being efficient to give an idea about the hydrodynamics. This approach stands in the descriptive area. Nevertheless, one can say that the clayey fraction is associated with a calm depositional environment and the sandy fraction is associated with more agitated depositional environment. This agitation is strongly commanded by the climatic factor; the more the climate is wet the more the sandy fraction is available in sediments; the filling of the playa by the erosion of the surrounding outcropping sediments is also related to wind activity; as it is shown in chapter four, the outcropping sediments in the vicinities of the playa are basically aeolian. Thus, the sandy fraction is more than likely coming from the surrounding sandy outcrops. The computed statistical parameters may refine the distinction of bands; thus, will be used in the correlation between cores.

Correlation of Cores From Sidi El Hani Clay Pan on the Basis of the Descriptive Grain Size Distribution

Correlation Between Cores From Sebkha Sidi El Hani (Stricto Sensu), Souassi and Dkhila

The aim of this correlation is to compare the different parts of the saline environment. Since the three parts have the same watershed, the difference persists in the erosion of the outcropping sediment and the subsidence in locations of cores. Core1 (sebkha Sidi El Hani) and core3 (Dkhila) are correlated by nine alternated bands (Fig.10); five clayey bands are separated by four sandy bands. Generally, the clayey bands are more developed in core3. The silty fraction remains approximately constant along all cores. The core1 contains more of sandy fraction; it may be influenced by the sand dune in its vicinity; it is also more subsiding since it collects more sediment than core3. On the other hand, core6 (sebkha Souassi) records only seven of the nine bands; four clayey bands are interrupted by three sandy bands. Since it is located near the Beta Island of OuledMoussa-Aalalcha, core6 is influenced by its aeolian sedimentary content.

Variation of the Descriptive Grain Size Distribution Along Cores

Figure 8. Schepard's classification, folk classification and verbal description of core3

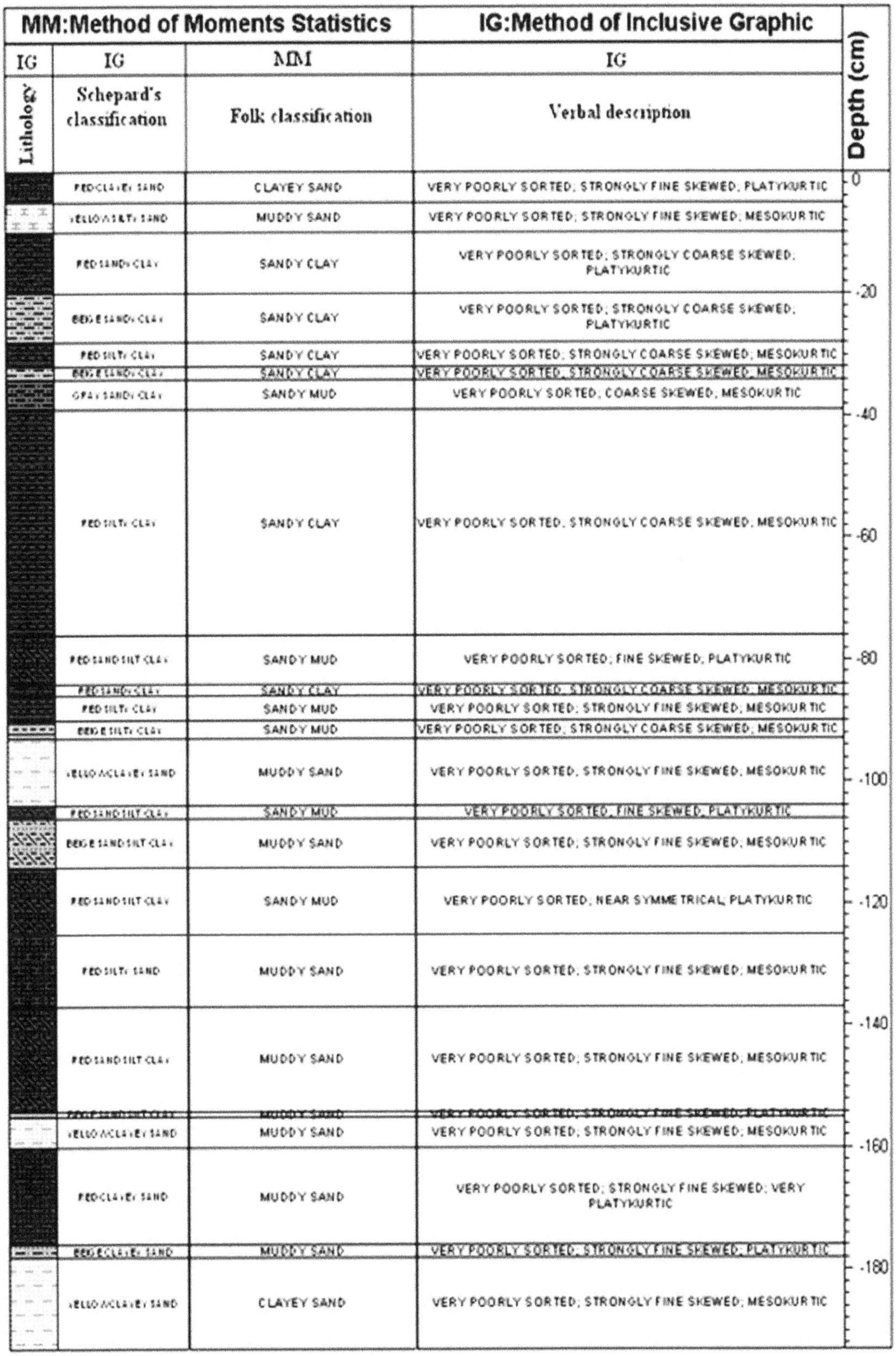

Figure 9. Computed grain size distribution along core3

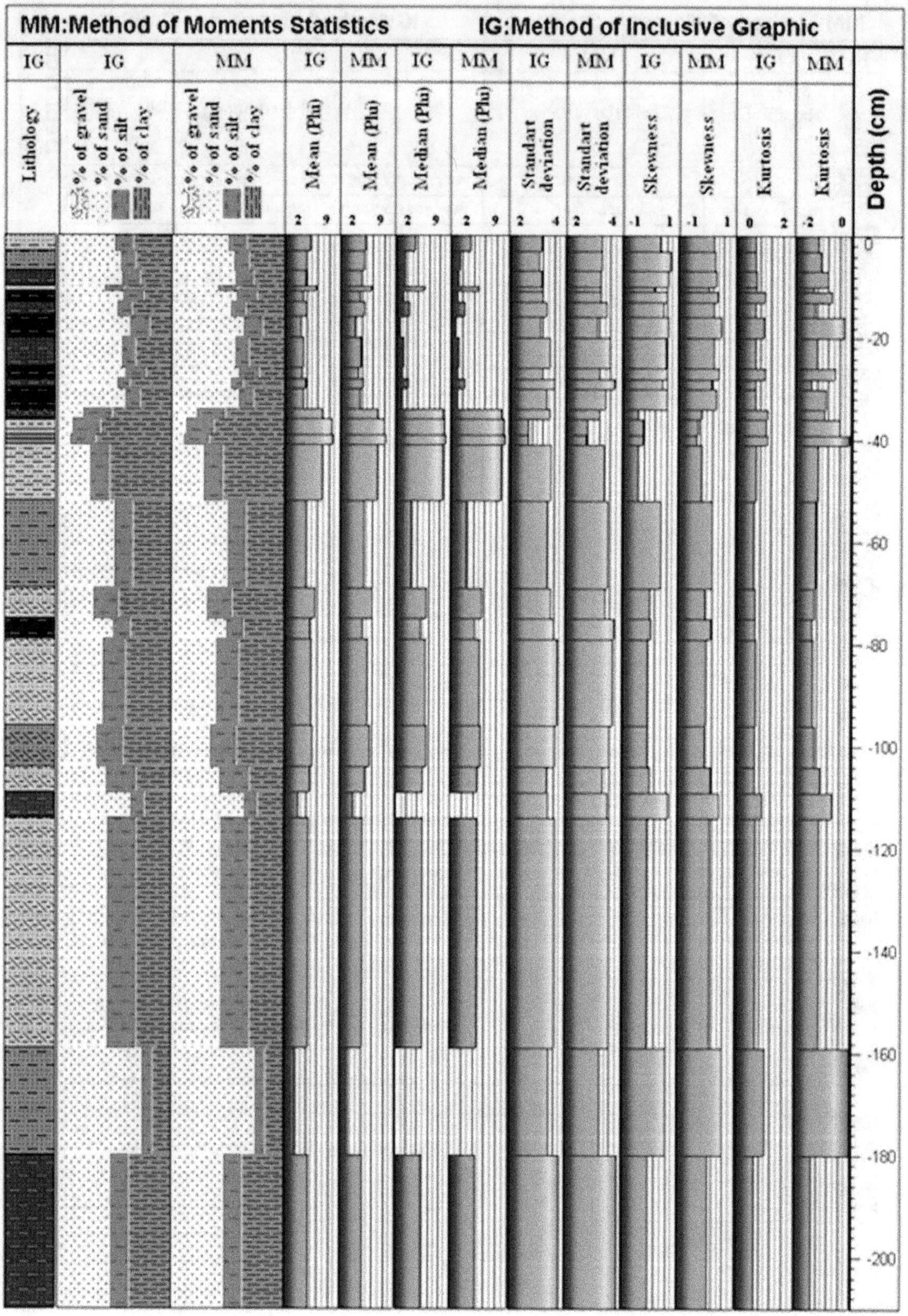

Variation of the Descriptive Grain Size Distribution Along Cores

Figure 10. Correlation between cores from Sidi El Hani (stricto sensu), Souassi and Dkhila on the basis of sand/silt/clay percentages

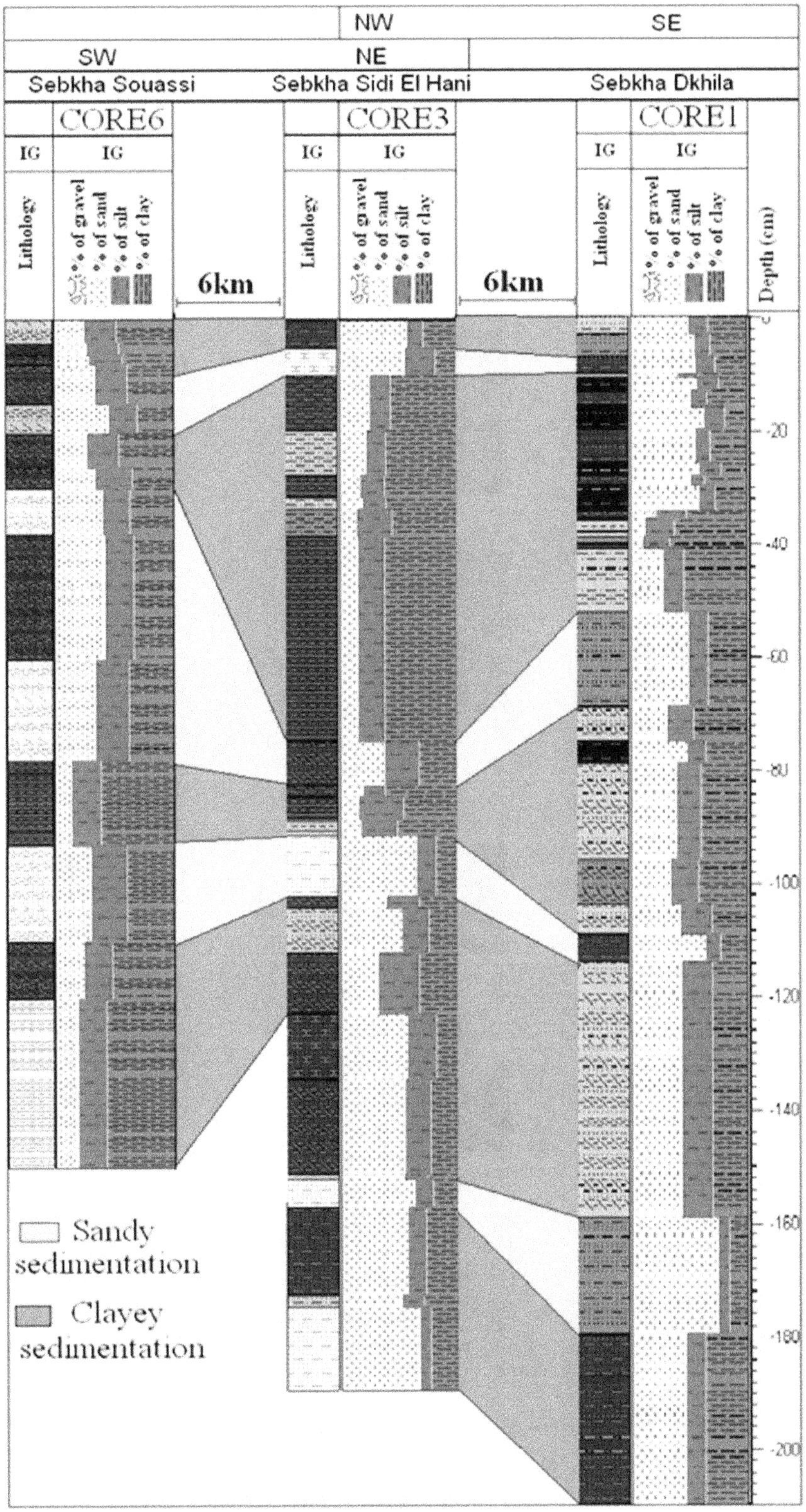

Figure 11. Correlation between cores from the margin Beta Island-clay pan on the basis of sand/silt/clay percentages

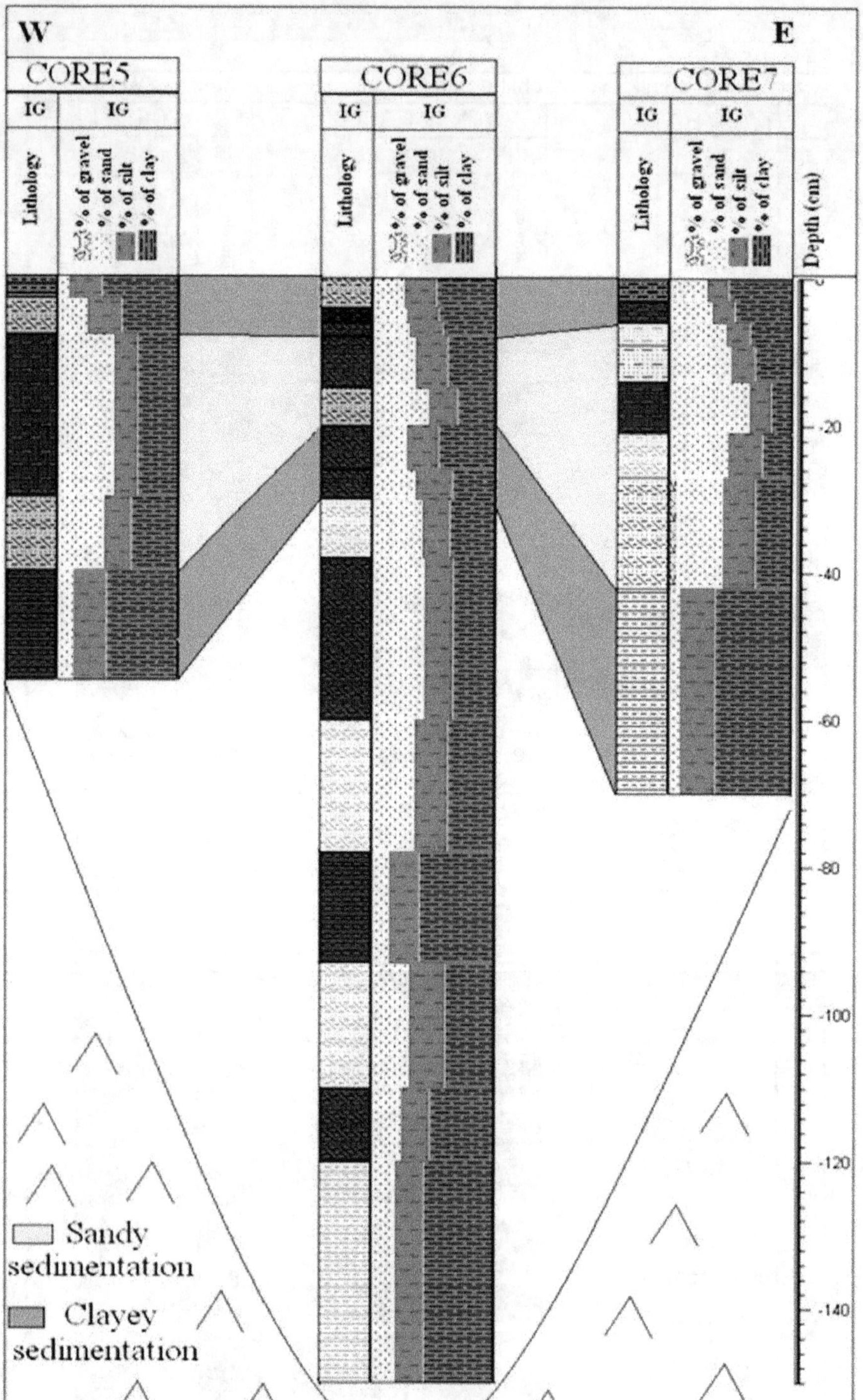

Correlation Between Cores From the Margin Between the Beta Island and Sidi El Hani Clay Pan

The Beta Island may be formed by a wind deflation or a tectonic uplift; seeing the geological context of Sidi El Hani clay pan, both hypotheses are possible. On the one hand, the wind activity may transport grains of sand from far zones such as the so-called Le Grand Erg Oriental and from near zones in Sidi El Hani watershed. On the other hand, the compressive tectonics like the Post-Villafranchian NW-SE¶ ¦compression may cause the uplift of such geological features. The Beta Island of OuledMoussa-Aalalcha is located westward in the Figure 11. Only three bands of the nine bands are recognized in these three cores. Concerning the fibrous gypsum layers, its surface is undulated. It outcrops at the edge of the Beta Island.

CONCLUSION

In spite of the difficulty to separate the fine fraction and the failure of classification methods to differentiate bands, the grain size distribution has given some results. The descriptive grain size distribution within cores of Sidi El Hani clay pan is dominated by clayey and silty fraction; but sandy sedimentation bands represent reference bands for correlation between cores. The correlation points out nine depositional periods. Five clayey periods are interrupted by four sandy periods. This correlation may have no significance, neither in space nor in time. Chronologically, at the very same time, different sedimentations may take place in different locations.

REFERENCES

Ben, J. M. A. A. H. (2008). *Le système endoréique de Sidi Al Hani, Chrita et Mechertate: Paléoenvironnement et dynamique récente* (Thèse de doctorat). Université de Tunis.

Besbes, M. (1975). *Etude hydrogéologique de la plaine de Kairouan sur modèle mathématique. RAPP. CIG*. EMP/DRE.

Blair, T. C., & Mcpherson, J. G. (1999). Grain-size and textural classification of coarse sedimentary particles. *Journal of Sedimentary Research*, *69*(1), 619. doi:10.2110/jsr.69.6

Cailleux, A., & Tricart, J. (1962). Le modèle glaciaire et nival. SEDES.

Collias, E. E., Rona, M. R., Mcmanus, D. A., & Creager, J. S. (1963). *Machine processing of geological data*. University of Washington Technical Report Number 87.

Feuillette, S. (2001). *Vers une gestion de la demande sur une nappe en accès libre: exploration des interactions ressources usages par les systèmes multi-agents application à la nappe de Kairoun, Tunisie centrale* (Thèse de doctorat). Université de Montplier II.

Folk, R. L. (1966). A review of grain-size parameters. *Sedimentology*, *6*(2), 73–93. doi:10.1111/j.1365-3091.1966.tb01572.x

Folk, R. L. (1971). Longitudinal dunes of the northwestern edge of the Simpson Desert, Northern Territory, Australia, 1. Geomorphology and grain size relationships. *Sedimentology*, *16*(1-2), 5–54. doi:10.1111/j.1365-3091.1971.tb00217.x

Inman, D. L. (1952). Measures for describing size of sediments. *Journal of Sedimentary Petrology*, *19*(2), 125–145.

Jarraya-Horriche, F. (2004). *Contribution à l'analyse et à la rationalisation des réseaux piézométriques* (Thèse de doctorat). Ecole Nationale d'Ingénieurs de Tunis.

Kruger, C. J. C. (2003). *Constrained Cubic Spline Interpolation*. http://www.korf.co.uk/spline.pdf

Pejrup, M. (1988). *The triangular diagram used for classification of estuarine sediments: a new approach. Tide-influenced sedimentary environments and facies*. Reidel.

Poppe, L. J., Eliason, A. H., & Hastings, M. E. (2004). A Visual Basic program to generate sediment grain-size statistics and to extrapolate particle distributions. *Computers & Geosciences*, *30*(7), 791–795. doi:10.1016/j.cageo.2004.05.005

Reineck, H.E., & Siefert, W. (1980). Faktoren der Schlickbildung im Sahlenburger Watt und Neuwerker Watt. *Die KuK ste, 35*, 26-51.

Rodgers, J. (1950). The nomenclature and classification of sedimentary rocks. *American Journal of Science*, *248*(5), 297–311. doi:10.2475/ajs.248.5.297

Shepard, F. P. (1954). Nomenclature based on sand-silt-clay ratios. *Journal of Sedimentary Petrology*, *24*, 151–154.

Yechieli, Y., & Wood, W. W. (2002). Hydrogeologic processes in saline systems: Playas, sabkhas, and saline lakes. *Earth-Science Reviews*, *58*(3-4), 343–365. doi:10.1016/S0012-8252(02)00067-3

Chapter 13
Multidisciplinary Study of Cores From the Wetland of El Melah, Southwestern Tunisia: Climatic and Environmental Significance

Sana Bedoui
Higher Institute of Water Sciences and Techniques of Gabes, University of Gabes, Tunisia

Elhoucine Essefi
University of Gabes, Tunisia

Younes Hamed
University of Gafsa, Tunisia

ABSTRACT

Saline systems are candidates to be threatened by climatic change. In terms of methodology and materials, color identification, geochemistry, and mineralogy analysis were used. The spectral analysis of data of the amount of active K and Na firstly shows the individualization of a cycle stretching along with 1000 yr. Then, a less pronounced cycle of 2500 years is marked with a lower intensity. Also, on the spectral analysis of K data, the drowned cycles of 650 and 500 years are marked by weak intensities not reaching the threshold level. These two comparable cycles may be artifacts due to analysis errors or variability in the rate of sedimentation resulting in the bifurcation of two hybrid cycles from one real cycle. This work confirmed the cycles found by color studies through real analyses such as geochemical and magnetic measurements. As a matter of fact, the majority of cycles found out by the spectral analysis of colors data are confirmed through analyses.

DOI: 10.4018/978-1-7998-9498-8.ch013

INTRODUCTION

Sebkhas, or seasonally flooded salt depressions, can be found throughout North Africa (Marquer et al. 2008). The term "sebkha" refers to flat-bottomed depressions that usually flood and are devoid of vegetation due to salty soils. Depression lake is frequently mentioned because its water evaporates, leaving the salts it has dissolved in the watershed. Sometimes, like in the case of the sebkha el Melah Zarzis, these sebkhas are in contact with the sea. An exoreic saline system, sebkha El Melah, was chosen to deal with the coastal geological background. It is feasible to anticipate climate cyclicity based on the sediment record of some depressions, such as oceans (Goslin et al., 2018), seas, lakes (Sant et al., 2018), and sebkhas (Essefi et al., 2015; Bedoui (2016); Amal Gammoudi et al., 2019). This salt depression can be found along the dry coasts of North Africa, Arabia, Baja California, and Australia's Shark Bay (Warren, 2006). Various studies have discussed the effect of climate change on the sebkha (Jian et al., 2000; Jallali et al. 2016), their geological and morphological evolution, and climatic (Dillenburg et al. 2017) and Holocene climate variability (Wanner et al., 2008; Pena et al., 2010; Berner et al. 201; Berner et al., 2011). In the literature surveys, different researches have largely discussed the effect of climate change in the environment of Sebkha (Zaibi et al., 2012; Li et al., 2019; Ben Ameur et al., 2019, 2021).

Any paleoclimatic investigation should be timed so that the climatic phases may be assigned to well-known periods of time. Furthermore, precise dating is required for spectrum analysis for Holocene cyclostratigraphy. However, due to technical and economic constraints, dating is not always achievable. Furthermore, some anterior studies in the field may have conducted dating that is relevant to this research. As a result, facies identification allows for the development of an age model. We have levels for sebkha El Melah that are well-dated.

The investigation of the recorded paleoclimatic variability is still a subject of controversy. Nonetheless, according to the available sediments, one or another proxy may be more efficient at extracting information concerning the depositional environment and the climatic conditions. The saline environments in Tunisia are linked to the Mediterranean Sea's global oceanography, which is influenced by the Atlantic Oscillation. On the other hand, exoreic systems, have a significant influence over inland systems. In a similar vein to earlier research, the goal of this project is to conduct a multidisciplinary investigation into the record within a sebkha El Melah.

Study Area: Sebkha El Melah

Sebka El Melah is a Saharan salt flat located in southern Tunisia. It has an area of 150 km^2 (Fig.1). Its perfectly flat surface is slightly below sea level. This saline system occupies the bottom of a depression framed by trays of grown olive trees or left with sparse steppe. It is separated from the sea by a set of dune oolithic cords cut by a channel partly flooded by the sea, reflecting the former communication of the basin with the sea (Duplay et al., 2012). From a sedimentological point of view, the sebkha sediments essentially consist of several salt layers of rock salt and gypsum and polyhalite (Perthuisot 1971). Sebkha El Melah appeared at the beginning of the Wurmian Quaternary period (Perthuisot, 1975). Around 40,000 BP, the lagoon was highly desalinated by freshwater arrivals. At the upper Würm, seawater withdrew and the basin evolved into a temporary lake or continental sebkha.

Figure 1. Geological map of Sebkha El Melah and its vicinities
(Perthuisot and Florida, 1973)

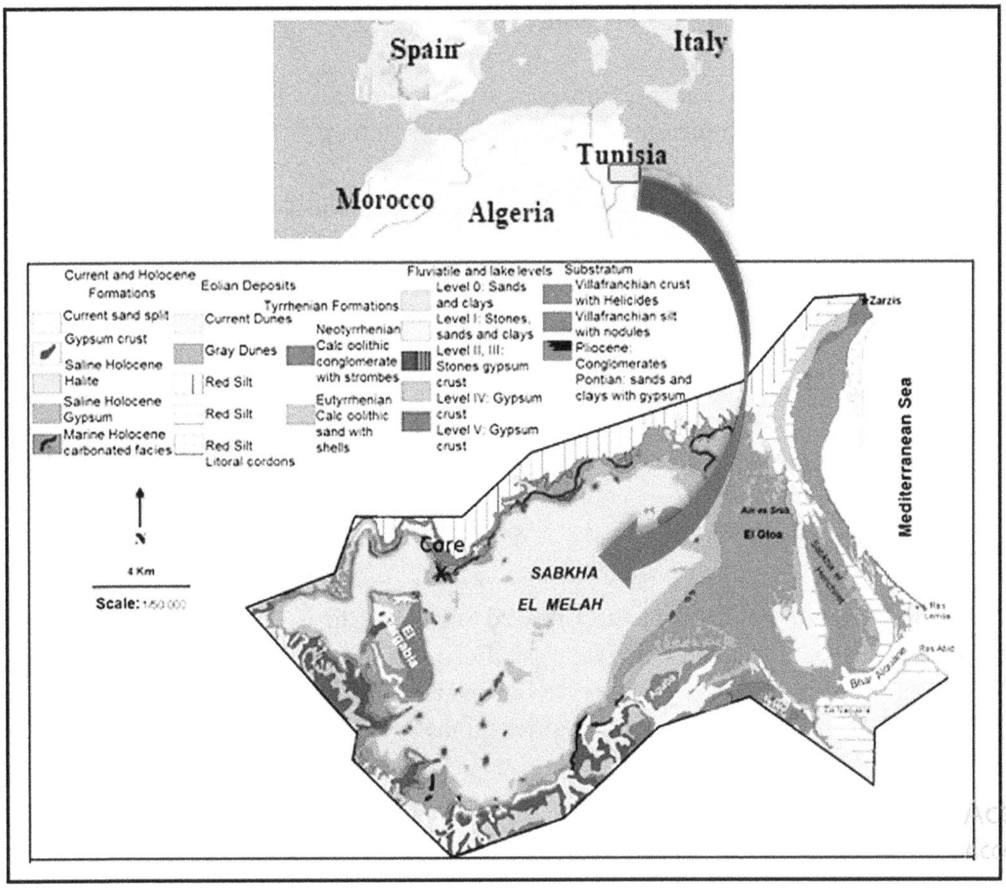

METHODS AND MATERIALS

Soil Sampling

In order to eliminate any potential contamination of the sedimentary fillings by metals, it was better to use a plastic tube for coring. During the field expedition, an 82 cm core was recuperated in the sebkha El Melah using a PVC tube of 3 mm in thickness and 63 mm in diameter. The core was then transported to the laboratory to be cut to a size of 1 cm for destructive (geochemical and mineralogical studies) and non-destructive analyses (color identification and their significance).

Non-Destructive Analyses

Color Evolution

Before doing any destructive analysis of the recuperated samples, the core underwent non-destructive analysis or the Visual Core Description (Essefi, 2009) using color identification and its significance.

Destructive Analyses

Geochemical and Geophysical Variability

Geochemistry applies tools and concepts from chemistry to the study of the Earth and sediments. Thus, Ca^{2+}, K^+, and Na^+ contents were measured using a flame photometer.

Data Presentation and Treatment: Depth-Parameters Plots

Once acquired, data should undergo a clear presentation along with the depth of the cores and a deep statistical and mathematical treatment. To visualize the variability of the 82 values of one parameter along with the cores, a plot with the appropriate scale was carried out. The scale should take into account the maximum and the minimum values of the parameter. Once acquired, data should undergo a clear presentation along with the depth of the cores and a deep statistical and mathematical treatment. Moreover, statistical and mathematical techniques such as spectral analysis were used to treat the obtained data.

Mineralogical Analyses

The diffraction of x-rays is a method universally used to identify the nature and structure of crystallized products. Indeed, this method applies only to crystalline mediums (rocks, crystals, minerals, pigments, clays...) showing the characteristics of the crystalline state, i.e. a periodic arrangement, ordered and in three-dimensional reticular plans of the constitutive atoms.

RESULTS AND DISCUSSIONS

Colors Evolution

Several authors have been interested in the evolution of Tunisian systems during the Holocene (Dearing et al. 2001; Zaîbi et al. 2011; Essefi et al. 2015). The grayscale is one of the adequate methods used to study the high-resolution climatic variability in sebkha Mhabeul southeast Tunisia (Marqueur et al. 2008). The plot of color variability along the core from sebkha El Melah (Fig.2) shows a change along with the core. This variability is due to a radical change in the geochemical and mineralogical compositions of sediment, which are interns due to changes in climatic conditions.

Spectral Analysis Based on Colors Evolution

As for data treatment, the simple spectral analysis was carried out through PAST software. Concerning the spectral analysis, PAST includes the Lomb periodogram algorithm for unevenly sampled data (Press et al., 1992), with time values given in the first column and dependent values in the second column. The data is automatically trended prior to analysis. The highest peak in the spectrum is presented with its frequency and power value, together with a probability that the peak could occur from random data. The 0.01 and 0.05 significance levels are shown as dashed lines. For the time series analysis, we need dating to find out temporal climatic cycles. Nonetheless, the rate of sedimentation in endorheic Tunisian saline systems may be estimated based on the dating of cores collected from sebkhas of the same region, namely the sebkhas of Mhabeul, Sidi Mansour, and El-Guettiate. For instance, tephrochronological dating indicated that the rate of sedimentation during the Late Holocene at sebkha Mhabeul was in the range of 0.33 mm/yr (Essefi et al., 2014) and 0.37 mm/yr (Marquer et al., 2008). Based on the previous dating, a 30 yr for each sediment cm in the sebkha of El Melah was adopted.

Figure 2. Plot of colors variability along the core from Sebkha El Melah

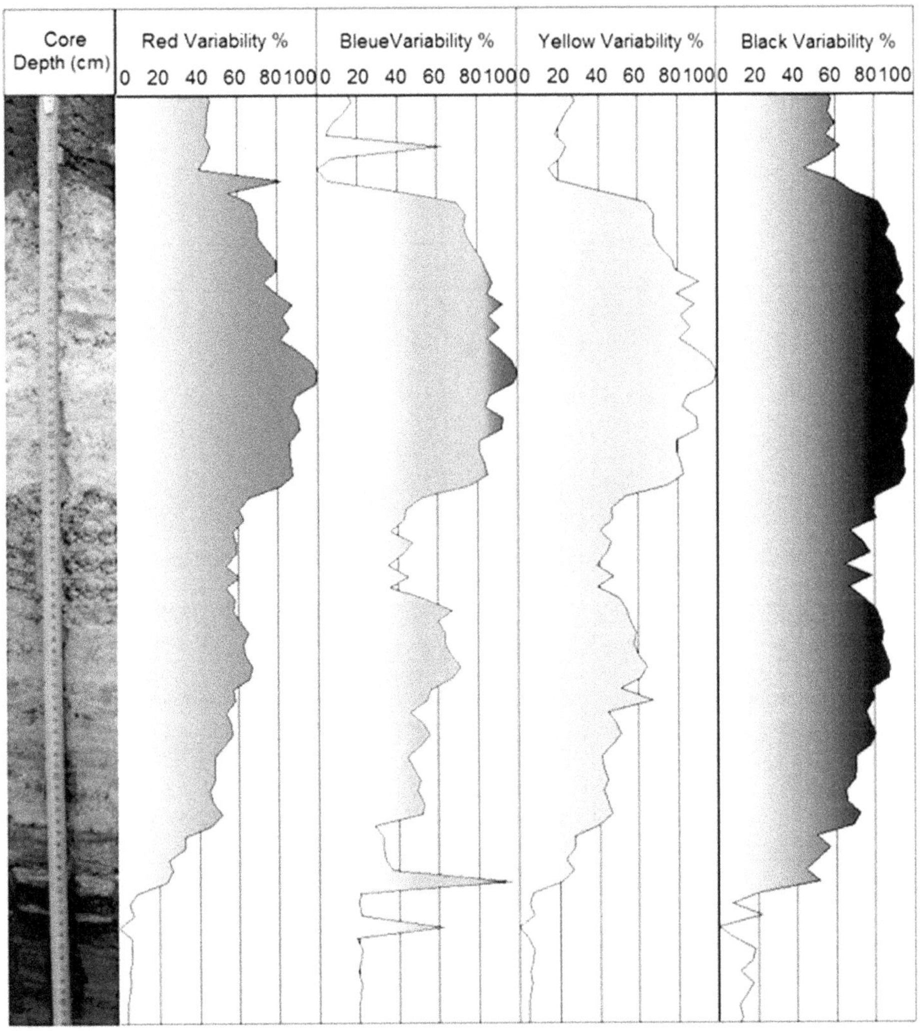

The spectral analysis of red and black data (Fig.3) shows the individualization of a first cycle centered around 2500 yr. Also, a second cycle that is centered on 1000 yr is well documented. On the other hand, the 500 yr cycle is less pronounced.

Figure 3. Spectral analysis of the black and the red

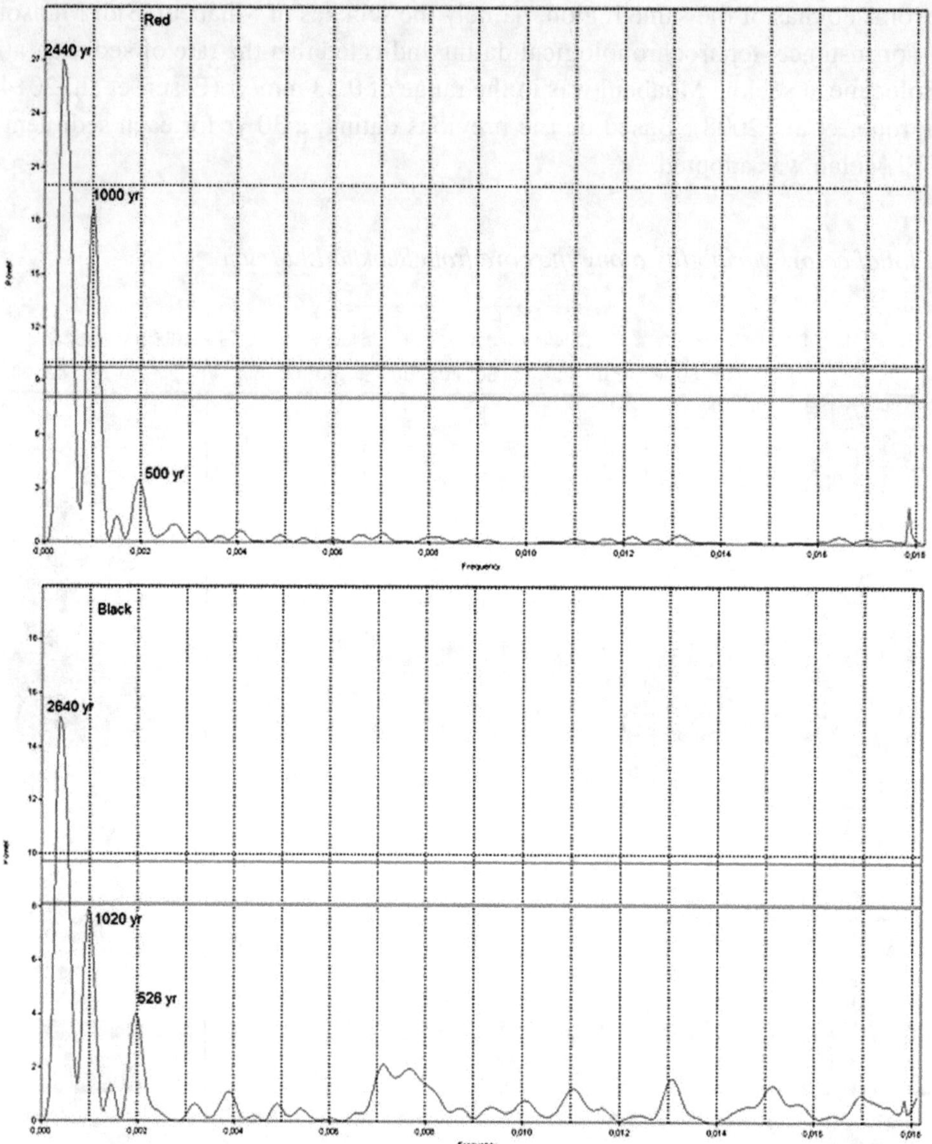

The spectral analysis of yellow and blue (Fig. 4) shows the individualization of a first cycle centered on 1000 yr. A second cycle centered on 2500 yr is well documented. The 500 yr cycle remains less pronounced.

Geochemical and Geophysical Variability

Evolution of Magnetic Susceptibility and Geochemical Elements

The study of the Sebkha El Melah saline system was based on the investigation of a core carried out in the southern part. In the laboratory, samples of core1 underwent analyses of magnetic susceptibility and chemical elements (Ca, Na and K). Values of the magnetic susceptibility along with the core show 2 obvious parts (Fig.5). The first stretche is 8 cm; it consists of continental sediment marked by a clayey dominance indicated by high values of active K and Na. This part is characterized by high values of magnetic susceptibility. Geochemically, it is also characterized by low values of active Ca, which is an indication of the presence of diamagnetic minerals such as calcite and gypsum.

Figure 4. Spectral analysis of the yellow and the blue

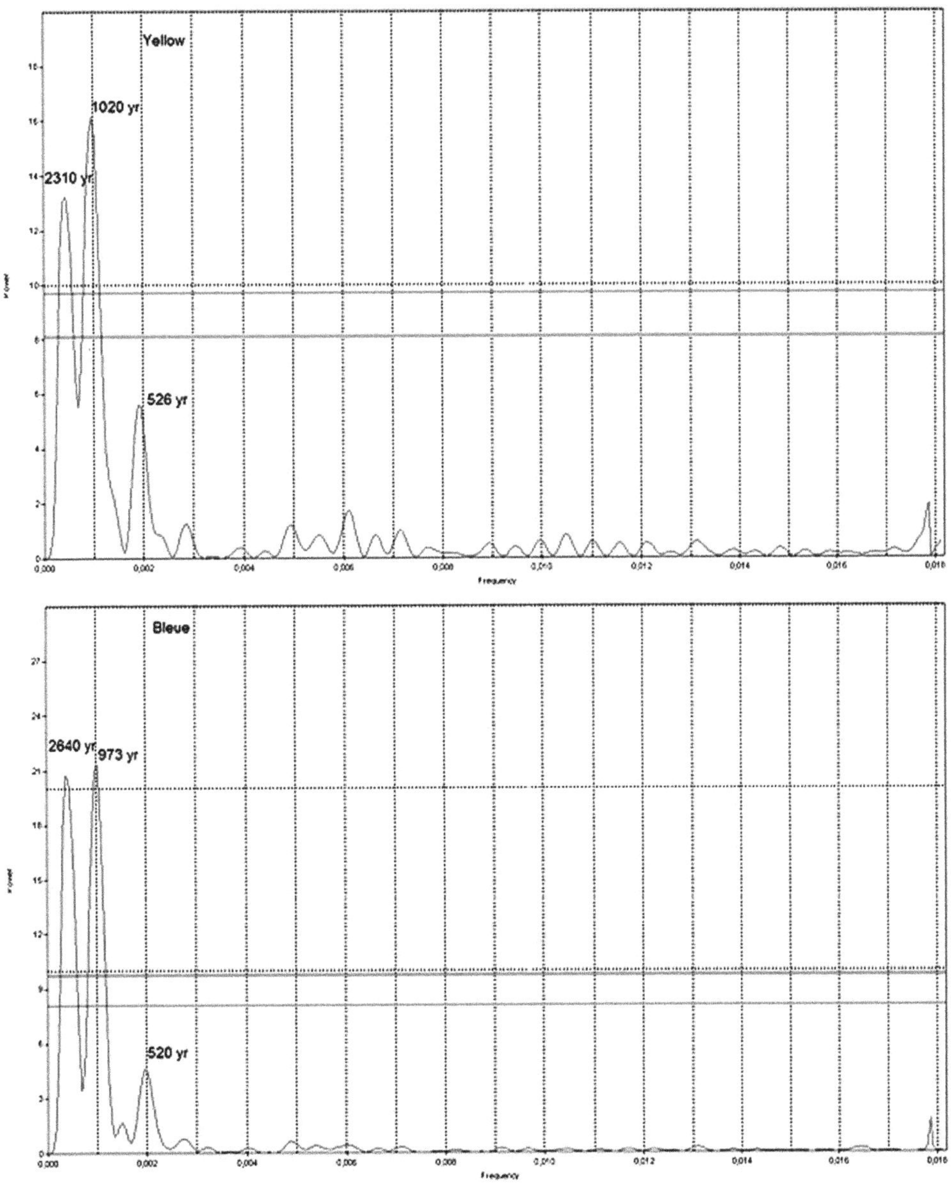

Figure 5. Evolution of magnetic susceptibility and geochemical elements along the core of the El Melah saline system

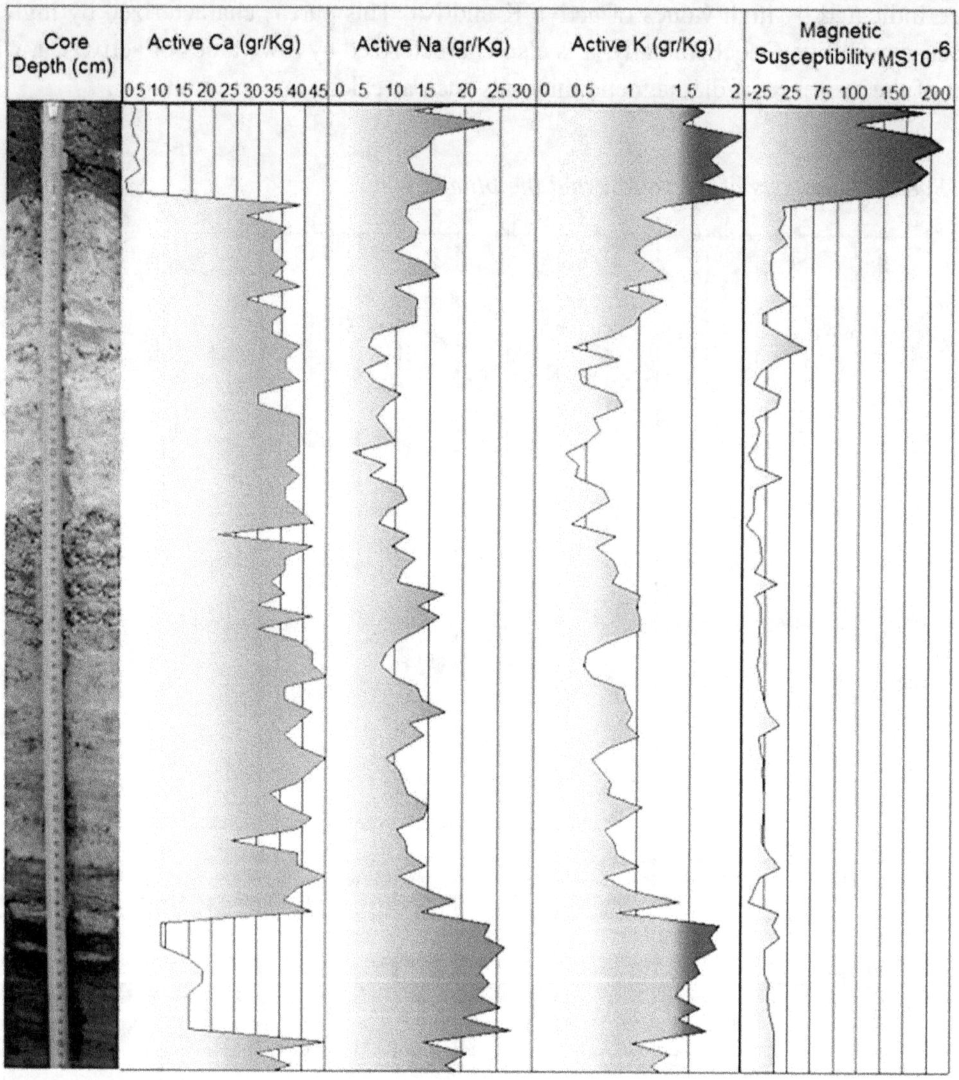

Spectral Analysis Based on the Magnetic Susceptibility and the Geochemistry

Exoreic saline systems in Tunisia, such as that of El Melah record climatic changes as well as eustatic variations (Lakhdar, 2009). Comparably to inland systems, they are the subject of concurrence between continental and marine effects. The oceanographic component directly intervenes in the shaping of the sedimentary cycles. The cyclicity obviously identified along the core may be found out through the spectral analysis of data of the different parameters. The spectral analysis of data of the magnetic susceptibility and the amount of active Ca shows the individualization of a cycle stretching over about 2500 yr (Fig.6). A less pronounced cycle of 1000 yr is marked with a lower intensity. Also, a drowned cycle of 500 yr is marked by a weak intensity not reaching the threshold level. Within sebkhas Mhabeul

(Essefi et al., 2014) and Sidi Mansour (Essefi et al., 2015), a cycle of about 1000 yr was found based on the magnetic susceptibility data.

On the other hand, the spectral analysis of data of the amount of active K and Na firstly shows the individualization of a cycle stretching along about 1000 yr (Fig.7). Then, a less pronounced cycle of 2500 yr is marked with a lower intensity. Also, on the spectral analysis of K data, the drowned cycles of 650 and 500 yr are marked by weak intensities not reaching the threshold level. These two comparable cycles may be artifacts due to analysis errors or variability in the rate of sedimentation resulting in the bifurcation of two hybrid cycles from one real cycle.

Figure 6. Spectral analysis based on the magnetic susceptibility and active Ca

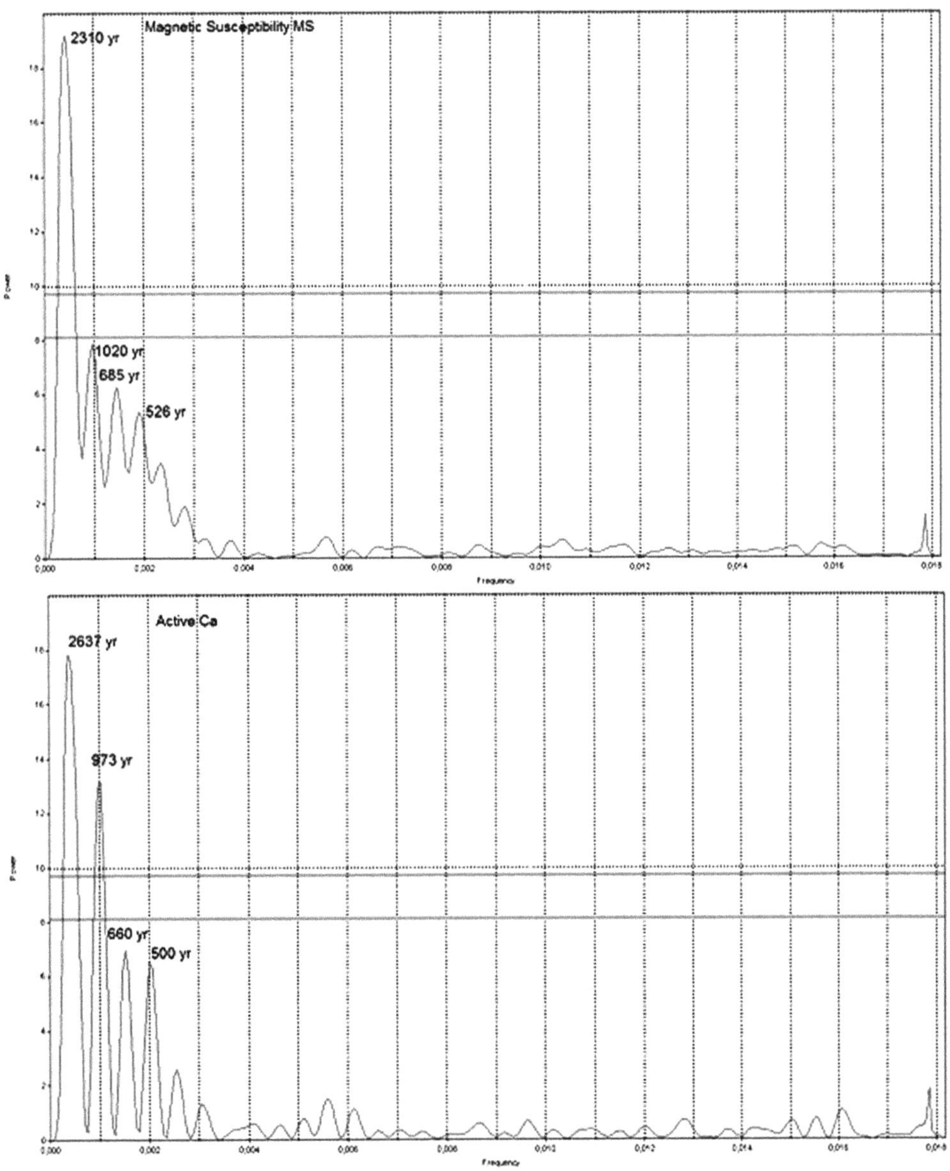

Figure 7. Spectral analysis based on active K and Na amounts

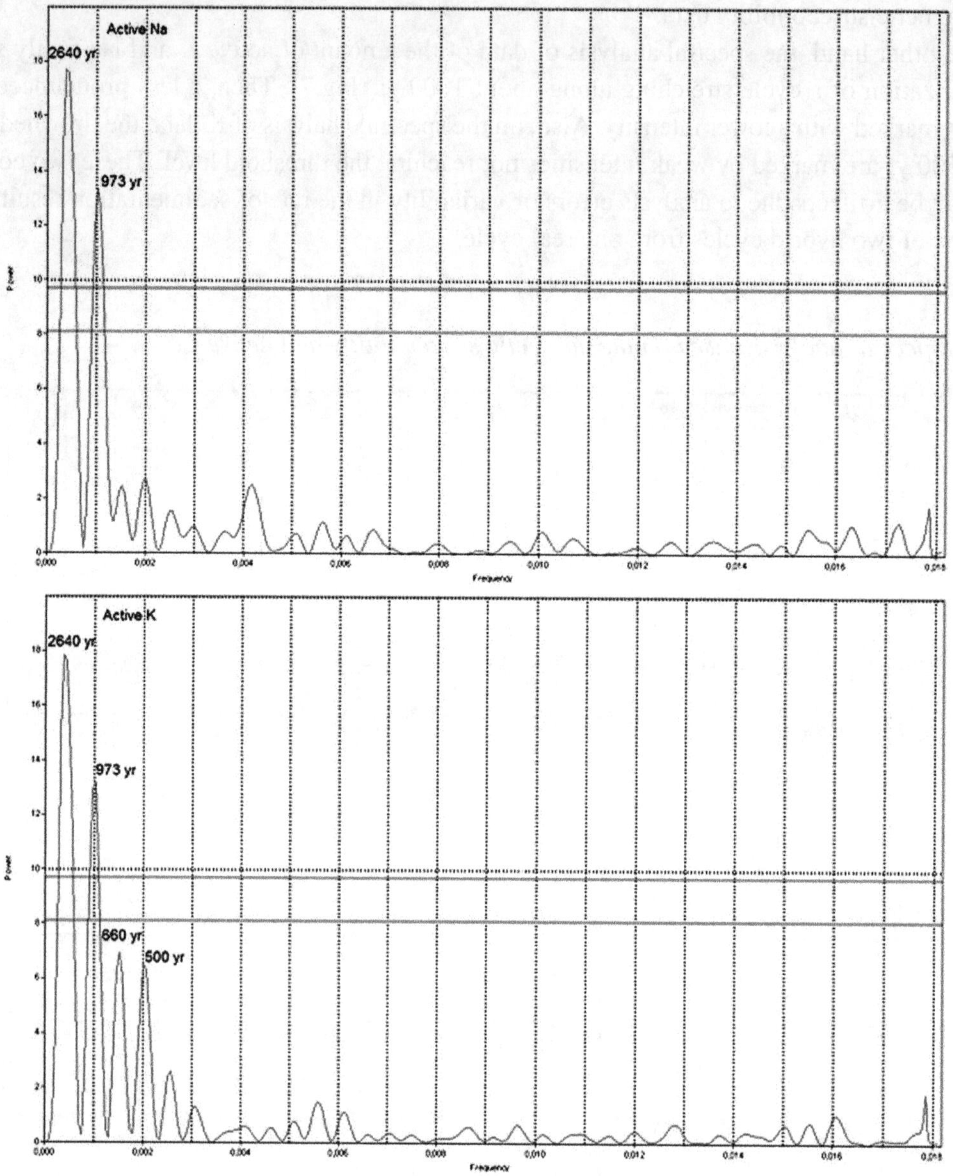

Mineralogical Study

The XRD pattern (Fig.8) of a 32 cm core sample from Sebkha El Melah shows the precipitation of primarily gypsum ($CaSO_4$, 2H2O) (7.59, 4.28, 3.06, 2.6, and 2.11) with traces of other minerals. According to the work of Ben Smida (2016) dealing with the precipitation of salt in sebkha El Melah, gypsum was not available in the present conditions. So, comparably to the present, more arid climatic conditions were responsible for the presence of gypsum.

Figure 8. X-ray diffraction patterns of the sample 32 cm of the core from Sebkha El Melah

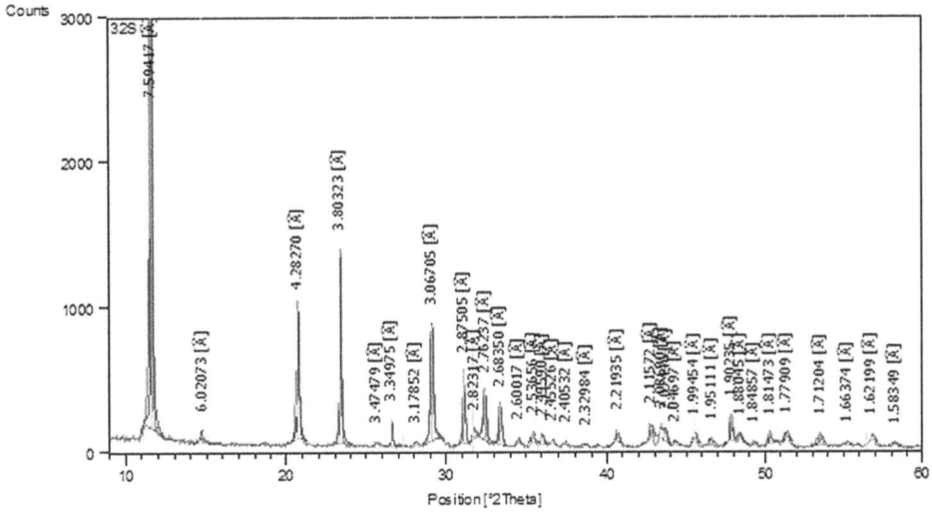

The investigation of the XRD pattern (Fig.9) revealed the precipitation of mainly gypsum (CaSO4, 2H2O) (7.59, 4.28, 3.06, 2.6, and 2.11) with some traces of other minerals for the sample 56 cm of the core from Sebkha El Melah. Comparably to the sample of 32cm (peak of gypsum is 3000), the peak of the sample of 56 cm (4000) is higher, indicating more arid climatic conditions. Various minerals such as halite, gypsum, dolomite, magnesite, carnallite and huntite were deposited at the sebkha of El Melah (Perthuisot 1971; Perthuisot et al. 1990).

Figure 9. X-ray diffraction patterns of the sample 56 cm of the core from sebkha El Melah

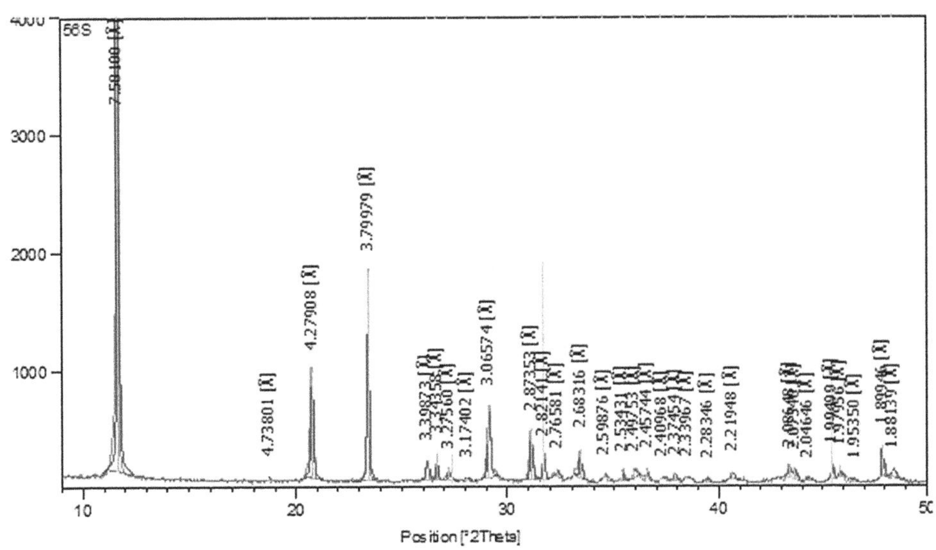

The XRD pattern (Fig.10) of the sample 79 cm of the core from Sebkha El Melah shows that the precipitation is primarily quartz with traces of halite (NaCl) (3.25, 2.82, 1.99, 1.62, and 1.41). Comparably, to the sample 56 cm, this sample indicates wetter climatic conditions.

Figure 10. X-ray diffraction patterns of the sample 79 cm of the core from sebkha El Melah

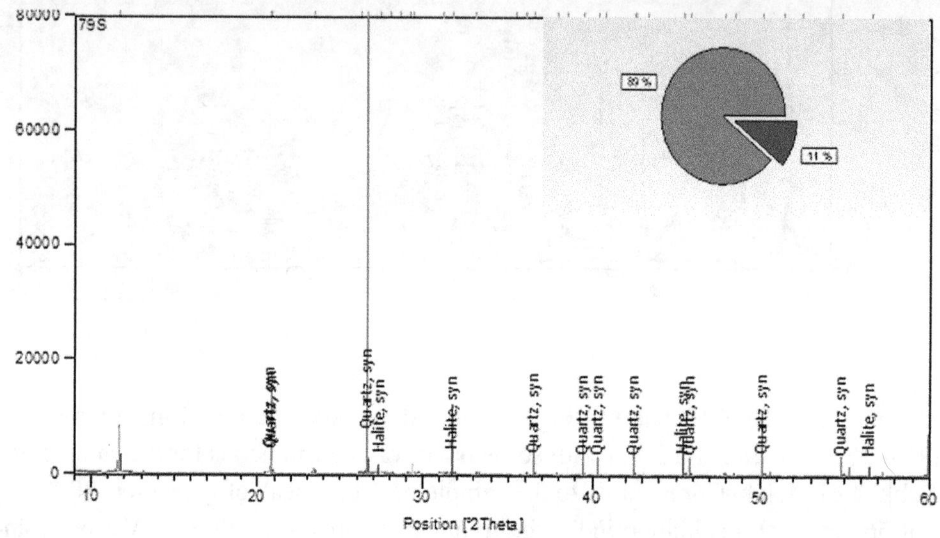

CONCLUSION

This work on El Melah allowed us the extraction of some millennial to decadal cycles. This Holocene cyclostratigraphy in inland systems is basically controlled by sun activity. The contribution of the marine component is definitely proven through the study of the soil by the mineralogy through the XRD. Indeed, the use of spectral analysis has made it possible to demonstrate that several Holocene paleoclimatic records include a continuous 2500-year and a 1000-year cycle during the Late Holocene, sometimes associated with a 1600-year cycle. This second spectral signature, not present in solar activity records, has been attributed to internal forcing, likely oceanic-driven as suggested by Broecker et al. (2001) and McManus et al. (1999). Work on saline systems is currently flourishing worldwide since they are more candidates to record climatic change during the Holocene. Further studies could only improve this understanding of these bodies. More sophisticated techniques and higher resolution may only improve the quality of this study. Concerning techniques, paleoclimatic studies need precise dating to be well established within time. In doing so, it can be possible to attribute each climatic phase to a well-known period. Added to cycles already found out by red and black data 1 cm and analyses, smaller cycles of 500 yr were extracted.

REFERENCES

Ameur, M. B., Masmoudi, S., Abichou, A., Medhioub, M., & Yaich, C. (2019). Use of the magnetic, geochemical, and sedimentary records in establishing paleoclimate change in the environment of Sebkha: Case of the Sebkha Mhabeul in southeastern Tunisia. *Comptes Rendus Geoscience*, *351*(7), 487–497. doi:10.1016/j.crte.2019.10.003

Ameur, M. B., Masmoudi, S., & Yaich, C. (2021). Flood and sandstorm events recorded in holocene sebkha deposits in Southeastern Tunisia: Evidence from magnetic and geochemical properties. *Quaternary International*, *571*, 46–57. doi:10.1016/j.quaint.2020.09.006

Bedoui, S. (2016). *Multidisciplinary study of endorheic (El Guettar) and exoreic (El Melah) saline systems*. Master thesis.

Ben Smida, N. (2016). *Comparative study of exorheic (Boujmal and El Melah) and endorheic (Mchiguig and Mhabeul) saline systems: a geo-economic approach*. Master thesis.

Berner, K. S., Koç, N., Godtliebsen, F., & Divine, D. (2011). Holocene climate variability of the Norwegian Atlantic Current during high and low solar insolation forcing. *Paleoceanography*, *26*(2), n/a. doi:10.1029/2010PA002002

Broecker, W. S., Lynch-Stieglitz, J., Clark, E., Hajdas, I., & Bonani, G. (2001). What caused the atmosphere's CO2 content to rise during the last 8000 years? *Geochemistry Geophysics Geosystems*, *2*(10), n/a. doi:10.1029/2001GC000177

Dearing, J. A., Livingstone, I. P., Bateman, M. D., & White, K. (2001). Palaeoclimate records from OIS 8.0–5.4 recorded in loess–palaeosol sequences on the Matmata Plateau, southern Tunisia, based on mineral magnetism and new luminescence dating. *Quaternary International*, *76*, 43–56. doi:10.1016/S1040-6182(00)00088-4

Dillenburg, S. R., Barboza, E. G., Rosa, M. L. C., Caron, F., & Sawakuchi, A. O. (2017). The complex prograded Cassino barrier in southern Brazil: Geological and morphological evolution and records of climatic, oceanographic and sea-level changes in the last 7–6 ka. *Marine Geology*, *390*, 106–119. doi:10.1016/j.margeo.2017.06.007

Duplay, J., Semhi, K., Bloundi, M. K., Khedhiri, S., Ahmed, R. B., & Darragi, F. (2012). Geochemical background and contamination level assessment in Tunisian and Moroccan Mediterranean coastal lagoon sediments. *Environmental Earth Sciences*, *67*(4), 1201–1214. doi:10.100712665-012-1566-6

Essefi, E., Gharsalli, N., & Yaich, C. (2015). Climatic significance of grain size parameters of the wet aeolian sedimentary filling of sebkha mhabeul, southeast tunisia: Contribution of the genetic approach. *Journal of Basic and Applied Research International*, *1*(4), 140–152.

Essefi, E., Mefteh, S., Medhioub, M., & Yaich, C. (2014). Magnetic study of the heated and unheated sedimentary fillings of Sebkha Mhabeul, Southeast Tunisia: A geophysical method for paleoclimatic investigation and tephrochronological dating. *International Journal of Geophysics*, *2014*, 2014. doi:10.1155/2014/908395

Gammoudi, A., Essefi, E., & Rigane, H. (2019). Variability of the grain-sized fractions during the last three millennia within the sedimentary fillings of the saline system of Mhabeul: The Holocene cyclostratigraphy. *Arabian Journal of Geosciences*, *12*(9), 1–11. doi:10.100712517-019-4472-3

Goslin, J., Fruergaard, M., Sander, L., Gałka, M., Menviel, L., Monkenbusch, J., & Clemmensen, L. B. (2018). Holocene centennial to millennial shifts in North-Atlantic storminess and ocean dynamics. *Scientific Reports*, *8*(1), 1–12. doi:10.103841598-018-29949-8 PMID:30143649

Jalali, B., Sicre, M. A., Bassetti, M. A., & Kallel, N. (2016). Holocene climate variability in the northwestern Mediterranean Sea (gulf of lions). *Climate of the Past*, *12*(1), 91–101. doi:10.5194/cp-12-91-2016

Jian, Z., Wang, P., Saito, Y., Wang, J., Pflaumann, U., Oba, T., & Cheng, X. (2000). Holocene variability of the Kuroshio current in the Okinawa Trough, northwestern Pacific Ocean. *Earth and Planetary Science Letters*, *184*(1), 305–319. doi:10.1016/S0012-821X(00)00321-6

Lakhdar, R., & Soussi, M. (2009). Les dépôts lagunaires holocènes du Sud-Est de la Tunisie: extension et organisation spatio-temporelle. Conférence Méditerranéenne Côtiere et Maritime, 115-118. doi:10.5150/cmcm.2009.029

Li, M., Huang, C., Ogg, J., Zhang, Y., Hinnov, L., Wu, H., Chen, Z.-Q., & Zou, Z. (2019). Paleoclimate proxies for cyclostratigraphy: Comparative analysis using a Lower Triassic marine section in South China. *Earth-Science Reviews*, *189*, 125–146. doi:10.1016/j.earscirev.2019.01.011

Marquer, L., Pomel, S., Abichou, A., Schulz, E., Kaniewski, D., & Van Campo, E. (2008). Late Holocene high resolution palaeoclimatic reconstruction inferred from Sebkha Mhabeul, southeast Tunisia. *Quaternary Research*, *70*(2), 240–250. doi:10.1016/j.yqres.2008.06.002

McManus, J. F., Oppo, D. W., & Cullen, J. L. (1999). A 0.5-million-year record of millennial scale climate variability in the North Atlantic. *Science*, *283*(5404), 971–975. doi:10.1126cience.283.5404.971 PMID:9974387

Pena, L. D., Francés, G., Diz, P., Esparza, M., Grimalt, J. O., Nombela, M. A., & Alejo, I. (2010). Climate fluctuations during the Holocene in NW Iberia: High and low latitude linkages. *Continental Shelf Research*, *30*(13), 1487–1496. doi:10.1016/j.csr.2010.05.009

Perthuisot, J. P. (1971). Recent polyhalite from Sebkha el Melah (Tunisia). *Nature. Physical Science (London)*, *232*(35), 186–187. doi:10.1038/physci232186a0

Perthuisot, J. P. (1974). *Les dépôts salins de la sebkha El Melah de Zarzis: conditions et modalites de la sedimentation evaporitique*. Academic Press.

Perthuisot, J. P., & Floridia, S. (1973). *Carte géologique de la Sebkha el Melah et de ses bordures 1: 50,000.* Laboratoire de géologie de l'Ecole normale supérieure.

Press, W. H., Teukolsky, S. A., Flannery, B. P., & Vetterling, W. T. (1992). Numerical recipes in Fortran 77: volume 1, volume 1 of Fortran numerical recipes: the art of scientific computing. Cambridge University Press.

Russell, J. M., & Johnson, T. C. (2005). A high-resolution geochemical record from Lake Edward, Uganda Congo and the timing and causes of tropical African drought during the late Holocene. *Quaternary Science Reviews*, *24*(12-13), 1375–1389. doi:10.1016/j.quascirev.2004.10.003

Sant, K., Mandic, O., Rundić, L., Kuiper, K. F., & Krijgsman, W. (2018). Age and evolution of the Serbian Lake System: Integrated results from middle Miocene Lake Popovac. *Newsletters on Stratigraphy*, *51*(1), 117–143. doi:10.1127/nos/2016/0360

Sawyer, S., & Tapia, A. (2005). The Sociotechnical Nature of Mobile Computing Work: Evidence from a Study of Policing in the United State. *International Journal of Technology and Human Interaction*, *1*(3), 1–14. doi:10.4018/jthi.2005070101

Wanner, H., Beer, J., Bütikofer, J., Crowley, T. J., Cubasch, U., Flückiger, J., & Widmann, M. (2008). Mid-to Late Holocene climate change: An overview. *Quaternary Science Reviews*, *27*(19-20), 1791–1828. doi:10.1016/j.quascirev.2008.06.013

Warren, J. K. (2006). *Evaporites: sediments, resources and hydrocarbons*. Springer Science & Business Media. doi:10.1007/3-540-32344-9

Zaïbi, C., Carbonel, P., Kamoun, F., Fontugne, M., Azri, C., Jedoui, Y., & Montacer, M. (2012). Evolution of the sebkha Dreîaa (South-Eastern Tunisia, Gulf of Gabes) during the late Holocene: Response of ostracod assemblages. *Revue de Micropaleontologie*, *55*(3), 83–97. doi:10.1016/j.revmic.2012.03.003

Zaîbi, C., Kamoun, F., Carbonel, P., & Montacer, M. (2011). Distribution des ostracodes dans les sédiments de subsurface de la Sebkha El-Guettiate (Skhira, golfe de Gabès). Intérêt pour la reconstitution des paléo-environnements de l'Holocène. *Carnets de géologie*.

KEY TERMS AND DEFINITIONS

Exoreic Saline System: It can be a Sebkha which was in communication with the sea, it is the case of the Sebkha El Melah.

Flame Photometer: Is used to carry out the analyses of active major elements (Ca^{2+}, Na^{+} and K^{+}).

Geochemistry: Applies tools and concepts of chemistry to the study of the Earth and sediments.

Mineral: It is a crystalline chemical substance formed by a geological process.

Mineralogy: It is a scientific discipline that is concerned the chemical composition, internal crystal structure of minerals.

Spectral Analysis: This technique allows the tissue impulse response function to be derived with minimal modeling assumptions.

X-Ray Diffraction: It is a method universally used to identify the nature and the structure of the crystallized products.

Chapter 14
Millennial to Centennial Cyclicity Within the Exoreic Saline System of Boujmel, Southern Tunisia

Elhoucine Essefi
University of Gabes, Tunisia

ABSTRACT

Previous works proposed different age models of sedimentation in Sebkha Lagoon of Boujmel leading to the setting of controversial interpretations of eustatic and climatic phases. The aim of this work is carrying out a geological correlation and an astrochronological calibration based on the Holocene cyclostratigraphy leading to the setting of an age model satisfying dates of climatic and eustatic phases identified in southern Tunisia, including the Anthropocene and the Great Acceleration. Along a 130 cm core, four major climatic phases were upward recorded.

INTRODUCTION

Cyclostratigraphy is a branch of stratigraphy dealing with the interpretation of cyclic variations in the stratigraphic record. The most important, and also the first to be found out, of these cycles are the Earth's orbital cycles of precession, obliquity, and eccentricity (Milankovitch cycles), which result from perturbations of the Earth's orbit and its rotational axis. These cycles translate (via orbital-induced changes in insolation) into climatic, oceanographic, sedimentary, and biological changes that are potentially recorded in the sedimentary archives through geologic time. Nonetheless, the Milancovitch Theory explaining the ancient cyclostratigraphy does not file a case for the Holocene cyclostratigraphy. For instance, the decadal to millennial scales cycles of the Late Holocene sediment in lakes and sebkhas recorded based on geophysical and geochemical parameters were possibly not explained by this Theory. Recent studies (e.g., Baumgarten and Wonik 2015; Timothy and Swindles, 2014; Martinez-Ruiz et al., 2015) have

DOI: 10.4018/978-1-7998-9498-8.ch014

highlighted other astronomical (such as the solar activity) and oceanographic (such as the thermohaline circulation) causes (e.g., Pérez-Asensio et al., 2020) coupled with atmospheric dynamics (Shimizu et al., 2020) to explain short climatic cycles.

As a matter of fact, the Sun activity and its effect on the climatic cyclicity have gained univocal popularity. The short term cyclicity of Sun activity is inferred through sunspots amountand activity. The Schawbe cycle (9 to 14 yr with an average of 11 yr) (Deng et al., 2015; Miyake et al., 2015; Ahluwalia et al., 2015) is governed by the Sun internal dynamics whereas the 22 yr Hale cycle is linked to solar convection (Sun et al., 2015). The Gleissberg cycle (87 yr) is due to the variability in Sun diameter (Kim et al., 2015). Other cycles such as that of Vriès (240yr) are still insufficiently explained (Herrera et al., 2015). These cycles are directly seen through telescopic sunspots and historical naked-eye sunspots observations of solar activity. Other proxies including the cosmogenic isotope ^{10}Be, cosmogenic isotope ^{14}C and the climate record allows inferring the longer cyclicity of the Sun activity (Herrera et al., 2015 and references therein). Nonetheless, the climatic cyclicities are not always related to the Sun activity. Instead, other oceanographic and atmospheric mechanisms and processes created hybrid cycles (Martinez-Ruiz et al., 2015). For instance, in the westernmost Mediterranean Sea, major periodicities at 1300 yr and 1515 yr, linked tothe North Atlantic climate variability and the African monsoon system were found out along the past 20,000 yr. Other major periodicities of 2000 yr and 5000 yr are related to the solar activity. Minor secondary millennial to centennial cycles stretching along 650 yr, 1087 yr, and 3000 yr are still debatable (Rodrigo-Gámiz et al., 2014). Cyclostratigraphy proves vital to give precision to ages found by absolute and relative dating techniques through the astrochronological calibration. Astrochronology is the dating of sedimentary units by calibration with astronomically tuned timescales, such as Milankovitch cycles (Krijgsman et al., 2001; Meyers, 2008) and sunspot cycles (Fairbridge, 1996; Liritzis & Fairbridge, 2003). When used in concern with radiometric dating, it allows the resolution of timescales to a high degree of accuracy. In the case of Milankovitch cyclostratigraphy, the dating error can be lower than 21000 years. As for the application of astrochronology for Holocene sediments, the use of astronomic cycles seems promising to give precision to radiocarbon dating and setting limit of some recent global modification on the planet such as the Anthropocene and the Great Acceleration.

Recently, the natural cyclicity governed by the interplay of the astronomic, oceanographic, and atmospheric factors have been interrupted by an abrupt and radical change related to the human polluting activities. Worldwide, this pollution dates back to the setting of industrial activities during the Anthropocene (Essefi, 2020). The so-called the Great Acceleration (Waters et al., 2014, 2016, 2018) is also marked by more pronounced radical changes of climatic conditions (Hébert et al., 2016; Hébert & Lovejoy 2017). Actually, the Anthropocene is an under discussion geologic era dating back to the setting of noticeable human impact on the terrestrial environments and the subsequent biotic conditions (Yusoff, 2013; Young et al., 2016). It has recently occupied the heart of the Earth Sciences (Crutzen, 2006). In spite of the existence of signs of micropaleontological species characterizing this epoch (Wilkinson et al., 2014), until August 2016, neither the International Commission of Stratigraphy nor the International Union of Geological Sciences has yet officially approved the term as a standalone subdivision of the geological scale (Castree, 2017). In Tunisian sebkha, apart from Chwaya's thesis (2017), there are no studies dealing with the Anthropocene. Nonetheless, the careful investigation of the few uppermost centimeters of cores from Tunisian sebkhas such as Boujmal (Essefi et al., 2015), Halk El Menjel (Gharsalli et al., 2017; 2020), and Mhabeul (Essefi et al., 2014, Essefi, 2020) remarked the radical change compared to the remaining part of the cores.core, including increasing magnetic susceptibility, heavy metals content, and organic matter amount. Nonetheless, no dating was carried out to set the limit of these extreme events.

Due to its location along the Mediterranean Sea already related to the Atlantic Ocean, North Africa is climatically fragile and it is the first to be damaged by any climatic change in the Mediterranean basin (Abd-Elmabod et al., 2020). Indeed, when the climate changes, water and air circulations worldwide change in the Atlantic Ocean and subsequently in the Mediterranean basin. Nonetheless, some regions (inland systems) immediately response; others (exoreic systems) should wait laps of time to modify the record of their depositional environments (Lim et al., 2019). The Saline systems have been an important field of study for academic and applied purposes. Sedimentologically, they give a lot of information about the sedimentary history and the depositional environment (e.g., Gammoudi et al., 2019 and references therein). To climatologists, the sedimentary records of these systems could be a subject for paleoclimatic researches (e.g., Essefi et al., 2013). Depending on the seawater contribution in the water budge, we can distinguish between exoreic and endorheic systems. Having close relation with the sea, exoreic systems receive seawater contribution in the sum of the water collected within the saline system. The exoreic system may be marine- or inland-influenced according to the dominance of marine or continental contribution in the water and salt budgets. This interplay between marine and continental influence may be variable with time due to the change of climatic and eustatic conditions. Tunisia is marked by the setting of numerous saline systems. Some of them are endorheic (Essefi, 2009; 2013); others are exoreic (Lakhdar, 2009). In this work, the saline systems will be introduced in terms of geographic, geologic, and climatic settings. These systems are reasonable candidate to record the climatic and eustatic variability due totheir setting at the outlets of drainage basins and the limits of marine and continental environments. Subsequently, these saline systems should be studied in their Mediterranean framework. Then, global phenomena and their effect on North Africa such as the Mediterranean Oscillation (MO), the North Atlantic Oscillation (NAO), and the Atlantic Multidecadal Oscillation (AMO) were introduced to understand the change of Boujmel saline system with a global line of thinking.

Based on geochemical (Ca, K and Na content) and geophysical (magnetic susceptibility) parameters, this work aimed to infer the cyclostratigraphy within exoreic system of Boujmel during the Holocene, southern Tunisia. Cycles were integrated in their Mediterranean context related to the global oceanography and atmospheric circulation. Also, cycles were used to give precision to dating of previous studies. In doing so, we may infer the setting of the Holocene-Anthropocene-Great Acceleration transitions in the area.

STUDY AREA

Geology and Paleoclimatology

The sebkha-lagoon of Boujmel (Fig.1), located 10 km of south sebkha El Melah and 28km from sebkha Mhabeul constitutes the largest evaporitic appendix of Bahira el Bibane with about 48 km^2 surface area. Currently, this area corresponds to a small modern coastal marine–tidal flat system characterized by a warm and arid climate, with only about a 200 mm yearly precipitation. Sebkha-lagoon of Boujmel is episodically supplied by seawater (Mediterranean Sea) through Bahira el Bibane and el Makhada channels (Lakhdar et al., 2006). The hydrological regime of the sebkha was divided into two main parts: coastal (marine) and continental part (Wasli, 2016).

Figure 1. (a) Setting of sebkha-lagoon of Boujmel within Mediterranean context, MV: Mejerda valley; (b) Position of sebkha-lagoon of Boujmel relative to Bibane lagoon, Sebkhas El Melah and Mhabeul (Jaouadi et al., 2016): Cores locations of Gammoudi et al. (2019) Mah1, Jaouadi et al. (2016) BJM2 and this work BJM1

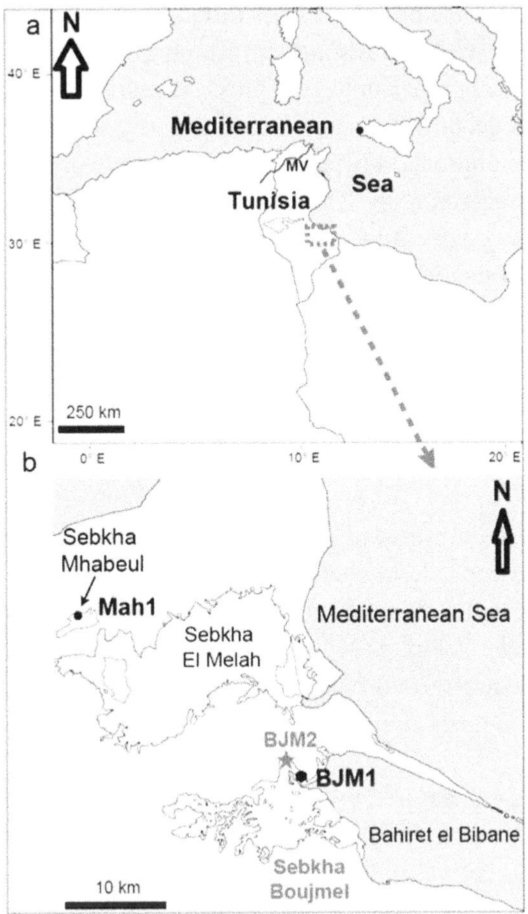

Climatic Phases and Holocene Cyclostratigraphy in Southern Tunisia

Recent paleoclimatic studies on sebkhas (e.g., Boujmel and Mhabeul, located 28 km from the sebkha-lagoon of Boujmel; Fig.1b) in southern Tunisia seem controversial in defining climatic stages. On the one hand, based on pollen study, Jaouadi et al. (2016) recorded at sebkha Boujmel four aridity pulses along a 160 cm core (distanced 3 Km from the core of this work) that punctuate the Late Holocene (4500 BP). According to the age model, dry (arid) events occur at ca., 3.70–3.25 (B4a), 3.00–2.00 (B3a), 1.40–1.10 (B2a) and 0.40–0.00 (B1a) ka BP and alternate with four more humid phases, (B4h) ca. 4.60-3.70, (B3h) ca. 3.25-3.00, (B2h) ca. 2.00-1.40 and (B1h) ca. 1.10-0.40 ka BP. Nonetheless, other imbedded climatic phases were not identified. For instance, within the B1a, which is an arid phase, two flooding event were identified correlated with those found by Gammoudi et al. (2019) at sebkha Mhabeul. Also, it encompasses the Great Acceleration (0-70 yr), the Anthropocene (70 yr-300 yr) and the Late Little Ice Age already recorded within sekkha Mhabeul (Gammoudi et al., 2019). Arid conditions are re-established in southern Tunisia and are contemporary with the Little Ice Age (LIA, ca. 1450–1850 AD). This arid climatic episode during the LIA coincides with an increase in clastic deposits at the top of the sedimentary fill from nearby sebkha Mhabeul (Marquer et al., 2008) and with the renewal of river activity in the Medjerda Valley (Faust et al., 2004) and is also contemporary with the most recent Bond event (NAC 0) at around 0.4 ka BP. The absence of these events in Jaoudi et al. (2016) is quite understood. On the one hand, Jaoudi et al. (2016) used lower resolution (sampling each 2.5 cm versus 0.5 cm in our work). On the other hand, they used a micropaleontogical proxy, which is not able to detect the high frequency climatic signal. However, the geochemical proxy, such as the tephrochronological used by Marquer et al. (2008), is able to reach aresolution as low as few millimeters. In addition, radiocarbon dating has big errors comparably with tephrochronology. Therefore, the two studies carried out on different sebkhas proposed two different succession of climatic phase for southern Tunisia. It is quite evident that climatic phases of Jaouadi et al. (2016) should be corrected by the ones of Marquer et al. (2008). Moreover, dating of Lakhdar et al. (2006) and Lakhdar (2009) already having big errors should be corrected based on climatic phases found by tephras dating and the Holocene cyclostratigraphy.

METHODS AND MATERIALS

During field expeditions, a 130 cm core was collected in the sebkha-lagoon of Boujmel (Fig.1). Within laboratory, the sampling was carried out each 5 mm to obtain 260 samples. The mineralogical composition of three samples located at 3 cm, 61 cm and 121.5 cm was determined by X-Ray Diffraction (XRD). The used diffractometer is Philips X-PERT with a Cu anticathode (Ka). The recording and the digital processing of the data are carried out using the software X' PERT HighScore Plus®. For the 260 samples, we carried out the measurement of the low frequency magnetic susceptibility LFMS and the amount of Ca, Na and K.

The low frequency MS were measured by the Bartington MS2B probe. As geophysical parameters, magnetic susceptibility (MS) is differently used in earth and environmental sciences. In coastal sebkha-lagoons such as of Boujmel, high values of the magnetic susceptibility are attributed to a continental depositional of ferrimagnetic and ferromagnetic minerals such as clayey minerals.Low values of the magnetic susceptibility due to the sedimentation of diamagnetic minerals such as carbonate and quartz, and thus are generally related to an arid climate. (e.g., Essefi, 2013). In terms of climatic conditions, high values of the magnetic susceptibility are attributed to a wet climate. Low values of the magnetic susceptibility are due to an arid climate. Nonetheless, in coastal sebkha-lagoons such as of Boujmel with variable depositional environment oscillating between marine and continental settings, the interpretation of MS is not straightforward. Instead, the competition and reciprocal effect between climate and depositional environment in registering the magnetic signal lead to controversial interpretation of MS values. For instance, high values of the magnetic susceptibility are attributed to a continental deposition of ferrimagnetic (e.g., magnetite) and ferromagnetic minerals such as clayey minerals (related to marine regression and/or wet conditions increasing the sedimentary flux). Low values of the magnetic susceptibility are attributed to marine deposition related to marine transgression and/or arid conditions decreasing the sedimentary flux and resulting in the crystallization of diamagnetic minerals such as carbonate, gypsum and halite (Wasli, 2016). In other words, high MS values may be contradictory interpreted as continental/arid as well as marine/wet conditions. For this reason, other proxies such as geochemical major elements are needed to be more decisive.

Chemical elements including Ca, Na and K were measured by Flame Photometer. When plotting the evolution of parameters along the core, we dealt on the variability based on the following formula Variability = (measured value/(maximum value – minimum value)) x 100.The use of variability allowed us toconvolute parameters with different scales and units of measurement.

Radiocarbon dating was carried along the sebkha by previous studies. On the one hand, Lakhdar et al. (2006) dated 8 samples from different cores laterally dispersed along the sebkha-lagoon with an average error of 300 yr. On the other hand, Jaouadi et al. (2016) dated 11 samples from a 160 cm core (distanced 3 Km from the core of this work) with an average error of 120 yr. Concerning the spectral analysis, PAST includes the Lomb periodogram algorithm for unevenly sampled data (Press et al., 1992), with time values given in the first column and dependent values in the second column. The data were automatically detrended prior to analysis. The highest peak in the spectrum was presented with its frequency and power value, together with a probability that the peak could occur from random data. The 0.01 and 0.05 significance levels were shown as dashed lines. With the big dating error reaching hundreds of years, cycles are also with big errors. In this work, we used an astrochronological calibration based on cycles already recorded in southern Tunisia and other sites. The Holocene cycles were used to calibrate the radiocarbon dating along the core For this reason, we dealt on well-identified cycles (Gammoudi et al., 2019), namely those in relation to sun activity, based on a precise tephrochronological dating at sebkha Mhabeul (Marquer et al., 2008; Essefi et al., 2114) to carry out an astrochronogical calibration. Based on discrete tephra layers, tephrochronology is a geochronological technique creating a chronological framework.The major advantage of this technique is that tephra layers are deposited relatively instantaneously over a large spatial area. In doing so, they provide accurate temporal marker layers which can be used to verify other dating techniques. In terms of statistical analysis, the principal component analysis PCA of all parameters and their possible convolution was used.

RESULTS AND DISCUSSION

Geological Correlation and Facies Visual Identification

Prior to any destructive analysis, the visual core description (VCD) proves relevant to identify obvious facies along the cores (Essefi, 2009). Based on the visual identification, facies already identified and dated by Lakhdar et al. (2006) were recognized along our core. Going from the bottom upward, five well-dated facies were recognized. From 130 cm to 120 cm, suggests a sufficiently confined marine environment. Along our core, the facies are quite influenced by continental sedimentation. The relative sea level of this facies dating to 4000 yr is similar to the current level. At ~4.0 ka BP, RSL dropped to the present datum and remained within ~-0.4 m and ~0.4 m in the late Holocene (Pirazzoli, 2005). From 120 cm to 80 cm, facies (F2) overcome by oolitic carbonated sands. The top of the F2 dated 2500 yr coincides with the RSL 0.1 ± 0.6 m while its bottom dating ~3.1 the RSL was at 0 ± 0.6 m (Pirazzoli, 2005). From 80 cm to 40 cm, a Facies (F3) with an age of 1800 yr contains a diversified fauna represented primarily by ostracods, lamellibranchs, gastropods and benthic foraminifera indicate a brackish lagoon environment (Carbonel & Pinson 1980; Saubade & Zaouali, 1986). Sedimentation continues with silty carbonated sands. At ~1.8 ka BP, the estimated the RSL is at 0.4 ± 0.6 m (Pirazzoli, 2005). From 40 cm to 20 cm, facies (F4) containing a faunistic assembly similar to that of the subjacent horizon. However, the individuals of small size indicate increasingly confined conditions inside the lagoon of Boujmel and enrichment in terrigenous detrital grains testifying the increased influence of the continental sedimentation. From 20 cm to 0 cm, the F5 is primarily eolic detritic (heavy quartz, feldspar, clays and minerals). The associated carbonated fraction consists of rare oolits, pellets and lithoclastes. These silts, containing authigenous gypsum, are poorly sorted and covered by a very thin rock salt film. This beginning of the F5 may coincide with ~0.6 ka BP when the RSL was at -0.1 ± 0.6 m (Pirazzoli, 2005). According to this correlation, the continental facies F1 (the top (123 cm) aged of 4000 yr) and F5 (the bottom (35 cm) aged of 600 yr) may be used as reference to validate the age model.

Age Model

At the level of sebkha-lagoonof Boujmel, dating was carried out in different sites and at different depths by Lakhdar et al. (2006). Afterwards, Jaouadi et al. (2016) dated samples along 160 cm core. In this work, we use all the previous dates to set an age model for the Holocene cyclostratigraphy and the Anthropocene setting. Radiocarbon dating is marked by high errors exceeding hundreds of years. Dates with errors exceeding 100 yr cannot be relevant for Holocene cyclostratigraphy having cycles of centennial orders. Also, these errors cannot set the limit of the Anthropocene-Holocene transition. To set a precise age model, all these dates of Lakhdar et al. (2006) and Jaouadi et al. (2016) were collected in a same plot (Fig.2). Then, a second order polynomial correlation was carried out to calculate age for each 0.5 cm. The polynomial model of ages within sebkha-lagoonof Boujmel based on the 8 datings of Lakhdar et al. (2006) (circles in Fig.2a) and the 11 ages of Jaouadi et al. (2016) (squares in Fig.2a) represents a good correlation (R^2= 0.91).Errors of Lakhdar (2006) ranged between 360 yr and 620 yr; these errors are really handicap for Holocene cyclostratigraphy and Anthropocene setting. On the other hand, errors of dating of Jaouadi et al. (2016) ranged between 53 yr and 133 yr. Taking into account errors in Jaouadi et al. (2016), we carry out third order polynomial age models based on maximum (Fig.2b) and minimum (Fig.2c) values.

Facies Mineralogy and Geochemistry

Mineralogical study of XRD patterns of samples along the core were studied at different depths (3 cm F5, 61 cm F3 and 121.5 cm F1). As indicated in Tab.1, under the current conditions of arid climate and continental sedimentation, the mineralogical assemblage of F5 (300 yr Anthropocene) is basically made up of quartz, halite and calcite. Quartz is basically of Quartz has basically detrital origin; high amount of potassium may be due to increasing clayey dust related to eolian activity recorded in southern Tunisia (Gammoudi, 2020). Calcite is the first precipitated mineral from the salty water due to the availability of calcium. Calcium may be combined in different mineral species such calcite and gypsum, which are precipitated from salt water of the sebkha-lagoon. Currently, salt budget of sebkha-lagoon Boujmel (99 g L^{-1}) is mainly governed by continental water (214 g L^{-1}) at the expense of seawater (35 g L^{-1}) (Ben Smida, 2016).Thus, halite precipitates as indication of seawater contribution. The mineralogical assemblage of F3 (1800 yr) is basically made up of quartz and halite. Quartz is basically of marine origin as it is indicated by the weak amount of potassium. At the bottom of the core, F1 (4000 yr) is marked by the absence of evaporitic minerals. The continental sedimentation had taken place under wet climate.

Figure 2. Age model estimation of sebkha-lagoon of Boujmel: (a) Model based on 19 samples dating of Lakhdar et al. (2006) and Jaouadi et al. (2016). (b) Model based on maximum values of Jaouadi et al. (2016). (c) Model based on minimum values of Jaouadi et al. (2016)

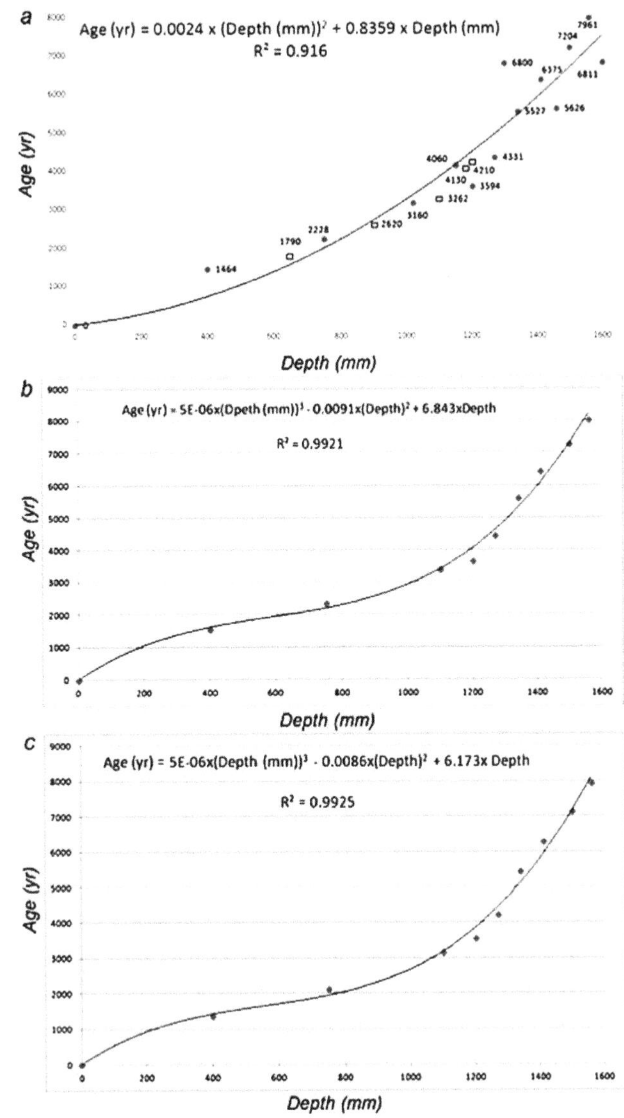

Table 1. Mineralogy and geochemistry of continental arid (F5) and wet (F1) facies at the top and middle of the core compared with the marine facies (F5) at the bottom(F3)

Depth (cm)	Halite [%]	Quartz [%]	Calcite [%]	Calcium [g/Kg]	Sodium [g/Kg]	Potassium [g/Kg]
3 (F5)	14	66	20	4.5	31	1.85
61 (F3)	8	92	0	6	24	0.75
121.5(F1)	0	100	0	3.5	11.4	1.3

The geochemistry of sediment may be used as proxy for the depositional environment, the climatic conditions and anthropogenic activities. (1) Potassium may be from the precipitation of sylvite, from continental K-clays leaching under wet climate, and from anthropogenic origin. Fertilizers are the main anthropogenic source (e.g., Gray et al., 2019, Essefi, 2020). Considering the influence of the seawater in the sebkha-lagoon brine geochemistry (Ben Smida, 2016), the hypothesis of sylvite precipitation may not be excluded. Therefore, high amount of potassium is an indication of continental and/or wet sedimentation. Alternatively,hand, the same high amount of potassium is an indication of sallow marine sedimentation under arid climate.

(2) High amount of Ca is an indication of arid climate resulting in gypsum precipitation and/or shallow water resulting in calcite/aragonite precipitation. Weak amount of Ca is an indication of wet continental sedimentation.

(3) High amount of Na is an indication of seawater flooding related to marine transgression as well as halite-type-minerals in the sebkha-lagoon.

The magnetic susceptibility is high due to continental sedimentation. Nonetheless, no straightforward interpretation of the MS related to climatic change. Due to the combination of climatic and eustatic effects, this already complicated interpretation of separate parameters within the sebkha-lagoon of Boujmel is further complicated due to the specific geologic setting of this saline system .system, namely its location under the influence of marine and continental conditions allowing the interplay of climatic variability and eustatic variation in modifying the sedimentary record. In sebkha-lagoon of Boujmel, the interplay of climate and depositional environment proposes a intricatecombination of arid/continental, arid/marine, wet/continental and wet/marine influences. In addition, other puzzlingmixed cases make the straightforward interpretation of one parameter next to impossible. For this reason, the use of all possible combination of all parameters may be useful to infer the simultaneous variability of climate and depositional environment conditions.The principal componment analysis PCA of the combination of different parameters (Fig.3) shows a first collection (Na, Na + K, Na + K + MS, Na + MS, K + MS) related to Na. Due to its availability in seawater and sebkha-lagoon brine, sodium is not the best proxy to record the weak variability of climate and depostional environment.Only major climatic and eustatic phases may be recorded by these parameters. The second collection (Ca, Ca + MS, Ca + K, Ca + Na + MS) is related to Ca amount. This collection may be used to infer weak variability of sedimentary conditions including the climate and depostional environment.

Identification of Major Climatic and Eustatic Phases

The plot of some parameters of the first collection related to Na (Fig.4) shows slight variability. However, major phases may be found out due to radical change of climate and depositional environment. From the bottom upwards, 5 major facies may be identified. (1) The sebkha continental facies started from the bottom of the core to 123 cm (4000 yr); this facies is marked by a weak influence of seawater. (2) From 123 cm to 54.5 cm (1000 yr), the sedimentation is marine with weak variability of parameters giving no idea about the climatic variability. (3) From 54.5 cm to 33.5 (600 yr), the sedimentation remains marine with a humid climate. This humid phase extends to change the depositional environment toward marine influenced continental sebkha. (4) From 33.5 cm (400 yr) upward, an arid phase is recorded. (5) At the mid of the B1a (200 yr), the depositional environments varied from a marine-influenced to pure continental sebkha.

Figure 3. Principal componnent analysis PCA of the combination of different parameters

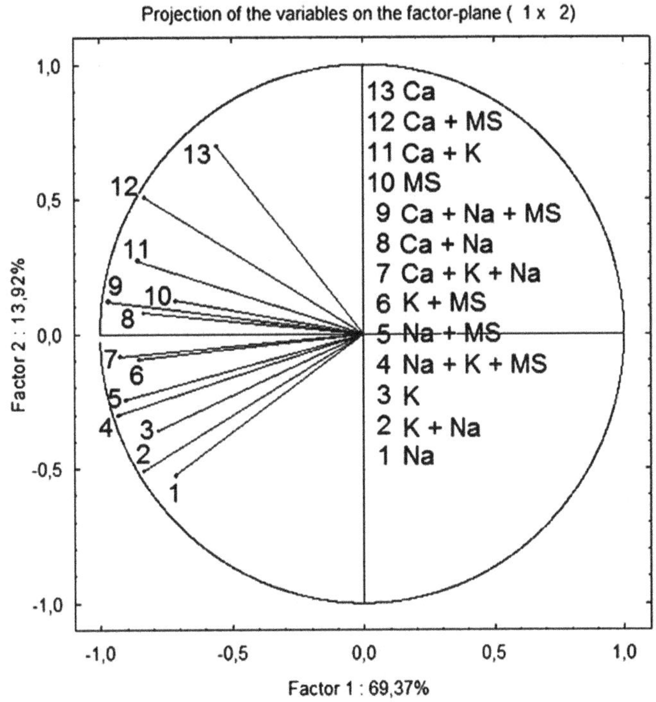

The plot of the collection related to Ca (Fig.5) shows clearer variability and more obvious climatic and eustatic stages. The interplay of climate and depositional environment resulted in convoluted sedimentary facies. From the bottom upwards, 11 facies (indicated in parenthesis) may be distinguished in four great sedimentary packages detailed above, three of them are natural and the uppermost is anthropogenic-related.

Natural Alternation Between a Continental Sebkha and a Marine Lagoon Between ca. 4400 Year and 1600 Year

From 4400 yr to 4000 yr, (1) we recorded the sedimentation of humid continental Sebkha (HCS) facies (Fig.5). With low values of Ca-related parameters, this facies recorded a marine regression and the setting of wetter conditions (B4h). At this time sebkha of Boujmel collected sediment during humid phase with weak amounts of Na, K and Ca. Sediment brought by water is richer inquartz rather than in clay and ferromagnetic minerals. (2) From 4000 yr to 2850 yr, a coupled change of climate and the depositional environment had taken place resulting in the sedimentation of the arid continental-influenced marine lagoon facies (ACIML) (Fig.5). All Ca-related parameters dramatically increased within the sedimentary record of Boujmel lagoon due to arid climate (B4a) enhancing evaporation and bringing eolian dust rich with ferromagnetic minerals. (3) The sedimentation of humid continental-influenced marine lagoon facies (HCIML) (Fig.5) had taken place between 2850 yr and 2700 yr due to a wet climate (B3h) and decreasing, but remaining, continental influence. (4) From 2700 yr to 1600 yr, a coeval change of climate and depositional environment resulted in the sedimentation of arid marine lagoon facies (AML). The arid phase B3a increased the evaporitic minerals and the magnetic dust within the lagoon of Boujmel. The continental influence decrease is due to decrease of rains bringing sediments to the lagoon.

Figure 4. Identification of major climatic and eustatic phases

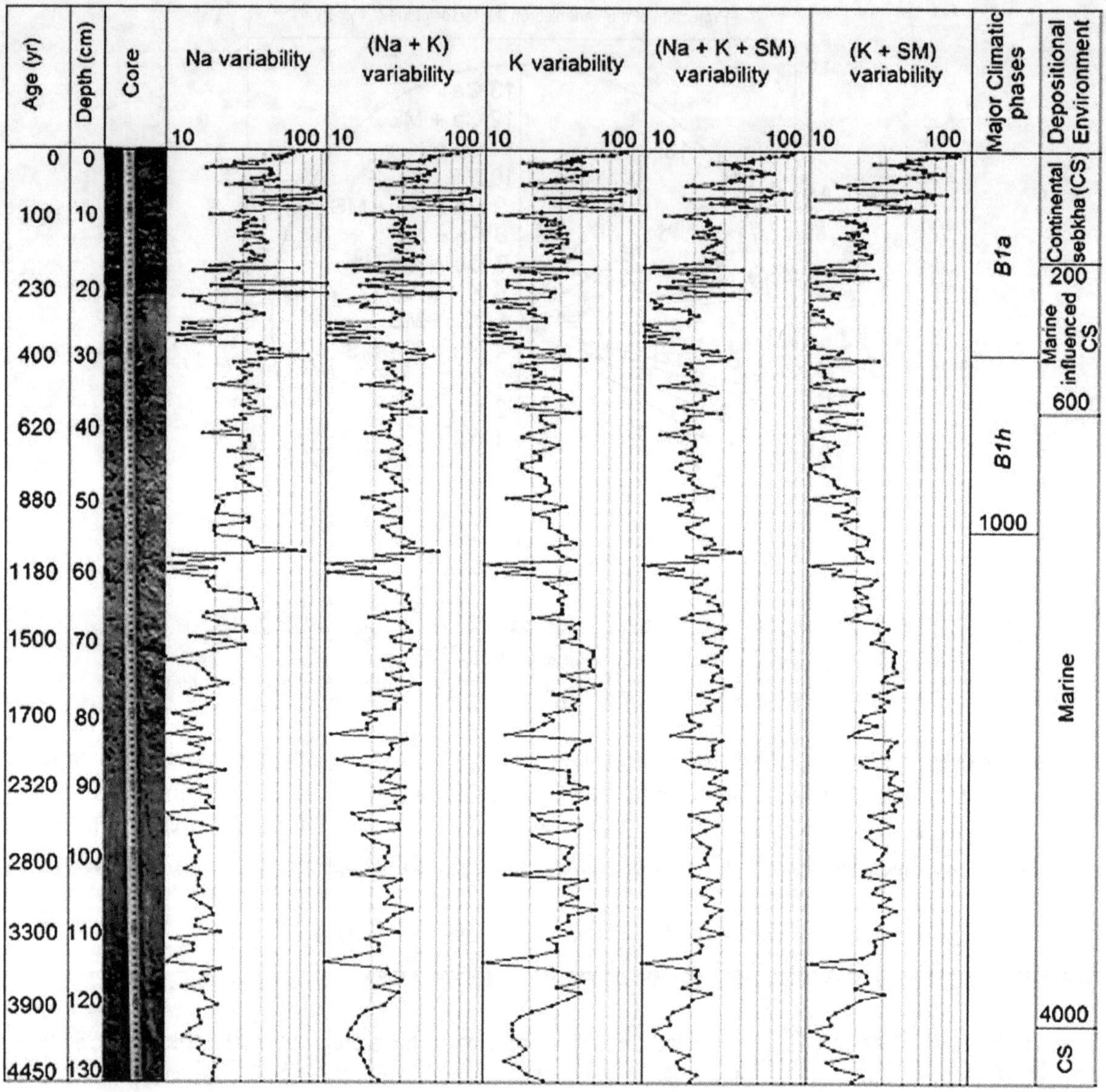

Natural Alternation Between Arid and Wet Climate Within Marine Sedimentation From 1600 yr to 600 yr

Due to marine transgression, the lagoon became a constant marine depositional environment with a variation of the climate. (5) From 1600 yr to 1400 yr, a humid phase (B2h=RWP) (B2h=Roman Warm Period) resulted in the sedimentation of humid marine facies (HM) with moderate ions content and weak magnetic susceptibility. (6) From 1400 yr to 1000 yr, an arid phase (B2a=DA) resulted in the sedimentation of an arid marine (AM) facies. (7) From 1000 yr to 600 yr, another longer humid marine (HM) facies is the result of the Medieval Climatic Anomaly (MCA) (part of the B1h) humid phase.

Alternation Between Marine Influenced and Sensu Stricto Continental Sebkha 600 yr and 200 yr

From 600 yr to 400 yr, (8) we recorded the sedimentation of humid continental sebkha (HMCS) facies (Fig.5). This facies recorded a relativedrop of the sea level and the setting of wetter conditions of the Early Little Ice Age (ELIA) (upper part of the Bh1). (9) From 400 yr to 200 yr, without any change of the depositional environment, the arid conditions of Late Little Ice Age (LLIA) (base of the Ba1) resulted in the deposition of the arid marine-influenced continental sebkha (AMICS).

Holocene- Anthropocene-Great Acceleration Transitions

At the top of the sebkha Boujmel core (Fig.5), (10) an abrupt variability is recorded from 24 cm to the top, indicating hence the setting of the Anthropocene (300 yr ago). Due to a very arid climatic phase (upper part of the B1a) and the setting of a continental sebkha, a very arid continental sebkha (VACS) facies rapidly deposited with high rate of sedimentation due to an increasing sedimentary flux. (11) At the depth of 7 cm (70 yr), all parameters greatly accelerate as indication of the setting of the Great Acceleration (1950). During this noticeable phase extreme arid conditions resulted in the deposition of the extremely arid continental sebkha (EACS) facies. This radical change in sebkha-lagoon of Boujmel was already recorded using a micropaleontological proxy (Jaouadi et al., 2015). The setting and the limits of the Anthropocene and the Great Acceleration still a matter of debate between scholars (Waters et al., 2018). Nonetheless, both of them are genetically linked and do have some common features including climate change pollution growth and increasing sedimentary flux. The increase of Ca, Na and K contents is also a direct indicator of an increasing aridity and/or sea level fall/marine regression. The change in the climatic conditions and sea level on the one hand and human induced activity on the other would have direct repercussions on the geochemistry of the sedimentary record. The magnetic susceptibility may be due to enrichment with heavy metals at the top of the core (Fig.5). Also, the setting of a novel microbiology (Gillings & Paulsen, 2014) related to the Anthropocene include the development of the magnetobacteria, which are responsible for the increasing magnetic susceptibility. Using the magnetic measurement in wetlands sediments as a marker for the start of the Anthropocene revealed itself relevant. In this vein, special attention is given to the importance of developing and disappearance of magnetic minerals. Actually, the starting of the Anthropocene is worldwide marked by an abrupt increase inthe magnetic susceptibility (Oldfield, 2015). This change deals also with wind erosion and perturbation of surface geochemistry during the Anthropocene (Marx et al., 2014). Toward the inland, Anthropocene conditions change the response of Wadi feeding the sebkha (Meybeck, 2001; Syvitski et al 2011). The sedimentary flux during the Anthropocene witnessed a radical change, especially during the Great Acceleration (Syvitskiet al., 2011). Actually, scholars (e.g., Ribot, 2014) remain dissonantconcerning causes and responses of climate during the Anthropocene due to the overlap of the natural and the human causes. The plot of the sedimentary fractions shows an upward tendency towards a decrease of clay percentage at the expense of silt and sand percentages, indicating hence an increase of sediment flux, which is related to some human-induced activities worldwide taking place during the Anthropocene (Syvitski et al., 2011):

1. Deforestation and its associated role in soil erosion, slope failure and downstream sedimentation; the soil erosion within drainage basins of southern Tunisia has recently increased due to an increasing aridity.
2. Agricultural activities in the southern Tunisia, including tillage, terracing, irrigation systems and subsurface water extraction, leading, respectively, to increased soil erosion, creep, siltation and subsidence. The coastal management takes place along the wetland through groynes, jetties, seawalls, breakwaters and harbours, leading to unnatural coastal erosion or sedimentation, wetland, mangrove and dune alterations.

Figure 5. Identification of facies related to the interplay of the climate and the depositional environment: Humid Continental Sebkha (HCS), Arid Continental-Influenced Marine Lagoon (ACIML), Humid Continental-Influenced Marine Lagoon (HCIML), Arid Marine Lagoon (AML), Humid Marine (HM), Arid Marine (AM), Very Arid Continental Sebkha (VACS), Extremely Arid Continental Sebkha (EACS)

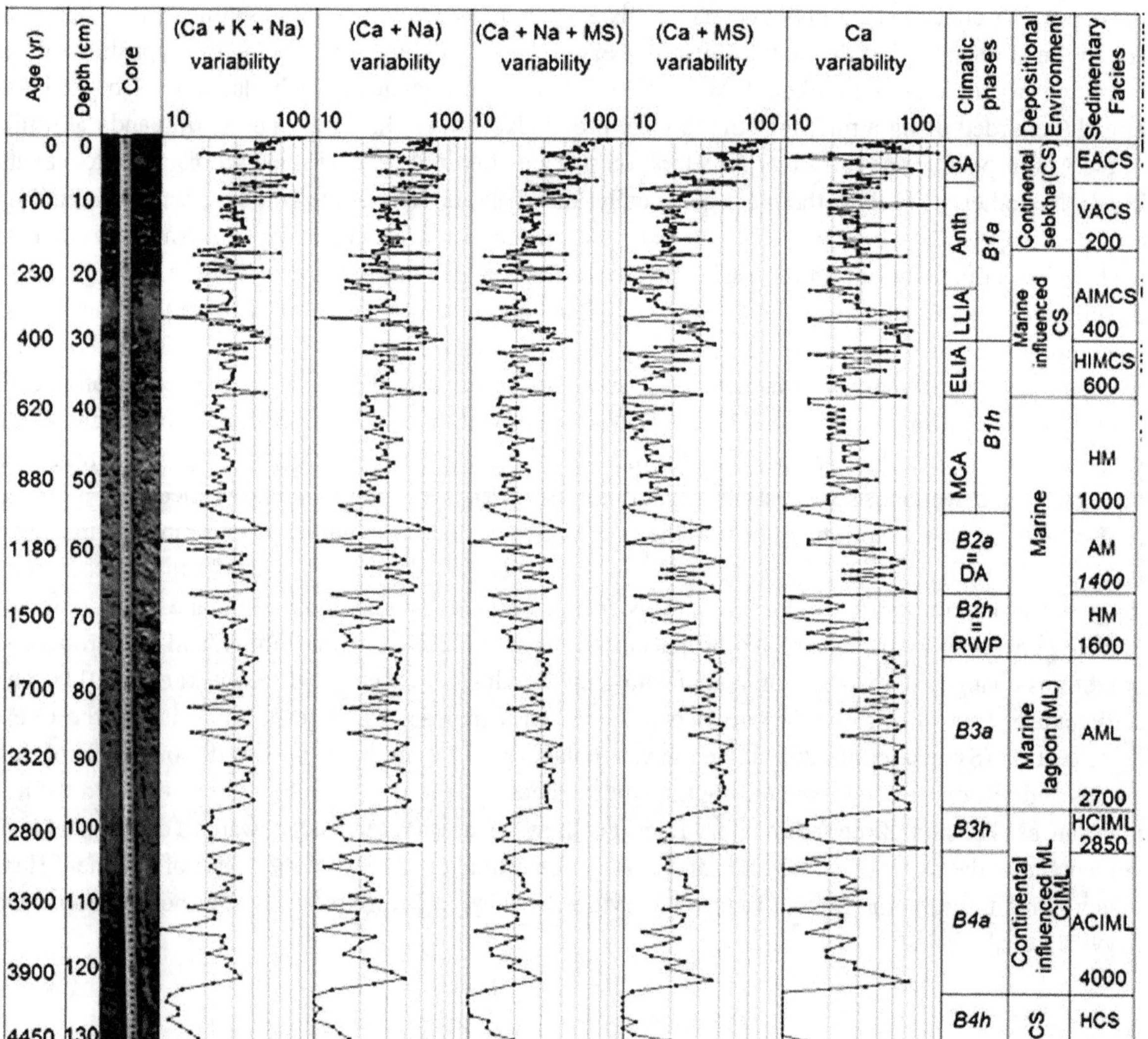

Spectral Analysis and Astrochronological Calibration

Spectral Analysis

Based on facies analysis (including mineralogy and facies identification), sebkha-lagoon Boujmel recorded two phases of shallowing (600 yr and 4000 yr) separated by 3400 yr. This well-defined eustatic cycle recorded in the literature (Pirazzoli, 2005) may be used as a reference to test the validity and to select the more adequate model. Using their fitting curves, ages at different depths were calculated (Table 2). Comparably to model based on maximum values of Jaouadi et al. (2016). (Table 2. C) and Model based on minimum values of Jaouadi et al. (2016) (Table 2. D), the mean of all models (Tab.2, B) seems the more adequate since it meets the two dates of minimum of the sea level (600 yr and 4000 yr). Also, this model should satisfy dates of the already identified climatic and eustatic phases. The corrected model (Table 2. A) (0.86) satisfies all the major and minor climatic phases recorded in sebkha Mhabeul. Also, the eustatic phases also recorded at coast of Tunisia. This model should be further validated based on the well-defined climatic cycles (Essefi et al., 2014; Gammoudi et al., 2019).

The spectral analysis of data of Na variability firstly shows the individualization of a cycle stretching along about ≈250 yr (Fig.6a). Then, a less pronounced cycle of 5000 yr is marked with a lower intensity. Also, the spectral analysis of K + Na variability datashows well-developed cycles of ≈3400 yr, 2000 yr, ≈1000 yr, and ≈250 yr (Fig.6b). The spectral analysis of Na and MS convolution shows other generations of cycles of 4000 yr, 2000 yr, 1500 yr, ≈1000 yr, ≈750 yr, and 600 yr (Fig.6c).

Table 2. Simplified age models: (A) Model corrected based on eustatic and climatic phases (B) Model based on 19 samples dating of Lakhdar et al. (2006) and Jaouadi et al. (2016). (C) Model based on maximum values of Jaouadi et al. (2016). (D) Model based on minimum values of Jaouadi et al. (2016)

Depth (cm)	A (yr)	B (yr)	C (yr)	D (yr)
10	100	536	598	414
20	230	930	1044	746
30	400	1212	1368	1016
35	600	1212	1368	1016
40	620	1413	1601	1244
50	880	1561	1771	1450
60	1180	1687	1909	1654
70	1500	1822	2046	1876
80	1700	1994	2210	2136
90	2320	2234	2432	2454
100	2800	2573	2743	2850
110	3300	3039	3171	3344
120	3900	3663	3747	3956
130	4450	4475	4501	4706

Figure 6. Spectral analysis based on the variability of some Na-related parameters

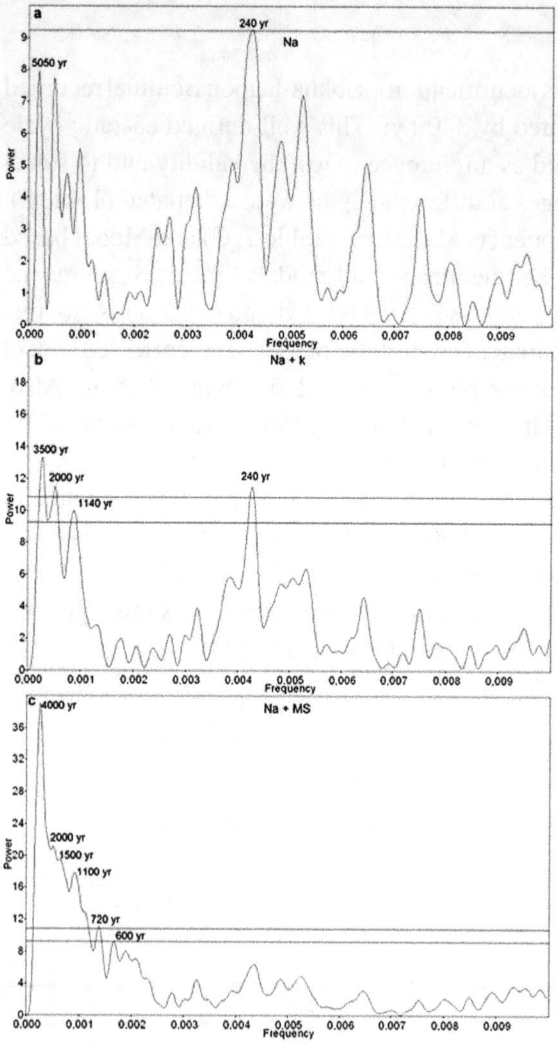

Figure 7. Spectral Analysis based on the variability of some Na-related parameters

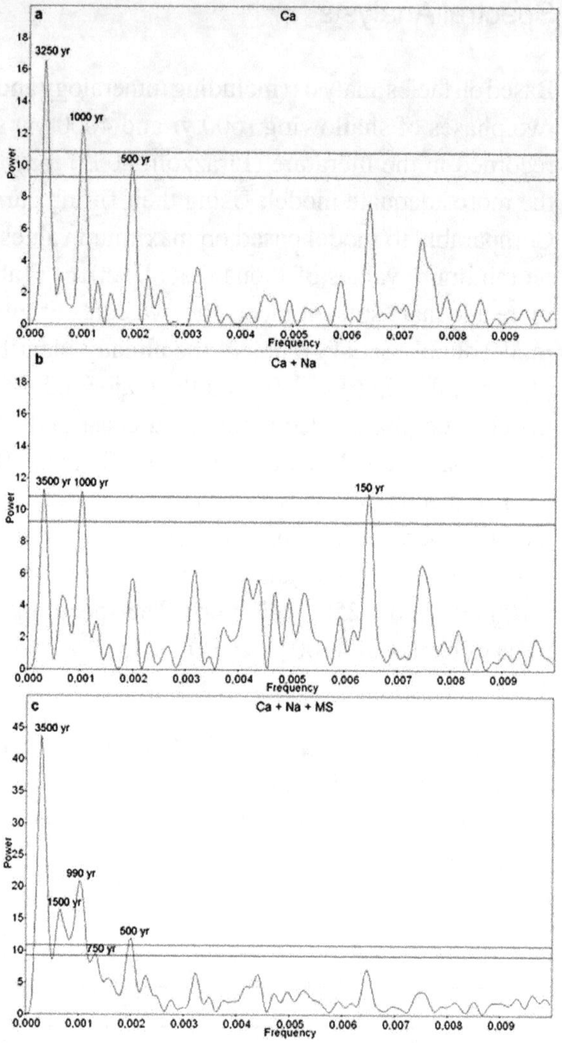

The average variability of calcium is basically related to water shallowing. The spectral analysis of calcium variability (Fig.7a) shows the setting of ≈3400 yr cycle. No comparable cycle was worldwide recorded neither in coastal nor in inland saline systems. On the other hand, cycles of 1000 yr and 500 yr were recorded. These cycles were already recorded at sebkha Mhabeul (Essefi et al., 2014; Gammoudi et al., 2019; Gammoudi, 2020). Added to cycles of ≈3400 yr and 1000 yr, the spectral analysis of Na + Ca variability (Fig.7b) shows the setting of 150 yr cycle. Adding MS variability (Fig.7c) allows the identification of ≈3400 yr, 1500 yr, ≈1000 yr, 750 yr and 500 yr cycles.

To validate the age model, all combinations of parameters were studied in order to find all possible cycles (Table 3). A sum of 60 cycles was found; most of them are already reported in the literature involved southern Tunisia and other sites of the world as result of an integrated astronomic, oceanographic and atmospheric forcing. Some of these cycles barely appear in one or two proxy; others are omnipresent for all proxies.

Table 3. Cycles within the sebkha-lagoon of Boujmel according to the used proxy

Proxy	Cycles (yr)
Na	240
K + Na	3500, 2000, 1140, 240
K	3250, 1800, 1200, 250, 98
Na + K + MS	3500, 1860, 1140
Na + MS	4000, 2000, 1500, 1100, 720, 600
K + MS	3500, 1700, 1140, 930, 750, 590
Ca + K + Na	3500, 960, 250
Ca + Na	3500, 1000, 150
Ca + Na + MS	3500, 1500, 990, 750, 500
MS	3500, 1600, 1140, 907, 730, 600, 530, 470
Ca + K	3250, 960
Ca + MS	3500, 1600, 960, 760, 500
Ca + K + MS	3500, 1600, 960, 760, 500, 430
Ca	3250, 1000, 500

Table 4. Occurrence of cycles within sebkha Boujmel and their identification in sebkha Mhabeul

Cycle With an Error of 5%	Occurrence in Sebkha Boujmel Record	Identification Within Sebkha Mhabeul Record
1000	12	Yes
3400	12	No
750	6	Yes
500	5	Yes
1500	5	Yes
250	4	Yes
1750	3	Yes
2000	3	Yes
600	3	Yes
900	2	Yes
4000	1	No
100	1	Yes
150	1	Yes

In terms of Holocene cyclostratigraphy, the sebkha-lagoon of Boujmel and sebkha Mhabeul must have similarities as well as differences. On the one hand, both saline systems are located in the same geological and climatic context in southern Tunisia. A distance of 28km between the two systems is not enough to cause radical change of the climatic forcing within the two systems. For this reason, climatic cycles cause by astronomic, atmospheric and oceanographic forcing should properly be recorded within both systems. As it is shown in Table 4, all cycles related to external forcing recorded in sebkha Mhabeul are already recorded within the sebkha-lagoon of Boujmel. On the other hand, eustatic variability controlling the sebkha-lagoon of Boujmel has no effect on the inland sebkha of Mhabeul. The 3400 yr cycle related to variation of sea level is absent in the case of sebkha Mhabeul. Therefore, added to its identification of climatic and eustatic phases, the adopted age model is relevant for finding out the Holocene cyclicity.

Astrochronological Calibration

Based on the age model validating the Holocene cyclostratigraphy already proved in southern Tunisia at sebkha Mhabeul and in other sites around the world, we are able to set the astrochronological calibration of radiocarbon dates of climatic and eustatic phases. As for ages of Lakhdar et al. (2006) and Lakhdar (2009), all values are overestimated. This is due to the technical handicaps including the traditional dating technique and geological complexity due to the interplay of marine and continental depositional environment as well as human activities during the Anthropocene causing a decreasein the radiocarbon dating accuracy. For this reason, the errors are higher for recent sediment.

Table 5.

Depth (cm)	Age yr (BP)	Errors (yr)	Errors [%]	Age (yr) After Astrochronological Calibration
40	1464	± 98	6.7	620
75	2228	± 124	5.6	1650
110	3262	± 124	3.5	3300
120	3594	± 69	1.9	3900
127	4331	± 133	3	4250

The use of the AMS radiocarbon technique allowed Jaouadi et al. (2016) obtaining ages with lower errors. However, ages of the upper part of the core remain overestimated. This is due to the worldwide polluting activities during the Anthropocene. According to this calibration, our results are in concordance with Jaouadi et al. (2016) for the first two climatic phases. The first arid phase B1a (0-400 yr) may be divided to different climatic phases including the Great Acceleration (0-70 yr), the Anthropocene (70 yr-300 yr), and the Late Little Ice Age (300 yr-400 yr). The second arid phase B2a (1000 yr-1400 yr) is equivalent to the MiddleAge. The third B3a (1600 yr-2700 yr) and fourth (2850yr-4000 yr) arid phases are not equivalent to that found by Jaouadi et al. (2016): BJ3 (2000-3000) and BJ4 (2850 yr-4000 yr).

Table 6. Proposed calibration of Jaouadi et al. (2016) ages

Depth (cm)	Age yr (BP)	Errors (yr)	Errors [%]	Age (yr) After Astrochronological Calibration
65	1790	± 260	14.5	1340
90	2630	± 620	23.5	2320
102	3160	± 290	9.2	2655
118	4060	± 380	9	3722

SUMMARY AND CONCLUSION

Previous studies (Lakhdar et al., 2006; Lakhdar, 2009; Jaouadi et al., 2016) on the sebkhas-lagoon of Boujmel seem controversial in defining variability in climatic phases and the depositional environment. Two approaches ascribed to different dating techniques were employed:. Lakhdaret al. (2006) used the conventional radiocarbon technique needing more quantity of the marine sample, leading hence to bigger error, whereaserror. Jaouadi et al. (2016) used the AMS technique on more continental influence sample, therefore achieving higher accuracy of dates. However, the increase of the continental contribution may increase dates errors. In terms of resolution and proxy, Lakhdaret al. (2006) and Lakhdar (2009) used facies sedimentology (sampling of many centimeters). This facies were geochemically and mineralogically studied. The focus in these studies was on depositional environment rather on the climate. Based on pollen and mineralogical investigation, Jaouadi et al. (2016) used higher resolution of 2.5 cm. They recorded at sebkha Boujmel four aridity pulses that punctuate the Late Holocene (4500 BP). Nonetheless, the work was climate oriented but the variability of the depositional environment was overlooked.

In our study, the 130 cm core was meant to be in a mid-way between the two previous studies with comparable length. In addition, we increased the resolution to 0.5 cm and we used the geochemical and geophysical proxies associated with facies description to set the climate as well as the depositional environment variability. In spite of the absence of our own dating, the choice of the location of the core allowed us using all dates of Lakhdar et al. (2006) and Jaouadi et al. (2016) to set a mean model satisfying dates of the climatic and eustatic phases recorded in southern Tunisia. This model was used to set the Holocene cyclostratigraphy and the astrochronological calibration in the region.

Based on the age model, four major climatic phases were upward recorded: (1) natural alternation between a continental sebkha and a marine lagoon between ca. 4400 yr and 1600 yr,followed by a (2) natural alternation between arid and wet climate within marine sedimentation from 1600 yr to 600 yr; (3) an alternation between marine-influenced and *sensu stricto* continental sebkha between 600 yr and 200 yrthat was interrupted by (4) Holocene-Anthropocene and Anthropocene-Great Acceleration transitions at 300 yr and 70 yr ago respectively. Based on the age model and the spectral analysis, a sum of 60 cycles was found. Most of them are already reported in the literature in southern Tunisia or other sites in the world and are related to integrated astronomic, oceanographic, and atmospheric forcing. The most occurring cycles are 1000 yr and 3400 yr. The first is related to Sun activity and is well reported in the literature. The second has not previously discussed neither in the study region nor in other sites in the world. It is probably related to eustatic phase recorded in southern Tunisia. These cycles served to set the astrochronological calibration of previous radiocarbon dates of climatic and eustatic phases. In addition, the limits of the Anthropocene and the Great Acceleration were determined in the region thus providing significant material to discussion regarding these novel periods of time.

REFERENCES

Abd-Elmabod, S. K., Muñoz-Rojas, M., Jordán, A., Anaya-Romero, M., Phillips, J. D., Laurence, J., & de la Rosa, D. (2020). Climate change impacts on agricultural suitability and yield reduction in a Mediterranean region. *Geoderma*, *374*, 114453. doi:10.1016/j.geoderma.2020.114453

Ahluwalia, H.S., & Ygbuhay. R.C. (2015). Cosmic ray 11-year modulation for sunspot cycle 24. *Solar Physics, 290*(2), 635-43.

Baumgarten, H., & Wonik, T. (2015). Cyclostratigraphic studies of sediments from Lake Van (Turkey) based on their uranium contents obtained from downhole logging and paleoclimatic implications. *International Journal of Earth Sciences, 104*(6), 1639–1654. doi:10.100700531-014-1082-x

Ben Smida, N. (2016). *Comparative geochemical study of exorheic (Boujmal and Melah) and endorheic (Mcheguig and Mhabeul) saline systems: a geo-economical approach* (Master Thesis). Higher Institute of Applied Sciences and Technology of Gabes.

Carbonel, P., & Pinson, J. (1980). Les Cyprideis témoins de l'évolution des sels en milieux laguno-lacustres sous climat semi-aride. In N. Nrstic (Ed.), *Taxonomy, Biostratigraphy and Distribution of Ostracodes* (pp. 211–217). Serbs KO Geolochko.

Chwaya, M. (2017). *Etude comparative entre la sebkha endoréique de Boujmal, centre tunisien et la sebkha exoréique de Jallabia, Sud-Est tunisien: changements globaux au cours de la transition Holocène-Anthropocène* (Master Thesis). Faculty of Sciences of Gafsa.

Crutzen, P. J. (2006). The "anthropocene". In Earth system science in the Anthropocene. Springer.

Deng, W. & Wei, G. (2015). Decoupling of seasonal temperature and precipitation over the western Pacific during the early mid-Holocene. *International Journal of Climatology, 35*(5), 794-800.

Essefi, E. (2009). *Multidisciplinary study of Sidi El Hani Saline environment: The history and the climatic variability* [MS thesis]. Faculty of sciences of Sfax, University of Sfax.

Essefi, E. (2013). *Wet Aeolian Sedimentology and Sequence Stratigraphy within the Terrestrial Analogues in Eastern Tunisia: Implications for Wet Aeolian Sedimentology and Sequence Stratigraphy on Mars* (Doctoral dissertation). National Engineering School of Sfax, Sfax, Tunisia.

Essefi, E. (2020). On the Geochemistry and Mineralogy of the Anthropocene. *Int J Water Wastewater Treat, 6*(2).

Essefi, E., Ben Smida, N., Jandoubi, I., Othmani, M. A., & Tagorti, M. A. (2020). Progressive evaporation of brine of sebkha Mchiguig, central Tunisia: A geo-economical comparative study of salt and brine. *Carbonates and Evaporites, 35*(2), 59. doi:10.100713146-020-00592-7

Essefi, E., Gharsalli, N., & Yaich, C. (2015). Geophysical and Geochemical Study of the Silico-evaporitic Sedimentary Filling of Boujmal wetland, Eastern Tunisia: Inferring the Climatic Signal within Groundwater Noise. *Journal of Basic and Applied Research International*, 14-25.

Essefi, E., Mefteh, S., Medhioub, M., & Yaich C. (2014) Magnetic Study of the Heated and Unheated Sedimentary Fillings of Sebkha Mhabeul, Southeast Tunisia: A Geophysical Method for Paleoclimatic Investigation and Tephrochronological Dating. *International Journal of Geophysics*. . doi:10.1155/2014/908395

Essefi, E., Touir, J., Tagorti, M. A., & Yaich, C. (2013). Record of the climatic variability and the sedimentary dynamics during the last two millennia at sebkha Dkhila, eastern Tunisia (2013). *ISRN Geology*.

Gammoudi, A. (2020). *Etude géochimique et cyclostratigraphie holocène des sédiments de la sebkha Mhabeul au Sud-Est de la Tunisie* (PhD Thesis). Higher Institute of Applied Sciences and Technology of Gabes.

Gammoudi, A., Essefi, E., & Rigane, H. (2019). Variability of the grain-sized fractions during the last three millennia within the sedimentary fillings of the saline system of Mhabeul: The Holocene cyclostratigraphy. *Arabian Journal of Geosciences*, *12*(9), 296. doi:10.100712517-019-4472-3

Gharsalli, N., Essefi, E., Baydoun, R., & Yaich, C. (2017). Filling Sebkha Halk el Menjel, Eastern Tunisia. *The 1st Atlas Georesources International Congress*.

Gharsalli, N., Essefi, E., Baydoun, R., & Yaich, C. (2020). The Anthropocene and Great Acceleration as Controversial Epoch Of Human-Induced Activities: Case Study Of The Halk El Menjel Wetland, Eastern Tunisia. *Applied Ecology and Environmental Research*, *18*(3), 4137–4166. doi:10.15666/aeer/1803_41374166

Gillings, M. R., & Paulsen, I. T. (2014). Microbiology of the Anthropocene. *Anthropocene*, *5*, 1–8. doi:10.1016/j.ancene.2014.06.004

Gray, M. H., & Jeffrey, D. M. (2019). A review of research on the fertiliser and lime requirements of pasture on the eastern coast of the North Island. *Journal of New Zealand Grasslands*, *81*, 21–32. doi:10.33584/jnzg.2019.81.373

Hébert, R., & Lovejoy, S. A. (2017). Scaling Model for the Anthropocene Climate Variability with Projections to 2100. EGU General Assembly Conference Abstracts, 19, 9943.

Herrera, V. V., Mendoza, B., & Herrera, G. V. (2015). Reconstruction and prediction of the total solar irradiance: From the Medieval Warm Period to the 21st century. *New Astronomy*, *34*, 221–233. doi:10.1016/j.newast.2014.07.009

Jaouadi, S., Lebreton, V., Bout-Roumazeilles, V., Siani, G., Lakhdar, R., Boussoffara, R., & Combourieu-Nebout, N. (2016). *Environmental changes, climate and anthropogenic impact in southern-eastern Tunisia during the last 8 kyr*. Academic Press.

Jaouadi, S., Lebreton, V., Mannai-Tayech, B., Lakhdar, R., & Soussi, M. (2015). Apport de l'analyse pollinique des sédiments de la sebkha Boujmel (Sud-Est tunisien) à la reconstitution des paléopysages et paléoclimats holocènes en milieu aride. *Méditerranée*, *125*(125), 39–49. doi:10.4000/mediterranee.7891

Lakhdar, R. (2009). *Les sédiments holocènes et les tapis microbiens du littoral du Sud-Est de la Tunisie: sédimentologie et paléoenvironnements* (PhD). Université de Sfax.

Lakhdar, R., Soussi, M., Ismail, M. H. B., & M'Rabet, A. (2006). A Mediterranean Holocene restricted coastal lagoon under arid climate: Case of the sedimentary record of Sabkha Boujmel (SE Tunisia). *Palaeogeography, Palaeoclimatology, Palaeoecology*, *241*(2), 177–191. doi:10.1016/j.palaeo.2006.02.014

Lim, J., Lee, J. Y., Hong, S. S., Park, S., Lee, E., & Yi, S. (2019). Holocene coastal environmental change and ENSO-driven hydroclimatic variability in East Asia. *Quaternary Science Reviews*, *220*, 75–86. doi:10.1016/j.quascirev.2019.07.041

Martinez-Ruiz, F., Kastner, M., Gallego-Torres, D., Rodrigo-Gámiz, M., Nieto-Moreno, V., & Ortega-Huertas, M. (2015). Paleoclimate and paleoceanography over the past 20,000 yr in the Mediterranean Sea Basins as indicated by sediment elemental proxies. *Quaternary Science Reviews*, *107*, 25–46. doi:10.1016/j.quascirev.2014.09.018

Marx, S. K., McGowan, H. A., Kamber, B. S., Knight, J. M., Denholm, J., & Zawadzki, A. (2014). Unprecedented wind erosion and perturbation of surface geochemistry marks the Anthropocene in Australia. *Journal of Geophysical Research. Earth Surface*, *119*(1), 45–61. doi:10.1002/2013JF002948

Meybeck, M. (2001). River basin under Anthropocene conditions. In *Science and Integrated Basin Management Dahlem workshop series* (pp. 275–294). Wiley.

Miyake, F., Suzuki, A., Masuda, K., Horiuchi, K., Motoyama, H., Matsuzaki, H., Motizuki, Y., Takahashi, K., & Nakai. Y. (2015). Cosmic ray event of AD 774–775 shown in quasi-annual 10Be data from the Antarctic Dome Fuji ice core. *Geophysical Research Letters, 42*(1), 84-9.

Oldfield, F. (2015). Can the magnetic signatures from inorganic fly ash be used to mark the onset of the Anthropocene? *The Anthropocene Review*, *2*(1), 3–13. doi:10.1177/2053019614534402

Pérez-Asensio, J. N., Frigola, J., Pena, L. D., Sierro, F. J., Reguera, M. I., Rodríguez-Tovar, F. J., & Cacho, I. (2020). Changes in western Mediterranean thermohaline circulation in association with a deglacial Organic Rich Layer formation in the Alboran Sea. *Quaternary Science Reviews*, *228*, 106075. doi:10.1016/j.quascirev.2019.106075

Pirazzoli, P. A. (2005). A review of possible eustatic, isostatic and tectonic contributions in eight late-Holocene relative sea-level histories from the Mediterranean area. *Quaternary Science Reviews*, *24*(18-19), 1989–2001. doi:10.1016/j.quascirev.2004.06.026

Press, W. H., Teukolsky, S. A., Vetterling, W. T., & Flannery, B. P. (1992). Numerical Recipes. Cambridge University Press.

Saubade, A. M., & Zaouali, L. J. (1986). Malacofaune subfaussile de la Bahira el Bibane, Tunisie méridionale. *Bull. Inst. Géol. Bassin d'Aquitaine, Bordeaux*, *40*, 23–33.

Shimizu, M. H., Sampaio, G., Venancio, I. M., & Maksic, J. (2020). Seasonal changes of the South American monsoon system during the Mid-Holocene in the CMIP5 simulations. *Climate Dynamics*, *54*(5), 2697–2712. doi:10.100700382-020-05137-1

Syvitski & Kettner. (2011). Sediment flux and the Anthropocene. *Philosophical Transactions of the Royal Society A: Mathematical, Physical and Engineering Sciences, 369*(1938), 957-975.

Timothy, P. R., & Swindles, G. T. (2014). Influence of ocean–atmospheric oscillations on lake ice phenology in eastern North America. *Climate Dynamics*, *45*(9), 2293–2308. doi:10.100700382-014-2415-y

Wasli, A. (2016). *Holocene cyclostratigraphy within the sedimentary record of exoreic (Boujmel) and endorheic (Mhabeul) saline systems: Eustatic and climatic controls* (Master Thesis). Higher Institute of Applied Sciences and Technology of Gabes.

Waters, C. N., Zalasiewicz, J., Summerhayes, C., Barnosky, A. D., Poirier, C., Gałuszka, A., & Jeandel, C. (2016). The Anthropocene is functionally and stratigraphically distinct from the Holocene. *Science*, *351*(6269).

Waters, C. N., Zalasiewicz, J., Summerhayes, C., Fairchild, I. J., Rose, N. J., Loader, W., ... Edgeworth, M. (2018). Global Boundary Stratotype Section and Point (GSSP) for the Anthropocene Series: Where and how to look for potential candidates (2018). *Earth-Science Reviews*, *178*, 379–42.

Waters, C. N., Zalasiewicz, J. A., Williams, M., Ellis, M. A., & Snelling, A. M. (2014). A stratigraphical basis for the Anthropocene? *Geological Society*, *395*(1), 1-21.

Wilkinson, I.P., Poirier, C., Head, M. J., Sayer, C. D., & Tibby, J. (2014). Microbiotic signatures of the Anthropocene in marginal marine and freshwater palaeoenvironments (2014). *Geological Society*, *395*(1), 185-219.

Young, H. S., McCauley, D. J., Galetti, M., & Dirzo, R. (2016). Patterns, causes, and consequences of anthropocene defaunation (2016). *Annual Review of Ecology, Evolution, and Systematics*, *47*(1), 333–358. doi:10.1146/annurev-ecolsys-112414-054142

Yusoff, K. (2013). Geologic life: Prehistory, climate, futures in theAnthropocene. *Environment and Planning. D, Society & Space*, *31*(5), 779–795. doi:10.1068/d11512

Chapter 15
Record of the Sedimentary Dynamics and Climatic Variability Within the Sedimentary Filling of Mchiguig Wetland During the Late Holocene:
Central Tunisia – Astronomic, Atmospheric, and Oceanographic Forces

Elhoucine Essefi
University of Gabes, Tunisia

ABSTRACT

This work aimed to study the cyclicity of the geochemical chemical parameters and the carbonate percentages along a 59 cm core from the sebkha of Mchiguig, Central Tunisia. In fact, from the bottom upwards, six climatic phases were recorded including the Warming Present (Great Acceleration), the Late Little Ice Age (Anthropocene), the Early Little Ice Age, the Medieval Climatic Anomaly, the Dark Age, and the Roman Warm Period. In fact, the spectral analysis of the studied parameters visualized many cycles. Those cycles are related to sun activity, oceanographic, and atmospheric factors. Solar activity generated 500 yr cycles; however, the oceanographic circulation generated other cycles of 1500 yr and 700-800 yr. The 1500 yr cycle may be the result of the solar activity and NAO-like circulation.

DOI: 10.4018/978-1-7998-9498-8.ch015

INTRODUCTION

The Holocene, the second time of the Quaternary period, is a relatively warm interval. Following the last glacial of the Pleistocene, this climatic period is one of Quaternary interglacials (Holcová et al., 2020). Many studies of Holocene climate oscillations suggest that this interglacial period, manifested by persistent unstable climatic oscillations and immediate climate events, (Aaby, 1976; Meese et al, 1994; Bond et al., 1997, 2001; Bianchi and McCave, 1999; McDermott et al., 2001; Mayewski et al., 2004; Andresen et al., 2005; Wanner et al., 2008; Fletcher et al., 2013; Karpinski et al., 2020; Njagi et al., 2021).

Actually, Mediterranean area is considered as an important study region for investigating Holocene climate variations owing to: (1) its sensitivity to global climate changes (Martines et al., 2015; Pepi & Focardi, 2021); (2) its latitudinal position where southern and northern climate regimes interact (Lionello et al., 2006; Lionello, 2012) and (3) its semi-enclosed nature (Martines et al., 2015). Therefore, Mediterranean region is one of the famous influenced regions to past (Jaouadi et al., 2016) and future climate variations (IPCC, 2014). In fact, a considerable amount of work has been carried out on the Mediterranean region which has indicated important information on global and regional scale patterns of climate changes in this region (e.g., Abrantes et al., 2012; Lionello, 2012; Malanotte-Rizzoli et al., 2014; Martines et al., 2015 Bazzicalupo et al., 2020). Besides, some studies of eastern Mediterranean regions show that sediments of saline lakes record climatic variability (e.g. Lakhdar et al., 2006; Pomel et al., 2006; Marquer et al., 2008; Essefi et al., 2013, 2014; 2015b,c; Jaouadi et al., 2016; Gammoudi et al., 2019; Essefi, 2021a). There are many saline systems located at the coast of the Mediterranean Sea, more precisely in Tunisia. Indeed, in this area, we quote the sebkhas of Mhabeul, Sidi El Hani, Ennouel, Mchiguig, Sidi Mansour, Halk El Menjel, (e.g. Schulz et al., 1995, 2002; Abichou, 2002; Pomel et al., 2006; Marquer et al., 2008; Essefi et al., 2014; Gammoudi et al., 2019; Essefi, 2021a and reference therein).

In this paper, the aim was to investigate the clamtic variability within the sedimentary filling of sebkha Mchiguig. Further, the cyclic evolution of the climate was investigated based on the spectral analysis of some parameters along the sedimentary filling of the Late Holocene. To do so, we used the carbonate percentage and the geochemical parameters (K, pH, and electric conductivity) to identify climatic phases and the Holocene cyclostratigraphy along a 59 cm core of sebkha Mchiguig during the Late Holocene.

STUDY AREA

Located at the crossroads between three governorates of Sfax, Kairouan and Sidi Bouzid, central Tunisia, the Sebkha of Mchiguig (also called sebkha Sidi Dhaher) is a NNE-SSW oriented depression (Fig.1b). It enormously reserves visible salt flush with the surface of the sebkha (Essefi et al., 2020). Located east of the Goubrar sheet between the North and Fatnassa bled Sidi Dhaher south, the Mchiguig sebkha approximately occupies a surface of 7 Km^2 along 13 km length and 500 m wide. Along the periphery of its watershed, a well-developed hydrographic network does not reach the sebkha (Fig.1b), due to weak rain in central Tunisia. However, extreme flooding events do reach the outlet in the playa and feed the surface by amounts of detritic sediment. In terms of topography, its background has an approxi-

Figure 1. Study area of sebkha Mchiguig: (a) Location of sebkha Mchiguig, (b) (Smida 2006)

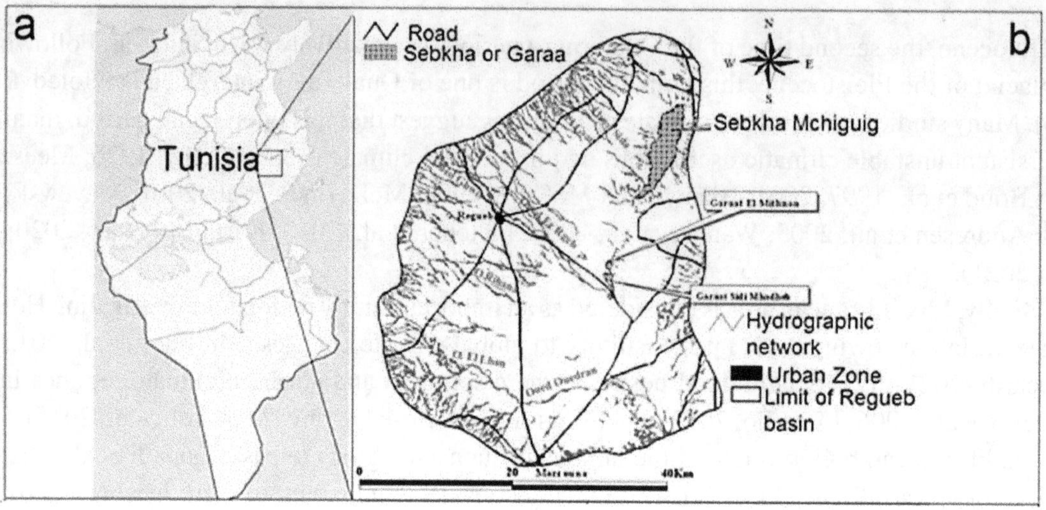

mate altitude of 113-114m, whereas the surrounding watershed approximately reaches the attitude of 500 m. Also, the watershed is marked by a fragile lihtology candidate to be dissolved by the extended hydrographic network. The hydrological balance between evaporation reaching the climax in July and a precipitation barely exceeding 30 mm/month enhances the evaporation and the precipitation of salt. Except some geoeconomics studies (Essefi et al., 2020), no paleoclimatic study had investigated the sedimentary filing of this wetland.

METHODS AND MATERIALS

A 59 cm core was carried out within the saline system of Mchiguig during the field mission of June 2017 (Fig. 1). This coring was realized by the penetration of an empty tube (6.3 cm diameter) and this penetration was determined by rotation and slight push on sediment. In the laboratory, the sampling was carried out continuously each 2 mm to obtain 290 samples. At the chemical laboratory, pH and electric conductivity were measured to investigate the response to climate variability. In addition, potassium was measured by Flame Photometer.

Up to now, classical Fourier spectral analysis remains the most common tool for studying the frequency pattern of a signal. For the processing of data, we are based on the PAST software to carry out the spectral analysis of the obtained data. Indeed, PAST software includes the Lomb periodogram algorithm for unequally sampled data (Press et al., 1992). As input to the PAST program, we enter the given time values in the first column and the dependent values in the second column. In the spectrum, the highest peak is presented with its power and frequency value. As probability, the peak could occur from random data (Essefi et al., 2015c). Besides, to find out temporal climatic cycles, we should deal on dating. In this work and in the previous work of Gammoudi et al. (2019), after detecting the climatic phases, we are based on the age-depth model of Marquer et al. (2008) in sebkha Mhabeul, determinate by the tephrochronological dating.

RESULTS

Climatic Phases

At the top of the sebkha Mchiguig core (Fig.2), a dramatic increase of all parameters has taken place from 11 cm to the top; it is an indication of the setting of the Anthropocene (300 yr ago). Climatically speaking, a very arid climatic phase in this continental sebkha has been recorded. So, a very arid continental sebkha (VACS) facies rapidly deposited with high rate of sedimentation related to an increasing sedimentary flux. At the depth of 3 cm (70 yr), all parameters further accelerate indicating hence the setting of the Great Acceleration (1950). During this noticeable phase, already recorded in the meteorological archive, extreme arid conditions resulted in the deposition of the extremely arid continental sebkha (EACS) facies. The measures of $CaCO_3$ percentages may indicate whether the sediment contains limestone or not (Fig. 2). Along the core, the percentage of $CaCO_3$ values range between 0 and 20% (Fig. 2). These values remain below 20% which is probably due to the dolomitization of calcium carbonates. The sediment along the core does not contain limestone. Indeed, the first 3 centimeters are characterized by a high percentage of carbonate (between 6% and 20%). The second interval stretches until 12 cm is characterized by a decrease in carbonate percentage (0% - 1.5%). The interval ranging from 12 cm to 46 cm shows the replication of the first two parts. Finally, the interval ranging from 46 cm to 59 cm is characterized by low percentage of $CaCO_3$ (between 0% and 2%). This variation in carbonate percentages is probably due to sun activity (Essefi et al., 2014). Indeed, the high carbonate contents indicate chemical conditions which are characterized by high primary productivity and hot temperatures. So, it is an indicator of arid climate. However, low carbonate levels is an indicator of humid climate. The results of geochemical study prove that potassium exists with minor proportions. The proportion of potassium (K), along the core, presents low values that range from 0 g/kg and 3 g/kg with slight fluctuation (Fig. 2). This fraction is probably due to the existence of KCl (Essefi et al., 2015a). The pH rate shows an obvious oscillation and values range between 6 g/kg and 9 g/kg, with average of 7 g/kg (Fig. 2). We can deduce that the rate of pH could be linked to the precipitation of salt mineral (NaCl). Hence, the fluctuation of pH rate is possibly due to the variation of climatic conditions. Also, it may be related to the periodicity of solar activity (Essefi et al., 2015a). Indeed, there are two periods: a sunny and arid period marked by the precipitation of high amount of (NaCl) and a wet period marked by a decrease in the salt level. The arid period is also recognized by a significant rise of the fine fraction which is explicated by the drop of the stream and the river activity (Essefi et al., 2015a). The climatic phases along the Mchiguig core may be relevant for the setting of an age model. First, in the uppermost part of the core (3cm), the stage Warming Present (WP) or the Great Acceleration stretches on ca. ≈70yr; the establishment of modern conditions is characterised by stable conditions. Added to a small salt crust, this period is dominated by a clayey sedimentation. This period is marked by increasing carbonate percentages, potassium, pH, and electric conductivity. Second, the stage C4 is located between 3 cm and 11 cm; it is called the Late Little Ice Age (Late LIA) (Wiles et al., 2008) or the Anthropocene (Essefi, 2021a,b,c,d); it stretches between the 80yrBP and 400yrBP, i.e. 320yr. The clayey sedimentation makes up the twofold and threefold laminates. With low amounts of carbonate and potassium, this period is marked by high values of pH and electric conductivity. Third, from ca. 400yrBP to 600yrBP, during the 200yr of cycle C3, which is called Early Little Ice Age (ELIA), This stage ending at the depth of 19 cm is marked by low values of pH. Fourth, the passage to the stage C2, which is called the Medieval Climatic Anomaly (MCA), is marked by a sharp increase of pH values at the transition with ELIA; then, pH returns to a decrease. The

Figure 2. Evolution of carbonate percentages and geochemical elements along the core of Sebkha Mchiguig

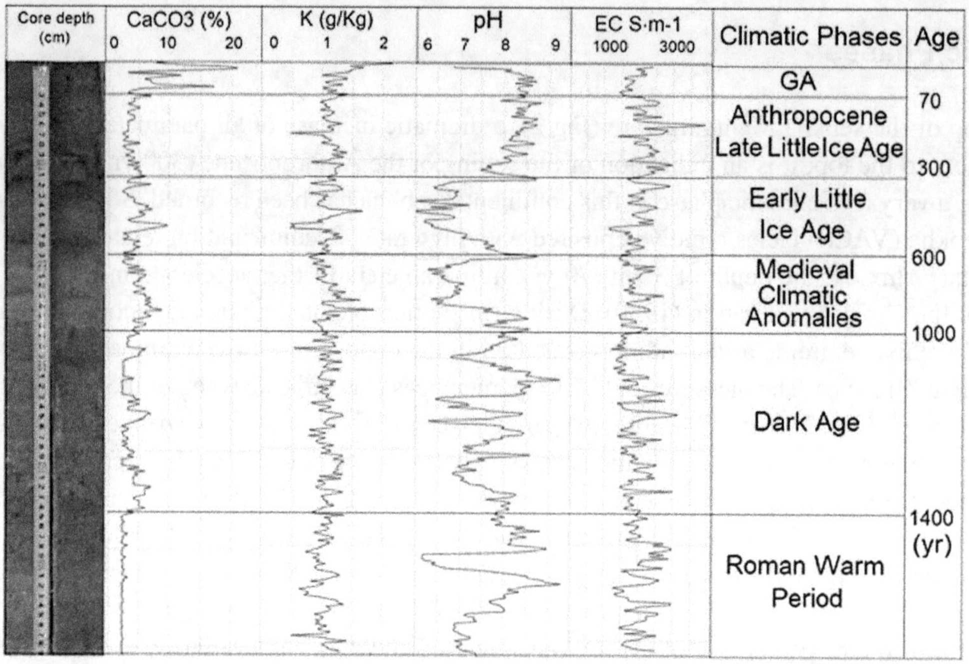

MCA spanned between ca. 600yrBP to 1000yrBP, i.e. ≈400yr. Fifth, during the stage (C1) of the Dark Age (DA), which stretches from 27 cm to 46 cm, a shift from light to dark sediments is recorded between ca. 1000yrBP and 1400yrBP, i.e. 400yr. As for the grain size distribution, the second reference sandy band is detected in the top of the DA. Sixth, the 300yr Roman Warm Period (RWP) stretches between 1400yrBP and 1750yrBP. During the (RWP), it suggests stable climatic conditions with small and/or scarce flood events; this period is recorded in the core of sebkha Mchiguig between 46 cm and 59 cm.

The Principal Component Analysis (PCA) is using as a treatment for carbonate percentage ($CaCO_3$), and geochemical (pH and K) parameters. The PCA was carried out through "XLStat" software. In this study, 290 samples and 3 variables were treated based on the multivariate statistical techniques. The correlation matrix (Table 1) was used to define the relationship existing between these variables (Vallès and Bourgeat, 1989). It shows a good correlation between pH and K ($r = 0.8$) and a good correlation between K and $CaCO_3$ ($r = 0.7$). Besides, the two principal axes of PCA explain 68.2% of the sampling variance of different parameters. The PCA of $CaCO_3$, pH and K rates of sebkha Mchiguig proves that the carbonate percentages is positive correlated with rates of pH and potassium. This has proven that climatic variability shown through the variation of pH and potassium rates along the core may have a relation with the oscillation of carbonate percentages. It is an indicator of climate change. In fact, during the wet period, the sediment supply coming through hydrographic network causes the dissolution of carbonates ($CaCO_3$) and salt minerals. Though, arid period is characterized by the rise of the $CaCO_3$ percentage and halite mineral rate (NaCl) (Essefi et al., 2015a).

Table 1. Pearson's correlation matrix. Bold values indicate a significant correlation factor (with significance level, $p < 0.05$).

Variables	pH	K (g/Kg)	$CaCO_3$ (%)
pH	1		
K (g/Kg)	**0.8**	1	
$CaCO_3$ (%)	0.50	**0.7**	1

Figure 3. Age-depth model within sebkha Mchiguig

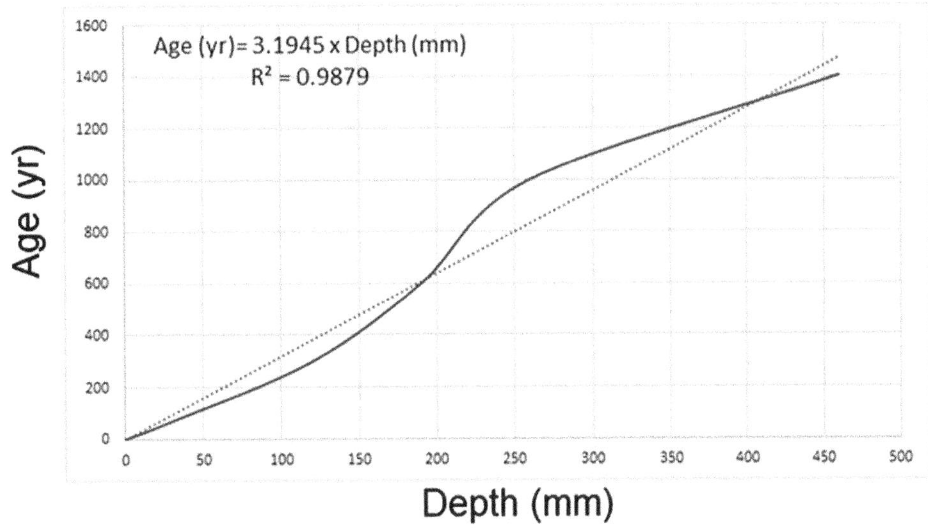

Age Model

Based on the climatic phases, we can estimate the age of each sample by interpolation to construct an age-depth model (Fig. 3). As result, the rate of sedimentation during the Late Holocene at sebkha Mchiguig was in the range of 3.19 yr/mm. Slight modifications of the sedimentation rate would have taken place due to variation in climatic conditions, but they do not the general influence the overall evolution of the sedimentary record.

Spectral Analysis

In order to find out climatic cycles along a core of sebkha Mchiguig, dating was a requirement. In this chapter, based on the age-depth model (Fig. 2), the extrapolated age at the bottom of the core is about 1600 yr. To define the cyclicity during the late Holocene, the obtained results were treated by Past software. First, the spectral analysis of data of the carbonate percentages shows the individualization of 4 cycles of ≈ 706 yr, 440-360 yr and 250 yr (Fig. 4). These cyclicities were worldwide recorded. They must be the result of different parameters such as the solar activity, atmospheric and/or oceanographic forcing.

The spectral analysis of data of the electric conductivity shows the individualization of 4 cycles of ≈ 500 yr (Fig. 5). On the other hand, it shows the individualization of cycles of 1500 yr, (Fig. 5). The

Figure 4. Spectral analysis based on $CaCO_3$ amount

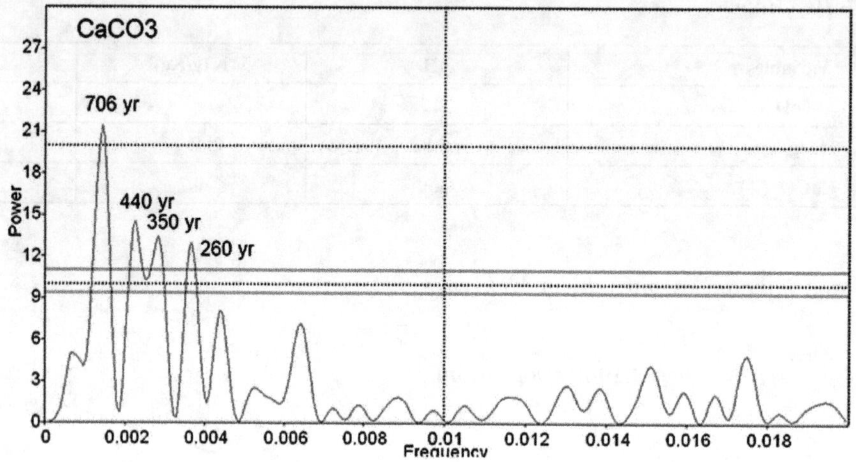

Figure 5. Spectral analysis based on electric conductivity

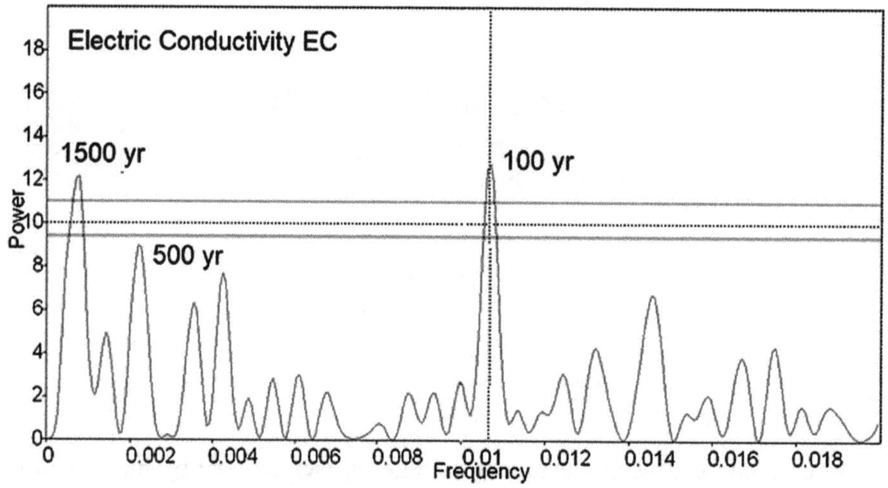

Figure 6. Spectral analysis based on pH values

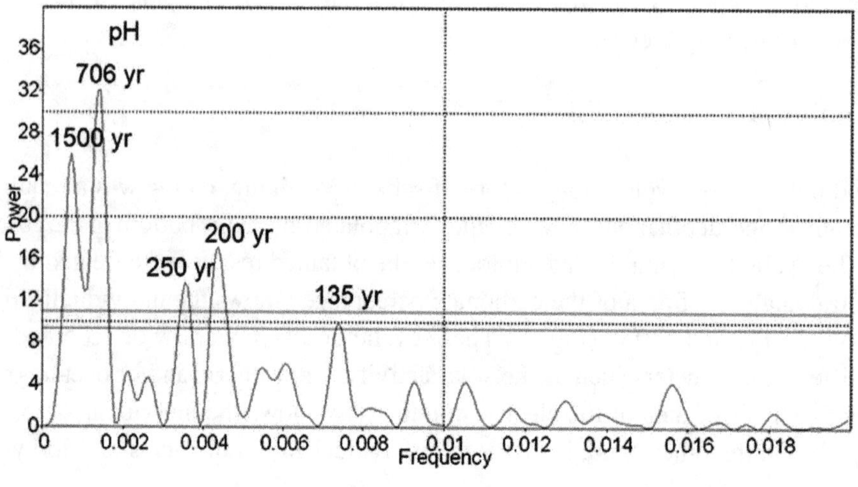

spectral analysis visualized 100 yr (Fig. 5b). These cyclicities were worldwide recorded. They must be the result of different parameters such as the solar activity, atmospheric and/or oceanographic forcing.

First, the spectral analysis of data of pH shows the individualization of 1500 yr cycle (Fig. 6). A cycle of 560 yr is not well developed and it does reach the critical level. These cyclicities were worldwide recorded. They must be the result of different parameters such as the solar activity, atmospheric and/or oceanographic forcing.

DISCUSSION

Climatic Conditions During the Last Three Millennia

Our studied core includes six major climatic events (Fig. 3): Warming Present (WP) stretching along the uppermost 3 cm; Late Little Ice Age (Late LIA) located between 3 cm and 11 cm; Early Little Ice Age (Early LIA) (Wiles et al., 2008) located between 12 cm and 19 cm, Medieval Climatic Anomaly (MCA) spanned from 19 cm to 27 cm; Dark Age (DA) located between 27 cm and 46 cm and Roman Warm Period (RWP) spanned from 46 cm to 59 cm. In fact, the WP is manifested by stable climatic conditions with high values of potassium, carbonate, and electric conductivity and low pH values. It is marked by the rise of the pH rates, which indicates dry climatic conditions during this period. Besides, the Late LIA is classified as moderate stage (Essefi el al., 2015a). The bottom Late LIA is marked by a sandy sedimentation (Fig. 3), revealing a transition period between the dry conditions (the end of the WP) and the wet conditions (the Early LIA) (Essefi el al., 2015a; Gammoudi et al., 2019). Moreover, the Early LIA is manifested by high grey scale values and low sand fraction percentages, approving the wet climatic conditions. Furthermore, the passage to the Medieval Climatic Anomaly (MCA) (Bradley et al., 2003), is marked by an apparent decline in the grey scale intensity, revealing a dry period. It is an unstable and dry climatic stage (Essefi el al., 2015a). In addition, during the DA (Desprat et al., 2003), an increase of the grey scale values was noted and the top of this stage is manifested by a sandy sedimentation. It is an indicator of wet climatic conditions. In addition, the bottom of the RWP (Reale & Dirmeyer, 2000) is marked by high sand fraction percentages. This stage is subdivided into two parts. Indeed, the first part (from 27 down to 46 cm) is marked by low grey scale values indicating dry climate. The second part (from 46 down to 59 cm) is marked by high grey scale values as signs of stable and wet climatic conditions (Gammoudi et al., 2019).

Holocene Cyclostratigraphy

The spectral analyses of different studied parameters show the presence of significant cyclicities (Fig. 5). In fact, the high-frequency oscillations of the saline system of Mchiguig during the Late Holocene show periodicities of 1500 yr, and 700-800 yr. These cycles are recorded in many other studies. Although there are many studies discussing the climatic variability during the Holocene, the factors involved in the Holocene cyclostratigraphy are still not well understood. The 1500 yr cycle, recorded in many studies of the North Atlantic and Mediterranean regions (Bond et al., 2001; Moreno et al., 2005; Debret et al., 2007, 2009; Pena et al., 2010; Rodrigo-Gámiz et al., 2014), may has a probable link with oceanographic circulation and solar activity. Indeed, the cycle of 1500 yr can be a sub-harmonic of the 1000 yr cycle (Debret et al., 2009) or the result of an immense reorganization of North Atlantic thermohaline circulation

(Colin et al., 2010). Based on the study of the effective humidity and the runoff in southern Spain during the Holocene, Jiménez-Espejo et al. (2014), confirmed that the 1500 yr cycle have been a probable link with the solar activity and the NAO (North Atlantic Oscillation) -like circulation. Besides, the 700-800 yr cycle may form a second harmonic of the 1500 yr cycle, possibly related to the oceanic thermohaline circulation, namely in the North Atlantic Ocean (Jian et al., 2000). The cycle of 500 yr, which was recorded in other studies (Carcaillet et al., 2001), is related to solar activity (Essefi et al., 2015a). Essefi et al. (2015a) have recently demonstrated that oceanographic and atmospheric factors could generate centennial cycles (493 yr and 329 yr). Our centennial cycles have been recorded in many studies. In fact, Wanner et al. (2008), Sejrup et al. (2011), Knudsen et al. (2012) and Turney et al. (2016) affirmed that the about 453 yr (≈500 yr), 360 yr (≈400 yr), 346 yr (≈300 yr) and 293 yr (≈200 yr) are related to the fluctuations in the total solar irradiance. Goslin et al. (2018) suggested that the solar irradiance variability is transmitted to atmospheric and surface wind circulation. In this context, the changes in the thermohaline circulation (Stuiver and Brazuinas, 1993) and the solar heating of the sea-surface transferring to the atmosphere (Meehl et al., 2009) can be the main factors intervening on the amplification of solar variability. Consequently, the centennial cycles could be the result of the interaction between oceanographic and atmospheric forcing and solar activity (Gammoudi et al., 2019).

CONCLUSION

The study of the grey scale of Sebkha Mchiguig core shows the presence of six climatic stages. It helps to investigate the climatic variability recorded in this sebkha during the last Two millennia. Indeed, there are two types of climate. Although, Warming Present, Medieval Climatic Anomaly and the first part of Roman Warm Period were categorized as dry climatic stages, Early Little Ice Age, Dark Age and the second part of Roman Warm Period were considered as wet climatic stages. Then, Late Little Ice Age has moderate climatic conditions. Also, the spectral analysis based on the study of carbonate percentage, grey scale intensity, sand fraction percentages and geochemical parameters, visualized some significant cycles of 1500 yr, 800 yr, 600 yr, 560 yr, 544 yr, 453 yr, 360 yr, 346 yr and 293 yr. Solar activity generated 500 yr cycles, however, the oceanographic circulation generated other cycles of 1500 yr and 700-800 yr. The 1500 yr cycle may be the result of the solar activity and NAO-like circulation. Moreover, the interaction of the atmospheric and oceanographic factors with the sun activity could engender the rest of cycles. Most previous studies of Holocene cycles have shown the presence of this cyclicity, regardless of the geographical position. Consequently, the Holocene cyclostratigraphy is related to solar activity and to atmospheric and oceanographic forcing.

REFERENCES

Aaby, B. (1976). Cyclic climatic variations in climate over the past 5500 yr reflected in raised bogs. *Nature, 263*(5575), 281–284. doi:10.1038/263281a0

Abichou, A. (2002). *Les changements de paysages du bassin-versant de l'oued-Tataouine Fessi (sud-est tunisien). Étude multiscalaire et micromorphologie des remplissages des sebkhas et des états de surface des sols* (Thèse Géographie). Univ. Michel de Montaigne.

Abrantes, F., Voelker, A., Sierro, F. J., Naughton, F., Rodrigues, T., Cacho, I., Ariztegui, D., Brayshaw, D., Sicre, M. A., & Batista, L. (2012). Paleoclimate variability in the Mediterranean region. In P. Lionello (Ed.), *The Climate of the Mediterranean Region* (pp. 1–86). E-Publishing Elsevier. doi:10.1016/B978-0-12-416042-2.00001-X

Andresen, C. S., Bond, G., Kuijpers, A., Knutz, P. C., & Bjoôrck, S. (2005). Holocene climate variability at multidecadal time scales detected by sedimentological indicators in a shelf core NW off Iceland. *Marine Geology*, *214*(4), 323–338. doi:10.1016/j.margeo.2004.11.010

Bazzicalupo, P., Maiorano, P., Girone, A., Marino, M., Combourieu-Nebout, N., Pelosi, N., & Incarbona, A. (2020). Holocene climate variability of the Western Mediterranean: Surface water dynamics inferred from calcareous plankton assemblages. *The Holocene*, *30*(5), 691–708. doi:10.1177/0959683619895580

Bianchi, G. G., & McCave, I. N. (1999). Holocene periodicity in North Atlantic climate and deep-ocean flow south of Iceland. *Nature*, *397*(6719), 515–517. doi:10.1038/17362

Bond, G., Kromer, B., Beer, J., Muscheler, R., Evans, M. N., Showers, W., Hoffmann, S., Lotti-Bond, R., Hajdas, I., & Bonani, G. (2001). Persistent solar influence on North Atlantic climate during the Holocene. *Science*, *294*(5549), 2130–2136. doi:10.1126cience.1065680 PMID:11739949

Bond, G., Showers, W., Cheseby, M., Lotti, R., Almasi, P., deMenocal, P., Priore, P., Cullen, H., Hajdas, I., & Bonani, G. (1997). A pervasive millenial-scale cycle in North Atlantic: Holocene and glacial climates. *Science*, *278*(5341), 1257–1266. doi:10.1126cience.278.5341.1257

Bradley, R. S., Hughes, M. K., & Diaz, H. F. (2003). Climate in Medieval Time. *Science*, *302*(5644), 404–405. doi:10.1126cience.1090372 PMID:14563996

Carcaillet, C., Bouvier, M., Fréchette, B., Larouche, A. C., & Richard, P. J. H. (2001). Comparison of pollen-slide and sieving methods in lacustrine charcoal analyses for local and regional fire history. *The Holocene*, *11*(4), 467–476. doi:10.1191/095968301678302904

Colin, C., Frank, N., Copard, K., & Douville, E. (2010). Neodymium isotopic composition of deep-sea corals from the NE Atlantic: Implications for past hydrological changes during the Holocene. *Quaternary Science Reviews*, *29*(19-20), 2509–2517. doi:10.1016/j.quascirev.2010.05.012

Debret, M., Bout-Roumazeilles, V., Grousset, F., Desmet, M., McManus, J. F., Massei, N., Sebag, D., Petit, J. R., Copard, Y., & Trentesaux, A. (2007). The origin of the 1500-year climate cycles in Holocene North-Atlantic records. *Climate of the Past*, *3*(4), 569–575. doi:10.5194/cp-3-569-2007

Debret, M., Sebag, D., Costra, X., Massei, N., Petit, J. R., Chapron, E., & Bout-Roumazeilles, V. (2009). Evidence from wavelet analysis for a mid-Holocene transition in global climate forcing. *Quaternary Science Reviews*, *28*(25-26), 2675–2688. doi:10.1016/j.quascirev.2009.06.005

Desprat, S., Sanchez Goni, M. F., & Loutre, M.-F. (2003). Revealing climatic variability of the last three millennia in northwestern Iberia using pollen influx data. *Earth and Planetary Science Letters*, *213*(1-2), 63–78. doi:10.1016/S0012-821X(03)00292-9

Essefi, E. (2021c). Parallel evolution of energy consumption and environmental pollution during the Anthropocene. *Journal of Global Ecology and Environment*, 36-42.

Essefi, E. (2020). On the Geochemistry and Mineralogy of the Anthropocene. *International Journal of Water and Wastewater Treatment*, *6*(2), 1–14.

Essefi, E. (2021a). *Apport des Approches Géochimique et Sédimentologique au Fonctionnement des Systèmes Endoréiques et Exoréiques de l'Holocène à l'Anthropocène: Intérêts Environnemental et Géoéconomique*. Institut Supérieur des Sciences et Techniques des Eaux de Gabès.

Essefi, E. (2021b). Record of the Anthropocene-Great Acceleration along a core from the coast of Sfax, southeastern Tunisia. *Turkish Journal of Earth Science*, 1-16.

Essefi, E. (2021d). Homo Sapiens Sapiens Progressive Defaunation During The Great Acceleration: The Cli-Fi Apocalypse Hypothesis. *International Journal of Toxicology and Toxicity Assessment*, *1*(1), 1–6.

Essefi, E., Gharsalli, N., Kalabi, S., Ben Ameur, M., & Yaich, C. (2015d). Spectral analysis of a core from the Sebkha of Sidi Mansour, Southern Tunisia: The Holocene cyclostratigraphy. *J Geophys Remote Sens*, *4*(02), 141. doi:10.4172/2169-0049.1000141

Essefi, E., Gharsalli, N., Salhi, S., & Yaich, C. (2015b). Cyclicity of grain size parameters along a Core from the Gleysol of Sebkha Ennoual, southeastern Tunisia. *J Basic Appl Res Int, 15*(5–3), 124–134.

Essefi, E., Gharsalli, N., & Yaich, C. (2015a). Climatic significance aeolian sedimentary filling of sebkha Mhabeul, southeastern Tunisia: Contribution of the genetic approach. *J Basic Appl Res Int*, *15*(2–4), 140–152.

Essefi, E., Gharsalli, N., & Yaich, C. (2015c). Geophysical and geochemical study of the silico-evaporitic sedimentary filling of Boujmal wetland, eastern Tunisia: inferring the climatic signal within groundwater noise. *J Basic Appl Res Int, 15*(2–1), 14–25.

Essefi, E., Jmaa, H. B., Touir, J., Tagorti, M. A., & Yaich, C. (2013). Variability to sedimentary dynamics and climatic conditions during the last two millennia at sebkha Souassi in eastern Tunisia. *Papers on Global Change IGBP*, *20*(1), 29–45. doi:10.2478/igbp-2013-0001

Essefi, E., Mefteh, S., Medhioub, M., & Yaich, C. (2014). Magnetic study of the heated and unheated sedimentary fillings of Sebkha Mhabeul, Southeast Tunisia: A geophysical method for paleoclimatic investigation and tephrochronological dating. *International Journal of Geophysics*, *2014*, 7–7. doi:10.1155/2014/908395

Essefi, E., Smida, N. B., Jandoubi, I., Othmani, M. A., & Tagorti, M. A. (2020). Progressive evaporation of brine of sebkha Mchiguig, central Tunisia: A geo-economical comparative study of salt and brine. *Carbonates and Evaporites*, *35*(2), 1–10. doi:10.100713146-020-00592-7

Fletcher, W. J., & Zielhofer, C. (2013). Fragility of Western Mediterranean landscapes during Holocene Rapid Climate Changes. *Catena*, *103*, 16–29. doi:10.1016/j.catena.2011.05.001

Goslin, J., Fruergaard, M., Sander, L., Gałka, M., Menviel, L., Monkenbusch, J., Thibault, N., & Clemmensen, L. B. (2018). Holocene centennial to millennial shifts in North-Atlantic storminess and ocean dynamics. *Scientific Reports*, *8*(1), 12778. doi:10.103841598-018-29949-8 PMID:30143649

Holcová, K., Suchánek, V., Heřmanová, Z., & Vondrák, D. (2020). The First Evidence of Bioerosion in Chitinous Lacustrine Microfossils from Glacial Lakes (late Pleistocene and Holocene, Bohemian Forest). *Palaios, 35*(8), 317–326. doi:10.2110/palo.2020.008

IPCC. (2014). *Climate Change 2014: Synthesis Report.* In R. K. Pachauri & L. A. Meyer (Eds.), *Contribution of Working Groups I, II and III to the Fifth Assessment Report of the Intergovernmental Panel on Climate Change* (p. 151). IPCC.

Jaouadi, S., Lebreton, V., Bout-Roumazeilles, V., Siani, G., Lakhdar, R., Boussoffara, R., Dezileau, L., Kallel, N., Mannai-Tayech, B., & Combourieu-Nebout, N. (2016). Environmental changes, climate and anthropogenic impact in south-east Tunisia during the last 8 kyr. *Climate of the Past Discussions, 12*(6), 1339–1359. doi:10.5194/cp-12-1339-2016

Jian, Z., Wang, P., Saito, Y., Wang, J., Plaumann, U., Oba, T., & Cheng, X. (2000). Holocene variability of the Kuroshio Current in the Okinawa Trough, northwestern Pacifc Ocean. *Earth and Planetary Science Letters, 184*(1), 305–319. doi:10.1016/S0012-821X(00)00321-6

Jiménez-Espejo, F. J., García-Alix, A., Jiménez-Moreno, G., Rodrigo-Gámiz, M., Anderson, R. S., Rodríguez-Tovar, F. J., Martínez-Ruiz, F., Giralt, S., Delgado Huertas, A., & Pardo-Igúzquiza, E. (2014). Saharan aeolian input and effective humidity variations over western Europe during the Holocene from a high altitude record. *Chemical Geology, 374-375*, 1–12. doi:10.1016/j.chemgeo.2014.03.001

Karpinski, E., Hackenberger, D., Zazula, G., Widga, C., Duggan, A. T., Golding, G. B., & Poinar, H. N. (2020). American mastodon mitochondrial genomes suggest multiple dispersal events in response to Pleistocene climate oscillations. *Nature Communications, 11*(1), 1–9. doi:10.103841467-020-17893-z PMID:32873779

Knudsen, M. F., Jacobsen, B. H., Riisager, P., Olsen, J., & Seidenkrantz, M. S. (2012). Evidence of Suess solar-cycle bursts in subtropical Holocene speleothem? 18O records. *The Holocene, 22*(5), 597–602. doi:10.1177/0959683611427331

Knudsen, M. F., Riisager, P., Jacobsen, B. H., Muscheler, R., Snowball, I., & Seidenkrantz, M. S. (2009). Taking the pulse of the Sun during the Holocene by joint analysis of 14C and 10Be. *Geophysical Research Letters, 36*(16), L16701. Advance online publication. doi:10.1029/2009GL039439

Knudsen, M. F., Seidenkrantz, M. S., Jacobsen, B. H., & Kuijpers, A. (2011). Tracking the Atlantic Multidecadal Oscillation through the last 8,000 years. *Nature Communications, 2*(1), 178. doi:10.1038/ncomms1186 PMID:21285956

Lakhdar, R., Soussi, M., Ben Ismail, M. H., & M'Rabet, A. (2006). A Mediterranean Holocene restricted coastal lagoon under arid climate: Case of the sedimentary record of Sabkha Boujmel (SE Tunisia). *Palaeogeography, Palaeoclimatology, Palaeoecology, 241*(2), 177–191. doi:10.1016/j.palaeo.2006.02.014

Lionello, P. (2012). *The Climate of the Mediterranean Region. From the Past to the Future*. Elsevier.

Lionello, P., Malanotte-Rizzoli, P., Boscolo, R., Alpert, P., Artale, V., Li, L., Luterbacher, J., May, W., Trigo, R., Tsimplis, M., Ulbrich, U., & Xoplaki, E. (2006). The Mediterranean climate: An overview of the main characteristics and issues. Mediterranean Climate Variability. Developments in Earth and Environmental Sciences, 4, 1-26.

Malanotte-Rizzoli, P., Artale, V., Borzelli-Eusebi, G. L., Brenner, S., Crise, A., Gacic, M., Kress, N., Marullo, S., Ribera d'Alcalà, M., Sofianos, S., Tanhua, T., Theocharis, A., Alvarez, M., Ashkenazy, Y., Bergamasco, A., Cardin, V., Carniel, S., Civitarese, G., D'Ortenzio, F., ... Triantafyllou, G. (2014). Physical forcing and physical/biochemical variability of the Mediterranean Sea: A review of unresolved issues and directions for future research. *Ocean Science*, *10*(3), 281–322. doi:10.5194/os-10-281-2014

Marquer, L., Pomel, S., Abichou, A., Schulz, E., Kaniewski, D., & Van Campo, E. (2008). Late Holocene high resolution palaeoclimatic reconstruction inferred from Sabkha Mhabeul, southeast Tunisia. *Quaternary Research*, *70*(2), 240–250. doi:10.1016/j.yqres.2008.06.002

Martinez-Ruiz, F., Kastner, M., Gallego-Torres, D., Rodrigo-Gámiz, M., Nieto-Moreno, V., & Ortega-Huertas, M. (2015). Paleoclimate and paleoceanography over the past 20,000 yr in the Mediterranean Sea Basins as indicated by sediment elemental proxies. *Quaternary Science Reviews*, *107*, 25–46. doi:10.1016/j.quascirev.2014.09.018

Mayewski, P. A., Rohling, E. E., Stager, J. C., Karlén, W., Maasch, K. A., Meeker, L. D., Meyerson, E. A., Gasse, F., van Kreveld, S., Holmgren, K., Lee-Thorp, J., Rosqvist, G., Rack, F., Staubwasser, M., Schneider, R. R., & Steig, E. J. (2004). Holocene climate variability. *Quaternary Research*, *62*(3), 243–255. doi:10.1016/j.yqres.2004.07.001

McDermott, F., Mattey, D. P., & Hawkesworth, C. (2001). Centenial scale Holocene climate variability revealed by a high-resolution speleothem y18O record from SW Ireland. *Science*, *294*(5545), 1328–1331. doi:10.1126cience.1063678 PMID:11701925

Meehl, G. A., Arblaster, J. M., Matthes, K., Sassi, F., & Van Loon, H. (2009). Amplifying the Pacific Climate System Response to a Small 11-Year Solar Cycle Forcing. *Science*, *325*(5944), 1114–1118. doi:10.1126cience.1172872 PMID:19713524

Meese, D. A., Gow, A. J., Grootes, P., Mayewski, P. A., Ram, M., Stuiver, M., Taylor, K. M., Waddington, K. C., & Zielinski, G. A. (1994). The accumulation record from the GISP2 record as an indicator of climate change throughout the Holocene. *Science*, *266*(5191), 1680–1682. doi:10.1126cience.266.5191.1680 PMID:17775628

Moreno, A., Cacho, I., Canals, M., Grimalt, J. O., Sánchez Goñi, M. F., Shackleton, N., & Sierro, F. J. (2005). Links between marine and atmospheric processes oscillating on a millennial time-scale. A multi-proxy study of the last 50,000 yr from the Alboran Sea (Western Mediterranean Sea). *Quaternary Science Reviews*, *24*(14-15), 1623–1636. doi:10.1016/j.quascirev.2004.06.018

Njagi, D. M., Routh, J., Olago, D., & Gayantha, K. (2021). A multi-proxy reconstruction of the late Holocene climate evolution in the Kapsabet Swamp, Kenya (East Africa). *Palaeogeography, Palaeoclimatology, Palaeoecology*, *574*, 110475. doi:10.1016/j.palaeo.2021.110475

Pena, L. D., Francés, G., Diz, P., Esparza, M., Grimalt, J. O., Nombela, M. A., & Alejo, I. (2010). Climate fluctuations during the Holocene in NW Iberia: High and low latitude linkages. *Continental Shelf Research*, *30*(13), 1487–1496. doi:10.1016/j.csr.2010.05.009

Pepi, M., & Focardi, S. (2021). Antibiotic-Resistant Bacteria in Aquaculture and Climate Change: A Challenge for Health in the Mediterranean Area. *International Journal of Environmental Research and Public Health, 18*(11), 5723. doi:10.3390/ijerph18115723 PMID:34073520

Pomel, S., Abichou, A., & Schulz, E. (2006). La signification paléo-impluviométrique des sédiments de la sebkha Mhabeul (Sud-Est de la Tunisie). *L'érosion entre Société, Climat et Paléoenvironnement: Table ronde en l'honneur du Professeur René Neboit-Guilhot*, 435-442.

Press, W. H., Teukolsky, S. A., Vetterling, W. T., & Flannery, B. P. (1992). *Numerical Recipes in C*. Cambridge University Press.

Reale, O., & Dirmeyer, P. (2000). Modeling the effects of vegetation on Mediterranean climate during the Roman Classical Period. Part I: Climate history and model sensitivity. *Global and Planetary Change, 25*(3-4), 163–184. doi:10.1016/S0921-8181(00)00002-3

Rodbell, D. T., Seltzer, G. O., Anderson, D. M., Abbott, M. B., Enfield, D. B., & Newman, J. H. (1999). An 15,000-year record of El Nino-driven alluviation in southwestern Ecuador. *Science, 283*(5401), 516–520. doi:10.1126cience.283.5401.516 PMID:9915694

Rodrigo-Gámiz, M., Martínez-Ruiz, F., Rodríguez-Tovar, F. J., Jiménez-Espejo, F. J., & Pardo-Igúzquiza, E. (2014). Millennial-to centennial-scale climate periodicities and forcing mechanisms in the westernmost Mediterranean for the past 20,000 yr. *Quaternary Research, 81*(1), 78–93. doi:10.1016/j.yqres.2013.10.009

Schulz, E., Abichou, A., Hachicha, T., Pomel, S., Salzmann, U., & Zouari, K. (2002). Sebkhas as ecological archives and the vegetation and landscape history of southeastern Tunisia during the last two millennia. *Journal of African Earth Sciences, 34*(3-4), 223–229. doi:10.1016/S0899-5362(02)00021-0

Schulz, E., Smykatz-Kloss, W., & Abichou H., (1995). Zaderg environmental history of the semidesert region in southernTunisia. *Zentralblatt fur Geologie und Paleontology, 1*(3-4), 423-440.

Sejrup, H. P., Haflidason, H., & Andrews, J. T. (2011). A Holocene North Atlantic SST record and regional climate variability. *Quaternary Science Reviews, 30*(21-22), 3181–3195. doi:10.1016/j.quascirev.2011.07.025

Stengele, F., & Smykatz-Kloss, W. (1995). Mineralogical and geochemical study of Holocene sebkha sediments in southeastern Tunisia. *Chemie der Erde-Geochemistry, 55*(4), 241–256.

Stuiver, M., & Brazuinas, T. (1993). Modeling atmospheric 14C influences and 14C ages of marine samples to 10,000 bc. In M. Stuiver, A. Long, & R. S. Kra (Eds.), *Calibration, Radiocarbon*. doi:10.1017/S0033822200013874

Turney, C. S. M., Jones, R. T., Fogwill, C., Hatton, J., Williams, A. N., Hogg, A., Thomas, Z. A., Palmer, J., Mooney, S., & Reimer, R. W. (2016). A 250-year periodicity in Southern Hemisphere westerly winds over the last 2600 years. *Climate of the Past, 12*(2), 189–200. doi:10.5194/cp-12-189-2016

Vallès, V., & Bourgeat, F. (1989). Détermination géochimique de l'exigence de plâtre d'une culture des sols sodiques. I. Le développement du modèle thermodynamique GYPSOL simulant l'interaction chimique de l'eau d'irrigation-sol. *Recherche sur les sols aides et de readaptation, 44*(3), 157-163.

Wanner, H., Beer, J., Bütikofer, J., Crowley, T. J., Cubasch, U., Flückiger, J., Goose, H., & Grosjean, M. (2008). Mid- to Late Holocene climate change: An overview. *Quaternary Science Reviews*, *27*(19-20), 1791–1828. doi:10.1016/j.quascirev.2008.06.013

Wiles, G. C., Barclay, D. J., Calkin, P. E., & Lowell, T. V. (2008). Century to millennial-scale temperature variations for the last two thousand years indicated from glacial geologic records of Southern Alaska. *Global and Planetary Change*, *60*(1-2), 115–125. doi:10.1016/j.gloplacha.2006.07.036

Chapter 16
The Geobiology of Bacteria Within the Extreme Conditions of Sidi El Hani Wetland, Eastern Tunisia:
Environmental and Astrobiological Significance

Tagorti Mohamed Ali
Univeristy of Sousse, Tunisia

Elhoucine Essefi
University of Gabes, Tunisia

ABSTRACT

Extreme conditions in wetlands may be the niche of development of colonies of bacteria. In this chapter, the authors study the gnotobiology of moderate bacteria within the saline wetland of Sidi El Hani. In terms of geology, the coring within the sedimentary filling of the wetland shows color variation from white, grey, black to red related to the variability of bacteria species. On the other hand, in terms of microbiological investigations, isolates of bacteria show a variability from Gram+ to Gram-, from oxidase+ to oxidase-, from catalase+ to catalase-. This geobiological variability is related to radical change in climatic conditions. In doing so, the wetland of Sidi El Hani may record the climatic variability during the Late Holocene. On the other hand, it may be considered as terrestrial analogue with a development of extremophiles.

DOI: 10.4018/978-1-7998-9498-8.ch016

INTRODUCTION

As combination of geology and biology, geobiology encompasses many disciplines including paleoenvironmental modeling, genomics, paleontology and geochemistry (Visconti, 2021). Geobiology is a scientific discipline applying principles and tools of biology to study of the Earth. Geobiology targets the use of physics and chemistry methods. It dates back to the 1970s when the Gaia hypothesis was formulated. It may bay be used for environmental and paleoclimatic (Phillips et al., 2021) as well as astrobiological (Perl et al., 2021) studies.

Past climatic changes may be inferred based on the sedimentary record of some depressions such as seas (Zhang et al., 2021), lakes (Fedorova al., 2021), oceans (Goosse et al., 2021), and sebkhas (Essefi, 2021 and references therein). Worldwide, a number of sites have been studied for their application to understand geological and potential biological processes on Mars (e.g., Essefi, 2013; Essefi et al., 2014a). To name but a few, the Channeled Scablands (Gallagher and Bahia, 2021), Australia (Van Kranendonk et al., 2021), Sahara Desert (Savijärvi & Harri, 2021), Tunisia (Essefi et al., 2014b), the Dead Sea (e.g., Wentworth & Morris, 2001), and the Arctic and Antarctica (Coleine et al., 2021, and references therein) are sites sharing similarities with Mars. The key word for all these features of similarities is life. For instance, life in extreme environments (e.g., Fernández-Martínez et al., 2021) is the actual analogue of martian life if ever existed. Consequently, scholars insisted on detailed studies to understand the whys and the wherefores of the extreme life on terrestrial analogues. During the recent decades, research in the field of geomicrobiology flourished to advocate the astrobiological implications of life in extreme environments (Barbieri, 2013 and references therein). In this chapter, we study the gnotobiology of moderate bacteria within the saline wetland of Sidi El Hani. First, we investigate the sedimentary filling of the wetland and the recorded environmental changes during the last two millennia. Then, moderate bacteria were studied to show their astrobiological significance.

STUDY AREA

The sebkha of Sidi El Hani is a NW-SE lengthened depression in the Sahel area in Tunisia) with a relatively small drainage basin. It is by no means communicating with the sea. But it contains huge quantities of salt spanned on tens of kilometres. It is fed by temporary (Oued Maktaa) and (Fig: 1) and permanent (Oued Chrita) hydrologic networks. It represents the basal part of the endorheic system Sidi El Hani-Chrita-Mechertat (Ben Jmaa, 2008). Moreover, eventhough it has always been treated as a single unit, the sebkha of Sidi El Hani as such is actually made up of three communicated playas; from north to the south one finds: the playas of Sidi El Hani (sensu stricto), the playa of Souassi and the playa of Dkhila. Administratively, it stretches on four governmental subdivisions (Fig: 1). Its northeasternmost belongs to Sousse state; its northwesternmost is distant 13 km of Kairouan center; a minor part of Sidi El Hani playa is located in Mounastir state. The Mahdia state includes the southeasternmost of Dkhila sebkha (i.e. the southeasternmost of Sidi El Hani saline environment). This administrative subdivision and its large geographic extension are handicaps to create a clear database about the sebkha; thus, the dispersed and incomplete data are hard to collect. Topographically, Sidi El Hani playa is limited by various hills that barely reach more than 100m of altitude. Westward, one finds the undulation of Aïn El Ksar, Aïn Tarfaïa and El Guessaat; southward, the undulation of Bou Thadi and Eljem and towards the East the Zarmdine anticline. The basin of Sidi El Hani is an eastern diverticulum, of the Kairouan

Figure 1. Administrative location and topography of Sidi El Hani playa
(Leduc et al., 2004; Ben Jmaa, 2008)

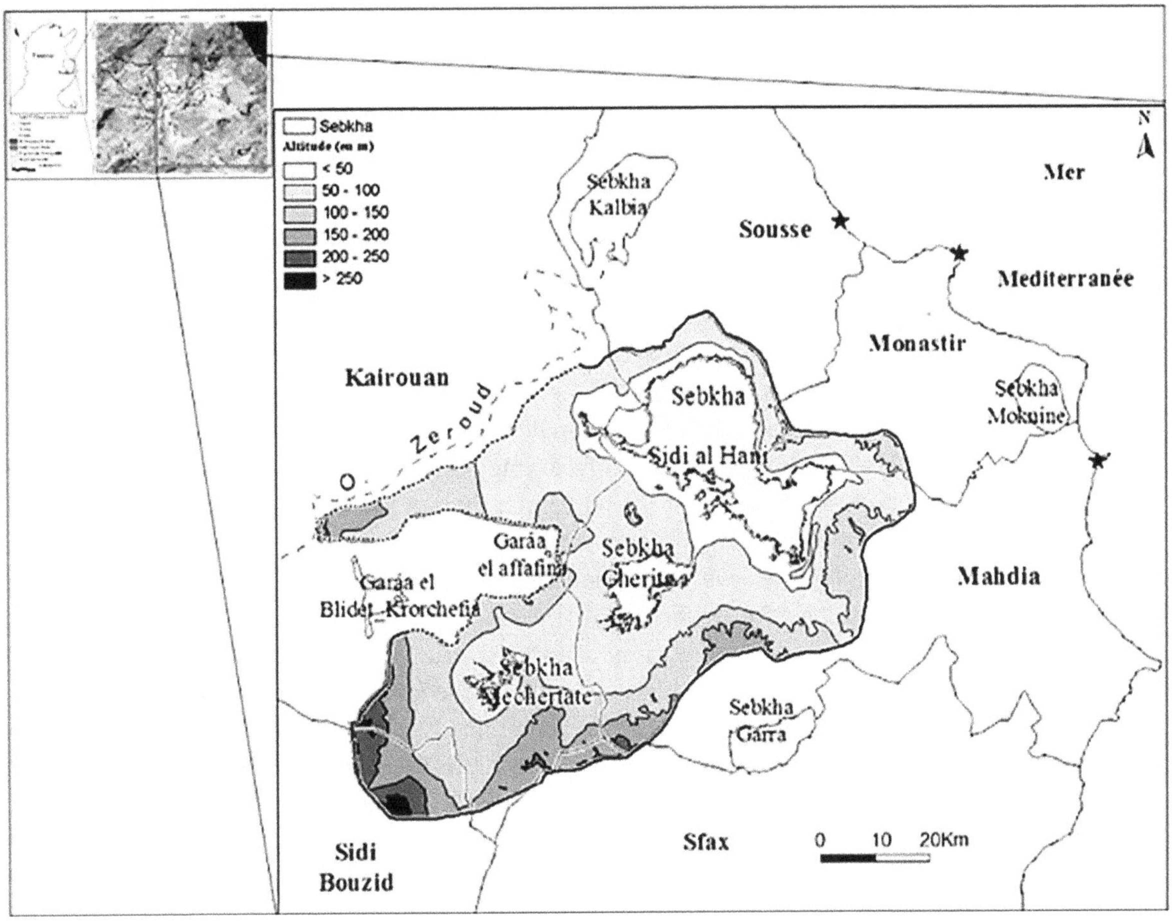

plain, which penetrates toward a coastal folded upland. This topographic contrast helps to create a developed hydrological network within the 3300 km^2 of its hydrological basin, on which many dams are constructed (Fig: 1). The three components of sebkha Sidi El Hani have different orientations. Sidi El Hani (sensu stricto) and Dkhila playas are oriented N 170; the playa of Souassi is oriented N 90. The bottom of the playa of Sidi El Hani (sensu stricto) hangs slightly towards the East and that of Souassi towards the North. The surface of Sidi El Hani playa is approximately 370 Km^2 and its average depth of water is 0.4m; but it can reach in some locations 0.8m. The principal highest point, which is located in North East of the depression, can reach 33m of altitude.

MATERIALS AND METHODS

Twelve sediment samples were taken from three ever flooded stations in the sebkha of Sidi El Hani (S1, S2 and S3). The moderate halophilous bacteria isolated from the surface sediment from the sebkha were purified by adding 34 g/l NaCl. Incubation of 1g of sample of sediment in a test tube containing 20 ml of the nutritive bubble added to NaCl 34g/l during 7 days at a temperature of 37 °C. By the method of the scratches, 0.1 ml of the samples of were taken and spread out over limp of Petri containing a nutritive gelose added to NaCl 34g/l. The dominant colonies presenting different aspect and form are mended on others limp of Petri containing the same field. Then, they are incubated thereafter during 24 hours in the drying oven at 37 °C. This last operation is repeated until obtaining pure stocks. The pure stocks were preserved at - 20 °C in a liquid field containing 34 g/l NaCl and 10 to 15% of glycerol in tubes of conservation. The purified strains underwent a series of tests for their identification. The Gram test was carried out according to the method described by Fluharty et al. (1967). The gelation, after 5 to 60 second of application of the bacterial colony in the presence of some drops of potassium hydroxide to 3%, indicates that the bacterium is Gram-ve otherwise, the bacterium is Gram+ve. This fast technique did not give an idea about the morphology of the bacterium. Nevertheless, the coloring of Gram is an essential stage for the characterization of the bacterial strains. Parallel to the determination of the reaction of Gram, it gave an idea about the aspect of the colonies, their forms like their modes of grouping. The analyzed bacterial strains related to a blade with a handle are red. Then, we dilute a colony in a drop of physiological water. The following stage consists of a drying at an ambient air and fixing heat, by slowly passing out three times on the higher part of the flame of a Bunsen burner. Then, we let the blade cool and we colour according to the following protocol: we cover the blade with the purple Gentian solution during 1minute. We abundantly rinse with water and we cover the blade with Lugol solution to fix the purple during 1minute. Washing with alcohol lasts 10 second (until it is not formed any more coloured cloud) and colouring with the solution of Fushine 30 second. The blades are carried then to the microscopic observation. For the micro-organisms with Gram+ve, the iodized coloured complex cannot any more be dissolved out of the cell by a product of discolouration like alcohol or acetone. The cell remains coloured in purple blue then. For the micro-organisms with Gram-ve, the iodized colored complex is dissolved and the cell is colored in pink or red. The oxidase test is highlighted in the bacteria by the reagent Tetramethyl P Phenylenediamine dihydrochloride at 1%. The appearance of a violet coloration, after 1 to 5 mn of application to the bacterial colony in the presence of this reagent, indicates a production of the oxidase (bacteria positive oxidase). The catalase is the enzyme of the oxygenated water degradation resulting from the respiratory chain. Its presence is tested by putting the bacterial colonies in the presence of some hydrogen peroxide drops (H_2O_2) at 10%. The appearance of an effervescence indicates the production of the catalase by the tested strain. In order to highlight the morphological variations of the colonial shapes of the various isolated strains and to eliminate the colonies which are repeated on the level of each station, an examination in a fresh state under the photonic microscope and the magnifying glass was carried out. The diffusion on field Mueller Hinton Agar (MH) is a method employed to study the antimicrobic activity (Piddok, 1990). The pathogenic bacteria used are *Vibrio alginolyticus* isolated starting from fish infected like previously described by Ben Kahla-Nakbi et al.. (2006), *Staphylococcus aureus* ATCC 25923 and *Salmonella typhymurium* ATCC1408. These later were revivified in 10 ml of nutritive bubble and cultivated during 24 hours on a nutritive gelose at 30 °C the common colonies of the pure culture were suspended in 10 ml of physiological field and agitated during 5 mn. 1 ml of the suspensions prepared of each pathogenic was spread out over a solid field of Mueller Hinton (MH) added with NaCl 34g/l and incubated during 30 mn at 37 °C. The isolated stocks were incubated in a nutritive bubble containing 34g/l NaCl (24 H, 37 °C). Wells were dug fruitlessly in the field (MH) and were filled by 100 µl of each bacterial suspension enriched in the nutritive bubble. The presence of an antimicrobic activity is revealed by the presence of a zone of inhibition around the well. The diameter of the zone of inhibition around the well measures the degree of inhibition (Parish & Davidson, 1993).

The identification of the biochemical characteristics of the isolates was carried out by the biochemical galleries. Only the strains presenting a good antibacterial activity were tested. To identify the bacilli with Gram+ve, we used the galleries 50 CHB (Biomerieux. France). The Api the 50 CHB, encompasses 50 the biochemical tests allowing the study of the fermentation of substrate pertaining to the family of carbohydrates and derived (uronic heterosides, polyalcohols, acids). After 24 and 48 hours of incubation, the reactions were read using a software (Api Web, Biomeriaux, France). Api the 20 E comprise 20 microtubes containing dehydrated substrates which are inoculated by a bacterial suspension which reconstitutes the tests. The reactions produced during 24 hours of incubation result in coloured turns spontaneous or revealed by the addition of reagents.These reactions were read using a software (Api Web, Biomeriaux, France). The detection of the capacity of biofilm formation of the isolated bacteria for purpose probiotic is carried out by cultivating the bacteria on limp of Cetri containing Red Agar Congo (CRA) like previously described by Freeman et al. (1989). This field (CRA), was prepared by mixing 0.8 g of Red of Congo with 36 g of saccharose (Sigma) in 1litre of gelose heart brain (Biorad) added with 34g/l of NaCl. Limp then are incubated during 24 H at 37 °C under conditions of aerobe and are left one night at an ambient temperature. The strains known as positive "biofilm" appear in the form of irregular black colonies on gelose CRA, while the strains known as negative "biofilm" develop a red rosy color on this field (Vasudevan et al., 2003).

Detection of the Capacity of Formation of Biofilm on a Glass Surface

The capacity of the isolated strains to adhere to abiotic surfaces (glass) was tested on tubes out of glass of ten-millilitres (0.5 cm diameters) like previously described by Wolfe et al. (2004). All the tested strains were enriched in a bubble SWT (sea-water-tryptone-yeast extract) prepared by the mixture by 5 g bactotryptone, 3 g of yeast extract and 3 ml of glycerol in 1 liter of water (700 ml of filtered sea water + 300 ml of distilled water). The glass tubes were incubated during one night at 37 °C with agitation. 100 ml of this pre-enriched culture were added to new glass tubes containing the same field and incubated at 37 °C during 10 hours without agitation. All the glass tubes were colored with a Crystal Violet solution 1% (Merck, France) during 15 min, then washed with distilled water. The strains forming a purple strip on the wall of the tubes were regarded as biofilm positive.

Quantitative Study of the Adhesion of the Isolated Stocks

This study was determined by a quantitative semi technique in the plates with 96-well (Nunc, Roskilde, Denmark) like previously described by Christensen and al. (1985) and Mahdhi et al. (2010). The strains were incubated in bubble Tryptocase-soya-Broth (TSB) with 34g/L of NaCl, (Biorad) at 37 °C during 24 heurs. Optical density of suspensions at 595 Nm (DO595) was measured and adjusted. Thereafter, these suspensions were diluted to the 1:100 in a bubble TSB with 2% of glucose (w/v). A total volume of 200 µl of each suspension transferred in plates from polystyrene 96-well in the form of U (Nunclon; Nunc, Denmark). Each strain was tested three times. A well containing sterile bubble TSB only was useful like controls negative. The plate was incubated thereafter in aerobiosis during 24 h at 37 °C. Then, the bubble was aspired and the plates was rinsed three times with saline plug phosphates (PBS) (7 mm Na2 HPO4, 3mM NaH2 PO4, 130 mm NaCl, pH 7,4), in order to eliminate the nonadherent strains. The plate was dried thereafter.

The adherent strains were fixed by an ethanol solution at 95% and were coloured with 100 µl of a purple crystal solution with 1% (Merck) during 5 min. The surplus of dye was eliminated; then the plate was washed three times with 300 µl of sterile distilled water. Then, it was emptied and dried. The optical density of each cup was measured to 570 Nm by using a reader ELISA (automated Multiskan reader, Italy). The capacity of the strains to form a biofilm was classified according to three categories:

- Strongly adherent strains: ($DO570 \geq 1$)
- Slightly adherent strains: ($0.1 \leq DO570 \leq 1$)
- Nonadherent strains ($DO570 < 0.1$)

Each test was carried out in triple.

The search for certain enzymes such as amylase, caseinase, lecithinase and lipase was made on Trytocase-Soya-Agar (TSA 1%) field added to 0.2% of starch, 1% skimmed milk, 5% egg yolk, 1% of tween 80 and 34g/L of NaCl. The strains were mended on the surface of these fields. After incubation at 37 °C during 24 to 48 H, a positive reaction is highlighted by a clear halation all around the colony. Only for the starch, a positive reaction is highlighted after addition of a Lugol solution.

RESULTS

Geology

The correlation between core1 (210cm) from Sidi El Hani and a core from sebkha Mhabeul (65cm) (Fig.2) shows that the climatic variability recorded during the last two millennia in 65cm is recorded in core1 of Sidi El Hani discharge in 114cm. As it is proved in core10 location, the rate of sedimentation is higher in this location of Sidi El Hani than in sebkha Mhabeul. Similarly, the sedimentation in this location of Sidi El Hani clay pan follows a chaotic evolution. By following the major grey scale variability and the three reference bands along both cores, the six climatic stages recognized in sebkha Mhabeul and core10 were found out also along core1. First, the Warming Present (WP) stretches along the uppermost 3cm of core1 with its high grey scale as sign of a dry climate. Second, the Late Little Ice Age (Late LIA) is located between 3cm and 7cm in core1. During this stage, the too low rate of sedimentation in this location of Sidi El Hani clay pan forbids the distinction between intermediate GS values and high GS values. Nevertheless, this period may be classified climatologically dry. As in sebkha Mhabeul core and core10, this period is limited the bottom by the first reference sandy band. Third, Early Little Ice Age (ELIA) is located in core1 between 7cm and 28cm represents the. Climatologically, the intermediate values of GS indicate that this stage may be classified as moderate. Fourth, the Medieval Climatic Anomaly (MCA) spanning from 28cm to 40cm in core1 is marked by a sharp increase of the GS revealing a wet period. Fifth, Dark Age (DA) appears along the part between 40cm and 79cm, a shift from light to dark sediments is recorded. As for the grain size distribution, the second reference sandy band is detected in the top of the DA. Sixth, the Roman Warm Period (RWP) appears between 79cm and 114cm. According to the major grey scale, this period may be divided into sub-stages. The first (from 79cm to 109cm in core1) is marked by high GS values associated with dark sediments suggesting stable climatic conditions. The second (from 109cm to 114cm in core1) is recorded in the bottom is marked by a third sandy reference band and low GS values as signs of wet climate.

Figure 2. Correlation between Sebkha Mhabeul core and core1 from Sidi El Hani clay pan

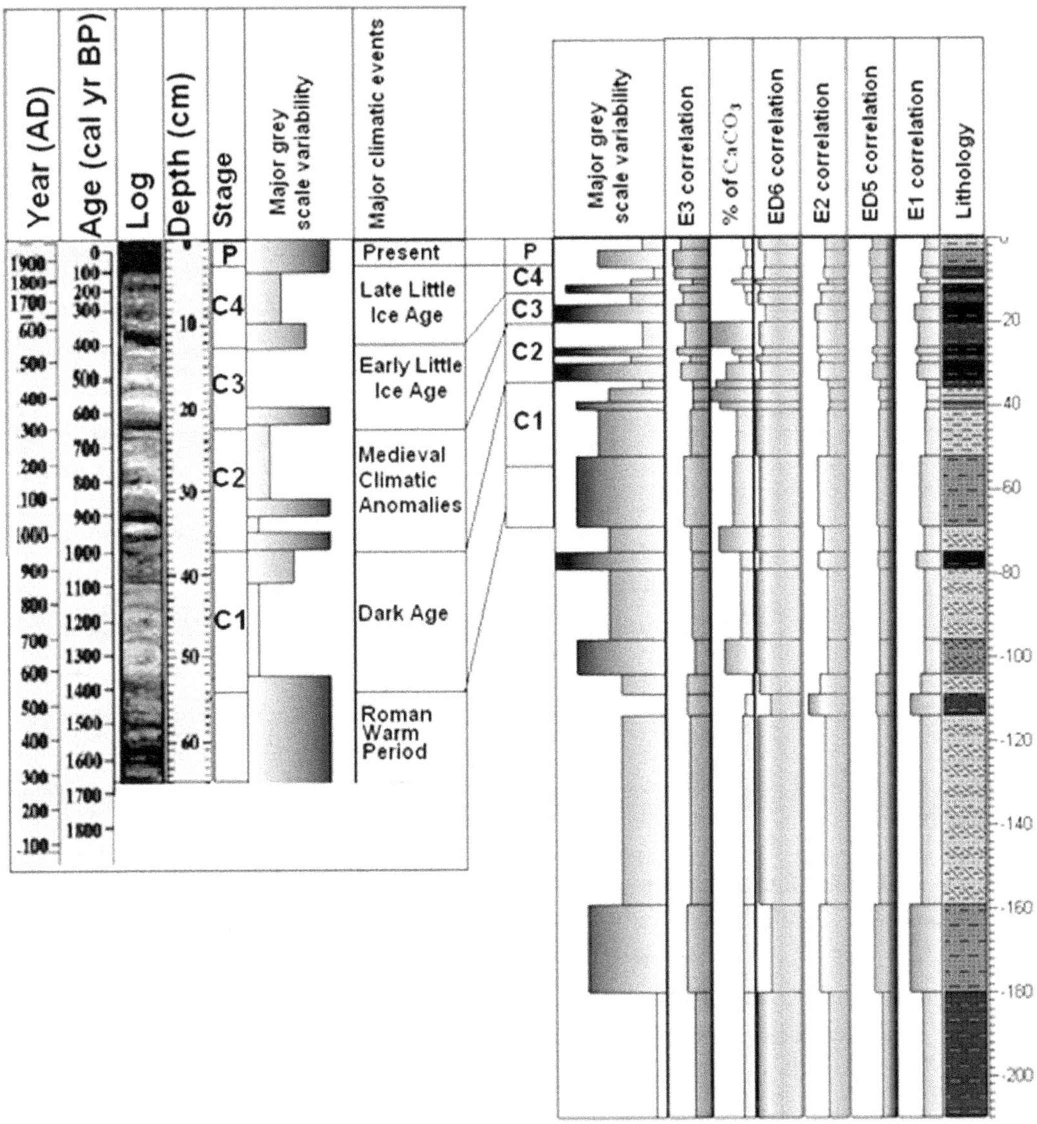

The remaining part (96cm) of core1 is made of the alternation of humid and arid periods. From 114cm to 180cm the high major grey scale indicates a stable climatic condition. To the end of the core, the darkness slightly decreases to reach minimum values in the bottom of the core (210cm). It is worth to be noted that this core is located in the outlet of Wadi Enmira where the occasional rain in the eastern part of the Sidi El Hani watershed could cause the coming of quantities of detritic sediments mixed with fresh water.

Physicochemical characteristics of sampling sites (dissolved oxygen, pH, temperature, and salinity) evolute from moderate to extreme geological conditions (Tab.1). The major aim of this part is the temporal follow-up of some physicochemical factors of surface water of the stations of S1 samplings, S2 and S3 at end to characterize this biotope supporting the development of the halophilous bacterial strains. It

Table 1. Physicochemical characteristics of sampling sites

Date	Campaign	Station	Oxygen (mg/l)	pH	Temperature (°C)	Salinity (g/l)
(24.12.2009)	C1	S1	4.4	7.3	14.7	198
		S2	3.8	7.3	15.6	199
		S3	3.75	7.3	16.4	203
(05.02.2010)	C2	S1	5.7	7.5	9	196.5
		S2	6.2	7.5	9	204.9
		S3	5.6	7.5	10.5	199.54
(25.02.2010)	C3	S1	5.7	7.5	10	204.8
		S2	6.2	7.5	10	205
		S3	5.6	7.5	11.5	210.6
(11.03.2010)	C4	S1	5.4	7.4	14	222.6
		S2	5.5	7.4	14	214.3
		S3	5.1	7.5	14	230.6
(26.03.2010)	C5	S1	4	7.4	21	221
		S2	4.8	7.3	22	225
		S3	3.8	7.4	23	223
(13.04.2010)	C6	S1	4.9	6.2	20	225
		S2	4.6	6.1	25	224
		S3	4.2	6.1	23	221

worth noting that all examined samples represent extreme ecosystems due to the very reduced presence of vegetation and high salinity. During the Winter, the salinity varies between 196.5 and 210.6 g / L and between 214.3 and 230.6 g / L during the Spring. The temperature of overlying water lies between 9 and 16.5 °C in winter. It reaches 25 °C in spring (Table 6). The lowest values are always measured on the level of the S1 station, being in the North of the sebkha. The high temperature of water supports the development of the halophilous micro-organisms. The variation of the pH shows slightly basic neutral contents, these values lie between 6.1 and 7.5.

The values of dissolved oxygen measured during the Winter season vary between 3.75 and 6.2 mg / L and between 3.8 to 5.5 mg / L during the Spring. Generally, in winter and in spring, a significant fall of the values of the dissolved oxygen was records, comparably to the Summer, which is characterized by a high content of the dissolved O_2 indicating an algal proliferation . This fall can be explained by a significant consumption of the dissolved O_2.

These fluctuations on the level of the physicochemical parameters closely depend on the weather factors which are determinant due to the low depth of water in the sebkha.

Biology

Detailed biological studies (Sellami, 2020a,b,c) has proven the presence of a well-developed microbiology marked by the presence of different species. For the identification of the isolated strains, isolated

Figure 3. Photos of the principal colonial shapes of the halophilous bacteria isolated from the surface sediment of the sebkha of Sidi El Hani

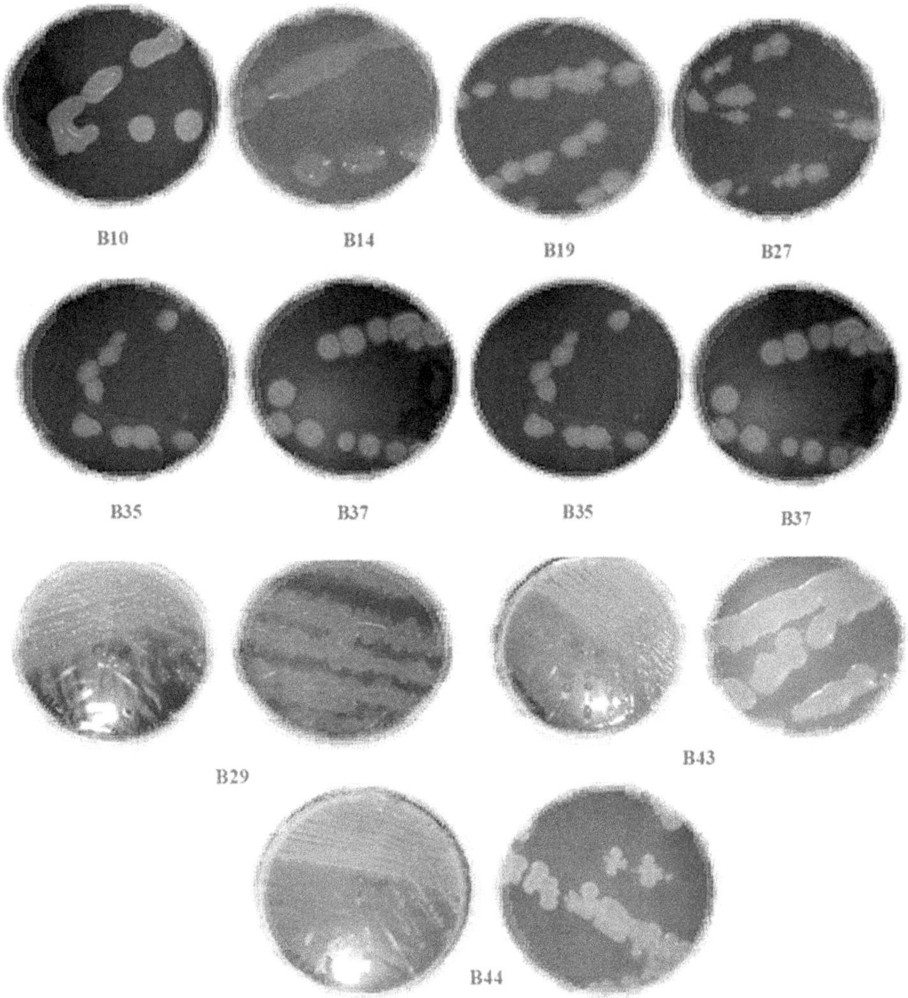

halophilous strains are characterized. In our study, we analyzed the morphology of the purified colonies and determined certain biochemical characteristics. To investigate the morphology of the isolated colonies, after enrichment of the samples in a nutritive bubble containing 34 g/L of Na Cl and incubation during 7 days at 37 °C under aerobic conditions, the bacteria develop making the bubble turbid. The successive planting of samples on the nutritive gelose containing 34 g/l Na Cl allows the isolation of the pure colonies, these last are located according to their morphological characters (form, aspect, color…). The majority of the purified colonies are round, circular of a diameter from approximately 1 to 2 mm, with smooth contours, convex, and white cream color (examples: B10; B14 and B 37). Other colonies are more or less lengthened or stretched, of smooth contours (examples: B 35, B19, and B25) or irregular (examples: B44 and B27) having a pigmentation which varies from cream white to yellow pale. Certain colonies well spreading out over the nutritive gelose (Fig.3), are sticky, creamy more or less translucent, presenting irregular (example: B28 and B29) or regular (B43) contours.

Table 2. Biochemical characteristics of isolates

Compaigns	Stations	Bacteria	Form	Gram	Oxidase	Catalase
C1	S1	B1	Bacille	+	-	+
		B2	Bacille	+	-	+
C2	S1	B3	Bacille	+	-	+
		B4	Bacille	+	-	+
		B5	Bacille	+	-	+
		B6	Bacille	+	-	+
		B7	Bacille	-	+	+
	S2	B8	Bacille	+	+	+
		B9	Bacille	+	+	+
		B10	Bacille	+	-	+
		B11	Bacille	+	+	+
		B12	Bacille	-	+	+
		B13	Bacille	-	+	+
		B14	Bacille	-	+	+
		B15	Bacille	-	+	+
		B16	Bacille	-	+	+
		B17	Bacille	+	+	+
		B18	Bacille	+	+	+
	S3	B19	Bacille	+	-	+
		B20	Bacille	+	-	+
		B21	Bacille	+	+	+
C3	S1	B23	Bacille	-	-	+
		B24	Bacille	+	+	+
		B25	Bacille	+	-	+
		B26	Bacille	+	-	+
		B27	Bacille	+	-	+
		B28	Bacille	-	+	+
		B29	Bacille	-	+	+
		B30	Bacille	-	+	+
	S3	B31	Bacille	+	+	+
		B32	Bacille	+	+	+
C4	S1	B33	Bacille	+	-	+
		B34	Bacille	+	+	+
		B35	Bacille	+	-	
	S3	B36	Bacille	+	-	+

continued on following page

Table 2. Continued

Compaigns	Stations	Bacteria	Form	Gram	Oxidase	Catalase
C5	S2	B37	Bacille	+	-	+
		B38	Bacille	-	+	+
		B39	Bacille	+	+	+
		B40	Bacille	+	-	+
	S3	B41	Bacille	+	-	+
		B42	Bacille	+	-	+
C6	S1	B43	Bacille	+	-	+
	S2	B44	Bacille	+	-	+

Identification of the Isolated Strains

A total of forty three stocks were purified. The microscopic observation, in a fresh state and after coloring of Gram, revealed that among the forty three stocks thirty-two are with Gram+ve coloring, the remainder of the stocks are of Gram-ve. All the isolates present a bacillar form. All the strains are able to produce the catalase enzyme of the degradation of the oxygenated water scour resulting from the respiratory chain. Oxidase tests revealed that among the forty-three stocks tested, twenty-two are oxidase positive whereas twenty-one are oxidase negative.

CONCLUSION

This study shows the setting of an extreme environment at Sidi El Hani wetland. Harsh conditions with increasing salinity resulted in the flourishing of bacteria. In extreme environment the climatic signal is not frank. Instead, the saline water generated a noise leading to the absence of an obvious signal. To overcome this difficulty, we used the gray scale variability to identify all climatic phases along a core from the wetland of Sidi El Hani. On the other hand, the biologic filling of the wetland, which considered as terrestrial analogues, is marked by the development of extremophiles. These extremophiles are candidate to develop in other planets such as Mars.

REFERENCES

Ben Jmaa, H. (2008). *The endorheic system of Sidi El Hani, Chrita and Mechertate: Paleoenvironment and recent dynamics* (Doctoral dissertation). University of Tunis.

Coleine, C., Biagioli, F., de Vera, J. P., Onofri, S., & Selbmann, L. (2021). Endolithic microbial composition in Helliwell Hills, a newly investigated Mars-like area in Antarctica. *Environmental Microbiology*, *23*(7), 4002–4016. doi:10.1111/1462-2920.15419 PMID:33538384

Essefi, E. (2021a). *Apport des Approches Géochimique et Sédimentologique au Fonctionnement des Systèmes Endoréiques et Exoréiques de l'Holocène à l'Anthropocène: Intérêts Environnemental et Géoéconomique*. Institut Supérieur des Sciences et Techniques des Eaux de Gabès.

Essefi, E., Komatsu, G., Fairén, A. G., Chan, M. A., & Yaich, C. (2014a). Groundwater influence on the aeolian sequence stratigraphy of the Mechertate–Chrita–Sidi El Hani system, Tunisian Sahel: Analogies to the wet–dry aeolian sequence stratigraphy at Meridiani Planum, Terby crater, and Gale crater, Mars. *Planetary and Space Science*, *95*, 56–78. doi:10.1016/j.pss.2013.05.010

Essefi, E., Komatsu, G., Fairén, A. G., Chan, M. A., & Yaich, C. (2014b). Models of formation and activity of spring mounds in the Mechertate-Chrita-Sidi El Hani system, eastern Tunisia: Implications for the habitability of Mars. *Life (Chicago, Ill.)*, *4*(3), 386–432. doi:10.3390/life4030386 PMID:25370379

Fedorova, I., Zdorovennov, R., Kadutskiy, V., Fedorov, G., Shestakova, E., Zdorovennova, G., & Nigamatzyanova, G. (2021, April). Geochemical sensitivity of lacustrine ecosystems of Yamal Peninsula (Russian Arctic) to climate change. In EGU General Assembly Conference Abstracts (pp. EGU21-15069). Academic Press.

Fernández-Martínez, M. Á., García-Villadangos, M., Paz, M. M., Gangloff, V., Carrizo, D., Blanco, Y., & Fairén, A. (2021). Geomicrobiological heterogeneity of lithic habitats in the extreme environment of Antarctic Nunataks: A potential early Mars analog. *Frontiers in Microbiology*, *12*, 1568. doi:10.3389/fmicb.2021.670982 PMID:34276605

Fluharty, R. G., Simpson, F. B., Russell, G. J., & Morris, R. H. (1967). *A Proposal for a Repetitively Pulsed Test Facility (RPTF) (No. IN-1149)*. Idaho Nuclear Corp. doi:10.2172/4557773

Freeman, D. J., Falkiner, F. R., & Keane, C. T. (1989). New method for detecting slime production by coagulase negative staphylococci. *Journal of Clinical Pathology*, *42*(8), 872–874. doi:10.1136/jcp.42.8.872 PMID:2475530

Gallagher, C. J., & Bahia, R. (2021). Outflow channels on Mars. In *Mars Geological Enigmas* (pp. 13–40). Elsevier. doi:10.1016/B978-0-12-820245-6.00002-1

Goosse, H., Dalaiden, Q., Cavitte, M. G., & Zhang, L. (2021). Can we reconstruct the formation of large open-ocean polynyas in the Southern Ocean using ice core records? *Climate of the Past*, *17*(1), 111–131. doi:10.5194/cp-17-111-2021

Kahla-Nakbi, A. B., Chaieb, K., Besbes, A., Zmantar, T., & Bakhrouf, A. (2006). Virulence and enterobacterial repetitive intergenic consensus PCR of Vibrio alginolyticus strains isolated from Tunisian cultured gilthead sea bream and sea bass outbreaks. *Veterinary Microbiology*, *117*(2-4), 321–327. doi:10.1016/j.vetmic.2006.06.012 PMID:16870360

Leduc, C., Calvez, R., Beji, R., Nazoumou, Y., Lacombe, G., & Aouadi, C. (2004). Evolution de la ressource en eau dans la vallée du Merguellil (Tunisie centrale). *Actes du Séminaire Modernisation de l'Agriculture Irriguée Rabat.*

Perl, S. M., Celestian, A. J., Cockell, C. S., Corsetti, F. A., Barge, L. M., Bottjer, D., & Melwani Daswani, M. (2021). A proposed geobiology-driven nomenclature for astrobiological in situ observations and sample analyses. *Astrobiology*, *21*(8), 954–967. doi:10.1089/ast.2020.2318 PMID:34357788

Phillips, A. A., Speth, D. R., Miller, L. G., Wang, X. T., Wu, F., Medeiros, P. M., & Sessions, A. L. (2021). *Microbial succession and dynamics in meromictic Mono Lake*. Geobiology. doi:10.1111/gbi.12437

Piddock, L. J. (1990). Techniques used for the determination of antimicrobial resistance and sensitivity in bacteria. *The Journal of Applied Bacteriology*, *68*(4), 307–318. doi:10.1111/j.1365-2672.1990.tb02880.x PMID:2190965

Savijärvi, H. I., & Harri, A. M. (2021). Water vapor adsorption on Mars. *Icarus*, *357*, 114270. doi:10.1016/j.icarus.2020.114270

Sellami, I., Charmantier, G., Naceur, H. B., Kacem, A., & Lorin-Nebel, C. (2020a). Osmoregulatory performance and immunolocalization of Na+/K+-ATPase in the branchiopod Artemia salina from the Sebkha of Sidi El Hani (Tunisia). *Tissue & Cell*, *63*, 101340. doi:10.1016/j.tice.2020.101340 PMID:32223951

Sellami, I., Naceur, H. B., & Kacem, A. (2020c). Study of Cysts Biometry and Hatching Percentage of the Brine Shrimp Artemia salina (Linnaeus, 1758) from the Sebkha of Sidi El Hani (Tunisia) According to Successive Generations. *Aquaculture Studies*, *21*(1), 41–46. doi:10.4194/2618-6381-v21_1_05

Sellami, I., Naceur, H. B., & Kacem, A. (2021b). Reproductive performance in successive generations of the brine shrimp Artemia salina (Crustacea: Anostraca) from the Sebkha of Sidi El Hani (Tunisia). *Animal Reproduction Science*, *225*, 106692. doi:10.1016/j.anireprosci.2021.106692 PMID:33450437

Van Kranendonk, M. J., Djokic, T., Baumgartner, R., Bontognali, T. R., Sugitani, K., Kiyokawa, S., & Walter, M. R. (2021). Life analog sites for Mars from early Earth: diverse habitats from the Pilbara Craton and Mount Bruce Supergroup, Western Australia. In Mars Geological Enigmas (pp. 357-403). Elsevier.

Vasudevan, P., Nair, M. K. M., Annamalai, T., & Venkitanarayanan, K. S. (2003). Phenotypic and genotypic characterization of bovine mastitis isolates of Staphylococcus aureus for biofilm formation. *Veterinary Microbiology*, *92*(1-2), 179–185. doi:10.1016/S0378-1135(02)00360-7 PMID:12488081

Visconti, G. (2021). Geobiology. In *Climate, Planetary and Evolutionary Sciences* (pp. 317–329). Springer. doi:10.1007/978-3-030-74713-8_9

Zhang, K., Li, A., Liu, X., Chen, M. T., Lu, J., Zhang, J., & Wang, H. (2021). Heavy mineral record from the east China sea inner shelf: Implications for provenance and climate changes over the past 1500 years. *Continental Shelf Research*, *226*, 104488. doi:10.1016/j.csr.2021.104488

Chapter 17
Public Participation in Restoration and Sustainable Use of Wetland Ecosystem Services in India

Pallavi Mitra
Centre for Distance and Online Education, Rabindra Bharati University, India

Anwesha Haldar
East Calcutta Girls' College, India

Priya Banerjee
https://orcid.org/0000-0001-6695-5588
Centre for Distance and Online Education, Rabindra Bharati University, India

ABSTRACT

Ecosystem services include conditions and processes that make up natural ecosystems and the species present therein. According to recent studies, wetland ecosystems provide the maximum service value per area by playing a significant role in regulating and purifying water supplies, controlling flood, acting as carbon-sinks, and sustaining human lives and livelihoods. Challenges like wetland loss and degradation, declining freshwater resources, and probable consequences of climate change have attracted significant scientific and public attention towards wetland conservation and restoration. Despite diligent conservation efforts, the global status of wetland security is still alarming. Long-term sustainable management and use of wetlands necessitate active public participation from all sectors. This study reviews the current status of different wetlands in India. It also provides a detailed discussion of different aspects of economic evaluation of ecosystem services, wetland restoration, and public participation for improving wetland policies and governance.

DOI: 10.4018/978-1-7998-9498-8.ch017

INTRODUCTION

Ecosystem services (ES) is a concept that highlights the magnanimous impact of nature on mankind. Besides referring to the benefits obtained by humans from ecology, ES also implies quality life for mankind in terms of recreation or the cultural and inspirational values of natural wonders which in turn improves cognitive thinking as well as emotional perspectives (West, 2015). The ES categories suggested by the Millennium Ecosystem Assessment have been enlisted in Table 1. ES has attracted increasing attention of the scientific community owing to its significance and relevance for practical management of different types of ecosystems (Müller & Burkhard, 2012; Salata et al., 2017). According to previous reports, wetland ecosystems present in different types of agricultural and urban landscapes yield the highest service value per area by playing significant roles in water supply and purification, control of floods and maintenance of people's livelihoods (Cimon-Morin & Poulin 2018). However, existing studies are yet to meet the increasing demand of data and robust evidence of policy-decision makers, thereby hindering the transformation of research findings to actual management practice (Martinez-Harms et al., 2015; Förster et al., 2015).

Global wetlands account for an approximate area of 7–10 million km^2, which is equivalent to 5–8% of the global land area. 23.2% of the entire global ES value (US$125 trillion yr^{-1}) is yielded by the wetland ecosystems of this world (Costanza et al., 2014; Xu et al., 2018). Wetland research is carried out in terms of hydrology, origin and development, and linkages of the same to society (Pal & Debanshi, 2021). Wetland ES has been categorized as provisioning (e.g., fresh water provision or services include products obtained from ecosystems, such as food, water, timber, fibre, or genetic resources), regulating (e.g., water purification, flood regulation, air quality regulation, climatic regulation, disease regulation, pest regulation, pollination, and natural hazard regulation), cultural (include benefits that people obtain from ecosystems related to spiritual enrichment, recreation, ecotourism, aesthetics, formal and informal education, inspiration, and cultural heritage) and supporting (include habitat for wildlife and basic ecosystem processes of nutrient cycling and primary productivity that may, in turn, lead to the other three services listed above) (de Groot et al., 2012; MEA, 2005; William and James, 2015).

Owing to its ecological and economic importance, several initiatives like the China's National Wetland Conservation Program, the Nature Conservancy, the Wetlands Conservancy, the Ramsar Convention, etc. have been implemented for conservation of the same (Wang et al., 2012; Xu et al., 2018). Policymakers are aware of the significance of healthy wetlands for achieving sustainable

Table 1. Categories of ecosystem services.

ES Categories	Summary Description
Provisioning	Physical and other resources extracted from ecosystems to support a diversity of human needs, such as food, fibre and natural medicines
Regulating	Regulatory processes within ecosystems maintaining balance, such as pollination, water purification and climate moderation
Cultural	Aspects of ecosystems providing non-material benefits enriching society, such as those supporting tourism, recreation and spiritual interests
Supporting	Processes within ecosystems that maintain overall functioning and resilience, such as soil formation, photosynthetic production of oxygen and habitat for wildlife

Reproduced with permission from Everard et al. (2019).

development goals (SDGs) and are concerned regarding their loss. However, policy-makers and practitioners are yet to identify ES as a significant component for overcoming challenges facing efficient wetland management. Unavailability of data has always been a deterrent in this case. Despite conservation efforts undertaken by environmental non-governmental organizations (NGOs) and scientists, the global biosecurity of wetlands is still alarming (Eskandari Damaneh et al., 2019). Therefore, restoration and conservation of wetlands should be prioritized in environmental policies on an urgent basis. The present chapter considers the concept of community participation in wetland management and conservation as of central importance in achieving socially and politically legitimate environmental management under conditions of environmental change, institutional complexity and stakeholder conflict.

50 YEARS OF RAMSAR– PRESENT STATUS OF WETLANDS IN INDIA

Since its inception in 1971, Ramsar Convention has been instrumental in highlighting and thereby conserving all types of natural or artificial, permanent or temporary, static or flowing water, fresh or brackish water, inland or marine wetlands, provided the depth of the concerned wetland does not exceed six meters at low tide. The convention entered into force in India on 1 February 1982. However, interests in wetlands have been more in the domain of research than in public purview. In India, wetland resources consist of a number of perennial river systems, besides reservoirs, tanks, estuaries, backwater swamp, lakes and floodplains, sloughs and ox-bow lakes. The available estimates about the areal extent of wetlands in India vary widely from a minimum of 1% to a maximum of 5% of geographical area, but

Figure 1. Expansion of settlements into urban wetlands. A-C: Urban wetlands of London, the restored Xin'an River waterfront of Huangshan City, China and the Văcărești Nature Park, Bucharest, Romania, respectively
(Ramsar Convention, 2018)

do support nearly 20% of the recorded biodiversity (Talukdar & Rajbongshi, 2018). These wetlands are distributed in different geographical regions ranging from Himalayan foothills to the islands and coastal plains. Extensive study of wetland ecosystems has become essential, especially in the urban fringe zones where the wetlands are being lost due to severe settlement pressures, as shown in Figure 1 (Nengroo et al., 2017). Even today wetlands are commonly considered as marshes that are best suited for reclamation or dumping of waste materials, which in turn degrades the water quality due to upstream developments and over-exploitation. Monitoring initiatives and documentation of the same has rarely been carried out post restoration. However, restoration activities carried out at Pashan Lake in Pune in 2005 was found to improve both the abiotic and biotic aspects of the lake (Bake & Sharma, 2019; Yardi et al., 2019). Efforts of restoration of this wetland included lake drainage, silt and debris extraction, island development for birds and planting of indigenous trees. Moreover, long term benefits in terms of chemical quality and ecological health were assessed by volunteers from neighbouring schools and communities who had been trained to assess the chosen parameters regularly as per internationally recognized and standardized procedures.

In 2020, India has added 2 more wetlands selected under Ramsar Convention taking total number of Ramsar wetlands in the country to 46 covering an area of about 1,083,322 hectares across 15 different Indian States and two Union Territories (UTs). Only one Ramsar site of India, namely the Upper Ganga River (Brijghat to Narora Stretch) is a river. In terms of size the four wetlands in East India cover 6170 sq km area which is about seven times more than the areas under 20 Ramsar sites of North India. The largest Ramsar sites are Sundarban deltaic region (4230 sq km) followed by Chilika Lake area (1165 sq km). List of wetlands designated as Ramsar Sites have been shown in Table 2.

In recent decades, the science of environmental flows has advanced greatly, with global recognition of the importance of flow regimes for maintaining wetland biodiversity, maintaining ecological integrity, and restoring or enhancing the delivery of ecosystem services. Despite their ecological and economic values, about two thirds of the world's natural wetlands have disappeared since 1900, based on existing data (Davidson, 2014). This loss is continuing, mostly for agricultural and urban development (Dixon et al., 2016). Remaining wetlands are threatened by hydrological change and nutrient enrichment, driven by population growth, economic development and climate change (Craig et al., 2017).

ECONOMIC VALUATION OF WETLAND SERVICES

A clearer focus on the basic economics of ecosystem services can help clarify their values and help us understand how to devise defensible estimates of those values. Valuation is majorly done over the natural asset—the ecosystem, and its components—that contributes to the production of something. Economic valuation (EV) of wetland ES facilitates efficient management and sustainable use of environmental resources. EV assigns quantitative values to wetland goods and ES in absence of any market price for the same. EV of wetland ES or good is generally estimated in terms of what the respondent/consumer is willing to pay when the exact cost of supplying the same is undetermined or unavailable. When an ES is available for free, *willingness to pay* determines the value of this resource irrespective of whether any actual payment is done. This questions the need for assigning a quantitative value to such resources and whether the process of evaluation is appropriate for ensuring wise use of these resources.

Table 2. Ramsar Wetlands sites of India

S. No.	Name of Site	State Location	Date of Declaration	Area (*in Sq. km.*)
1	Kolleru Lake	Andhra Pradesh	19.08.2002	901
2	Deepor Beel	Assam	19.08.2002	40
3	Kabartal Wetland	Bihar	21.07.2020	26.20
4	Nalsarovar Bird Sanctuary	Gujarat	24.09.2012	120
5	Thol Lake Wildlife Sanctuary	Gujarat	05.04.2021	6.99
6	Wadhvana Wetland	Gujarat	05.04.2021	6.30
7	Sultanpur National Park	Haryana	25.05.2021	1.425
8	Bhindawas Wildlife Sanctuary	Haryana	25.05.2021	4.12
9	Chandertal Wetland	Himachal Pradesh	08.11.2005	0.49
10	Pong Dam Lake	Himachal Pradesh	19.08.2002	156.62
11	Renuka Wetland	Himachal Pradesh	08.11.2005	0.2
12	Wular Lake	Jammu & Kashmir	23.03.1990	189
13	Hokera Wetland	Jammu & Kashmir	08.11.2005	13.75
14	Surinsar-Mansar Lakes	Jammu & Kashmir	08.11.2005	3.5
15	Tsomoriri Lake	Jammu & Kashmir	19.08.2002	120
16	Asthamudi Wetland	Kerala	19.08.2002	614
17	Sasthamkotta Lake	Kerala	19.08.2002	3.73
18	Vembanad Kol Wetland	Kerala	19.08.2002	1512.5
19	Tso Kar Wetland Complex	Ladakh	17.11.2020	95.77
20	Bhoj Wetlands	Madhya Pradesh	19.08.2002	32.01
21	Lonar Lake	Maharashtra	22.07.2020	4.27
22	Nandur Madhameshwar	Maharashtra	21.06.2019	14.37
23	Loktak Lake	Manipur	23.03.1990	266
24	Bhitarkanika Mangroves	Orissa	19.08.2002	650
25	Chilka Lake	Orissa	01.10.1981	1165
26	Beas Conservation Reserve	Punjab	26.09.2019	64.289
27	Harike Lake	Punjab	23.03.1990	41
28	Kanjli Lake	Punjab	22.01.2002	1.83
29	Keshopur-Miani Community Reserve	Punjab	26.09.2019	3.439
30	Nangal Wildlife Sanctuary	Punjab	26.09.2019	1.16
31	Ropar Lake	Punjab	22.01.2002	13.65
32	Keoladeo Ghana NP	Rajasthan	01.10.1981	28.73
33	Sambhar Lake	Rajasthan	23.03.1990	240
34	Point Calimere Wildlife and Bird Sanctuary	Tamil Nadu	19.08.2002	385
35	Rudrasagar Lake	Tripura	08.11.2005	2.4
36	Nawabganj Bird Sanctuary	Uttar Pradesh	19.09.2019	2.246

continued on following page

Table 2. Continued

S. No.	Name of Site	State Location	Date of Declaration	Area (*in Sq. km.*)
37	Parvati Agra Bird Sanctuary	Uttar Pradesh	02.12.2019	7.22
38	Saman Bird Sanctuary	Uttar Pradesh	02.12.2019	52.63
39	Samaspur Bird Sanctuary	Uttar Pradesh	03.10.2019	79.94
40	Sandi Bird Sanctuary	Uttar Pradesh	26.09.2019	30.85
41	Sarsai Nawar Jheel	Uttar Pradesh	19.09.2019	16.13
42	Sur Sarovar	Uttar Pradesh	21.08.2020	4.31
43	Upper Ganga River (Brijghat to Narora Stretch)	Uttar Pradesh	08.11.2005	265.9
44	Asan Conservation Reserve	Uttarakhand	21.07.2020	4.444
45	East Kolkata Wetlands	West Bengal	19.08.2002	125
46	Sunderbans Wetland	West Bengal	30.01.2019	4230

(ENVIS–Resource Partner on Wildlife and Protected Areas, 2021)

The Rapid Assessment of Wetland Ecosystem Services (RAWES) is another method of EV of wetland ES (McInnes & Everard, 2017). It is a rapid method developed through extensive field trials conducted in wetlands of Chad, India, Ireland and United Kingdom. This method is applicable for different types of wetlands and considers time and money resource constraints as significant parameters (McInnes & Everard, 2017). Wetland ES evaluated by the RAWES approach have been shown in Table 3.

However, EV only addresses wetland ES for human welfare (Campos et al., 2021). However, few wetland services have existential values which exceed the values of their utility for humanity. In such cases, conservation of wetlands becomes a moral obligation rather than efficient resource allocation. Political considerations may also motivate initiatives of wetland management. Hence, EV is one of the parameters and not the sole required for efficient wetland management. Efficient utilization of EV for wetland decision and policy making necessitates a framework (like total economic value; TEV) for identifying and grouping these values as required (Thapa et al., 2020). TEV consists of the components described as follows.

1. *The use values* are of the following types:
   a. *Direct use values* (DUV) including the benefits obtained from agriculture, dyes, fish, fruits, fuel wood, peat/energy, recreation, transport, vegetable oils, wildlife harvesting, etc.;
   b. *Indirect use values* (IUV) including the indirect benefits obtained from different wetland ES like external ecosystem support, flood control, groundwater recharge, micro-climatic stabilization, nutrient retention, shoreline stabilization, storm protection, etc.
   c. *Option values* (OV) obtained by an individual from ensuring that a resource will be available in future for use.
2. *The non-use values* (NUV) are current or potential values obtained as a result of ensuring continued existence of resources (like bequest, biodiversity, cultural heritage and, religious sites) and are unrelated to use. Environmentalists supporting the pure intrinsic value of nature advocate for NUVs.

Table 3. Wetland ecosystem services (ES) evaluated by the RAWES approach.

Categories of ES	Specific ES	Example	Examples of Questions Assessors Can Ask Regarding This ES
Provisioning	Provision of fresh water	Water used for domestic drinking supply, for irrigation, for livestock, etc.	• Does the wetland provide a source of fresh water? • Does the wetland store fresh water for human use? • Is the wetland a net source of pollution, degrading fresh water provision?
	Provision of food	Crops, fruit, fish, etc.	• What is grown in the wetland, either formally or from informal harvesting? • Are animals are harvested from the wetland? • Are livestock using the wetland?
	Provision of fibre	Timber for building, wool for clothing, etc.	• Are any natural materials such as wood, fibre, straw, animal fibre (wool/hide/ sinew/antler/other) taken from the wetland?
	Provision of fuel	Fuelwood, peat, etc.	• Is any material taken from the wetland and used as fuel for domestic or other uses?
	Provision of genetic resources	Rare breeds used for crop/ stock breeding, etc.	• Are any native or rare strains of plants and animals, wild and domesticated, which could contribute genetic diversity for human uses (for instance for drug manufacture, improving resilience of domestic animals and plants, horticultural trade, etc.)
	Provision of natural medicines and pharmaceuticals	Plants used as traditional medicines, etc.	• Are there any plants, animals or their parts derived from the wetland which are harvested and used for their medicinal properties?
	Provision of ornamental resources	Collection of shells, flowers, etc.	• Are there any plants, animals or their parts are derived from wetland that are collected and used/sold for their ornamental properties?
	Clay, mineral, aggregate harvesting	Sand and gravel extracted for building use, clay extracted for brick-making, etc.	• What substances are extracted or dug up from the wetland for construction or other human uses?
	Waste disposal	Dumping of solid waste, discharge of waste water, etc.	• Does the wetland provide a location for the disposal of liquid, solid or other waste materials?
	Energy harvesting from natural air and water flows	Water wheels driven by flowing water, windmills driven by the wind, etc.	• Are any technologies (water wheels, wind turbines, etc.) used to capture natural flows of energy through or across the wetland?
Regulating	Air quality regulation	Removal of airborne particles from the exhaust of cars, chimneys of industry, dust from agricultural land, etc.	• Is there a source for airborne pollutants? • Does the wetland habitat structure help to settle out airborne pollutants? • Does the state of the wetland make it a source of air pollutants (microbial, particulate or chemical)?
	Local climate regulation	Regulation of the local microclimate, through shading, reducing air temperature, etc.	• Does the wetland habitat structure provide shade for humans? • Does the wetland have areas of standing water with or without vegetation that will be generating evapotranspiration and consequently reducing air temperatures?
	Global climate regulation	Regulation of the global climate through control in greenhouse gas emissions, the sequestration of carbon, etc.	• Does the wetland store and/or sequester carbon? • Does this balance with generation of methane and other greenhouse gases?

continued on following page

Table 3. Continued

Categories of ES	Specific ES	Example	Examples of Questions Assessors Can Ask Regarding This ES
	Water regulation	Regulation of flows of surface water during high and low flows, regulation of recharge of groundwater, etc.	• Do the topography, permeability and roughness of the wetland enable it to store water during high rainfall/ discharge and top slowly release it back to surface waters or to groundwater? • Does the wetland regulate discharges during dry periods to buffer low flows during dry weather?
	Flood hazard regulation	Regulation and storage of flood water, regulation of intense rainfall events, etc.	• Does the wetland regulate, store and retain floodwaters? • Does the wetland store rainfall and surface water that might contribute to flooding and damage to property or ecosystems downstream?
	Storm hazard regulation	Regulation of tidal or storm surges, regulation of extreme winds, etc.	• Does the complexity of habitat, particularly trees, tall reeds and other vegetation and surface topography, absorb energy from extreme events such as storms and waves that might otherwise damage property or adjacent ecosystems?
	Pest regulation	Control of pest species such as mosquitoes, rats, flies, etc.	• Do natural predation and other ecological processes in the wetland regulate and control pest organisms? • Is the wetland a source of pests (for example rats thriving in dirty water systems)?
Regulating	Regulation of human diseases	Presence of species that control the species (vectors) that transmit human diseases such as malaria, West Nile fever, dengue fever, Zika virus, leptospirosis, schistosomiasis, etc.	• Do natural predation and other ecological processes in the wetland regulate organisms that may cause human diseases? • Are faecal deposits, bacteria or other potentially pathogenic microbes immobilised by processes in the wetland? • Is the condition of the wetland contributing to the negative spread of populations of disease vectors (such as mosquitoes)?
	Regulation of diseases affecting livestock	Presence of species that control the species (vectors) that transmit diseases to livestock such as leptospirosis, schistosomiasis, duck virus enteritis, highly pathogenic avian influenza, tick-borne diseases, etc.	• Do natural predation and other ecological processes in the wetland regulate organisms that may cause diseases in livestock? • Are faecal deposits, bacteria or other potentially pathogenic microbes immobilised by processes in the wetland? • Is the condition of the wetland contributing to the negative spread of populations of disease vectors (such as mosquitoes or snails)?
	Erosion regulation	Regulation of energy environment to reduce risk of erosion, presence of dense vegetation protecting soils, etc.	• Does the wetland vegetation provide protection from erosion for the soils? • Are there any signs of erosion, such as bare earth, in the wetland?
	Water purification	Cleaning of water, improvement of water quality, deposition of silts, trapping of contaminants and pollutants, etc.	• Do physico-chemical (sunlight exposure in shallow waters, detention of water in aerobic and anaerobic microhabitats) and biological processes in the wetland result in the breakdown of organic, microbial and other pollutants in the water passing though? • Are suspended solids deposited? • Is there a noticeable change in the quality, such as the turbidity, of water entering and leaving the wetland?

continued on following page

Table 3. Continued

Categories of ES	Specific ES	Example	Examples of Questions Assessors Can Ask Regarding This ES
Regulating	Pollination	Pollination of plants and crops by pollinators such as bees, butterflies, wasps, etc.	• Do populations of pollinating organisms (butterflies, wasps, bees, bats, etc.) in the wetland contribute to pollination within the wetland? • Do pollinators using the wetland also help to pollinate nearby crops, gardens, allotments, etc.?
	Salinity regulation	Freshwater in the wetland provides a barrier to saline waters.	• Does the hydrology of the wetland help prevent saline water contaminating freshwaters? • Doe the presence of freshwater in the wetland prevents the salinisation of soils?
	Fire regulation	Providing physical barriers to the spread of fire, maintaining wet conditions to prevent fires spreading, etc.	• Does the configuration of waterbodies (ditches, streams, etc.) help to prevent the spread of fires? • Is there water at or near the soil surface that restricts the spread of fire? • Are organic rich or peat soils drained and susceptible to fire and burning?
	Noise and visual buffering	Wetland trees or tall reeds absorbing and buffering the impact of noise.	• Is there a source (busy road, industry, construction, etc.) and receptor (houses, wildlife, etc.) for noise pollution? • Does wetland ecosystem structure, particularly tall trees and reeds, provide visual screening as well as suppress noise transmission?
Cultural	Cultural heritage	Importance of the wetland for historical or archaeological value, as an example of traditional uses or management practices, as a cultural landscape, etc.	• Does the wetland system have cultural importance, either due to its natural character or traditional uses?
	Recreation and tourism	Importance of the wetland for providing a location for recreation such as fishing, water sports or swimming, or as a tourism destination, etc.	• Is the wetland used for organised or informal recreational purposes? • Is there infrastructure provided for access and recreation? • Are their wider tourism/ecotourism benefits flowing from these uses?
	Aesthetic value	The wetland is overlooked by properties, is part of an of known area of natural beauty, is used as a subject for painters and artists, etc.	• Does the wetland provide aesthetic benefits through the desirability of siting houses of commercial development adjacent to it? • Does the presence of a wetland have a significant impact on property prices? • Is the wetland depicted in many works of art?
	Spiritual and religious value	The wetland holds plays a role in local religious festivals. The wetland is considered as a sacred site, the wetland forms part of a traditional belief system, etc.	• What spiritual and/or religious values do people derive from the wetland? • Does the wetland hold any important spiritual or cultural value to people? • Does the wetland play any part in traditional religious ceremonies? • Are there any traditional wetland management practices (such as the timing of planting and cropping of rice to Buddhist or other traditions and teachings) associated with the wetland?

continued on following page

Table 3. Continued

Categories of ES	Specific ES	Example	Examples of Questions Assessors Can Ask Regarding This ES
Cultural	Inspirational value	Presence of local myths or stories relating to the wetland, traditional oral or written histories about the wetland or wetland animals, creation of different art forms associated with the wetland, development of distinct architecture based on the wetland, etc.	• Are there any particular myths or other folklore associated with the wetland? • Do any wetland animals appear or are featured in local stories and myths? • Does the wetland inspire people to create music or other forms of art? • Have particularly ways of designing and building developed which reflect the wetland?
	Social relations	Presence of fishing, grazing or cropping communities which have developed within and around the wetland.	• Have communities formed around the wetland and its uses, including for example fishing (subsistence, commercial and recreational), cropping or stock management, walking and jogging, birdwatching and photography, etc?
	Educational and research	Use of the wetland by local school children for education, site of long-term research and monitoring, site visited by organised educational study tours, etc.	• Is the wetland used for any educational purposes, organised or informal, ranging from school-level visits to university research and teaching? • Are there any public awareness or educational materials present?
Supporting	Soil formation	Deposition of sediment, accumulation of organic matter, etc.	• Do accretion processes (both sedimentation of mineral material and the build-up of organic material) on the wetland result in the formation of soils?
	Primary production	Presence of primary producers such as plants, algae, etc.	• Do photosynthetic processes on the wetland produce organic matter and store energy in biochemical form?
	Nutrient cycling	Source of nutrients present from inputs from agricultural land, internal cycling of plant material, inputs of nutrients from floodwaters, presence of fauna to recycling nutrients, etc.	• Do wetland processes biochemically transform nutrients (for example nitrification/denitrification)? • Are nutrients settled out in particulate forms, changing the characteristics of water passing through the system? • Are there abundant invertebrates and detritivores that are decomposing and cycling organic material?
	Water recycling	Presence of wetland vegetation and open water result in evapotranspiration and local recycling of water, relatively closed canopies and low exposure to winds retains water in local cycles, sandy or coarse substrates allow exchange with groundwaters, etc.	• Does the structure of the wetland retain water in tight cycles (for example recapture of vapour produced by evapotranspiration)? • Does the wetland enable exchanges with groundwater (either discharge or recharge)?
	Provision of habitat	Presence of locally important habitats and species, presence of species and habitats of conservation concern, etc.	• Does the wetland support a diversity of representative biodiversity (plants and animals) locally? • Does the wetland support species which humans consider of conservation concern or as charismatic interests?

Reproduced with permission from McInnes & Everard (2017).

PROBLEMS AFFECTING PRESERVATION OF WETLANDS

Assessing the importance of wetlands in sustenance of environment and humans, the basic issues of wetlands must be identified before holistic management procedures can be undertaken (Mahyuni & Syahrin, 2021). The term 'wetland loss' deals with issues like agricultural conversion, direct deforestation, hydrologic alteration, inundation, defoliation, altered upper watersheds, accumulative water demands, water quality degradation, wetland consolidation, global climate change, ground-water depletion, exotic species and biodiversity loss (Day & Rybczyk, 2019; Foote et al. 1996; Werners et al., 2015). One of the significant reasons for granting a Ramsar recognition is to halt the worldwide loss of wetlands and to conserve, through wise use and management as these ecosystems are not only unique but are most congenial for migratory birds and endangered animal species. Yet, owing to increasing pollution, siltation, disturbing human activities, avian numbers in most Ramsar sites in India are sharply declining. Some species, such as cranes, have declined by up to 47 per cent from population levels recorded in the 1980s. In the Asia-Pacific region, the decline of water bird populations has been particularly severe (Delany, 2010). It is observed that most of these wetlands are in close proximity and are in direct connected to the river channels. As a result, the riparian and aquatic ecosystem has a strong interrelationship with the wetland ecosystem. The construction of dams, barrages, canals without credible impact assessments, options assessment or democratic decision making for the dependent communities and their faulty operation have been detrimental for the ecological flows of the wetlands connected to these reservoirs in many Ramsar sites in India. The booming population pressures on land and resources coupled with expansion of urban territories. According to MoEF records, 34,000 ha. of the water spread area of the Kolleru lake (Andhra Pradesh) have been reclaimed for agriculture and between 1973 and 2007, Greater Bengaluru Region lost 66 wetlands with a water spread area of around 1100 ha due to urban sprawl (Ramachandra & Kumar, 2008).

Moreover, changes in water management policy and practice in India towards supply-based water policies that have increased water insecurity also pose a long-term systemic threat to wetlands: such as facilitating deep-pumped well drilling, inter-basin water transfer, and dam construction. According to the 'Wire' report, the introduction of Loktak Lake (Protection) Act, 2006 has affected the livelihoods of thousands of fishermen since they were prohibited them from fishing in the core area of the lake. Even after continued protests, in 2011 most of the fishermen were removed from the lake by the Loktak Development Authority (LDA), after a report by Wetlands International claimed detrimental fishing practices as a cause behind degradation of the lake. Challenges faced by wetlands have been discussed as follows.

Urban Encroachments

The progressions and rapidity of wetland losses in urban spaces vary greatly among countries, particularly in South Asian cities, where they result in urban floods, water scarcity, and livelihood losses (Mondal et al., 2017). In many megacities in lower income countries in the Global South, where social inequalities and weak governance aggravate the consequences of environmental pressure, the patterns and trends of urbanization are significantly different from those in post-industrial cities in industrialized countries (Nagendra et al. 2018), and wetland degradation is more complex. This is particularly true for large urban agglomerations in India, where population growth and city expansion are exceptionally high (Grimm et al., 2008; Taubenböck et al., 2009) due to rapid industrial and economic development, leading to the high extinction rate of natural habitats, vegetated areas, and wetlands (Dasgupta, 2007). Wular Lake was once spread across 20,200 hectares, now remains restricted to 2,400 hectares. A survey conducted by South Asian Forum for Environment (SAFE) over a period of eight months showed a 46% reduction in the water bodies of Kolkata since 2006 (Dasgupta & Salam, 2018). From 3,874 ponds, lakes and canals 10 years ago, the number has dwindled to 1,670 (Dasgupta & Salam, 2018).

Increased Pollutants

According to the reports of South Asia Network on Dams, Rivers and People (SANDRP, 2017) Indian wetlands are severely compromised due to different anthropogenic activities. Gujarat presently has the maximum number of wetlands along its coastline accounting for 22.7% of the total wetland area of this country (Suthar et al., 2019). The Khajadiya Bird Sanctuary located in Gujarat have been converted to a picnic spot. Wetlands of Uttar Pradesh inhabited by the Sarus (State Bird of Uttar Pradesh) are threatened by increasing real estate encroachments (Choudhury et al., 2021). A drastic reduction in the number of wetlands has also been reported by this study. Similar encroachment has also been observed in Haryana (Solanki & Joshi, 2017). Unplanned construction and deforestation activities carried out in these wetlands in the name of beautification has also affected natural nesting sites of migratory birds. In Maharashtra, rampant dumping of solid waste has caused major degradation of wetlands (Chouhan et al., 2018). In Kolkata, units of plastic recycling and leather tanneries have illegally mushroomed on wetlands banks, polluting the same (Banerjee & Dey, 2017; Dasgupta & Salam, 2018; Vicziany et al., 2017). Deepor Beel, a freshwater lake designated as the first Ramsar Site of Assam, has been choked by the dumping of garbage generated by the entire Kamrup city located on its edge (Sur & Mandal, 2019). Moreover, illegal and unplanned construction has disrupted the lake's link with the river Brahmaputra. Untreated sewage dumped in this lake has also led to its eutrophication. Absence of sewage network has also led to disposal of 42% of sewage generated in Bengaluru city into its lakes (Ramachandra et al., 2017). A large number of developmental activities of Kerala, like construction of the Kochi metro, the LNG Terminal (Puthu Vypeen) and the NH bypass etc. has been carried out on wetlands (Rawat & Yadav, 2021; Roshnath & Sashikuma, 2019). According to them, wetland boundaries have not been scientifically demarcated. These has led to disappearance of significant aquatic species from the wetlands of Kerala. Many of the state governments, though aware of the situation, have failed to form Wetland Conservation Committees mandated by the National Wetland Conservation Programme (NWCP). Inspite of rapid decline in wetland health, states have not undertaken significant efforts for conservation of the same (Ragavan et al., 2020). Environmentalists have also provided repeated warnings against reclamation of wetlands and destruction of mangroves due to the irreplaceable ESs offered by them.

WETLAND MANAGEMENT – CONSERVATION AND RESTORATION POLICIES

The case for preventing further wetland loss and degradation has been made for decades. The signing of the Ramsar Convention on Wetlands signalled the development of a global commitment for maintaining the ecological character of all wetlands. However, the Millennium Ecosystem Assessment (2005) and more recent analyses (Finlayson et al., 2019; Gardner et al. 2015) have shown that this commitment has not been fulfilled. The reasons are many and complex, including the perception that conserving wetlands conflicts with efforts to improve the immediate, short-term personal and economic wellbeing of people.

Given that the Convention has existed for about five decades, the failure to generate major public and government support signals that an alternative approach is needed. National governments have been unable to address and solve wetland issues of this country and on-site (Finlayson et al., 2019). Decision makers should consider inputs of researchers and practitioners for forming policies that will define the future of wetlands for meeting short and mid-term climate goals and long-term biosphere requirements in a sustainable manner (Moomaw et al., 2018).

These efforts are largely targeted to elevate wetland issues in the international considerations of climate change. Scientists should vividly portray the important role of wetlands in the welfare of society and economy to policy makers and managers. Proper communication with the managers, decision makers and public necessitate the aligning of public mindset and concern with wetland science and aspects of ESs offered by wetlands. The initiatives required for sustainable management of wetlands are given as follows (Finlayson et al., 2019; Ripple et al., 2017):

1. Conversion of a significant proportion of the world's wetlands to interconnected, well-managed and well-funded networks of protected areas should be prioritised;
2. Wetland ESs should be restored and maintained by checking the conversion of wetlands to other forms of land use;
3. Large scale restoration of indigenous plant communities of wetlands;
4. Ecological processes and dynamics of wetlands should be rejuvenated with indigenous species (including top-level predators);
5. Development and adoption of sufficient policies for restoring the wetland animals lost/ threatened by exploitation, over-fishing, poaching and trade;
6. Development of better education and infrastructure for reducing the wastage of food derived from wetlands;
7. Promotion of alternative food sources for cattle for preventing overgrazing on wetlands;
8. Enhancement of wetland education for both children and adults;
9. Greater involvement of local and indigenous communities in wetland monitoring and management;
10. Development of ecologically-sound financial investments in place of ecologically destructive investments for ensuring positive environmental change in wetlands;
11. Development and promotion of eco-friendly technologies and clean energy that does not exert any detrimental impacts on the environment;
12. Reduction of economic inequality and promotion of prices, taxation, and incentive systems that consider the real costs incurred by patterns of wetland ES consumption.

PUBLIC PARTICIPATION IN WETLAND RESTORATION

Participatory management is different from conventional management practice. It considers the interests, synergistic effects, level of participation, and level of vulnerability of all stakeholders for sustainable use of natural resources and empowerment of local institutional capacity (Devarani & Basu, 2009). Since its formulation in the Rio Declaration of 1992 (Principle 10), public participation has become an integral component of how environmental law should work. The strength of international environmental laws to restore wetlands depends upon the extent of public participation. In the context of protected areas, public participation was once a 'subsidiary issue' but has now moved to 'the centre stage'. Whilst there have been several studies that have looked at the Ramsar Convention in India, there have been relatively few which have evaluated the legal aspects of the principle as it applies (or ought to apply) under Ramsar. Water bodies have always been the means of livelihood for the local community, more so the ethnic culture that they sustain. However, a gradual shift of their interests from the consumptive to the non-consumptive practices have been observed (Lamsal et al., 2015). Public participation in ecosystem restoration has been shown in Figure 2.

There has been increasing interest in recent years in community-based management of natural resources or where the government is responsible for management and encouraging public participation. Community involvement can thus range from autonomous management by the resource community or area to some form of shared responsibility in collaboration with the development programmes or projects of the state. *Participatory development* generally implies developmental activities carried out through a participatory approach. This approach promotes public participation as a significant parameter for ensuring the success of the concerned activity. Participatory approach has reportedly achieved sustainable management of wetlands and tangible benefits for the local communities participating in the same. Consideration of the cultural values of wetland resources may facilitate a better understanding of public attitudes (support/opposition) and expectations thereby enhancing their participation in wetland restoration. Moreover, the RAWES approach for EV of wetland ES has further encouraged public participation by recognizing the needs and sentiments of local communities (McInnes & Everard, 2017). This approach trains the local people to identify field indicators for assessing several wetland ESs. The simple results obtained through this process also facilitates decision makers to interlink ES with management policies.

Figure 2. Public participation in wetland restoration. Figures A and B show restoration of mangrove forests along the Tallo River Basin.
Reproduced with permission from Surya et al. (2020)

Gender and Ecological Restoration

Restoration efforts are not gender neutral in nature. However, for achieving success, these efforts require the participation of both men and women at different levels. Restoration activities should be screened for a proper understanding of probable gender-differentiated impacts. Involvement of both genders may be ensured through capacity building, promoting participation and using quotas. Gender dimension of environmental issues have been evident in case of climate change, loss of biodiversity, energy and water security, desertification, etc. (MacGregor, 2017). Proper understanding of the crucial role played by women as agents of change requires the consideration of multiple issues of gender inequality. The success of the joint forest management (JFM) initiatives practiced in different Indian states have highlighted the significance of participation of both genders in restoration efforts (Rai et al., 2018). Restoration activities of the Attappady Wasteland Comprehensive Environmental Conservation Project were carried out with a participatory approach where women were involved in project preparation and implementation through use of tribal hamlets, quotas and JFM committees (Thomas et al., 2021). Results obtained in both initiatives established the need for both genders to perform their roles in restoration activities.

Participation of Tribal and Rural Communities

Rural people inhabiting the watershed of the wetlands rely upon wetland ES. Hence, they are the actual 'managers' of that ecosystem (Ghoochani et al., 2020). They play a significant role in conservation efforts due to their familiarity with the environment gained through continuous interactions, recreation and practices of resource use. Participation of rural communities is specifically significant in actions of local development as well as stages, scales and procedures for achieving restoration goals. However, larger focus on restoration requires farmers to reduce their croplands for reaching conservation goals. Alignment of wetland conservation and agricultural practices is essential for achieving socially and politically legitimate outcomes of environmental management even under detrimental scenarios of environmental change, complexity of institutions and conflicting interests of stakeholders (Cotton & Mahroos-Alsaiari, 2015; Eskandari-Damaneh et al., 2020; Millar et al., 2019). However, lack of participation of these people in the management and resource supervision often results in failed attempts of environmental management. Alienation of these rural communities from processes of decision-making and improper implementation of sustainable agricultural practices also leads to the failure of management endeavours. Therefore, proper understanding of broad parameters influencing rural participation and specific ones guiding wetland management is crucial for achieving both wetland conservation and sustainable livelihoods (Savari et al., 2020). Similar participatory approach carried out in the Sunderbans Biosphere Reserve of West Bengal, India had resulted in the overall improvement of wetland biodiversity and promoted wise use of ESs (D'Souza et al., 2017). Participatory management of East Kolkata Wetlands (West Bengal, India) also revealed similar improvements and wise use of wetland resources (Chakraborty & Halder, 2019; Kundu & Chakraborty, 2017; Roy-Basu et al., 2020).

Role of Civil Societies

Public participation in conversation practices may decline if citizens are unaware of the details of the project, especially local inhabitants and regular visitors. Conservation projects should prioritize ES of greater value to these participants like water purification, flood protection, etc. Landscape changes, removal of landmarks and project associated restrictions imposed on resource usage may also discourage public participation in conservation projects. Surveys conducted with open-ended questions are more appropriate for recording the perspectives of these residents regarding values and cultural aspects of resources. These perspectives often remain unacknowledged in studies carried out with predefined categories of ecosystem values and services (Ruoso et al. 2015). Members of the civil society should be included in processes of decision-making for achieving restoration goals. Moreover, tourism activities encourage further involvement of society in wetland conservation. Hence, recreational activities should be carried out in parallel to conservation efforts though they are commonly considered to be in conflict with each other (Chakraborty et al., 2020).

Positive Outcomes Recorded in India

Positive outcomes of public participation in restoration activities (shown in Figure 3) have been compiled by the South Asia Network on Dams, Rivers and People (SANDRP, 2020). As per their reports, individual initiatives have been undertaken for cleaning of lakes and water bodies from Chennai, greater Noida, and Udaipur. Citizens and local communities of Maharashtra, Goa and Kerala have also succeeded in wetland restoration efforts. New wetlands have been designated as protected areas by the state governments of Punjab and Uttar Pradesh. State Wetlands Authorities have been constituted by the states of Delhi, Rajasthan, Telangana and Uttarakhand. Moreover, wetland committees of Maharashtra and Delhi have reportedly carried out impressive wetland restoration activities. In East Kolkata Wetlands (EKW),

Figure 3. Improvement in ecological health of Indian wetlands achieved through public participation in restoration activities
(SANDRP, 2020)

the state secretariat has made it imperative for the Housing Infrastructure Development Corporation (HIDCO) to engage a consulting agency for developing an Environment Impact Assessment report as well as an Environment Management Plan for minimizing the impact of developmental activities on the wetland. As per the verdict of the National Green Tribunal, the city administration has ordered for the demolition of all illegal encroachments in and around the EKW since 2012. Similar initiatives have been reported from other states as well.

FACTORS EFFECTING LONG TERM SUSTAINABILITY OF RESTORATION EFFORTS

It is not always that application of ES to gainful practices is a boon to the ecology. There are a number of direct and indirect drivers that negatively impacting the availability of ESs, thus impacting the livelihoods of the people. Conservation of the same includes several complex challenges posed by one or more of ecological, economic, social and/or cultural issues (Root-Bernstein & Frascaroli, 2016). Over-exploitation leading to resource depletion and the weak enactment of laws and the lack of a management plan for the area are important factors contributing to the wetland's degradation (Bhatta et al., 2016). Factors exerting positive long-term impacts on sustainability of restoration efforts include ecological, aesthetic, recreational ones along with those specifically considered/rejected for the project (Pueyo-Ros et al., 2019). Factors exerting negative long-term impacts on are accessibility, recreational, technical and economic in nature (Pueyo-Ros et al., 2019).

LINKING ECOSYSTEM SERVICES TO PUBLIC POLICY

ES have been a popular theme in conservation policy. It is widely known that preserving or restoring areas of natural habitat, boosts the ecosystem service of procuring clean air and water, flood control, and crop pollination will be provided to society. ESs and goods if properly evaluated, may even yield sufficient economy to meet the cost of conserving the ecosystems that produce them. Any ES that has not been considered by 'the market' is now given due importance in public and private decision-making. Recognizing the importance of ES is expected to encourage public participation in restoration efforts and identify areas of research for government and private sectors.

Current Practice and Future Directions Towards Wetland Governance

Wetlands and wildlife dependant on the same are globally protected using legal agreements like the Ramsar Convention, the Bonn Convention (on Migratory Species) and the Convention on Biological Diversity. EV of wetland ES has been efficiently determined using the *willingness-to-pay* and *willingness to accept* approaches. Valuable parameters considered in these approaches include acreage, age, benefits, cultural values, distance between the wetland and the settlement, economic incentives, education, household size, location, etc. Despite several initiatives undertaken by the scientific community and non-governmental organizations all over the world, biosecurity of the same is in an alarming condition (Eskandari Damaneh et al., 2019).

Almost half of the global wetlands have been lost due to anthropogenic activities. Wetlands are threatened by competing land uses of the agricultural, industrial and waste disposal sectors. Decision makers and management officials are simultaneously focusing on controlled use and conservation of water and increasing the availability of the same for public use (Ghoochani et al., 2017, 2015). However, excessive water consumption for household activities, commercial manufacturing, construction of dams and energy production has also exerted detrimental impacts on the wetlands (Fung et al., 2019; Pazvakawambwa, 2018; Sok et al., 2019; Zou et al., 2018). Dams can cause wetlands to dry up (Kingsford, 2000). Construction of dams alter land morphology and use, (Zhao et al., 2010) and result in large scale migration of population. It also causes sediment build-up, changes in water quality, and destruction of wildlife and forests downstream (Benchimol & Peres, 2015; Kondolf et al., 2014; Yang et al., 2005). Therefore, in recent attempts of policy making for integrated management of wetlands, larger focus is devoted to environmental and socio-economic consequences of integrated water management through construction of dams (Del Bene et al., 2018; Miles & Ebrey, 2017; Newell et al., 2019; Tahmiscioğlu et al., 2007; Yang et al., 2011).

CONCLUSION

Despite several global initiatives, condition of global wetlands is on the decline. Participation of indigenous rural/ tribal communities in all stages of wetland management is being considered mandatory for ensuring successful restoration and sustainable use of these wetlands. For ensuring the success of all efforts of conservation, ES should be considered as a significant aspect of policy making. EV of ES has led to a better understanding of its importance. However, more research is required in this regard to understand the attitude and sentiment of people towards wetland ES and encourage their participation in conservation and sustainable use of wetlands.

REFERENCES

Aryal, K., Ojha, B. R., & Maraseni, T. (2021). Perceived importance and economic valuation of ecosystem services in Ghodaghodi wetland of Nepal. *Land Use Policy*, *106*, 105450. doi:10.1016/j.landusepol.2021.105450

Bake, H., & Sharma, R. (2019). Estimating Visitors' Willingness to Pay for Lake Conservation: A Contingent Valuation Study of Pashan Lake, Pune, India. In *Environmental Impacts of Tourism in Developing Nations* (pp. 234–251). IGI Global. doi:10.4018/978-1-5225-5843-9.ch013

Banerjee, S., & Dey, D. (2017). Eco-system complementarities and urban encroachment: A SWOT analysis of the East Kolkata Wetlands, India. *Cities and the Environment*, *10*(1), 2.

Benchimol, M., & Peres, C. A. (2015). Widespread forest vertebrate extinctions induced by a mega hydroelectric dam in lowland Amazonia. *PLoS One*, *10*(7), e0129818. doi:10.1371/journal.pone.0129818 PMID:26132139

Campos, F. S., David, J., Lourenço-de-Moraes, R., Rodrigues, P., Silva, B., da Silva, C. V., & Cabral, P. (2021). The economic and ecological benefits of saving ecosystems to protect services. *Journal of Cleaner Production*, *311*, 127551. doi:10.1016/j.jclepro.2021.127551

Chakraborty, A., Halder, S., Nazneen, S., & Dey, S. K. (2019). East Kolkata Wetlands (EKW), India: A Unique Example of Resource Recovery. In Wastewater Reuse and Watershed Management (pp. 25-40). Apple Academic Press.

Chakraborty, S., Saha, S. K., & Selim, S. A. (2020). Recreational services in tourism dominated coastal ecosystems: Bringing the non-economic values into focus. *Journal of Outdoor Recreation and Tourism*, *30*, 100279. doi:10.1016/j.jort.2020.100279

Choudhury, M., Sharma, A., Singh, P., & Kumar, D. (2021). Impact of climate change on wetlands, concerning Son Beel, the largest wetland of North East, India. In *Global Climate Change* (pp. 393–414). Elsevier. doi:10.1016/B978-0-12-822928-6.00006-X

Chouhan, H. A., Parthasarathy, D., & Pattanaik, S. (2018). Urban at the edges: Mumbai's coastline urbanisms. In *Sustainable urbanization in India* (pp. 279–293). Springer. doi:10.1007/978-981-10-4932-3_15

Cimon-Morin, J., & Poulin, M. (2018). Setting conservation priorities in cities: Approaches, targets and planning units adapted to wetland biodiversity and ecosystem services. *Landscape Ecology*, *33*(11), 1975–1995. doi:10.100710980-018-0707-z

Costanza, R., De Groot, R., Sutton, P., Van der Ploeg, S., Anderson, S. J., Kubiszewski, I., & Turner, R. K. (2014). Changes in the global value of ecosystem services. *Global Environmental Change*, *26*, 152–158. doi:10.1016/j.gloenvcha.2014.04.002

Cotton, M. D., & Mahroos-Alsaiari, A. A. (2015). Key actor perspectives on stakeholder engagement in Omani Environmental Impact Assessment: An application of Q-Methodology. *Journal of Environmental Planning and Management*, *58*(1), 91–112. doi:10.1080/09640568.2013.847822

Craig, L. S., Olden, J. D., Arthington, A. H., Entrekin, S., Hawkins, C. P., Kelly, J. J., & Groffman, P. M. (2017). Meeting the challenge of interacting threats in freshwater ecosystems: A call to scientists and managers. *Elementa*, *5*, 5. doi:10.1525/elementa.256

D'Souza, N. M., Ishwar, N. M., Sumra, I., & Vyas, P. (2017). Participatory wetland management: a solution to conservation challenges in the sundarbans biosphere reserve. In *Wetland science* (pp. 575–587). Springer. doi:10.1007/978-81-322-3715-0_30

Dasgupta, A., & Salam, A. (2018). A City Reclaiming Assurance on Open Vulnerable Areas–A Dossier on Wetlands and Canals of Kolkata. *International Conference on Urban Sustainability: Emerging Trends, Themes, Concepts & Practices (ICUS)*. 10.2139srn.3236659

Davidson, N. C. (2014). How much wetland has the world lost? Long-term and recent trends in global wetland area. *Marine and Freshwater Research*, *65*(10), 934–941. doi:10.1071/MF14173

Day, J. W., & Rybczyk, J. M. (2019). Global change impacts on the future of coastal systems: perverse interactions among climate change, ecosystem degradation, energy scarcity, and population. In *Coasts and Estuaries* (pp. 621–639). Elsevier. doi:10.1016/B978-0-12-814003-1.00036-8 de Groot, R., Brander, L., Van Der Ploeg, S., Costanza, R., Bernard, F., Braat, L., & Van Beukering, P. (2012). Global estimates of the value of ecosystems and their services in monetary units. *Ecosystem Services*, *1*(1), 50–61. doi:10.1016/j.ecoser.2012.07.005

Del Bene, D., Scheidel, A., & Temper, L. (2018). More dams, more violence? A global analysis on resistances and repression around conflictive dams through co-produced knowledge. *Sustainability Science*, *13*(3), 617–633. doi:10.100711625-018-0558-1

Devarani, L., & Basu, D. (2009). Participatory wetland management in Loktak lake: A road to sustainable development. *Journal of Crop and Weed*, *5*(1), 178–190.

Dixon, M. J. R., Loh, J., Davidson, N. C., Beltrame, C., Freeman, R., & Walpole, M. (2016). Tracking global change in ecosystem area: The Wetland Extent Trends index. *Biological Conservation*, *193*, 27–35. doi:10.1016/j.biocon.2015.10.023

Ehrlich, P. R., & Mooney, H. A. (1983). Extinction, substitution, and ecosystem services. *Bioscience*, *33*(4), 248–254. doi:10.2307/1309037

ENVIS Centre on Wildlife & Protected Areas. (2021). *RAMSAR Wetland Sites*. http://www.wiienvis.nic.in/Database/ramsar_wetland_sites_8224.aspx

Eskandari Damaneh, H., Eskandari Damaneh, H., Khosravi, H., & Gholami, H. (2019). Analysis and monitoring of drought using NDVI index (Case study: The west basin of Jaz Murian wetland). *Rangeland*, *13*(3), 461–475.

Eskandari-Damaneh, H., Noroozi, H., Ghoochani, O. M., Taheri-Reykandeh, E., & Cotton, M. (2020). Evaluating rural participation in wetland management: A contingent valuation analysis of the set-aside policy in Iran. *The Science of the Total Environment*, *747*, 141127. doi:10.1016/j.scitotenv.2020.141127 PMID:32777492

Everard, M., Kangabam, R., Tiwari, M. K., McInnes, R., Kumar, R., Talukdar, G. H., & Das, L. (2019). Ecosystem service assessment of selected wetlands of Kolkata and the Indian Gangetic Delta: Multi-beneficial systems under differentiated management stress. *Wetlands Ecology and Management*, *27*(2), 405–426. doi:10.100711273-019-09668-1

Finlayson, C. M. (2012). Forty years of wetland conservation and wise use. *Aquatic Conservation*, *22*(2), 139–143. doi:10.1002/aqc.2233

Finlayson, C. M., Davies, G. T., Moomaw, W. R., Chmura, G. L., Natali, S. M., Perry, J. E., & Sutton-Grier, A. E. (2019). The second warning to humanity–providing a context for wetland management and policy. *Wetlands*, *39*(1), 1–5. doi:10.100713157-018-1064-z

Foote, A. L., Pandey, S., & Krogman, N. T. (1996). Processes of wetland loss in India. *Environmental Conservation*, *23*(1), 45–54. doi:10.1017/S0376892900038248

Förster, J., Barkmann, J., Fricke, R., Hotes, S., Kleyer, M., Kobbe, S., & Wittmer, H. (2015). Assessing ecosystem services for informing land-use decisions: A problem-oriented approach. *Ecology and Society*, *20*(3), 31. doi:10.5751/ES-07804-200331

Fung, Z., Pomun, T., Charles, K. J., & Kirchherr, J. (2019). Mapping the social impacts of small dams: The case of Thailand's Ing River basin. *Ambio*, *48*(2), 180–191. doi:10.100713280-018-1062-7 PMID:29797277

Gardner, R. C., Barchiesi, S., Beltrame, C., Finlayson, C., Galewski, T., Harrison, I., & Walpole, M. (2015). State of the world's wetlands and their services to people: a compilation of recent analyses. Ramsar Convention Secretariat, Ramsar scientific and technical briefing note no. 7, Gland, Switzerland.

Ghoochani, O. M., Ghanian, M., Khosravipour, B., & Crotts, J. C. (2020). Sustainable tourism development performance in the wetland areas: A proposed composite index. *Tourism Review*. Advance online publication. doi:10.1108/TR-02-2019-0061

Kondolf, G. M., Rubin, Z. K., & Minear, J. T. (2014). Dams on the Mekong: Cumulative sediment starvation. *Water Resources Research*, *50*(6), 5158–5169. doi:10.1002/2013WR014651

Kumar, M., Mukherjee, N., Sharma, G. P., & Raghubanshi, A. S. (2010). Land use patterns and urbanization in the holy city of Varanasi, India: A scenario. *Environmental Monitoring and Assessment*, *167*(1), 417–422. doi:10.100710661-009-1060-0 PMID:19562495

Kundu, N., & Chakraborty, A. (2017). Dependence on ecosystem goods and services: a case study on East Kolkata Wetlands, West Bengal, India. In *Wetland Science* (pp. 381–405). Springer. doi:10.1007/978-81-322-3715-0_20

Lamsal, P., Pant, K. P., Kumar, L., & Atreya, K. (2015). Sustainable livelihoods through conservation of wetland resources: A case of economic benefits from Ghodaghodi Lake, western Nepal. *Ecology and Society*, *20*(1), art10. doi:10.5751/ES-07172-200110

MacGregor, S. (Ed.). (2017). *Routledge handbook of gender and environment*. Taylor & Francis. doi:10.4324/9781315886572

Mahyuni, M., & Syahrin, M. N. (2021). Sister City Cooperation Model In Wetland Environmental Governance: Case Study In Banjarmasin, South of Kalimantan. *International Journal of Politic. Public Policy and Environmental Issues*, *1*(02), 115–125.

Martinez-Harms, M. J., Bryan, B. A., Balvanera, P., Law, E. A., Rhodes, J. R., Possingham, H. P., & Wilson, K. A. (2015). Making decisions for managing ecosystem services. *Biological Conservation*, *184*, 229–238. doi:10.1016/j.biocon.2015.01.024

McInnes, R. J., & Everard, M. (2017). Rapid assessment of wetland ecosystem services (RAWES): An example from Colombo, Sri Lanka. *Ecosystem Services*, *25*, 89–105. doi:10.1016/j.ecoser.2017.03.024

Miles, A., & Ebrey, J. (2017). The village in the city: Participation and cultural value on the urban periphery. *Cultural Trends*, *26*(1), 58–69. doi:10.1080/09548963.2017.1274360

Millar, J., Robinson, W., Baumgartner, L., Homsombath, K., Chittavong, M., Phommavong, T., & Singhanouvong, D. (2019). Local perceptions of changes in the use and management of floodplain fisheries commons: The case of Pak Peung wetland in Lao PDR. *Environment, Development and Sustainability*, *21*(4), 1835–1852. doi:10.100710668-018-0105-3

Mondal, B., Dolui, G., Pramanik, M., Maity, S., Biswas, S. S., & Pal, R. (2017). Urban expansion and wetland shrinkage estimation using a GIS-based model in the East Kolkata Wetland, India. *Ecological Indicators*, *83*, 62–73. doi:10.1016/j.ecolind.2017.07.037

Moomaw, W. R., Chmura, G. L., Davies, G. T., Finlayson, C. M., Middleton, B. A., Perry, J. E., & Sutton-Grier, A. E. (2018). The relationship between wetlands and a changing climate. *Wetlands*, *38*, 183–205. doi:10.100713157-018-1023-8

Müller, F., & Burkhard, B. (2012). The indicator side of ecosystem services. *Ecosystem Services*, *1*(1), 26–30. doi:10.1016/j.ecoser.2012.06.001

Nagendra, H., Bai, X., Brondizio, E. S., & Lwasa, S. (2018). The urban south and the predicament of global sustainability. *Nature Sustainability*, *1*(7), 341–349. doi:10.103841893-018-0101-5

Nengroo, Z. A., Shah, A. H., & Bhat, M. S. (2017). Dynamics of land use change in rural-urban fringe: A case study of Srinagar city. *Environmental Science: An Indian Journal*, *13*(4), 142.

Newell, J. P., Goldstein, B., & Foster, A. (2019). A 40-year review of food–energy–water nexus literature and its application to the urban scale. *Environmental Research Letters*, *14*(7), 073003. doi:10.1088/1748-9326/ab0767

Pal, S., & Debanshi, S. (2021). Developing wetland landscape insecurity and hydrological security models and measuring their spatial linkages. *Ecological Informatics*, *66*, 101461. doi:10.1016/j.ecoinf.2021.101461

Pazvakawambwa, G. T. (2018). *Water resources governance in the upper Swakop basin of Namibia* (Doctoral dissertation). University of Namibia.

Pueyo-Ros, J., Ribas, A., & Fraguell, R. M. (2019). A cultural approach to wetlands restoration to assess its public acceptance. *Restoration Ecology*, *27*(3), 626–637. doi:10.1111/rec.12896

Ragavan, P., Kathiresan, K., Mohan, P. M., Ravichandran, K., Jayaraj, R. S. C., & Rana, T. S. (2020). Ensuring the adaptive potential of Coastal wetlands of India-the need of the hour for sustainable management. *Wetlands Ecology and Management*, 1–12.

Rai, N. D., Bhasme, S., & Balaji, P. (2018). Power, inequality and rights: A political ecology of forest restoration. In *Forest landscape restoration* (pp. 47–62). Routledge. doi:10.4324/9781315111872-4

Ramachandra, T. V., & Kumar, U. (2008). Wetlands of greater Bangalore, India: Automatic delineation through pattern classifiers. *Electronic Green Journal*, *1*(26), 2. doi:10.5070/G312610729

Ramachandra, T. V., Mahapatra, D. M., Vinay, S., Varghese, S., Asulabha, K. S., Bhat, S. P., & Aithal, B. H. (2017). *Bellandur and Varthur Lakes rejuvenation blueprint*. ENVIS Technical Report 116, Environmental Information System, CES, Indian Institute of Science.

Ramsar Convention. (2018). *Urban wetlands: prized land, not wasteland*. https://www.ramsar.org/sites/default/files/documents/library/wwd18_infographic_engl.pdf

Rawat, K., & Yadav, A. K. (2021). Sustainable Practices for Conservation of Wetland Ecosystem. *Wetlands Conservation: Current Challenges and Future Strategies*, 129-139.

Ripple, W. J., Wolf, C., Newsome, T. M., Galetti, M., Alamgir, M., Crist, E., & Laurance, W. F. (2017). World Scientists' Warning to Humanity: A Second Notice. *Bioscience*, *67*(12), 10–1093. doi:10.1093/biosci/bix125

Root-Bernstein, M., & Frascaroli, F. (2016). Where the fish swim above the birds: Configurations and challenges of wetland restoration in the Po Delta, Italy. *Restoration Ecology*, *24*(6), 773–784. doi:10.1111/rec.12369

Roshnath, R., & Sashikuma, C. (2019). Conservation challenges of the heronries in Kerala. *Journal of the Bombay Natural History Society*, *116*, 63–67.

Roy-Basu, A., Bharat, G. K., Chakraborty, P., & Sarkar, S. K. (2020). Adaptive co-management model for the East Kolkata wetlands: A sustainable solution to manage the rapid ecological transformation of a peri-urban landscape. *The Science of the Total Environment*, *698*, 134203. doi:10.1016/j.scitotenv.2019.134203 PMID:31783445

Ruoso, L. E., Plant, R., Maurel, P., Dupaquier, C., Roche, P. K., & Bonin, M. (2015). Reading ecosystem services at the local scale through a territorial approach: The case of peri-urban agriculture in the Thau Lagoon, Southern France. *Ecology and Society*, *20*(3), art11. doi:10.5751/ES-07694-200311

Salata, S., Ronchi, S., & Arcidiacono, A. (2017). Mapping air filtering in urban areas. A land use regression model for ecosystem services assessment in planning. *Ecosystem Services*, *28*, 341–350. doi:10.1016/j.ecoser.2017.09.009

Savari, M., Damaneh, H. E., & Damaneh, H. E. (2020). Factors influencing local people's participation in sustainable forest management. *Arabian Journal of Geosciences*, *13*(13), 1–13. doi:10.100712517-020-05519-z

Sok, K., Méral, P., Pillot, D., & Defossez, S. (2019). Ecosystem Services from Tonle Sap Flood Pulse: Spatial and Economic Analysis in Aek Phnom and Sangkae Districts of Battambang Province, Cambodia. In *Water and Power* (pp. 123–151). Springer. doi:10.1007/978-3-319-90400-9_8

Solanki, V., & Joshi, A. (2017). Disappearing wetland: A study of Basai wetlands, Haryana (India). *International Journal of Economic Research*, *14*(21), 159–169.

South Asia Network on Dams, Rivers, and People (SANDRP). (2020). *World Wetlands Day 2020: Positive Stories from India*. https://sandrp.in/2020/02/01/world-wetlands-day-2020-positive-stories-from-india/

South Asia Network on Dams, Rivers, and People (SANDRP). (2021). *India's Wetlands 2016: Encroached and Polluted*. https://sandrp.in/2017/02/02/indias-wetlands-2016-encroached-and-polluted/

Sur, S., & Mandal, J. (2019). Deepor Beel: A Weeping and Blemishing Ramsar Site. *Jalaplavit*, *9*(3), 48–58.

Surya, B., Syafri, S., Sahban, H., & Sakti, H. H. (2020). Natural resource conservation based on community economic empowerment: Perspectives on watershed management and slum settlements in Makassar City, South Sulawesi, Indonesia. *Land (Basel)*, *9*(4), 104. doi:10.3390/land9040104

Suthar, A. M., Tatu, K., Gujar, R., & Kamboj, R. D. (2019). A Comparative Account of Diversity of Hydrophytes in Some Inland Wetlands (Pariej, Kanewal and Wadhwana) of Central Gujarat. *Research & Reviews. Journal of Life Science*, *9*(2), 39–43.

Tahmiscioğlu, M. S., Anul, N., Ekmekçi, F., & Durmuş, N. (2007, March). Positive and negative impacts of dams on the environment. In *International Congress on River Basin Management* (pp. 759-769). Academic Press.

Talukdar, J. K., & Rajbongshi, M. K. (2018). Ichthyofaunal Diversity and Conservation Status of Puthimari Beel of Barpeta, Assam, India. Academic Press.

Thapa, S., Wang, L., Koirala, A., Shrestha, S., Bhattarai, S., & Aye, W. N. (2020). Valuation of ecosystem services from an important wetland of Nepal: A Study from Begnas watershed system. *Wetlands*, *40*(5), 1071–1083. doi:10.100713157-020-01303-7

Thomas, S. T., Thomas, E. T., McLean, M., & Titus, T. T. (2021). Paving the way to achieving the United Nations Sustainable Development Goals for women from Indigenous communities: Lessons from Attappady, India. *Discover Sustainability*, *2*(1), 1–15. doi:10.100743621-021-00009-y

Vicziany, M., Chattopadhyay, D., & Bhattacharyya, S. (2017). Food from sewage: Fish from the east Kolkata wetlands and the limits of traditional knowledge. *South Asia*, *40*(3), 619–644. doi:10.1080/00856401.2017.1341038

Wang, Z., Wu, J., Madden, M., & Mao, D. (2012). China's wetlands: Conservation plans and policy impacts. *Ambio*, *41*(7), 782–786. doi:10.100713280-012-0280-7 PMID:22457078

Werners, S. E., Bos, E. J., Civic, K., Hlásny, T., Hulea, O., Jones-Walters, L. M., & Zsuffa, I. J. (2015). *Climate change vulnerability and ecosystem-based adaptation measures in the Carpathian region: final report-Integrated assessment of vulnerability of environmental resources and ecosystem-based adaptation measures* (No. 2572). Wageningen-UR.

West, A. (2015). Core concept: Ecosystem services. *Proceedings of the National Academy of Sciences of the United States of America*, *112*(24), 7337–7338. doi:10.1073/pnas.1503837112 PMID:26082537

Xu, X., Jiang, B., Tan, Y., Costanza, R., & Yang, G. (2018). Lake-wetland ecosystem services modeling and valuation: Progress, gaps and future directions. *Ecosystem Services*, *33*, 19–28. doi:10.1016/j.ecoser.2018.08.001

Yang, S. L., Zhang, J., Zhu, J., Smith, J. P., Dai, S. B., Gao, A., & Li, P. (2005). Impact of dams on Yangtze River sediment supply to the sea and delta intertidal wetland response. *Journal of Geophysical Research. Earth Surface*, *110*(F3).

Yardi, K. D., Bharucha, E., & Girade, S. (2019). Post-restoration monitoring of water quality and avifaunal diversity of Pashan Lake, Pune, India using a citizen science approach. *Freshwater Science*, *38*(2), 332–341. doi:10.1086/703440

Zhao, Q., Liu, S., & Dong, S. (2010). Effect of dam construction on spatial-temporal change of land use: A case study of Manwan, Lancang River, Yunnan, China. *Procedia Environmental Sciences*, *2*, 852–858. doi:10.1016/j.proenv.2010.10.096

Zou, Y., Duan, X., Xue, Z., Mingju, E., Sun, M., Lu, X., & Yu, X. (2018). Water use conflict between wetland and agriculture. *Journal of Environmental Management*, *224*, 140–146. doi:10.1016/j.jenvman.2018.07.052 PMID:30036808

Compilation of References

Aaby, B. (1976). Cyclic climatic variations in climate over the past 5500 yr reflected in raised bogs. *Nature*, *263*(5575), 281–284. doi:10.1038/263281a0

Abd-Elmabod, S. K., Muñoz-Rojas, M., Jordán, A., Anaya-Romero, M., Phillips, J. D., Laurence, J., & de la Rosa, D. (2020). Climate change impacts on agricultural suitability and yield reduction in a Mediterranean region. *Geoderma*, *374*, 114453. doi:10.1016/j.geoderma.2020.114453

Abe, K., Nomura, N., & Suzuki, S. (2020). Biofilms: hot spots of horizontal gene transfer (HGT) in aquatic environments, with a focus on a new HGT mechanism. *FEMS Microbiology Ecology, 96*(5), fiaa031. doi:10.1093/femsec/fiaa031

Abell, R., Lehner, B., Thieme, M., & Linke, S. (2016). Looking beyond the fenceline: Assessing protection gaps for the world's rivers. *Conservation Letters*. https://onlinelibrary.wiley.com/doi/10.1111/conl.12312/full

Abichou, A. (2002). *Les changements de paysages du bassin-versant de l'oued-Tataouine Fessi (sud-est tunisien). Étude multiscalaire et micromorphologie des remplissages des sebkhas et des états de surface des sols* (Thèse Géographie). Univ. Michel de Montaigne.

Abrantes, F., Voelker, A., Sierro, F. J., Naughton, F., Rodrigues, T., Cacho, I., Ariztegui, D., Brayshaw, D., Sicre, M. A., & Batista, L. (2012). Paleoclimate variability in the Mediterranean region. In P. Lionello (Ed.), *The Climate of the Mediterranean Region* (pp. 1–86). E-Publishing Elsevier. doi:10.1016/B978-0-12-416042-2.00001-X

Abrol, A. P., Yadav, J. S. D., & Massoud, F. I. (1988). Salt-affected soils and their management. FAO Soils Bulletin, 39.

Adams, W. A., & Akhtar, N. (1994). The possible consequences for herbage growth of waterlogging compacted pasture soils. *Plant and Soil*, *162*(1), 1–17. doi:10.1007/BF01416085

Africa Solid Waste Management Data Book. (2019). Available online at https://africancleancities.org/assets/data/JICA_databook_EN_web_20191218.pdf https://www.ramsar.org/sites/default/files/documents/library/info2007-01-e.pdf https://vedas.sac.gov.in/vedas/downloads/atlas/Wetlands/NWIA_National_atlas.pdf

Agamennone, V., Krul, C. A. M., Rijkers, G., & Kort, R. (2018). A practical guide for probiotics applied to the case of antibiotic-associated diarrhea in The Netherlands. *BMC Gastroenterology*, *18*(1), 103. doi:10.118612876-018-0831-x PMID:30078376

Agarwal, S., & Khan, S. (2020). Heavy Metal Phytotoxicity: DNA Damage. In Cellular and Molecular Phytotoxicity of Heavy Metals. Springer.

Aghalari, Z., Dahms, H. U., Sillanpää, M., Sosa-Hernandez, J. E., & Parra-Saldívar, R. (2020). Effectiveness of wastewater treatment systems in removing microbial agents: A systematic review. *Globalization and Health*, *16*(1), 13. doi:10.118612992-020-0546-y PMID:32013988

Ahluwalia, H.S., & Ygbuhay. R.C. (2015). Cosmic ray 11-year modulation for sunspot cycle 24. *Solar Physics, 290*(2), 635-43.

Ahmad, Z., Hussain, A., & Shakeel, A. (2019). *Economics Importance of Wetlands, their Benefits and Values Case of Pakistan*. Academic Press.

Ahmed, W., Angel, N., Edson, J., Bibby, K., Bivins, A., O'Brien, J. W., Choi, P. M., Kitajima, M., Simpson, S. L., Li, J., Tscharke, B., Verhagen, R., Smith, W., Zaugg, J., Dierens, L., Hugenholtz, P., Thomas, K. V., & Mueller, J. F. (2020). First confirmed detection of SARS-CoV-2 in untreated wastewater in Australia: A proof of concept for the wastewater surveillance of COVID-19 in the community. *The Science of the Total Environment, 728*, 138764. doi:10.1016/j.scitotenv.2020.138764 PMID:32387778

Ajiboye, T. O., Oyewo, O. A., & Onwudiwe, D. C. (2021). Simultaneous removal of organics and heavy metals from industrial wastewater: A review. *Chemosphere, 262*, 128379. doi:10.1016/j.chemosphere.2020.128379 PMID:33182079

Akhilesh, S. (2013). Indian Society: Issues and Problems. Academic Press.

Alegbeleye, O. O., & Sant'Ana, A. S. (2020). Manure-borne pathogens as an important source of water contamination: An update on the dynamics of pathogen survival/transport as well as practical risk mitigation strategies. *International Journal of Hygiene and Environmental Health, 227*, 113524. doi:10.1016/j.ijheh.2020.113524 PMID:32298989

Alibardi, L., & Cossu, R. (2016). Effects of carbohydrate, protein and lipid content of organic waste on hydrogen production and fermentation products. *Waste Management, 47*(Pt A), 69–77. doi:10.1016/j.wasman.2015.07.049

Al-Isawi, R., Scholz, M., Wang, Y., & Sani, A. (2015). Clogging of vertical-flow constructed wetlands treating urban wastewater contaminated with a diesel spill. *Environmental Science and Pollution Research International, 22*(17), 12779–12803. doi:10.100711356-014-3732-8 PMID:25339533

Alongi, D. M., Murdiyarso, D., Fourqurean, J. W., Kauffman, J. B., Hutahaean, A., Crooks, S., & Wagey, T. (2015). Indonesia's blue carbon: A globally significant and vulnerable sink for seagrass and mangrove carbon. *Springer. Wetlands Ecology and Management, 24*(1), 3–13. doi:10.100711273-015-9446-y

Alves, M. M., Pereira, M. A., Sousa, D. Z., Cavaleiro, A. J., Picavet, M., Smidt, H., & Stams, A. J. (2009). Waste lipids to energy: How to optimize methane production from long-chain fatty acids (LCFA). *Microbial Biotechnology, 2*(5), 538–550. doi:10.1111/j.1751-7915.2009.00100.x PMID:21255287

American Public Health Association (APHA). (2005). *Standard methods for the examination of water and waste water* (21st ed.). APHA.

Ameur, M. B., Masmoudi, S., Abichou, A., Medhioub, M., & Yaich, C. (2019). Use of the magnetic, geochemical, and sedimentary records in establishing paleoclimate change in the environment of Sebkha: Case of the Sebkha Mhabeul in southeastern Tunisia. *Comptes Rendus Geoscience, 351*(7), 487–497. doi:10.1016/j.crte.2019.10.003

Ameur, M. B., Masmoudi, S., & Yaich, C. (2021). Flood and sandstorm events recorded in holocene sebkha deposits in Southeastern Tunisia: Evidence from magnetic and geochemical properties. *Quaternary International, 571*, 46–57. doi:10.1016/j.quaint.2020.09.006

Andresen, C. S., Bond, G., Kuijpers, A., Knutz, P. C., & Bjoôrck, S. (2005). Holocene climate variability at multidecadal time scales detected by sedimentological indicators in a shelf core NW off Iceland. *Marine Geology, 214*(4), 323–338. doi:10.1016/j.margeo.2004.11.010

Angeler, D. G., Sanchez-Carrillo, S., Rodrigo, M. A., Viedma, O., & Alvarez-Cobelas, M. (2005). On the importance of water depth, macrophytes and fish in wetland picocyanobacteria regulation. *Hydrobiologia, 549*(1), 23–32. doi:10.100710750-005-1700-5

Anon. (1990). *World Resources 1990-91*. Oxford University Press.

Anon. (1991). *World Resources 1991-92*. Oxford University Press.

Antia, N. J., Harrison, P. J., & Oliveira, L. (1991). The role of dissolved organic nitrogen in phytoplankton nutrition, cell biology and ecology. *Phycologia, 30*(1), 1–89. doi:10.2216/i0031-8884-30-1-1.1

Ardón, M., Montanari, S., Morse, J. L., Doyle, M. W., & Bernhardt, E. S. (2010). Phosphorus export from a restored wetland ecosystem in response to natural and experimental hydrologic fluctuations. *Journal of Geophysical Research, 115*(G4), G04031. doi:10.1029/2009JG001169

Arslan, M., & Gamal El-Din, M. (2021). Removal of per- and poly-fluoroalkyl substances (PFASs) by wetlands: Prospects on plants, microbes and the interplay. *The Science of the Total Environment, 800*, 149570. doi:10.1016/j.scitotenv.2021.149570 PMID:34399352

Arthington, A.H., Bunn, S.E., Poff, N.L., & Naiman, R.J. (2006). The challenge of providing environmental flow rules to sustain river ecosystems. *Ecological Applications, 16*(4), 1311–1318.

Aryal, K., Ojha, B. R., & Maraseni, T. (2021). Perceived importance and economic valuation of ecosystem services in Ghodaghodi wetland of Nepal. *Land Use Policy, 106*, 105450. doi:10.1016/j.landusepol.2021.105450

Assessment, M. E. (2005). *Ecosystems and human well-being: wetlands and water*. World Resources Institute.

Australian Government, Department of Agriculture, Water and the Environment. (n.d.). *Wetlands and the Climate Change*. https://www.environment.gov.au/water/wetlands/publications/wetlands-climate-change

Babu, M. S. U., & Nautiyal, S. (2015). Conservation and Management of Forest Resources in India: Ancient and Current Perspectives. *Natural Resources, 6*(04), 256–272. doi:10.4236/nr.2015.64023

Badamfirooz, J., Mousazad, R., & Hamid, S. (2021). A proposed framework for economic valuation and assessment of damages cost to national wetlands ecosystem services using the benefit-transfer approach. *Environmental Challenges, 5*.

Bai, L., Liu, X. L., Hu, J., Li, J., Wang, Z. L., Han, G., Li, S. L., & Liu, C. Q. (2018). Heavy metal accumulation in common aquatic plants in rivers and lakes in the Taihu basin. *International Journal of Environmental Research and Public Health, 15*(12), 2857. doi:10.3390/ijerph15122857 PMID:30558148

Bake, H., & Sharma, R. (2019). Estimating Visitors' Willingness to Pay for Lake Conservation: A Contingent Valuation Study of Pashan Lake, Pune, India. In *Environmental Impacts of Tourism in Developing Nations* (pp. 234–251). IGI Global. doi:10.4018/978-1-5225-5843-9.ch013

Balakrishnan, M. K. vs. Union of India. Civil (W.P.) No. 230 of 2001. Supreme Court of India. https://main.sci.gov.in/jonew/courtnic/rop/2000/17346/rop_886913.pdf

Banerjee, S., & Dey, D. (2017). Eco-system complementarities and urban encroachment: A SWOT analysis of the East Kolkata Wetlands, India. *Cities and the Environment, 10*(1), 2.

Baral, S., Basnyat, B., Khanal, R., & Gauli, K. (2016). A total economic valuation of wetland ecosystem services: An evidence from Jagadishpur Ramsar site, Nepal. *TheScientificWorldJournal, 2016*, 2016. doi:10.1155/2016/2605609 PMID:27830175

Barbier, E. B., Acreman, M., & Knowler, D. (1997). *Economic Valuation of Wetlands: A Guide for Policy Makers and Planners*. Ramsar Convention Bureau.

Barcelo, D. (2020). An environmental and health perspective for COVID-19 outbreak: Meteorology and air quality influence, sewage epidemiology indicator, hospitals disinfection, drug therapies and recommendations. *Journal of Environmental Chemical Engineering*, *8*(4), 104006. doi:10.1016/j.jece.2020.104006 PMID:32373461

Barros, D. F., & Albernaz, A. L. M. (2014). Possible impacts of climate cane on wetlands and its biota in the Brazilian Amazon. *Brazilian Journal of Biology*, *74*(4), 810–820. doi:10.1590/1519-6984.04013 PMID:25627590

Bartolucci, N. N., Anderson, T. R., & Ballantine, K. A. (2021). Restoration of retired agricultural land to wetland mitigates greenhouse gas emissions. *Restoration Ecology*, *29*(3), e13314. doi:10.1111/rec.13314

Bassi, N., Kumar, M. D., Sharma, A., & Saradhi, P. P. (2014). Status of wetlands in India: A review of extent, ecosystem benefits, threats and management strategie. *Journal of Hydrology. Regional Studies*, *2*, 1–19. doi:10.1016/j.ejrh.2014.07.001

Baumgarten, H., & Wonik, T. (2015). Cyclostratigraphic studies of sediments from Lake Van (Turkey) based on their uranium contents obtained from downhole logging and paleoclimatic implications. *International Journal of Earth Sciences*, *104*(6), 1639–1654. doi:10.100700531-014-1082-x

Bazzicalupo, P., Maiorano, P., Girone, A., Marino, M., Combourieu-Nebout, N., Pelosi, N., & Incarbona, A. (2020). Holocene climate variability of the Western Mediterranean: Surface water dynamics inferred from calcareous plankton assemblages. *The Holocene*, *30*(5), 691–708. doi:10.1177/0959683619895580

Bedoui, S. (2016). *Multidisciplinary study of endorheic (El Guettar) and exoreic (El Melah) saline systems*. Master thesis.

Belyea, L. R., & Clymo, R. S. (2001). Feedback control of the rate of peat formation. *Proceedings of the Royal Society of London. Series B, Biological Sciences*, *268*(1473), 1315–1321. doi:10.1098/rspb.2001.1665 PMID:11410160

Belyea, L. R., & Malmer, N. (2004). Carbon sequestration in peatland: Patterns and mechanisms of response to climate change. *Global Change Biology*, *10*(7), 1043–1052. doi:10.1111/j.1529-8817.2003.00783.x

Ben Jmaa, H. (2008). *The endorheic system of Sidi El Hani, Chrita and Mechertate: Paleoenvironment and recent dynamics* (Doctoral dissertation). University of Tunis.

Ben Smida, N. (2016). *Comparative geochemical study of exorheic (Boujmal and Melah) and endorheic (Mcheguig and Mhabeul) saline systems: a geo-economical approach* (Master Thesis). Higher Institute of Applied Sciences and Technology of Gabes.

Ben Smida, N. (2016). *Comparative study of exorheic (Boujmal and El Melah) and endorheic (Mchiguig and Mhabeul) saline systems: a geo-economic approach*. Master thesis.

Ben, J. M. A. A. H. (2008). *Le système endoréique de Sidi Al Hani, Chrita et Mechertate: Paléoenvironnement et dynamique récente* (Thèse de doctorat). Université de Tunis.

Benchimol, M., & Peres, C. A. (2015). Widespread forest vertebrate extinctions induced by a mega hydroelectric dam in lowland Amazonia. *PLoS One*, *10*(7), e0129818. doi:10.1371/journal.pone.0129818 PMID:26132139

Berg, M., Stengel, C., Pham, T. K., Pham, H. V., Sampson, M. L., Leng, M., Samreth, S., & Fredericks, D. (2007). Magnitude of arsenic pollution in the Mekong and Red River Deltas—Cambodia and Vietnam. *The Science of the Total Environment*, *372*(2-3), 413–425. doi:10.1016/j.scitotenv.2006.09.010 PMID:17081593

Bernard, C., Escalas, A., Villeriot, N., Agogué, H., Hugoni, M., Duval, C., Carré, C., Got, P., Sarazin, G., Jézéquel, D., Leboulanger, C., Grossi, V., Ader, M., & Troussellier, M. (2019). Very Low Phytoplankton Diversity in a Tropical Saline-Alkaline Lake, with Co-dominance of *Arthrospira fusiformis* (Cyanobacteria) and *Picocystissalinarum* (Chlorophyta). *Microbial Ecology, 78*(3), 603–617. doi:10.100700248-019-01332-8 PMID:30729265

Berner, K. S., Koç, N., Godtliebsen, F., & Divine, D. (2011). Holocene climate variability of the Norwegian Atlantic Current during high and low solar insolation forcing. *Paleoceanography, 26*(2), n/a. doi:10.1029/2010PA002002

Berntsen, T. K., Fuglestvedt, J. S., Joshi, M. M., Shine, K. P., Stuber, N., Ponater, M., & Li, L. (2005). Response of climate to regional emissions of ozone precursors: Sensitivities and warming potentials. Tellus B. *Chemical and Physical Meteorology, 57*(4), 283–304. doi:10.3402/tellusb.v57i4.16549

Besbes, M. (1975). *Etude hydrogéologique de la plaine de Kairouan sur modèle mathématique. RAPP. CIG.* EMP/DRE.

Bhateria, R., & Jain, D. (2016). Water quality assessment of lake water: A review. *Sustainable Water Resources Management, 2*(2), 161–173. doi:10.100740899-015-0014-7

Bhat, R. A., Shafiq-ur-Rehman, M. M., Dervash, M. A., Mushtaq, N., Bhat, J. I. A., & Dar, G. H. (2017). Current status of nutrient load in Dal Lake of Kashmir Himalaya. *Journal of Pharmacognosy and Phytochemistry, 6*(6), 165–169.

Bhowmik, S. (2020). *Ecological and Economic Importance of Wetlands and Their Vulnerability: A Review, in Current State and Future Impacts of Climate Change on Biodiversity*. Retrieved October 9, 2021, from: https://reefresilience.org/blue-carbon/blue-carbon-introduction/blue-carbon-benefits/

Biagi, K. M., Clark, M. G., & Carey, S. K. (2021). Hydrological functioning of a constructed peatland watershed in the Athabasca oil sands region: Potential trajectories and lessons learned. *Ecological Engineering, 166*, 106236. doi:10.1016/j.ecoleng.2021.106236

Bianchi, G. G., & McCave, I. N. (1999). Holocene periodicity in North Atlantic climate and deep-ocean flow south of Iceland. *Nature, 397*(6719), 515–517. doi:10.1038/17362

Bigio & Bharat. (2004). *Urban Environment and Infrastructure Towards Liveable Cities*. The World Bank.

Billings, S. A., Lajtha, K., Malhotra, A., Berhe, A. A., de Graaff, M. A., Earl, S., & Wieder, W. (2021). Soil organic carbon is not just for soil scientists: Measurement recommendations for diverse practitioners. *Ecological Applications, 31*(3), e02290. doi:10.1002/eap.2290 PMID:33426701

Bisseux, M., Colombet, J., Mirand, A., Roque-Afonso, A. M., Abravanel, F., Izopet, J., Archimbaud, C., Peigue-Lafeuille, H., Debroas, D., Bailly, J. L., & Henquell, C. (2018). Monitoring human enteric viruses in wastewater and relevance to infections encountered in the clinical setting: a one-year experiment in central France, 2014 to 2015. *Euro surveillance: bulletin Europeen sur les maladies transmissibles = European communicable disease bulletin, 23*(7), 17-00237. doi:10.2807/1560-7917.ES.2018.23.7.17-00237

Bi, W., Weng, B., Yuan, Z., Ye, M., Zhang, C., Zhao, Y., Yan, D., & Xu, T. (2018). Evolution characteristics of surface water quality due to climate change and LUCC under scenario simulations: A Case study in the Luanhe River Basin. *International Journal of Environmental Research and Public Health, 15*(8), 1724. doi:10.3390/ijerph15081724 PMID:30103482

Blackall, L. L., & Marshall, K. C. (1989). The mechanism of stabilization of actinomycete foams and the prevention of foaming under laboratory conditions. *Journal of Industrial Microbiology, 4*(3), 181–187. doi:10.1007/BF01574075

Blackwell, M. S., & Pilgrim, E. S. (2011). Ecosystem services delivered by small-scale wetlands. *Hydrological Sciences Journal, 56*(8), 1467–1484. doi:10.1080/02626667.2011.630317

Blair, T. C., & Mcpherson, J. G. (1999). Grain-size and textural classification of coarse sedimentary particles. *Journal of Sedimentary Research*, *69*(1), 619. doi:10.2110/jsr.69.6

Blankespoor, B., Dasgupta, S., & Laplante, B. (2012). *Sea-level rise and coastal wetlands: Impacts and costs* [Policy Research Working Paper 6277]. The World Bank. doi:10.1596/1813-9450-6277

Bodor, A., Bounedjoum, N., Vincze, G. E., Erdeiné Kis, Á., Laczi, K., Bende, G., Szilágyi, Á., Kovács, T., Perei, K., & Rákhely, G. (2020). Challenges of unculturable bacteria: Environmental perspectives. *Reviews in Environmental Science and Biotechnology*, *19*(1), 1–22. doi:10.100711157-020-09522-4

Bonanno, G., Borg, J. A., & Di Martino, V. (2017). Levels of heavy metals in wetland and marine vascular plants and their biomonitoring potential: A comparative assessment. *The Science of the Total Environment*, *576*, 796–806. doi:10.1016/j.scitotenv.2016.10.171 PMID:27810764

Bond, G., Kromer, B., Beer, J., Muscheler, R., Evans, M. N., Showers, W., Hoffmann, S., Lotti-Bond, R., Hajdas, I., & Bonani, G. (2001). Persistent solar influence on North Atlantic climate during the Holocene. *Science*, *294*(5549), 2130–2136. doi:10.1126cience.1065680 PMID:11739949

Bond, G., Showers, W., Cheseby, M., Lotti, R., Almasi, P., deMenocal, P., Priore, P., Cullen, H., Hajdas, I., & Bonani, G. (1997). A pervasive millenial-scale cycle in North Atlantic: Holocene and glacial climates. *Science*, *278*(5341), 1257–1266. doi:10.1126cience.278.5341.1257

Boorman, L. A. (1990). Impact of sea level changes on coastal areas. In M. M. Boer & R. S. de Groot (Eds.), *Landscape ecological impacts of climate change* (pp. 379–391). IOS Press.

Boutwell, J. L., & Westra, J. V. (2016). The role of wetlands for mitigating economic damage from Hurricanes. *Journal of the American Water Resources Association*, *52*(6), 1472–1481. doi:10.1111/1752-1688.12473

Boyd, J., & Banzhaf, S. (2007). What are ecosystem services? The need for standardized environmental accounting units. *Ecological Economics*, *63*(2–3), 616–626. doi:10.1016/j.ecolecon.2007.01.002

Boyer, B. (2000). Institutional mechanisms for sustainable development: A look at national councils for sustainable development in Asia. *Glob Environment Change-Hum Policy Dimens*, *10*(2), 157–160. doi:10.1016/S0959-3780(00)00019-4

Bradley, R. S., Hughes, M. K., & Diaz, H. F. (2003). Climate in Medieval Time. *Science*, *302*(5644), 404–405. doi:10.1126cience.1090372 PMID:14563996

Bragg, O. M. (2002). Hydrology of peat-forming wetlands in Scotland. *The Science of the Total Environment*, *294*(1-3), 111–129. doi:10.1016/S0048-9697(02)00059-1 PMID:12169001

Bravo-Inclan, L. A., Saldana-Fabela, M. P., & Sanchez-Chavez, J. J. (2008). Long-term eutrophication diagnosis of a high altitude body of water, Zimapan Reservoir, Mexico. *Water Science and Technology*, *57*(11), 1843–1849. doi:10.2166/wst.2008.279 PMID:18547939

Breitwieser, F. P., & Salzberg, S. L. (2020). Pavian: Interactive analysis of metagenomics data for microbiome studies and pathogen identification. *Bioinformatics (Oxford, England)*, *36*(4), 1303–1304. doi:10.1093/bioinformatics/btz715 PMID:31553437

Bridgham, S. D., Cadillo-Quiroz, H., Keller, J. K., & Zhuang, Q. (2013). Methane emissions from wetlands: Biogeochemical, microbial, and modeling perspectives from local to global scales. *Global Change Biology*, *19*(5), 1325–1346. doi:10.1111/gcb.12131 PMID:23505021

Broecker, W. S., Lynch-Stieglitz, J., Clark, E., Hajdas, I., & Bonani, G. (2001). What caused the atmosphere's CO2 content to rise during the last 8000 years? *Geochemistry Geophysics Geosystems*, *2*(10), n/a. doi:10.1029/2001GC000177

Brown. (2021). Opportunities for improving recognition of coastal wetlands in global ecosystem assessment frameworks. *Ecological Indicators, 126*.

Brown, L., Flavinc, Q., & Pootel, S. (1991). *State of the World. A World Watch Institute report on progress towards a sustainable society*. World Watch Institute India.

Buckland, S. T., Anderson, D. R., Burnham, K. P., & Laake, J. L. (1993). *Distance Sampling: Estimating Abundance of Biological Populations*. Chapman and Hall.

Bürgmann, H., Frigon, D., H Gaze, W., M Manaia, C., Pruden, A., Singer, A. C., F Smets, B., & Zhang, T. (2018). Water and sanitation: an essential battlefront in the war on antimicrobial resistance. *FEMS Microbiology Ecology, 94*(9), . doi:10.1093/femsec/fiy101

Burton, P. J., Jentsch, A., & Walker, L. R. (2020). The ecology of disturbance interactions. *Bioscience, 70*(10), 854–870. doi:10.1093/biosci/biaa088

Butler, S. (Ed.). (2010). *Macquarie Concise Dictionary* (5th ed.). Macquarie Dictionary Publishers.

Bystrom, O., Andersson, H., & Gren, I. (2000). Economic criteria for using wetlands as nitrogen sinks under uncertainty. *Ecological Economics, 35*(1), 35–45. doi:10.1016/S0921-8009(00)00166-X

Byun, C., Nam, J. M., & Kim, J. G. (2017). Effects of flooding regime on wetland plant growth and species dominance in a mesocosm experiment. *Plant Ecology, 218*(5), 517–527. doi:10.100711258-017-0707-0

Cailleux, A., & Tricart, J. (1962). Le modèle glaciaire et nival. SEDES.

Calvert, G. M., Karnik, J., Mehler, L., Beckman, J., Morrissey, B., Sievert, J., Barrett, R., Lackovic, M., Mabee, L., Schwartz, A., Mitchell, Y., & Moraga-McHaley, S. (2008). Acute pesticide poisoning among agricultural workers in the United States, 1998-2005. *American Journal of Industrial Medicine, 51*(12), 883–898. doi:10.1002/ajim.20623 PMID:18666136

Camill, P., Lynch, J. A., Clark, J. S., Adams, J. B., & Jordan, B. (2001). Changes in biomass, above ground net primary production, and peat accumulation following permafrost thaw in the boreal peatlands of Manitoba, Canada. *Ecosystems (New York, N.Y.), 4*(5), 461–478. doi:10.100710021-001-0022-3

Campos, F. S., David, J., Lourenço-de-Moraes, R., Rodrigues, P., Silva, B., da Silva, C. V., & Cabral, P. (2021). The economic and ecological benefits of saving ecosystems to protect services. *Journal of Cleaner Production, 311*, 127551. doi:10.1016/j.jclepro.2021.127551

Cappuccino, J. G., & Sherman, N. (2005). Microbiology: A laboratory manual. *Benjamin Cummings: New York, 2005*, 125–179.

Carbonel, P., & Pinson, J. (1980). Les Cyprideis témoins de l'évolution des sels en milieux laguno-lacustres sous climat semi-aride. In N. Nrstic (Ed.), *Taxonomy, Biostratigraphy and Distribution of Ostracodes* (pp. 211–217). Serbs KO Geolochko.

Carcaillet, C., Bouvier, M., Fréchette, B., Larouche, A. C., & Richard, P. J. H. (2001). Comparison of pollen-slide and sieving methods in lacustrine charcoal analyses for local and regional fire history. *The Holocene, 11*(4), 467–476. doi:10.1191/095968301678302904

Carp, E. (Ed.). (1972). *Proceedings of the International Conference on the Conservation of Wetlands and Waterfowl*. Ramsar, Iran: IWRB.

Carpenter, S. R. (2005). Eutrophication of aquatic ecosystems: Biostability and soil phosphorus. *Proceedings of the National Academy of Sciences of the United States of America*, *102*(29), 10002–10005. doi:10.1073/pnas.0503959102 PMID:15972805

Carvalho, F. P. (2017). Pesticides, environment, and food safety. *Food and Energy Security*, *6*(2), 48–60. doi:10.1002/fes3.108

CBD-Convention of Biological Diversity. (2008). *Inland water biodiversity*. https://www.cbd.int/waters/inland-waters

Chai, M., Li, R., Tam, N., & Zan, Q. (2019). Effects of mangrove plant species on accumulation of heavy metals in sediment in a heavily polluted mangrove swamp in Pearl River Estuary, China. *Environmental Geochemistry and Health*, *41*(1), 175–189. doi:10.100710653-018-0107-y PMID:29876676

Chakraborty, A., Halder, S., Nazneen, S., & Dey, S. K. (2019). East Kolkata Wetlands (EKW), India: A Unique Example of Resource Recovery. In Wastewater Reuse and Watershed Management (pp. 25-40). Apple Academic Press.

Chakraborty, S., Saha, S. K., & Selim, S. A. (2020). Recreational services in tourism dominated coastal ecosystems: Bringing the non-economic values into focus. *Journal of Outdoor Recreation and Tourism*, *30*, 100279. doi:10.1016/j.jort.2020.100279

Chambers, L. G., Guevara, R., Boyer, J. N., Troxler, T. G., & Davis, S. E. (2016). Effects of salinity and inundation on microbial community structure and function in a mangrove peat soil. *Wetlands*, *36*(2), 361–371. doi:10.100713157-016-0745-8

Chapin, F. S. III, Woodwell, G. M., Randerson, J. T., Rastetter, E. B., Lovett, G. M., Baldocchi, D. D., & Matson, P. A. (2006). McGuire, AD, Melillo, JM, Mooney, HA, Neff, JC. Houghton, RA, Pace, ML, Ryan, MG, Running, SW, Sala, OE, Schelesinger, WH, and Schulze, ED: Reconciling carbon-cycle concepts, terminology, and methods. *Ecosystems (New York, N.Y.)*, *9*(7), 1041–1050. doi:10.100710021-005-0105-7

Chatrath, K. J. S. (1992). *Wetlands of India – Major threats faced by wetlands of India and their conservative measures*. Ashish Publication House.

Chatterjee, N. (1995). Social forestry in environmentally degraded regions of India: Case-study of the Mayurakshi Basin. *Environmental Conservation*, *22*, 20–30.

Chazarenc, F., Gagnon, V., & Korkusuz, E. A. (2015). Treatment of pollution in constructed wetlands: From the fundamental mechanisms to the full-scale applications. *Environmental Science and Pollution Research International*, *22*(17), 12775–12778. doi:10.100711356-015-4969-6 PMID:26160123

Chen, G., Shi, H., Tao, J., Chen, L., Liu, Y., Lei, G., Liu, X., & Smol, J. P. (2015). Industrial arsenic contamination causes catastrophic changes in freshwater ecosystems. *Scientific Reports*, *5*(1), 17419. doi:10.1038rep17419 PMID:26615891

Cheng, R., Zhu, H., Cheng, X., Shutes, B., & Yan, B. (2020). Saline and Alkaline tolerance of wetland plants—What are the most representative evaluation indicators? *Sustainability*, *12*(5), 1913. doi:10.3390u12051913

Chen, J., Li, J., Zhang, H., Shi, W., & Liu, Y. (2019). Bacterial Heavy-Metal and Antibiotic Resistance Genes in a Copper Tailing Dam Area in Northern China. *Frontiers in Microbiology*, *10*, 1916. doi:10.3389/fmicb.2019.01916 PMID:31481945

Chen, Y., Parvez, F., Gamble, M., Islam, T., Ahmed, A., Argos, M., Graziano, J. H., & Ahsan, H. (2009). Arsenic exposure at low-to-moderate levels and skin lesions, arsenic metabolism, neurological functions, and biomarkers for respiratory and cardiovascular diseases: Review of recent findings from the Health Effects of Arsenic Longitudinal Study (HEALS) in Bangladesh. *Toxicology and Applied Pharmacology*, *239*(2), 184–192. doi:10.1016/j.taap.2009.01.010 PMID:19371619

Cherry, J. A. (2011). Ecology of Wetland Ecosystems: Water, Substrate, and Life. *Nature Education Knowledge*, *3*(10), 16.

Chimner, R. A., & Cooper, D. J. (2003). Influence of water table levels on CO2 emissions in a Colorado subalpine fen: An in situ microcosm study. *Soil Biology & Biochemistry*, *35*(3), 345–351. doi:10.1016/S0038-0717(02)00284-5

Chin, A., Chu, J., Perera, M., Hui, K., Yen, H. L., Chan, M., Peiris, M., & Poon, L. (2020). Stability of SARS-CoV-2 in different environmental conditions. *The Lancet Microbe*, *1*(1), e10. doi:10.1016/S2666-5247(20)30003-3 PMID:32835322

Chislock, M. F., Doster, E., Zitomer, R. A., & Wilson, A. E. (2013). Eutrophication: Causes, consequences, and controls in aquatic ecosystems. *Nature Education Knowledge*, *4*(4), 10.

Chmura, G. L., Anisfeld, S. C., Cahoon, D. R., & Lynch, J. C. (2003). Global carbon sequestration in tidal, saline wetland soils. *Global Biogeochemical Cycles*, *17*(4), n/a. doi:10.1029/2002GB001917

Chopra, R. (1985). *The State of India's Environment*. Ambassador Press.

Choudhury, M., Sharma, A., Singh, P., & Kumar, D. (2021). Impact of climate change on wetlands, concerning Son Beel, the largest wetland of North East, India. In *Global Climate Change* (pp. 393–414). Elsevier. doi:10.1016/B978-0-12-822928-6.00006-X

Chouhan, H. A., Parthasarathy, D., & Pattanaik, S. (2018). Urban at the edges: Mumbai's coastline urbanisms. In *Sustainable urbanization in India* (pp. 279–293). Springer. doi:10.1007/978-981-10-4932-3_15

Christensen, V. G., Maki, R. P., Stelzer, E. A., Norland, J. E., & Khan, E. (2019). Phytoplankton community and algal toxicity at a recurring bloom in Sullivan Bay, Kabetogama Lake, Minnesota, USA. *Scientific Reports*, *9*(1), 1–11. doi:10.103841598-019-52639-y PMID:31695119

Chu, Z., Jin, X., Iwami, N., & Inamori, Y. (2007). The effect of temperature on growth characteristics and competitions of *Microcystis aeruginosa* and *Oscillatoria mougeotii* in a shallow, eutrophic lake simulator system. In B. Qin, Z. Liu, & K. Havens (Eds.), *Eutrophication of Shallow Lakes with Special Reference to Lake Taihu, China. Developments in Hydrobiology* (Vol. 194). Springer. doi:10.1007/978-1-4020-6158-5_24

Chwaya, M. (2017). *Etude comparative entre la sebkha endoréique de Boujmal, centre tunisien et la sebkha exoréique de Jallabia, Sud-Est tunisien: changements globaux au cours de la transition Holocène-Anthropocène* (Master Thesis). Faculty of Sciences of Gafsa.

Cimon-Morin, J., & Poulin, M. (2018). Setting conservation priorities in cities: Approaches, targets and planning units adapted to wetland biodiversity and ecosystem services. *Landscape Ecology*, *33*(11), 1975–1995. doi:10.100710980-018-0707-z

Clarkson, B. R., Ausseil, A. E., & Gerbeaux, P. (2013). Wetland ecosystem services. In J. R. Dymond (Ed.), *Ecosystem services in New Zealand - conditions and trends*. Manaaki Whenua Press.

Clarkson, B. R., Ausseil, A. G. E., & Gerbeaux, P. (2013). *Wetland ecosystem services. Ecosystem services in New Zealand: conditions and trends*. Manaaki Whenua Press.

Clarkson, B. R., Ausseil, A. G. E., & Gerbeaux, P. (2014). Wetland ecosystem services. In J. R. Dymond (Ed.), *Ecosystem Services in New Zealand* (pp. 192–202). Manaaki Whenua Press.

Clements, F. E. (1916). *Plant succession analysis of the development of vegetation*. Carnegie Institution of Washington Publication Sciences. doi:10.5962/bhl.title.56234

Coban, O., Kuschk, P., Wells, N. S., Strauch, G., & Knoeller, K. (2015). Microbial nitrogen transformation in constructed wetlands treating contaminated groundwater. *Environmental Science and Pollution Research International*, *22*(17), 12829–12839. doi:10.100711356-014-3575-3 PMID:25233917

Coleine, C., Biagioli, F., de Vera, J. P., Onofri, S., & Selbmann, L. (2021). Endolithic microbial composition in Helliwell Hills, a newly investigated Mars-like area in Antarctica. *Environmental Microbiology*, *23*(7), 4002–4016. doi:10.1111/1462-2920.15419 PMID:33538384

Colin, C., Frank, N., Copard, K., & Douville, E. (2010). Neodymium isotopic composition of deep-sea corals from the NE Atlantic: Implications for past hydrological changes during the Holocene. *Quaternary Science Reviews*, *29*(19-20), 2509–2517. doi:10.1016/j.quascirev.2010.05.012

Collias, E. E., Rona, M. R., Mcmanus, D. A., & Creager, J. S. (1963). *Machine processing of geological data*. University of Washington Technical Report Number 87.

Conner, B., & Skladany, M. (1991). Aquacultural development in tropical Asia: A re-evaluation. *Natural Resources Forum*, *15*(1), 66–73. doi:10.1111/j.1477-8947.1991.tb00111.x

Conner, W. H., & Day, J. W. (1982). The ecology of forested wetlands in the southeastern United States. In *Wetlands: Ecology and Management* (pp. 69–87). National Institute of Ecology and International Scientific Publications.

Connor, R. (2015). *The United Nations world water development report 2015: water for a sustainable world* (Vol. 1). UNESCO Publishing.

Cornell, J. A., Craft, C. B., & Megonigal, J. P. (2007). Ecosystem gas exchange across a created salt marsh chronosequence. *Wetlands*, *27*(2), 240–250. doi:10.1672/0277-5212(2007)27[240:EGEAAC]2.0.CO;2

Cornish, G., Mensah, E., & Ghesquire, P. (1999). *Water quality and peri-urban irrigation: an assessment of surface water quality for irrigation and its implications for human health in the peri-urban zone of Kumasi, Ghana*. Project Report. HR Wallingford Ltd.

Costanza, R. (2021). The global value of coastal wetlands for storm protection. *Global Environmental Change*, 70.

Costanza, R. W., Costanza, R., d'Arge, R., de Groot, R., Farber, S., Grasso, M., Hannon, B., Limburg, K., Naeem, S., O'Neill, R. V., Paruelo, J., Raskin, R. G., Sutton, P., & van den Belt, M. (1997). The value of the world's ecosystem services and natural capital. *Nature*, *387*(6630), 253–260. doi:10.1038/387253a0

Costanza, R., De Groot, R., Sutton, P., Van der Ploeg, S., Anderson, S. J., Kubiszewski, I., & Turner, R. K. (2014). Changes in the global value of ecosystem services. *Global Environmental Change*, *26*, 152–158. doi:10.1016/j.gloenvcha.2014.04.002

Costanza, R., P'erez-Maqueo, O., Martinez, M. L., Sutton, P., Anderson, S. J., & Mulder, K. (2008). The value of coastal wetlands for hurricane protection. *Ambio*, *37*(4), 241–248. doi:10.1579/0044-7447(2008)37[241:TVOCWF]2.0.CO;2 PMID:18686502

Cotton, M. D., & Mahroos-Alsaiari, A. A. (2015). Key actor perspectives on stakeholder engagement in Omani Environmental Impact Assessment: An application of Q-Methodology. *Journal of Environmental Planning and Management*, *58*(1), 91–112. doi:10.1080/09640568.2013.847822

Cowardin, L. M., & Carter, V. (1979). Classification of wetlands and deepwater habitats in the United States. U.S. Dept. Interior, Fish & Wildlife Service, FWS/OBS- 79/31.

Cowardin, L. M., Carter, V., Golet, F. C., & LaRoe, E. T. (1979). *Classification of Wetlands and Deep Water Habitats of the United States*. US Department of the Interior, Fish and Wildlife Service, Office of Biological Services.

Cowardin, L. M., Carter, V., Golet, F. C., & LaRoe, E. T. (1979). *Classification of Wetlands and Deepwater Habitats of the United States. FWS/OBS-79/31*. US Fish and Wildlife Service.

Cowardin, L. M., Carter, V., Golet, F. C., & LaRoe, E. T. (1992). *Classification of Wetlands and Deepwater Habitats of the United States*. U.S. Department of the Interior, Fish and Wildlife Service.

Craig, L. S., Olden, J. D., Arthington, A. H., Entrekin, S., Hawkins, C. P., Kelly, J. J., & Groffman, P. M. (2017). Meeting the challenge of interacting threats in freshwater ecosystems: A call to scientists and managers. *Elementa*, *5*, 5. doi:10.1525/elementa.256

Craven, D., Hall, J. S., Berlyn, G. P., Ashton, M. S., & van Breugel, M. (2018). Environmental filtering limits functional diversity during succession in a seasonally wet tropical secondary forest. *Journal of Vegetation Science*, *29*(3), 511–520. doi:10.1111/jvs.12632

Cronk, J. K., & Fennessy, M. S. (2009). Wetland Plants. In G. E. Likens (Ed.), *Encyclopedia of Inland Waters* (pp. 590–598). Academic Press. doi:10.1016/B978-012370626-3.00060-0

Crush, J. R., Briggs, L. R., Sprosen, J. M., & Nichols, S. N. (2008). Effect of irrigation with lake water containing microcystins on microcystin content and growth of ryegrass, clover, rape, and lettuce. *Environmental Toxicology*, *23*(2), 246–252. doi:10.1002/tox.20331 PMID:18214908

Crutzen, P. J. (2006). The “anthropocene”. In Earth system science in the Anthropocene. Springer.

Cui, B. S., & Yang, Z. F. (2006). Wetland Science. Beijing Normal University Press.

D’Costa, P. M., D’Silva, M. S., & Naik, R. K. (2017). Impact of pollution on Phytoplankton and implications for marine econiches. In M. Naik & S. Dubey (Eds.), *Marine Pollution and Microbial Remediation*. Springer. doi:10.1007/978-981-10-1044-6_13

D’Souza, N. M., Ishwar, N. M., Sumra, I., & Vyas, P. (2017). Participatory wetland management: a solution to conservation challenges in the sundarbans biosphere reserve. In *Wetland science* (pp. 575–587). Springer. doi:10.1007/978-81-322-3715-0_30

Danilov, R. A., & Ekelund, N. G. A. (2001). Comparative studies on the usefulness of seven ecological indices for the marine coastal monitoring close to the shore on the Swedish East coast. *Environmental Monitoring and Assessment*, *66*(3), 265–279. doi:10.1023/A:1006364317956 PMID:11281424

Daou, K., Wang, R. Z., & Xia, Z. Z. (2006). Desiccant cooling air conditioning: A review. *Renewable & Sustainable Energy Reviews*, *10*(2), 55–77. doi:10.1016/j.rser.2004.09.010

Dar, S. A., Bhat, S. U., & Rashid, I. (2021a). The Status of Current Knowledge, Distribution and Conservation Challenges of Wetland Ecosystems in Kashmir Himalaya, India. In Sharma, S. & Singh, P. (eds) Wetlands Conservation: Current Challenges and Future Strategies. https://doi.org/10.1002/9781119692621.ch10

Dar, S. A., Bhat, S. U., Aneaus, S., & Rashid, I. (2020b). A geospatial approach for limnological characterization of Nigeen Lake, Kashmir Himalaya. *Environmental Monitoring and Assessment*, *192*(2), 1–18. doi:10.100710661-020-8091-y PMID:31953600

Dar, S. A., Bhat, S. U., & Rashid, I. (2021d). Landscape transformations, morphometry and water quality of Anchar wetland in Kashmir Himalaya, India: Implications for urban wetland management. *Water, Air, and Soil Pollution*, *232*(462) 1-19. doi:10.100711270-021-05416-5

Dar, S. A., Bhat, S. U., Rashid, I., & Dar, S. A. (2020a). Current status of wetlands in Srinagar City: Threats, management strategies, and future perspectives. *Frontiers in Environmental Science*, *7*, 199. doi:10.3389/fenvs.2019.00199

Dar, S. A., Hamid, I., Bhat, S. U., & Rashid, I. (2021c). Identification of anthropogenic contribution to wetland degradation: Insights from the Environmetric Techniques. *Stochastic Environmental Research and Risk Assessment*, •••, 1–15. doi:10.100700477-021-02121-x

Dar, S. A., Rashid, I., & Bhat, S. U. (2021b). Land system transformations govern the trophic status of an urban wetland ecosystem: Perspectives from remote sensing and water quality analysis. *Land Degradation & Development*, *32*(14), 1–18. doi:10.1002/ldr.3924

Dar, S. A., Rashid, I., & Bhat, S. U. (2021e). Linking land system changes (1980-2017) with the trophic status of an urban wetland: Implications for wetland management. *Environmental Monitoring and Assessment*, 193(710), 1–17. https://doi.org/10.1007/s10661-021-09476-2

Dasgupta, A., & Salam, A. (2018). A City Reclaiming Assurance on Open Vulnerable Areas–A Dossier on Wetlands and Canals of Kolkata. *International Conference on Urban Sustainability: Emerging Trends, Themes, Concepts & Practices (ICUS)*. 10.2139srn.3236659

Das, H. S., Mishra, R. K., & Sinha, B. N. (1990). Some observations on Chilka wetland ecosystem. In B. N. Sinha (Ed.), *Ecosystem degradation in India* (pp. 85–93). Ashis Publishing House.

Datta, S., Rajnish, K. N., Samuel, M. S., Pugazhendhi, A., & Ethiraj, S. (2020). Metagenomic applications in microbial diversity, bioremediation, pollution monitoring, enzyme and drug discovery. A review. *Environmental Chemistry Letters*, *18*(4), 1229–1241. doi:10.100710311-020-01010-z

Datti, A. A., Nanda, S., Annadurai, R., & Sengupta, M. (2017). Geospatial estimation of chlorophyll-a in urban lakes. *Rasayan Journal of Chemistry*, *10*(1), 263–270.

Daughton, C. G. (2018). Monitoring wastewater for assessing community health: Sewage Chemical-Information Mining (SCIM). *The Science of the Total Environment*, *619-620*, 748–764. doi:10.1016/j.scitotenv.2017.11.102 PMID:29161600

Davidson, N. C. (2014). How much wetland has the world lost? Long-term and recent trends in global wetland area. *Marine and Freshwater Research*, *65*(10), 934. doi:10.1071/MF14173

Davies, G. (2018, May 24). *How will wetlands be affected by climate change?* Wisconsin Wetlands Association. https://www.wisconsinwetlands.org/updates/how-will-wetlands-be-affected-by-climate-change/

Davies, T. D. (2007). Sulphate toxicity to the aquatic moss, *Fontinalis antipyretica*. *Chemosphere*, *66*(3), 444–451. doi:10.1016/j.chemosphere.2006.06.021 PMID:16899275

Davis, J., Sim, L., & Chambers, J. (2010). Multiple stressors and regime shifts in shallow aquatic ecosystems in antipodean landscapes. *Freshwater Biology*, *55*(s1), 5–18. doi:10.1111/j.1365-2427.2009.02376.x

Davis, S. N., & DeWiest, R. J. (1966). *Hydrogeology*. Wiley.

Day, J. W., & Rybczyk, J. M. (2019). Global change impacts on the future of coastal systems: perverse interactions among climate change, ecosystem degradation, energy scarcity, and population. In *Coasts and Estuaries* (pp. 621–639). Elsevier. doi:10.1016/B978-0-12-814003-1.00036-8

Day, R. H., Williams, T. M., & Swarzenski, C. M. (2007). Hydrology of tidal freshwater forested wetlands of the southeastern United States. In *Ecology of tidal freshwater forested wetlands of the Southeastern United States* (pp. 29–63). Springer. doi:10.1007/978-1-4020-5095-4_2

de Groot, R., Brander, L., Van Der Ploeg, S., Costanza, R., Bernard, F., Braat, L., & Van Beukering, P. (2012). Global estimates of the value of ecosystems and their services in monetary units. *Ecosystem Services*, *1*(1), 50–61. doi:10.1016/j.ecoser.2012.07.005

de la Cruz, A. A. (1978). Production and transport of detritus in wetlands. In *Wetland Functions and Values: The State of our Understanding.* American Water Resources.

deAnda, J., & Shear, H. (2001). Nutrients and Eutrophication in Lake Chapala. In A. M. Hansen & M. van Afferden (Eds.), *The Lerma-Chapala Watershed.* Springer. doi:10.1007/978-1-4615-0545-7_8

Dearing, J. A., Livingstone, I. P., Bateman, M. D., & White, K. (2001). Palaeoclimate records from OIS 8.0–5.4 recorded in loess–palaeosol sequences on the Matmata Plateau, southern Tunisia, based on mineral magnetism and new luminescence dating. *Quaternary International, 76*, 43–56. doi:10.1016/S1040-6182(00)00088-4

Debén, S., Aboal, J., Carballeira, A., Cesa, M., Real, C., & Fernandez, J. (2015). Inland water quality monitoring with native bryophytes: A methodological review. *Ecological Indicators, 53*, 115–124. doi:10.1016/j.ecolind.2015.01.015

Debret, M., Bout-Roumazeilles, V., Grousset, F., Desmet, M., McManus, J. F., Massei, N., Sebag, D., Petit, J. R., Copard, Y., & Trentesaux, A. (2007). The origin of the 1500-year climate cycles in Holocene North-Atlantic records. *Climate of the Past, 3*(4), 569–575. doi:10.5194/cp-3-569-2007

Debret, M., Sebag, D., Costra, X., Massei, N., Petit, J. R., Chapron, E., & Bout-Roumazeilles, V. (2009). Evidence from wavelet analysis for a mid-Holocene transition in global climate forcing. *Quaternary Science Reviews, 28*(25-26), 2675–2688. doi:10.1016/j.quascirev.2009.06.005

Del Bene, D., Scheidel, A., & Temper, L. (2018). More dams, more violence? A global analysis on resistances and repression around conflictive dams through co-produced knowledge. *Sustainability Science, 13*(3), 617–633. doi:10.100711625-018-0558-1

Deng, W. & Wei, G. (2015). Decoupling of seasonal temperature and precipitation over the western Pacific during the early mid-Holocene. *International Journal of Climatology, 35*(5), 794-800.

Deng, J., Qin, B., Paerl, H. W., Zhang, Y., Wu, P., Ma, J., & Chen, Y. (2014). Effects of nutrients, temperature and their interactions on spring phytoplankton community succession in Lake Taihu, China. *PLoS One, 9*(12), e113960. doi:10.1371/journal.pone.0113960 PMID:25464517

Desai, A. (2018, May 24). *How can wetlands provide resilience against climate change?* Wisconsin Wetlands Association. https://www.wisconsinwetlands.org/updates/how-can-wetlands-provide-resilience-against-climate-change/

Deshpande, R., & Zou, C. (2020). Pseudomonas Aeruginosa Induced Cell Death in Acute Lung Injury and Acute Respiratory Distress Syndrome. *International Journal of Molecular Sciences, 21*(15), 5356. doi:10.3390/ijms21155356 PMID:32731491

Desprat, S., Sanchez Goni, M. F., & Loutre, M.-F. (2003). Revealing climatic variability of the last three millennia in northwestern Iberia using pollen influx data. *Earth and Planetary Science Letters, 213*(1-2), 63–78. doi:10.1016/S0012-821X(03)00292-9

Devarani, L., & Basu, D. (2009). Participatory wetland management in Loktak lake: A road to sustainable development. *Journal of Crop and Weed, 5*(1), 178–190.

Dhariwal, A., Chong, J., Habib, S., King, I. L., Agellon, L. B., & Xia, J. (2017). MicrobiomeAnalyst: A web-based tool for comprehensive statistical, visual and meta-analysis of microbiome data. *Nucleic Acids Research, 45*(W1), W180–W188. doi:10.1093/nar/gkx295 PMID:28449106

Díaz-Pardo, E., Vazquez, G., & López-López, E. (1998). The phytoplankton community as a bioindicator of health conditions of Atezca Lake, Mexico. *Aquatic Ecosystem Health & Management, 1*(3-4), 257–266. doi:10.1080/14634989808656922

Dillenburg, S. R., Barboza, E. G., Rosa, M. L. C., Caron, F., & Sawakuchi, A. O. (2017). The complex prograded Cassino barrier in southern Brazil: Geological and morphological evolution and records of climatic, oceanographic and sea-level changes in the last 7–6 ka. *Marine Geology, 390*, 106–119. doi:10.1016/j.margeo.2017.06.007

Dinh, Q. T. (2016). Vietnam-Mekong Delta Integrated Climate Resilience and Sustainable Livelihoods (MD-ICRSL) Project: Environmental assessment (No. SFG1760, pp. 1-209). The World Bank.

Dise, N. B. (2009). Peatland response to global change. *Science, 326*(5954), 810–811. doi:10.1126cience.1174268 PMID:19892972

Dixon, M. J. R., Loh, J., Davidson, N. C., Beltrame, C., Freeman, R., & Walpole, M. (2016). Tracking global change in ecosystem area: The Wetland Extent Trends index. *Biological Conservation, 193*, 27–35. doi:10.1016/j.biocon.2015.10.023

Dokulil, M. T., & Qian, K. (2021). Photosynthesis, carbon acquisition and primary productivity of phytoplankton: A review dedicated to *Colin Reynolds. Hydrobiologia, 848*(1), 77–94. doi:10.100710750-020-04321-y

Dormann, C. F., Bagnara, M., Boch, S., Hinderling, J., Janeiro-Otero, A., Schäfer, D., & Hartig, F. (2020). Plant species richness increases with light availability, but not variability, in temperate forests under storey. *BMC Ecology, 20*(1), 1–9. doi:10.118612898-020-00311-9 PMID:31941480

Dorney, J., Savage, R., Adamus, P., & Tiner, R. (Eds.). (2018). *Wetland and Stream Rapid Assessments: Development, Validation, and Application*. Academic Press.

Dudgeon, D. (2012). Threats to freshwater biodiversity globally and in the Indo-Burma Biodiversity *Hotspot*.Chair of Ecology and Biodiversity, School of Biological Sciences, The University of Hong Kong, Pok Fu Lam Road, Hong Kong SAR, China. *Environmental Science*.

Duplay, J., Semhi, K., Bloundi, M. K., Khedhiri, S., Ahmed, R. B., & Darragi, F. (2012). Geochemical background and contamination level assessment in Tunisian and Moroccan Mediterranean coastal lagoon sediments. *Environmental Earth Sciences, 67*(4), 1201–1214. doi:10.100712665-012-1566-6

Durand, P., Breuer, L., Johnes, P. J., Billen, G., Butturini, A., Pinay, G., ... Wright, R. (2011). *Nitrogen processes in aquatic ecosystems*. Academic Press.

Echeveste, P., Dachs, J., Galbán-Malagón, C., Berrojalbiz, N., & Agust, S. (2010). Decrease in the abundance and viability of oceanic phytoplankton due to trace levels of complex mixtures of organic pollutants. *Chemosphere, 81*(2), 161–168. doi:10.1016/j.chemosphere.2010.06.072 PMID:20673958

Edelkraut, K. A., & Güsewell, S. (2006). Progressive effects of shading on experimental wetland communities over three years. *Plant Ecology, 183*(2), 315–327. doi:10.100711258-005-9042-y

Edgar, R. C. (2004). MUSCLE: Multiple sequence alignment with high accuracy and high throughput. *Nucleic Acids Research, 32*(5), 1792–1797. doi:10.1093/nar/gkh340 PMID:15034147

Edokpayi, J. N., Odiyo, J. O., & Durowoju, O. S. (2017). *Impact of Wastewater on Surface Water Quality in Developing Countries: A Case Study of South Africa, Water Quality, Hlanganani Tutu*. IntechOpen. Available from: https://www.intechopen.com/chapters/53194 doi:10.5772/66561

Edwards, P. (1985). *Aquaculture: A Component of Low-Cost Sanitation Technology*. The World Bank.

Ehrlich, P. R., & Mooney, H. A. (1983). Extinction, substitution, and ecosystem services. *Bioscience, 33*(4), 248–254. doi:10.2307/1309037

EHSO. (2021). *Causes and Control of Water Pollution in Wetlands.* Retrieved June 25, 2021, from http://www.ehso.com/WaterPollution_Wetlands.htm

Eisler, R. (2003). Health risks of gold miners: A synoptic review. *Environmental Geochemistry and Health, 25*(3), 325–345. doi:10.1023/A:1024573701073 PMID:12971253

Ejtahed, H. S., Hoseini-Tavassol, Z., Khatami, S., Zangeneh, M., Behrouzi, A., Ahmadi Badi, S., Moshiri, A., Hasani-Ranjbar, S., Soroush, A. R., Vaziri, F., Fateh, A., Ghanei, M., Bouzari, S., Najar-Peerayeh, S., Siadat, S. D., & Larijani, B. (2020). Main gut bacterial composition differs between patients with type 1 and type 2 diabetes and non-diabetic adults. *Journal of Diabetes and Metabolic Disorders, 19*(1), 265–271. doi:10.100740200-020-00502-7 PMID:32550175

Eliot, C. (2007). Method and metaphysics in Clements's and Gleason's ecological explanations. *Studies in History and Philosophy of Science Part C Studies in History and Philosophy of Biological and Biomedical Sciences, 38*(1), 85–109. doi:10.1016/j.shpsc.2006.12.006 PMID:17324810

Emamverdian, A., Ding, Y., Mokhberdoran, F., & Xie, Y. (2015). Heavy metal stress and some mechanisms of plant defense response. *TheScientificWorldJournal, 2015*, 756120. doi:10.1155/2015/756120 PMID:25688377

Emery, S. (2010). Succession: A Closer Look. *Nature Education Knowledge, 3*(10), 45.

Emlen, J. T. (1977). Estimating breeding season bird densities from transect counts. *The Auk, 94*, 455–468.

Engin, M. S., Uyanik, A., & Kutbay, H. G. (2015). Accumulation of heavy metals in water, sediments and wetland plants of Kizilirmak delta (Samsun, Turkey). *International Journal of Phytoremediation, 17*(1-6), 66–75. doi:10.1080/15226514.2013.828019 PMID:25174426

Engle, V. D. (2011). Estimating the Provision of Ecosystem Services by Gulf of Mexico Coastal Wetlands.*Springer. Wetlands, 31*(1), 179–193. doi:10.100713157-010-0132-9

Environmental Protection Office of Wetlands. (2001). *Agency Oceans and Watersheds* (4502T). Author.

ENVIS Centre on Wildlife & Protected Areas. (2021). *RAMSAR Wetland Sites*. http://www.wiienvis.nic.in/Database/ramsar_wetland_sites_8224.aspx

EPA. (2001). Treats to wetlands. United States Office of Water. EPA 843-F-01-002d.

EPA. (2021). *Why are wetlands important?* https://www.epa.gov/wetlands/why-are-wetlands-important

Ervin, G. N., & Wetzel, R. G. (2002). Influence of a dominant macrophyte, Juncuseffusus, on wetland plant species richness, diversity, and community composition. *Oecologia, 130*(4), 626–636. doi:10.100700442-001-0844-x PMID:28547266

Erwin, K. L. (2009). Wetlands and global climate change: The role of wetland restoration in a changing world. *Wetlands Ecology and Management, 17*(1), 71–84. doi:10.100711273-008-9119-1

Eskandari Damaneh, H., Eskandari Damaneh, H., Khosravi, H., & Gholami, H. (2019). Analysis and monitoring of drought using NDVI index (Case study: The west basin of Jaz Murian wetland). *Rangeland, 13*(3), 461–475.

Eskandari-Damaneh, H., Noroozi, H., Ghoochani, O. M., Taheri-Reykandeh, E., & Cotton, M. (2020). Evaluating rural participation in wetland management: A contingent valuation analysis of the set-aside policy in Iran. *The Science of the Total Environment, 747*, 141127. doi:10.1016/j.scitotenv.2020.141127 PMID:32777492

Essefi, E. (2009). *Multidisciplinary study of Sidi El Hani Saline environment: The history and the climatic variability* [MS thesis]. Faculty of sciences of Sfax, University of Sfax.

Essefi, E. (2013). *Wet Aeolian Sedimentology and Sequence Stratigraphy within the Terrestrial Analogues in Eastern Tunisia: Implications for Wet Aeolian Sedimentology and Sequence Stratigraphy on Mars* (Doctoral dissertation). National Engineering School of Sfax, Sfax, Tunisia.

Essefi, E. (2021b). Record of the Anthropocene-Great Acceleration along a core from the coast of Sfax, southeastern Tunisia. *Turkish Journal of Earth Science*, 1-16.

Essefi, E. (2021c). Parallel evolution of energy consumption and environmental pollution during the Anthropocene. *Journal of Global Ecology and Environment*, 36-42.

Essefi, E., Gharsalli, N., & Yaich, C. (2015). Geophysical and Geochemical Study of the Silico-evaporitic Sedimentary Filling of Boujmal wetland, Eastern Tunisia: Inferring the Climatic Signal within Groundwater Noise. *Journal of Basic and Applied Research International*, 14-25.

Essefi, E., Gharsalli, N., & Yaich, C. (2015c). Geophysical and geochemical study of the silico-evaporitic sedimentary filling of Boujmal wetland, eastern Tunisia: inferring the climatic signal within groundwater noise. *J Basic Appl Res Int, 15*(2–1), 14–25.

Essefi, E., Gharsalli, N., Salhi, S., & Yaich, C. (2015b). Cyclicity of grain size parameters along a Core from the Gleysol of Sebkha Ennoual, southeastern Tunisia. *J Basic Appl Res Int, 15*(5–3), 124–134.

Essefi, E., Touir, J., Tagorti, M. A., & Yaich, C. (2013). Record of the climatic variability and the sedimentary dynamics during the last two millennia at sebkha Dkhila, eastern Tunisia (2013). *ISRN Geology*.

Essefi, E. (2020). On the Geochemistry and Mineralogy of the Anthropocene. *Int J Water Wastewater Treat*, *6*(2).

Essefi, E. (2020). On the Geochemistry and Mineralogy of the Anthropocene. *International Journal of Water and Wastewater Treatment*, *6*(2), 1–14.

Essefi, E. (2021a). *Apport des Approches Géochimique et Sédimentologique au Fonctionnement des Systèmes Endoréiques et Exoréiques de l'Holocène à l'Anthropocène: Intérêts Environnemental et Géoéconomique*. Institut Supérieur des Sciences et Techniques des Eaux de Gabès.

Essefi, E. (2021d). Homo Sapiens Sapiens Progressive Defaunation During The Great Acceleration: The Cli-Fi Apocalypse Hypothesis. *International Journal of Toxicology and Toxicity Assessment*, *1*(1), 1–6.

Essefi, E., Ben Smida, N., Jandoubi, I., Othmani, M. A., & Tagorti, M. A. (2020). Progressive evaporation of brine of sebkha Mchiguig, central Tunisia: A geo-economical comparative study of salt and brine. *Carbonates and Evaporites*, *35*(2), 59. doi:10.100713146-020-00592-7

Essefi, E., Gharsalli, N., Kalabi, S., Ben Ameur, M., & Yaich, C. (2015d). Spectral analysis of a core from the Sebkha of Sidi Mansour, Southern Tunisia: The Holocene cyclostratigraphy. *J Geophys Remote Sens*, *4*(02), 141. doi:10.4172/2169-0049.1000141

Essefi, E., Gharsalli, N., & Yaich, C. (2015). Climatic significance of grain size parameters of the wet aeolian sedimentary filling of sebkha mhabeul, southeast tunisia: Contribution of the genetic approach. *Journal of Basic and Applied Research International*, *1*(4), 140–152.

Essefi, E., Gharsalli, N., & Yaich, C. (2015a). Climatic significance aeolian sedimentary filling of sebkha Mhabeul, southeastern Tunisia: Contribution of the genetic approach. *J Basic Appl Res Int*, *15*(2–4), 140–152.

Essefi, E., Jmaa, H. B., Touir, J., Tagorti, M. A., & Yaich, C. (2013). Variability to sedimentary dynamics and climatic conditions during the last two millennia at sebkha Souassi in eastern Tunisia. *Papers on Global Change IGBP*, *20*(1), 29–45. doi:10.2478/igbp-2013-0001

Essefi, E., Komatsu, G., Fairén, A. G., Chan, M. A., & Yaich, C. (2014a). Groundwater influence on the aeolian sequence stratigraphy of the Mechertate–Chrita–Sidi El Hani system, Tunisian Sahel: Analogies to the wet–dry aeolian sequence stratigraphy at Meridiani Planum, Terby crater, and Gale crater, Mars. *Planetary and Space Science*, *95*, 56–78. doi:10.1016/j.pss.2013.05.010

Essefi, E., Komatsu, G., Fairén, A. G., Chan, M. A., & Yaich, C. (2014b). Models of formation and activity of spring mounds in the Mechertate-Chrita-Sidi El Hani system, eastern Tunisia: Implications for the habitability of Mars. *Life (Chicago, Ill.)*, *4*(3), 386–432. doi:10.3390/life4030386 PMID:25370379

Essefi, E., Mefteh, S., Medhioub, M., & Yaich, C. (2014). Magnetic study of the heated and unheated sedimentary fillings of Sebkha Mhabeul, Southeast Tunisia: A geophysical method for paleoclimatic investigation and tephrochronological dating. *International Journal of Geophysics*, *2014*, 2014. doi:10.1155/2014/908395

Eur. Environ. Agency. (EEA). (2000). *Management of contaminated sites in Western Europe: Topic report No.13/1999*. EEA. Retrieved June 24, 2021, from https://www.eea.europa.eu/publications/Topic_report_No_131999

European Medicines Agency and the European Centre for Disease Prevention and Control. (2009). *The bacterial challenge – time to react. A call to narrow the gap between multidrug-resistant bacteria in the EU and development of new antibacterial agents*. European Medicines Agency. Available from: https://ecdc.europa.eu/sites/portal/files/media/en/publications/Publications/0909_TER_The_Bacterial_Challenge_Time_to_React.pdf

Eurostat. (2021). *Municipal Waste Statistics*. Retrieved from https://ec.europa.eu/eurostat/statistics-explained/index.php?title=Municipal_waste_statistics#Municipal_waste_generation

Everard, M., Kangabam, R., Tiwari, M. K., McInnes, R., Kumar, R., Talukdar, G. H., & Das, L. (2019). Ecosystem service assessment of selected wetlands of Kolkata and the Indian Gangetic Delta: Multi-beneficial systems under differentiated management stress. *Wetlands Ecology and Management*, *27*(2), 405–426. doi:10.100711273-019-09668-1

Fagorite, V. I., Odundun, O. A., Iwueke, L. E., Nwaigbo, U. N., & Okeke, O. C. (2019). Wetlands: A review of their classification, significance and management for sustainable development. *International Journal of Advanced Academic Research, Sciences, Technology and Engineering*, *5*(3), 2488–9849.

FAO. (2013). Guidelines to control water pollution from agriculture in China. *Water Reports*, 40.

Farkhondeh, T., Naseri, K., Esform, A., Aramjoo, H., & Naghizadeh, A. (2020). Drinking water heavy metal toxicity and chronic kidney diseases: A systematic review. *Reviews on Environmental Health*, *36*(3), 359–366. doi:10.1515/reveh-2020-0110 PMID:33128529

Favas, P. J., Pratas, J., Rodrigues, N., D'Souza, R., Varun, M., & Paul, M. S. (2018). Metal (loid) accumulation in aquatic plants of a mining area: Potential for water quality biomonitoring and biogeochemical prospecting. *Chemosphere*, *194*, 158–170. doi:10.1016/j.chemosphere.2017.11.139 PMID:29202268

Fedorova, I., Zdorovennov, R., Kadutskiy, V., Fedorov, G., Shestakova, E., Zdorovennova, G., & Nigamatzyanova, G. (2021, April). Geochemical sensitivity of lacustrine ecosystems of Yamal Peninsula (Russian Arctic) to climate change. In EGU General Assembly Conference Abstracts (pp. EGU21-15069). Academic Press.

Felsenstein, J. (1981). Evolutionary trees from DNA sequences: A maximum likelihood approach. *Journal of Molecular Evolution*, *17*(6), 368–376. doi:10.1007/BF01734359 PMID:7288891

Felsenstein, J. (1985). Phylogenies and the Comparative Method. *American Naturalist*, *125*(1), 1–15. doi:10.1086/284325

Feng, L., Zhang, W., & Li, X. (2018). Monitoring of regional drug abuse through wastewater-based epidemiology—A critical review. *Science China. Earth Sciences*, *61*(3), 239–255. doi:10.100711430-017-9129-x

Fernández-Martínez, M. Á., García-Villadangos, M., Paz, M. M., Gangloff, V., Carrizo, D., Blanco, Y., & Fairén, A. (2021). Geomicrobiological heterogeneity of lithic habitats in the extreme environment of Antarctic Nunataks: A potential early Mars analog. *Frontiers in Microbiology*, *12*, 1568. doi:10.3389/fmicb.2021.670982 PMID:34276605

Fernando, C. H. (1977). Investigations on the aquatic fauna of tropical ricefields with special reference to southeast Asia. *Geo-Eco-Trop*, *1*, 166–188.

Ferrelli, F., Brendel, A. S., Perillo, G. M. E., & Piccolo, M. C. (2021). Warming signals emerging from the analysis of daily changes in extreme temperature events over Pampas (Argentina). *Environmental Earth Sciences*, *80*(12), 1–13. doi:10.100712665-021-09721-4

Ferronato, N., & Torretta, V. (2019). Waste Mismanagement in Developing Countries: A Review of Global Issues. *International Journal of Environmental Research and Public Health*, *16*(6), 1060. doi:10.3390/ijerph16061060 PMID:30909625

Feuillette, S. (2001). *Vers une gestion de la demande sur une nappe en accès libre: exploration des interactions ressources usages par les systèmes multi-agents application à la nappe de Kairoun, Tunisie centrale* (Thèse de doctorat). Université de Montplier II.

Finlayson, C. M., D'Cruz, R., Aladin, N., Barker, D.R., Beltram, G., Brouwer, J., Davidson N., Duker, L., Junk, W.J., Kaplowitz, M.D., Ketelaars, H., Kreuzberg-Mukhina, E., Espino, G.L., Lévéque, C., Lopez, A., Milton, R.G., Mirabzadeh, P., Pritchard, D., Revenga, C., Rivera, M., … Steinkamp, M. (2005). Inland Water Systems. In Ecosystems and Human Well-being: Current State and Trends (pp. 551-583). Washington, DC: Island Press.

Finlayson, C. M. (1991). Plant ecology and management of an internationally important wetland in monsoonal Australia. In *Proceedings of an international symposium on wetlands and river corridor management*. New York: Association of State Wetland Managers.

Finlayson, C. M. (2012). Forty years of wetland conservation and wise use. *Aquatic Conservation*, *22*(2), 139–143. doi:10.1002/aqc.2233

Finlayson, C. M., Davies, G. T., Moomaw, W. R., Chmura, G. L., Natali, S. M., Perry, J. E., & Sutton-Grier, A. E. (2019). The second warning to humanity–providing a context for wetland management and policy. *Wetlands*, *39*(1), 1–5. doi:10.100713157-018-1064-z

Finlayson, C. M., & Spiers, A. G. (1999). *A compendium of information for managing and monitoring wetlands in tropical Australia. Supervising Scientist Report, 148*. Supervising Scientist.

Finlayson, C. M., & Spiers, A. G. (Eds.). (1999). *Global Review of Wetland Resources and Priorities for Wetland Inventory*. Supervising Scientist.

Fleischer, S., Gustafson, A., Joelsson, A., Pansar, J., & Stibe, L. (1994). Nitrogen removal in created ponds. *Ambio*, *6*, 349–357.

Fletcher, W. J., & Zielhofer, C. (2013). Fragility of Western Mediterranean landscapes during Holocene Rapid Climate Changes. *Catena*, *103*, 16–29. doi:10.1016/j.catena.2011.05.001

Fluharty, R. G., Simpson, F. B., Russell, G. J., & Morris, R. H. (1967). *A Proposal for a Repetitively Pulsed Test Facility (RPTF) (No. IN-1149)*. Idaho Nuclear Corp. doi:10.2172/4557773

Foley, J. A., DeFries, R., Asner, G. P., Barford, C., Bonan, G., Carpenter, S. R., Chapin, F. S., Coe, M. T., Daily, G. C., Gibbs, H. K., Helkowski, J. H., Holloway, T., Howard, E. A., Kucharik, C. J., Monfreda, C., Patz, J. A., Prentice, I. C., Ramankutty, N., & Snyder, P. K. (2005). Global consequences of land use. *Science*, *309*(5734), 570–574. doi:10.1126cience.1111772 PMID:16040698

Folk, R. L. (1966). A review of grain-size parameters. *Sedimentology*, *6*(2), 73–93. doi:10.1111/j.1365-3091.1966.tb01572.x

Folk, R. L. (1971). Longitudinal dunes of the northwestern edge of the Simpson Desert, Northern Territory, Australia, 1. Geomorphology and grain size relationships. *Sedimentology*, *16*(1-2), 5–54. doi:10.1111/j.1365-3091.1971.tb00217.x

Foote, A. L., Pandey, S., & Krogman, N. T. (1996). Processes of wetland loss in India. *Environmental Conservation*, *23*(1), 45–54. doi:10.1017/S0376892900038248

Förster, J., Barkmann, J., Fricke, R., Hotes, S., Kleyer, M., Kobbe, S., & Wittmer, H. (2015). Assessing ecosystem services for informing land-use decisions: A problem-oriented approach. *Ecology and Society*, *20*(3), 31. doi:10.5751/ES-07804-200331

Francis-Floyd, R., Watson, C., Petty, D., & Pouder, D.B. (2009). *Ammonia in aquatic systems*. UF/IFAS University of Florida (UF)/Institute of Food and Agricultural Sciences (IFAS).

Freeman, C., Ostle, N., & Kang, H. (2001). An enzymic'latch'on a global carbon store. *Nature*, *409*(6817), 149–149. doi:10.1038/35051650 PMID:11196627

Freeman, D. J., Falkiner, F. R., & Keane, C. T. (1989). New method for detecting slime production by coagulase negative staphylococci. *Journal of Clinical Pathology*, *42*(8), 872–874. doi:10.1136/jcp.42.8.872 PMID:2475530

Freitas, M., Azevedo, J., Pinto, E., Neves, J., Campos, A., & Vasconcelos, V. (2015). Effects of microcystin-LR, cylindrospermopsin and a microcystin-LR/cylindrospermopsin mixture on growth, oxidative stress and mineral content in lettuce plants (*Lactucasativa* L.). *Ecotoxicology and Environmental Safety*, *116*, 59–67. doi:10.1016/j.ecoenv.2015.02.002 PMID:25768423

Frolking, S., & Roulet, N. T. (2007). Holocene radiative forcing impact of northern peatland carbon accumulation and methane emissions. *Global Change Biology*, *13*(5), 1079–1088. doi:10.1111/j.1365-2486.2007.01339.x

Frolking, S., Roulet, N., & Fuglestvedt, J. (2006). How northern peatlands influence the Earth's radiative budget: Sustained methane emission versus sustained carbon sequestration. *Journal of Geophysical Research. Biogeosciences*, *111*(G1).

Frost, F. J., Muller, T., Petersen, H. V., Thomson, B., & Tollestrup, K. (2003). Identifying US populations for the study of health effects related to drinking water arsenic. *Journal of Exposure Science & Environmental Epidemiology*, *13*(3), 231–239. doi:10.1038j.jea.7500275 PMID:12743617

Fung, Z., Pomun, T., Charles, K. J., & Kirchherr, J. (2019). Mapping the social impacts of small dams: The case of Thailand's Ing River basin. *Ambio*, *48*(2), 180–191. doi:10.100713280-018-1062-7 PMID:29797277

Gajewska, M., Jóźwiakowski, K., Ghrabi, A., & Masi, F. (2015). Impact of influent wastewater quality on nitrogen removal rates in multistage treatment wetlands. *Environmental Science and Pollution Research International*, *22*(17), 12840–12848. doi:10.100711356-014-3647-4 PMID:25300181

Galatowitsch, S. M. (2018). Natural and anthropogenic drivers of wetland change. In C. Finlayson, G. Milton, R. Prentice, & N. Davidson (Eds.), *The Wetland Book II: Distribution, Description, and Conservation* (pp. 359–367). Springer Nature. doi:10.1007/978-94-007-4001-3_217

Gallagher, C. J., & Bahia, R. (2021). Outflow channels on Mars. In *Mars Geological Enigmas* (pp. 13–40). Elsevier. doi:10.1016/B978-0-12-820245-6.00002-1

Gammoudi, A. (2020). *Etude géochimique et cyclostratigraphie holocène des sédiments de la sebkha Mhabeul au Sud-Est de la Tunisie* (PhD Thesis). Higher Institute of Applied Sciences and Technology of Gabes.

Gammoudi, A., Essefi, E., & Rigane, H. (2019). Variability of the grain-sized fractions during the last three millennia within the sedimentary fillings of the saline system of Mhabeul: The Holocene cyclostratigraphy. *Arabian Journal of Geosciences, 12*(9), 1–11. doi:10.100712517-019-4472-3

Ganesh, K., & Kanhaiya, A. (Eds.). (2006). *Environmental Issues of Development*. The Associated Publishers.

Garcias-Bonet, N., Arrieta, J. M., de Santana, C. N., Duarte, C. M., & Marbà, N. (2012). Endophytic bacterial community of a Mediterranean marine angiosperm (Posidonia oceanica). *Frontiers in Microbiology, 3*, 342. doi:10.3389/fmicb.2012.00342 PMID:23049528

Garden Jenni, G., Mcalpine Clive, A., Possingham Hugh, P., & Jones Darryl, N. (2007). Using multiple survey methods to detect terrestrial reptiles and mammals: What are the most successful and cost-efficient combinations? *Wildlife Research, 34*(3), 218–227. doi:10.1071/WR06111

Gardner, R. C., Barchiesi, S., Beltrame, C., Finlayson, C., Galewski, T., Harrison, I., & Walpole, M. (2015). State of the world's wetlands and their services to people: a compilation of recent analyses. Ramsar Convention Secretariat, Ramsar scientific and technical briefing note no. 7, Gland, Switzerland.

Garg, J. K. (2013). Wetland assessment, monitoring and management in India using geospatial techniques. *Journal of Environmental Management*, 1–12. doi:10.1016/j.jenvman.2013.12.018 PMID:24486190

Gauci, V., Matthews, E., Dise, N., Walter, B., Koch, D., Granberg, G., & Vile, M. (2004). Sulfur pollution suppression of the wetland methane source in the 20th and 21st centuries. *Proceedings of the National Academy of Sciences of the United States of America, 101*(34), 12583–12587. doi:10.1073/pnas.0404412101 PMID:15297612

Gecheva, G., Pall, K., & Hristeva, Y. (2017). Bryophyte communities' responses to environmental factors in highly seasonal rivers. *Botany Letters, 164*(1), 79–91. doi:10.1080/23818107.2016.1263238

Gecheva, G., Yurukova, L., & Ganeva, A. (2011). Assessment of pollution with aquatic bryophytes in Maritsa River (Bulgaria). *Bulletin of Environmental Contamination and Toxicology, 87*(4), 480–485. doi:10.100700128-011-0377-x PMID:21814844

Gerhardt, P., Murray, R. G. E., Wood, W. A., & Krieg, N. R. (1994). *Methods for general and molecular bacteriology*. American Society for Microbiology. https://www.ncbi.nlm.nih.gov/

Ghani, S. A. A. (2015). Trace metals in seawater, sediments and some fish species from MarsaMatrouh Beaches in north-western Mediterranean coast, Egypt. *Egyptian Journal of Aquatic Research, 41*(2), 145–154. doi:10.1016/j.ejar.2015.02.006

Gharsalli, N., Essefi, E., Baydoun, R., & Yaich, C. (2017). Filling Sebkha Halk el Menjel, Eastern Tunisia. *The 1st Atlas Georesources International Congress*.

Gharsalli, N., Essefi, E., Baydoun, R., & Yaich, C. (2020). The Anthropocene and Great Acceleration as Controversial Epoch Of Human-Induced Activities: Case Study Of The Halk El Menjel Wetland, Eastern Tunisia. *Applied Ecology and Environmental Research, 18*(3), 4137–4166. doi:10.15666/aeer/1803_41374166

Ghermandi, A., van den Bergh, J. C. J. M., Brander, L. M., & Nunes, P. A. L. D. (2008). *The Economic Value of Wetland Conservation and Creation: A Meta-Analysis* [Working Paper 79]. Fondazione Eni Enrico Mattei.

Ghoochani, O. M., Ghanian, M., Khosravipour, B., & Crotts, J. C. (2020). Sustainable tourism development performance in the wetland areas: A proposed composite index. *Tourism Review*. Advance online publication. doi:10.1108/TR-02-2019-0061

Gibbs, J. P. (2011). Wetland loss and biodiversity conservation. *Conservation Biology, 14*(1), 314–317. doi:10.1046/j.1523-1739.2000.98608.x

Gibbs, R. J. (1970). Mechanisms controlling world water chemistry. *Science*, *170*(3962), 1088–1089. doi:10.1126cience.170.3962.1088 PMID:17777828

Gilbert, G., Gibbons, D. W., & Evans, J. (1998). *Bird Monitoring Methods*. RSPB Sandy.

Gillings, M. R., & Paulsen, I. T. (2014). Microbiology of the Anthropocene. *Anthropocene*, *5*, 1–8. doi:10.1016/j.ancene.2014.06.004

Gimenez, F. V., & Mas, C. R. (2020). The Valuation of Recreational Use of Wetlands and the Impact of the Economic Crisis. *International Journal of Environmental Research and Public Health*, *17*(9), 3228–3242. doi:10.3390/ijerph17093228 PMID:32384709

Giusti, L. (2009). A review of waste management practices and their impact on human health. *Waste Management (New York, N.Y.)*, *29*(8), 2227–2239. doi:10.1016/j.wasman.2009.03.028 PMID:19401266

Gleason, H. A. (1926). The individualistic concept of the plant association. *Bulletin of the Torrey Botanical Club*, *53*(1), 7–26. doi:10.2307/2479933

Goosse, H., Dalaiden, Q., Cavitte, M. G., & Zhang, L. (2021). Can we reconstruct the formation of large open-ocean polynyas in the Southern Ocean using ice core records? *Climate of the Past*, *17*(1), 111–131. doi:10.5194/cp-17-111-2021

Gopal, B. (1992) Tropical wetlands: degradation and need for rehabilitation. In Ecosystem Rehabilitation Vol. 2: Ecosystem analysis and synthesis. Academic Publishing.

Gopal, B. (1994) Conservation of inland waters in India: an overview. *Verhandlungen der Internationalen Vereinigung fiir Theoretische und Angewandte Limndlogie, 25*, 2494-7.

Gopal, B. (2013). Future of wetlands in tropical and subtropical Asia, especially in the face of climate change. *Aquatic Sciences*, *75*(1), 39–61. doi:10.100700027-011-0247-y

Gopal, B., & Chauhan, M. (2018). The Transboundary Sundarbans Mangroves (India and Bangladesh). In C. Finlayson, G. Milton, R. Prentice, & N. Davidson (Eds.), *The Wetland Book*. Springer.

Gopal, B., & Junk, W. J. (2000). Biodiversity in wetlands: An introduction. In B. Gopal, W. J. Junk, & J. A. Davis (Eds.), *Biodiversity in Wetlands: Assessment, Function and Conservation* (Vol. 1, pp. 1–10). Island Press.

Gopal, B., & Junk, W. J. (2001). *Assessment, determinants, function and conservation of biodiversity in wetlands: Present status and future needs*. Academic Press.

Gopal, B., Junk, W. J., Kakuru, W., Turyahabwe, N., & Mugisha, J. (2013). Total economic value of wetlands products and Services in Uganda. *The Scientific World Journal*, 1–13.

Gopal, B., & Sah, M. (1993). Conservation and management of rivers in India: Case-study of the River Yamuna. *Environmental Conservation*, *20*(3), 243–254. doi:10.1017/S0376892900023031

Goslin, J., Fruergaard, M., Sander, L., Gałka, M., Menviel, L., Monkenbusch, J., & Clemmensen, L. B. (2018). Holocene centennial to millennial shifts in North-Atlantic storminess and ocean dynamics. *Scientific Reports*, *8*(1), 1–12. doi:10.103841598-018-29949-8 PMID:30143649

Gosselink, J. G., & Turner, R. E. (1978). The role of hydrology in freshwater wetland ecosystems. In *Freshwater wetlands Ecological processes and management potential*. Academic Press. https://www.epa.gov/sites/default/files/2016-02/documents/wetlandfunctionsvalues.pdf

Gossner, M. M. (2009). Light intensity affects spatial distribution of Heteroptera in deciduous forests. *European Journal of Entomology*, *106*(2), 241–252. doi:10.14411/eje.2009.032

Government of India. (1991). *India 1990: A Reference Annual.* Research and Reference Division, Ministry of Information and Broadcasting, Government of India.

Granek, E., & Ruttenberg, B. I. (2008). Changes in biotic and abiotic processes following mangrove clearing. *Estuarine, Coastal and Shelf Science, 80*(4), 555–562. doi:10.1016/j.ecss.2008.09.012

Graversgaard, M., Jacobsen, B. H., Hoffmann, C. C., Dalgaard, T., Odgaard, M. V., Kjaergaard, C., & Tonderski, K. (2021). Policies for wetlands implementation in Denmark and Sweden–historical lessons and emerging issues. *Land Use Policy, 101*, 105206. doi:10.1016/j.landusepol.2020.105206

Gray, M. H., & Jeffrey, D. M. (2019). A review of research on the fertiliser and lime requirements of pasture on the eastern coast of the North Island. *Journal of New Zealand Grasslands, 81*, 21–32. doi:10.33584/jnzg.2019.81.373

Greninger, A. L. (2018). The challenge of diagnostic metagenomics. *Expert Review of Molecular Diagnostics, 18*(7), 605–615. doi:10.1080/14737159.2018.1487292 PMID:29898605

Guerrero-Latorre, L., Ballesteros, I., Villacrés-Granda, I., Granda, M. G., Freire-Paspuel, B., & Ríos-Touma, B. (2020). SARS-CoV-2 in river water: Implications in low sanitation countries. *The Science of the Total Environment, 743*, 140832. doi:10.1016/j.scitotenv.2020.140832 PMID:32679506

Guo, C., Chen, Y., Xia, W., Qu, X., Yuan, H., Xie, S., & Lin, L. S. (2020). Eutrophication and heavy metal pollution patterns in the water suppling lakes of China's south-to-north water diversion project. *The Science of the Total Environment, 711*, 134543. doi:10.1016/j.scitotenv.2019.134543 PMID:31812425

Guo, L., Dai, L., Yang, K., Li, D., & Li, G. (2018). Physiological changes of submerged macrophytes in response to a floating filamentous green algae bloom in clear-water conditions. *Journal of Oceanology and Limnology, 36*(5), 1604–1614. doi:10.100700343-018-7139-0

Gwin, S. E., Kentula, M. E., & Shaffer, P. W. (1999). Evaluating the Effects of Wetland Regulation through Hydrogeomorphic Classification and Landscape Profiles. *Wetlands, 19*(3), 477–489. doi:10.1007/BF03161687

Häder, D. P., Banaszak, A. T., Villafañe, V. E., Narvarte, M. A., González, R. A., & Helbling, E. W. (2020). Anthropogenic pollution of aquatic ecosystems: Emerging problems with global implications. *The Science of the Total Environment, 713*, 136586. doi:10.1016/j.scitotenv.2020.136586 PMID:31955090

Hall, A. (2021). *The Influence of Anthropogenic Climate Change on the 2015-2017 Hydrological Drought in the South-Western Cape, South Africa* (Master's thesis).

Halls, A. J. E. (1997). *Wetlands, biodiversity and the Ramsar convention: the role of the convention on wetlands in the conservation and wise use of biodiversity* (Vol. 13). In Ramsar Convention Bureau.

Halpern, B. S., Walbridge, S., Selkoe, K. A., Kappel, C. V., Micheli, F., D'Agrosa, C., & Watson, R. (2008). A global map of human impact on marine ecosystems. *Science, 319*(5865), 948–952. doi:10.1126cience.1149345 PMID:18276889

Hampel, M., Blasco, J., & Segner, H. (2015). Molecular and cellular effects of contamination in aquatic ecosystems. *Environmental Science and Pollution Research International, 22*(22), 17261–17266. doi:10.100711356-015-5565-5 PMID:26507731

Hansen, V. D., & Nestlerode, J. A. (2014). Carbon sequestration in wetland soils of the northern Gulf of Mexico coastal region. *Wetlands Ecology and Management, 22*(3), 289–303. doi:10.100711273-013-9330-6

Haque, R., Mazumder, D. G., Samanta, S., Ghosh, N., Kalman, D., Smith, M. M., & Smith, A. H. (2003). Arsenic in drinking water and skin lesions: Dose-response data from West Bengal, India. *Epidemiology (Cambridge, Mass.), 14*(2), 174–182. doi:10.1097/01.EDE.0000040361.55051.54 PMID:12606883

Harada, J., Inoue, T., Kato, K., Uraie, N., & Sakuragi, H. (2015). Performance evaluation of hybrid treatment wetland for six years of operation in cold climate. *Environmental Science and Pollution Research International*, *22*(17), 12861–12869. doi:10.100711356-014-3843-2 PMID:25424033

Harris, N. L., Gibbs, D. A., Baccini, A., Birdsey, R. A., De Bruin, S., Farina, M., & Tyukavina, A. (2021). Global maps of twenty-first century forest carbon fluxes. *Nature Climate Change*, *11*(3), 234–240. doi:10.103841558-020-00976-6

Harriss, R. C., Sebacher, D. I., & Day, F. P. Jr. (1982). Methane flux in the great dismal swamp. *Nature*, *297*(5868), 673–674. doi:10.1038/297673a0

Hartmann, D. L., Tank, A. M. K., Rusticucci, M., Alexander, L. V., Brönnimann, S., Charabi, Y. A. R., & Zhai, P. (2013). Observations: atmosphere and surface. In Climate change 2013 the physical science basis: Working group I contribution to the fifth assessment report of the intergovernmental panel on climate change (pp. 159-254). Cambridge University Press.

Hasan, M. K., Shahriar, A., & Jim, K. U. (2019). Water pollution in Bangladesh and its impact on public health. *Heliyon*, *5*(8), e02145. doi:10.1016/j.heliyon.2019.e02145 PMID:31406938

Hashimet, A. M., & Catherine, S. M. P. (2013). A Laboratory Study on Wave Reduction by Mangrove Forests. *Elsevier. APCBEE Procedia*, *5*, 27–32. doi:10.1016/j.apcbee.2013.05.006

Hassan, B., Yatoo, A. M., Khan, I. S., Ali, M. N., & Qadri, H. (2020). Determination of some commonly used antibiotic residues in hospital effluents of kashmir valley. *Uttar Pradesh Journal of Zoology*, *41*(17), 1–7. http://www.mbimph.com/index.php/UPJOZ/article/view/1703

Hastings, K. L., Smith, L. E., Lindsey, M. L., Blotsky, L. C., Downing, G. R., Zellars, D. Q., Downing, J. K., & Corena-McLeod, M. (2014). Effect of microalgae application on soil algal species diversity, cation exchange capacity and organic matter after herbicide treatments. *F1000 Research*, *3*(281), 281. doi:10.12688/f1000research.4016.1

Hawting, G., & Mateo, N. (1987). Farming on high International Development Research Center, Markham, Ontario. *Report*, *16*(1), 4–5.

He, Y., Yuan, Q., Mathieu, J., Stadler, L., Senehi, N., Sun, R., & Alvarez, P. J. J. (2020). Antibiotic resistance genes from livestock waste: occurrence, dissemination, and treatment. *NPJ Clean Water*, *3*(4). doi:10.1038/s41545-020-0051-0

Hébert, R., & Lovejoy, S. A. (2017). Scaling Model for the Anthropocene Climate Variability with Projections to 2100. EGU General Assembly Conference Abstracts, 19, 9943.

Hefting, M. M., Bobbink, R., & de Caluwe, H. (2003). Nitrous oxide emission and denitrification in chronically nitrate-loaded riparian buffer zones. *Journal of Environmental Quality*, *32*(4), 1194–1203. doi:10.2134/jeq2003.1194 PMID:12931872

Helbig, M., Gerken, T., Beamesderfer, E. R., Baldocchi, D. D., Banerjee, T., Biraud, S. C., & Richardson, A. D. (2021). Integrating continuous atmospheric boundary layer and tower-based flux measurements to advance understanding of land-atmosphere interactions. *Agricultural and Forest Meteorology*, *307*, 108509. doi:10.1016/j.agrformet.2021.108509

Hendriksen, R. S., Munk, P., Njage, P., van Bunnik, B., McNally, L., Lukjancenko, O., Röder, T., Nieuwenhuijse, D., Pedersen, S. K., Kjeldgaard, J., Kaas, R. S., Clausen, P., Vogt, J. K., Leekitcharoenphon, P., van de Schans, M., Zuidema, T., de Roda Husman, A. M., Rasmussen, S., Petersen, B., ... Aarestrup, F. M. (2019). Global monitoring of antimicrobial resistance based on metagenomics analyses of urban sewage. *Nature Communications*, *10*(1), 1124. doi:10.103841467-019-08853-3 PMID:30850636

Herbert, E. R., Boon, P., Burgin, A. J., Neubauer, S. C., Franklin, R. B., Ardón, M., & Gell, P. (2015). A global perspective on wetland salinization: Ecological consequences of a growing threat to freshwater wetlands. *Ecosphere*, *6*(10), 1–43. doi:10.1890/ES14-00534.1

Herrera, V. V., Mendoza, B., & Herrera, G. V. (2015). Reconstruction and prediction of the total solar irradiance: From the Medieval Warm Period to the 21st century. *New Astronomy*, *34*, 221–233. doi:10.1016/j.newast.2014.07.009

Hinton, J. J., Veiga, M. M., & Veiga, A. T. C. (2003). Clean artisanal gold mining: A utopian approach? *Journal of Cleaner Production*, *11*(2), 99–115. doi:10.1016/S0959-6526(02)00031-8

Hites, R. A. (2004). Polybrominateddiphenyl ethers in the environment and in people: A meta-analysis of concentrations. *Environmental Science & Technology*, *38*(4), 945–956. doi:10.1021/es035082g PMID:14998004

Hladikova, K., Ruzickova, I., Klucova, P., & Wanner, J. (2002). An investigation into studying of the activated sludge foaming potential by using physicochemical parameters. *Water Science and Technology*, *46*(1-2), 525–528. doi:10.2166/wst.2002.0529 PMID:12216681

Hoffman, J. S., Wells, J. B., & Titus, J. G. (1986). Future global warming and sea level rise. In J. G. Titus (Ed.), Greenhouse effect, sea level rise and coastal wetlands. US Government Printing Office, EPA-230-05-86-0I3.

Hoffmann, C. C. H. (1998). *Nutrient retention in wet meadows and fens* (PhD thesis). Ministry of Environment and Energy, National Environmental Research Institute, Silkeborg, Denmark.

Holcová, K., Suchánek, V., Heřmanová, Z., & Vondrák, D. (2020). The First Evidence of Bioerosion in Chitinous Lacustrine Microfossils from Glacial Lakes (late Pleistocene and Holocene, Bohemian Forest). *Palaios*, *35*(8), 317–326. doi:10.2110/palo.2020.008

Holdaway, R. J., Richardson, S. J., Dickie, I. A., Peltzer, D. A., & Coomes, D. A. (2011). Species- and community-level patterns in fine root traits along a 120000-year soil chronosequence in temperate rain forest. *Journal of Ecology*, *99*(4), 954–963. doi:10.1111/j.1365-2745.2011.01821.x

Holmberg, T. (2019). Animal waste work. The case of urban sewage management in Sweden. *Contemporary Social Science*, *16*(1), 1–15. doi:10.1080/21582041.2019.1630669

Ho, M., & Richardson, C. J. (2013). A five-year study of floristic succession in a restored urban wetland. *Ecological Engineering*, *61*, 511–518. doi:10.1016/j.ecoleng.2013.05.001

Hong, Z., Zhao, Q., Chang, J., Peng, L., Wang, S., Hong, Y., & Ding, S. (2020). Evaluation of water quality and heavy metals in wetlands along the Yellow River in Henan Province. *Sustainability*, *12*(4), 1300. doi:10.3390u12041300

Hossain, M. S., Rahman, N. N., Balakrishnan, V., Puvanesuaran, V. R., Sarker, M. Z., & Kadir, M. O. (2013). Infectious risk assessment of unsafe handling practices and management of clinical solid waste. *International Journal of Environmental Research and Public Health*, *10*(2), 556–567. doi:10.3390/ijerph10020556 PMID:23435587

Ho, Y. C., Show, K. Y., Guo, X. X., Norli, I., Abbas, F. A., & Morad, N. (2012). Industrial discharge and their effect to the environment. In *Industrial waste*. Intech.

Huang, Y., Chen, X. S., Li, F., Hou, Z. Y., Li, X., Zeng, J., & Xie, Y. H. (2021). Community Trait Responses of Three Dominant Macrophytes to Variations in Flooding During 2011–2019 in a Yangtze River-Connected Floodplain Wetland (Dongting Lake, China). *Frontiers in Plant Science*, *12*, 957. doi:10.3389/fpls.2021.604677 PMID:34122461

Hulkarni, D. H. (1970). *Forestry in India*. Oregon State University Printing Dept.

Hussain, S. A., Badola, R., Gopi, G. V., Johnson, J. A., Shivakumar, K., Dasgupta, N., Barthwal, S., & Tuboi, C. (2021). Framework for Ecological Monitoring of Ramsar Sites and other Wetlands in India. Ganga Aqualife Conservation Monitoring Centre (GACMC). Wildlife Institute of India.

Imer, D., Merbold, L., Eugster, W., & Buchmann, N. (2013). Temporal and spatial variations of CO2, CH4 and N2O fluxes at three differently managed grasslands. *Biogeosciences Discussions*, *10*, 2635–2673.

Ingole, M. P. D. (2021). Environment and global warming in current scenario. *EPRA International Journal of Agriculture and Rural Economic Research*, *9*(6), 16–19.

Inman, D. L. (1952). Measures for describing size of sediments. *Journal of Sedimentary Petrology*, *19*(2), 125–145.

IPCC. (2002). Climate change and biodiversity. IPCC Technical Paper V.

IPCC. (2014). *Climate Change 2014: Synthesis Report*. In R. K. Pachauri & L. A. Meyer (Eds.), *Contribution of Working Groups I, II and III to the Fifth Assessment Report of the Intergovernmental Panel on Climate Change* (p. 151). IPCC.

Islam, Z., & Rahmani, A. R. (2008). Potential and Existing Ramsar Sites in India. Bombay Natural History Society, Oxford University Press.

Ivčević, A., Statzu, V., Satta, A., & Bertoldo, R. (2021). The future protection from the climate change-related hazards and the willingness to pay for home insurance in the coastal wetlands of West Sardinia, Italy. *International Journal of Disaster Risk Reduction*, *52*, 101956. doi:10.1016/j.ijdrr.2020.101956

Iyengar, S., Dolšak, S., & Prakash, A. (2019, December). Selectively Assertive: Interventions of India's Supreme Court to Enforce Environmental Laws. *Sustainability*, *11*(24), 7234. Advance online publication. doi:10.3390u11247234

Jain, S. K. (2005). Water Resources of India. Water Encyclopedia. doi:10.1002/047147844X.wr243

Jain, A. K. (2005). *A Sustainable Vision for Urban India*. Kalpaz Publications.

Jaishankar, M., Tseten, T., Anbalagan, N., Mathew, B. B., & Beeregowda, K. N. (2014). Toxicity, mechanism and health effects of some heavy metals. *Interdisciplinary Toxicology*, *7*(2), 60–72. doi:10.2478/intox-2014-0009 PMID:26109881

Jalali, B., Sicre, M. A., Bassetti, M. A., & Kallel, N. (2016). Holocene climate variability in the north-western Mediterranean Sea (gulf of lions). *Climate of the Past*, *12*(1), 91–101. doi:10.5194/cp-12-91-2016

Jansson, M., Andersson, R., Berggren, H., & Leonardson, L. (1994). Wetlands and lakes as nitrogen traps. *Ambio*, *23*, 320–325.

Jaouadi, S., Lebreton, V., Bout-Roumazeilles, V., Siani, G., Lakhdar, R., Boussoffara, R., & Combourieu-Nebout, N. (2016). *Environmental changes, climate and anthropogenic impact in southern-eastern Tunisia during the last 8 kyr*. Academic Press.

Jaouadi, S., Lebreton, V., Bout-Roumazeilles, V., Siani, G., Lakhdar, R., Boussoffara, R., Dezileau, L., Kallel, N., Mannai-Tayech, B., & Combourieu-Nebout, N. (2016). Environmental changes, climate and anthropogenic impact in south-east Tunisia during the last 8 kyr. *Climate of the Past Discussions*, *12*(6), 1339–1359. doi:10.5194/cp-12-1339-2016

Jaouadi, S., Lebreton, V., Mannai-Tayech, B., Lakhdar, R., & Soussi, M. (2015). Apport de l'analyse pollinique des sédiments de la sebkha Boujmel (Sud-Est tunisien) à la reconstitution des paléopysages et paléoclimats holocènes en milieu aride. *Méditerranée*, *125*(125), 39–49. doi:10.4000/mediterranee.7891

Jarraya-Horriche, F. (2004). *Contribution à l'analyse et à la rationalisation des réseaux piézométriques* (Thèse de doctorat). Ecole Nationale d'Ingénieurs de Tunis.

Järvenpää, M., & Lindström, K. (2004). Water turbidity by algal blooms causes mating system breakdown in a shallow-water fish, the sand go by *Pomatos chistusminutus*. *Proceedings. Biological Sciences*, *271*(1555), 2361–2365. doi:10.1098/rspb.2004.2870 PMID:15556888

Jha, S. K., Singh, R. K., Damodaran, T., Mishra, V. K., Sharma, D. K., & Rai, D. (2013). Fluoride in groundwater: Toxicological exposure and remedies. *Journal of Toxicology and Environmental Health. Part B, Critical Reviews*, *16*(1), 52–66. doi:10.1080/10937404.2013.769420 PMID:23573940

Jiang, A. L., Lee, M. C., Zhou, G., Zhong, D., Hawaria, D., Kibret, S., & Hsu, K. (2021). Predicting distribution of malaria vector larval habitats in Ethiopia by integrating distributed hydrologic modeling with remotely sensed data. *Scientific Reports*, *11*(1), 1–14. doi:10.103841598-021-89576-8 PMID:33980945

Jian, Z., Wang, P., Saito, Y., Wang, J., Pflaumann, U., Oba, T., & Cheng, X. (2000). Holocene variability of the Kuroshio current in the Okinawa Trough, northwestern Pacific Ocean. *Earth and Planetary Science Letters*, *184*(1), 305–319. doi:10.1016/S0012-821X(00)00321-6

Jia, Y., Chen, W., Zuo, Y., Lin, L., & Song, L. (2018). Heavy metal migration and risk transference associated with cyanobacterial blooms in eutrophic freshwater. *The Science of the Total Environment*, *613-614*, 1324–1330. doi:10.1016/j.scitotenv.2017.09.180 PMID:28968935

Jiménez-Espejo, F. J., García-Alix, A., Jiménez-Moreno, G., Rodrigo-Gámiz, M., Anderson, R. S., Rodríguez-Tovar, F. J., Martínez-Ruiz, F., Giralt, S., Delgado Huertas, A., & Pardo-Igúzquiza, E. (2014). Saharan aeolian input and effective humidity variations over western Europe during the Holocene from a high altitude record. *Chemical Geology*, *374-375*, 1–12. doi:10.1016/j.chemgeo.2014.03.001

Jodha, N. S. (1992). *Common Property Resources ~ A Missing Dimension of Development Strategies*. Report No. 169. The World Bank.

Johnson, M. S., Matthews, E., Bastviken, D., Deemer, B., Du, J., & Genovese, V. (2021). Spatiotemporal methane emission from global reservoirs. *Journal of Geophysical Research: Biogeosciences*.

Johnson, P., & Rogers, G. (2003). *Ephemeral wetlands and their turfs in New Zealand*. Department of Conservation.

Junk, W. J., da Cunha, C. N., Thomaz, S. M., Agostinho, A. A., Ferreira, F. A., de Souza Filho, E. E., & Kawakita, K. (2021). Macrohabitat classification of wetlands as a powerful tool for management and protection: The example of the Paraná River floodplain, Brazil. *Ecohydrology & Hydrobiology*.

Jurburg, S. D., Nunes, I., Stegen, J. C., Le Roux, X., Priemé, A., Sørensen, S. J., & Salles, J. F. (2017). Autogenic succession and deterministic recovery following disturbance in soil bacterial communities. *Scientific Reports*, *7*(1), 1–11. doi:10.1038rep45691 PMID:28383027

Kahla-Nakbi, A. B., Chaieb, K., Besbes, A., Zmantar, T., & Bakhrouf, A. (2006). Virulence and enterobacterial repetitive intergenic consensus PCR of Vibrio alginolyticus strains isolated from Tunisian cultured gilthead sea bream and sea bass outbreaks. *Veterinary Microbiology*, *117*(2-4), 321–327. doi:10.1016/j.vetmic.2006.06.012 PMID:16870360

Kakuru, W., Turyahabwe, N., & Mugisha, J. (2013). Total economic value of wetlands products and services in Uganda. *The Scientific World Journal*. doi:10.1155/2013/192656 PMID:24163614

Kang, S. (2021). Microbes' Many Roles in Climate Change: Contribution, Consequence, Mitigation, and Model System. *Microbes: The Foundation Stone of the Biosphere*, 187-194.

Kansiime, F., & Maimuna, N. (1999). *Wastewater treatment by a natural wetland: The Nakivubo swamp*. CRC Press.

Karkman, A., Do, T. T., Walsh, F., & Virta, M. (2018). Antibiotic-Resistance Genes in Waste Water. *Trends in Microbiology, 26*(3), 220–228. doi:10.1016/j.tim.2017.09.005 PMID:29033338

Karpinski, E., Hackenberger, D., Zazula, G., Widga, C., Duggan, A. T., Golding, G. B., & Poinar, H. N. (2020). American mastodon mitochondrial genomes suggest multiple dispersal events in response to Pleistocene climate oscillations. *Nature Communications, 11*(1), 1–9. doi:10.103841467-020-17893-z PMID:32873779

Kaul, S., & Kumar, R. (2019). Wetland Conservation Ethos. New Delhi: Wetlands International. *South Asia*, 64.

Kaushik, S., & Saksena, D. N. (1999). Physico-chemical limnology of certain water bodies of central India. In K. Vijay Kumar (Ed.), *Freshwater ecosystem of India*. Daya Publishing House Delhi.

Keddy, A. P. (2008). Freshwater Marshes. In *Encyclopaedia of Ecology*. Academic Press; doi:10.1017/CBO9780511778179

Keddy, P. A. (2010). *Wetland ecology: Principles and conservation.* Cambridge University Press.

Keddy, P. A., Fraser, L. H., Solomeshch, A. I., Junk, W. J., Campbell, D. R., Arroyo, M. T., & Alho, C. J. (2009). Wet and wonderful: The world's largest wetlands are conservation priorities. *Bioscience, 59*(1), 39–51. doi:10.1525/bio.2009.59.1.8

Keeling, C. D., Piper, S., Bollenbacher, A., & Walker, S. (2018). *The keeling curve: A daily record of atmospheric carbon dioxide from Scripps Institution of Oceanography at UC San Diego*. https://scripps. ucsd. edu/programs/keelingcurve/, nd.

Khan, F. A., Söderquist, B., & Jass, J. (2019). Prevalence and Diversity of Antibiotic Resistance Genes in Swedish Aquatic Environments Impacted by Household and Hospital Wastewater. *Frontiers in Microbiology, 10*, 688. doi:10.3389/fmicb.2019.00688 PMID:31019498

Khan, H., & Brush, G. S. (1994). Nutrient and metal accumulation in a freshwater tidal marsh. *Estuaries, 17*(2), 345–360. doi:10.2307/1352668

Khatri, N., & Tyagi, S. (2015). Influences of natural and anthropogenic factors on surface and groundwater quality in rural and urban areas. *Frontiers in Life Science, 8*(1), 23–39. doi:10.1080/21553769.2014.933716

King, S. L., Laubhan, M. K., Tashjian, P., Vradenburg, J., & Fredrickson, L. (2021). Wetland Conservation: Challenges Related to Water Law and Farm Policy. *Wetlands, 41*(5), 1–17. doi:10.100713157-021-01449-y

Kingsford, R. T., Bino, G., Finlayson, C. M., Falster, D., Fitzsimons, J. A., Gawlik, D. E., Murray, N. J., Grillas, P., Gardner, R. C., Regan, T. J., Roux, D. J., & Thomas, R. F. (2021). Ramsar wetlands of international importance–improving conservation outcomes. *Frontiers in Environmental Science, 9*, 53. doi:10.3389/fenvs.2021.643367

Kinuthia, G. K., Ngure, V., Beti, D., Lugalia, R., Wangila, A., & Kamau, L. (2020). Levels of heavy metals in wastewater and soil samples from open drainage channels in Nairobi, Kenya: Community health implication. *Scientific Reports, 10*(1), 1–13. doi:10.103841598-020-65359-5 PMID:31913322

Knudsen, M. F., Jacobsen, B. H., Riisager, P., Olsen, J., & Seidenkrantz, M. S. (2012). Evidence of Suess solar-cycle bursts in subtropical Holocene speleothem? 18O records. *The Holocene, 22*(5), 597–602. doi:10.1177/0959683611427331

Knudsen, M. F., Riisager, P., Jacobsen, B. H., Muscheler, R., Snowball, I., & Seidenkrantz, M. S. (2009). Taking the pulse of the Sun during the Holocene by joint analysis of 14C and 10Be. *Geophysical Research Letters, 36*(16), L16701. Advance online publication. doi:10.1029/2009GL039439

Knudsen, M. F., Seidenkrantz, M. S., Jacobsen, B. H., & Kuijpers, A. (2011). Tracking the Atlantic Multidecadal Oscillation through the last 8,000 years. *Nature Communications, 2*(1), 178. doi:10.1038/ncomms1186 PMID:21285956

Kocamemi, B. A., Kurt, H., Sait, A., Sarac, F., Saatci, A. M., & Pakdemirli, B. (2020). *SARS- CoV-2 detection in Istanbul wastewater treatment plant sludges.* doi:10.1101/2020.05.12.20099358

Kolde, R. (2019). *pheatmap: Pretty Heatmaps*. R package version 1.0.12, URL https://CRAN. R-project.org/package=pheatmap

Kondolf, G. M., Rubin, Z. K., & Minear, J. T. (2014). Dams on the Mekong: Cumulative sediment starvation. *Water Resources Research*, *50*(6), 5158–5169. doi:10.1002/2013WR014651

Kováčik, J., Babula, P., & Hedbavny, J. (2017). Comparison of vascular and non-vascular aquatic plant as indicators of cadmium toxicity. *Chemosphere*, *180*, 86–92. doi:10.1016/j.chemosphere.2017.04.002 PMID:28391156

Kraemer, S. A., Ramachandran, A., & Perron, G. G. (2019). Antibiotic Pollution in the Environment: From Microbial Ecology to Public Policy. *Microorganisms*, *7*(6), 180. doi:10.3390/microorganisms7060180 PMID:31234491

Kristoffersen, P., Rask, A. M., Grundy, A. C., Franzen, I., Kempenaar, C., Raisio, J., & Zarina, L. (2008). A review of pesticide policies and regulations for urban amenity areas in seven European countries. *Weed Research*, *48*(3), 201–214. doi:10.1111/j.1365-3180.2008.00619.x

Kroeger, K. D., Crooks, S., Moseman-Valtierra, S., & Tang, J. (2017). Restoring tides to reduce methane emissions in impounded wetlands: A new and potent Blue Carbon climate change intervention. *Scientific Reports*, *7*(1), 1–12. doi:10.103841598-017-12138-4 PMID:28931842

Kruger, C. J. C. (2003). *Constrained Cubic Spline Interpolation*. http://www.korf.co.uk/spline.pdf

Kumar, M. D., & Dinesh Kumar, M. (2020). Water management in India: the multiplicity of views and solutions. In Politics and Policies for Water Resources Management in India (pp. 1–15). doi:10.1201/9780429316234-1

Kumar, R., Sharma, P., & Bandyopadhyay, S. (2021). Evidence of microplastics in wetlands: Extraction and quantification in Freshwater and coastal ecosystems. *Journal of Water Process Engineering, 40.*

Kumar, M., Mukherjee, N., Sharma, G. P., & Raghubanshi, A. S. (2010). Land use patterns and urbanization in the holy city of Varanasi, India: A scenario. *Environmental Monitoring and Assessment*, *167*(1), 417–422. doi:10.100710661-009-1060-0 PMID:19562495

Kumar, R., & Patnaik, P. (2016). Wetlands of Mahanadi Delta (India). In C. Finlayson, G. Milton, R. Prentice, & N. Davidson (Eds.), *The Wetland Book*. Springer. doi:10.1007/978-94-007-6173-5_28-4

Kumar, V., Parihar, R. D., Sharma, A., Bakshi, P., Singh Sidhu, G. P., Bali, A. S., Karaouzas, I., Bhardwaj, R., Thukral, A. K., Gyasi-Agyei, Y., & Rodrigo-Comino, J. (2019). Global evaluation of heavy metal content in surface water bodies: A meta-analysis using heavy metal pollution indices and multivariate statistical analyses. *Chemosphere*, *236*, 124364. doi:10.1016/j.chemosphere.2019.124364 PMID:31326755

Kundu, N., & Chakraborty, A. (2017). Dependence on ecosystem goods and services: a case study on East Kolkata Wetlands, West Bengal, India. In *Wetland Science* (pp. 381–405). Springer. doi:10.1007/978-81-322-3715-0_20

Kundu, P. (2018). Review article on the assessment of wetland. *International Journal of Research and Analytical Reviews*, *5*(3), 343–346.

Kusler, J. A., & Kentula, M. E. (1989a). Wetland Creation and Restoration: The Status of The Science. Vol. I. Regional review. EPA/600/3–89/038. USEPA, Environmental Research Laboratory.

Kutilova, I., Medvecky, M., Leekitcharoenphon, P., Munk, P., Masarikova, M., Davidova-Gerzova, L., Jamborova, I., Bortolaia, V., Pamp, S. J., & Dolejska, M. (2021). Extended- spectrum beta-lactamase-producing Escherichia coli and antimicrobial resistance in municipal and hospital wastewaters in Czech Republic: Culture-based and metagenomic approaches. *Environmental Research*, *193*(110487). Advance online publication. doi:10.1016/j.envres.2020.110487 PMID:33232750

La Rosa, G., Iaconelli, M., Mancini, P., Bonanno Ferraro, G., Veneri, C., Bonadonna, L., Lucentini, L., & Suffredini, E. (2020). First detection of SARS-CoV-2 in untreated wastewaters in Italy. *The Science of the Total Environment*, *736*, 139652. doi:10.1016/j.scitotenv.2020.139652 PMID:32464333

Lakeman, J. A. (1999). Climate Change 1995 The Science of Climate Change. In Contribution of WGI to the Second Assessment Report of the Intergovernmental Panel on Climate Change (Vol. 30, Issue 3). Cambridge University Press. doi:10.1007/978-3-662-03925-0_1

Lakhdar, R., & Soussi, M. (2009). Les dépôts lagunaires holocènes du Sud-Est de la Tunisie: extension et organisation spatio-temporelle. Conférence Méditerranéenne Côtiere et Maritime, 115-118. doi:10.5150/cmcm.2009.029

Lakhdar, R. (2009). *Les sédiments holocènes et les tapis microbiens du littoral du Sud-Est de la Tunisie: sédimentologie et paléoenvironnements* (PhD). Université de Sfax.

Lakhdar, R., Soussi, M., Ismail, M. H. B., & M'Rabet, A. (2006). A Mediterranean Holocene restricted coastal lagoon under arid climate: Case of the sedimentary record of Sabkha Boujmel (SE Tunisia). *Palaeogeography, Palaeoclimatology, Palaeoecology*, *241*(2), 177–191. doi:10.1016/j.palaeo.2006.02.014

Lakitan, B., Kartika, K., Susilawati, S., & Wijaya, A. (2021). Acclimating leaf celery plant (Apium graveolens) via bottom wet culture for increasing its adaptability to tropical riparian wetland ecosystem. *Biodiversitas Journal of Biological Diversity, 22*(1).

Lamastra, L., Suciu, N. A., & Trevisan, M. (2018). Sewage sludge for sustainable agriculture: Contaminants' contents and potential use as fertilizer. *Chemical and Biological Technologies in Agriculture*, *5*(10), 10. Advance online publication. doi:10.118640538-018-0122-3

Lambrou, T. P., Anastasiou, C. C., & Panayiotou, C. G. (2009, September). A nephelometric turbidity system for monitoring residential drinking water quality. In *International Conference on Sensor Applications, Experimentation and Logistics* (pp. 43-55). Springer.

Lamsal, P., Pant, K. P., Kumar, L., & Atreya, K. (2015). Sustainable livelihoods through conservation of wetland resources: A case of economic benefits from Ghodaghodi Lake, western Nepal. *Ecology and Society*, *20*(1), art10. doi:10.5751/ES-07172-200110

Langford, B. J., So, M., Raybardhan, S., Leung, V., Soucy, J. R., Westwood, D., Daneman, N., & MacFadden, D. R. (2021). Antibiotic prescribing in patients with COVID-19: rapid review and meta-analysis. *Clinical Microbiology and Infection, 27*(4), 520–531. doi:10.1016/j.cmi.2020.12.018

Larcher, W. (2003). *Physiological plant ecology*. Springer. doi:10.1007/978-3-662-05214-3

Larson, C. A., & Belovsky, G. E. (2013). Salinity and nutrients influence species richness and evenness of phytoplankton communities in microcosm experiments from Great Salt Lake, Utah, USA. *Journal of Plankton Research*, *35*(5), 1154–1166. doi:10.1093/plankt/fbt053

Lauridsen, T. L., Mønster, T., Raundrup, K., Nymand, J., & Olesen, B. (2020). Macrophyte performance in a low arctic lake: Effects of temperature, light and nutrients on growth and depth distribution. *Aquatic Sciences*, *82*(1), 18. doi:10.100700027-019-0692-6

Lawniczak, A. E., Zbierska, J., Nowak, B., Achtenberg, K., Grześkowiak, A., & Kanas, K. (2016). Impact of agriculture and land use on nitrate contamination in groundwater and running waters in central-west Poland. *Environmental Monitoring and Assessment*, *188*(3), 172. doi:10.100710661-016-5167-9 PMID:26887311

Leduc, C., Calvez, R., Beji, R., Nazoumou, Y., Lacombe, G., & Aouadi, C. (2004). Evolution de la ressource en eau dans la vallée du Merguellil (Tunisie centrale). *Actes du Séminaire Modernisation de l'Agriculture Irriguée Rabat.*

Leonardson, L., Bengtson, L., Davidsson, T., Person, T., & Emanuelsson, U. (1994). Nitrogen retention in artificially flooded meadows. *Ambio*, *23*, 332–341.

Leoni, B., Patelli, M., Soler, V., & Nava, V. (2018). Ammonium transformation in 14 lakes along a trophic gradient. *Water (Basel)*, *10*(3), 265. doi:10.3390/w10030265

Lewis, R. R., III. (1989). Wetland restoration/creation/enhancement terminology: Suggestions for standardization. Wetland Creation and Restoration: The Status of the Science, Vol. II. EPA 600/3/89/038B. U.S. Environmental Protection Agency.

Li, D., Tian, P., Luo, Y., Dong, B., Cui, Y., & Khan, S. (n.d.). Importance of stopping groundwater irrigation for balancing agriculture and wetland ecosystem. *Ecological Indicators, 127.* http:// doi:10.1016/j.ecolind.2021.107747

Li, C., Quan, Q., Gan, Y., Dong, J., Fang, J., Wang, L., & Liu, J. (2020). Effects of heavy metals on microbial communities in sediments and establishment of bioindicators based on microbial taxa and function for environmental monitoring and management. *The Science of the Total Environment*, *749*, 141555. doi:10.1016/j.scitotenv.2020.141555 PMID:32841857

Li, M., Huang, C., Ogg, J., Zhang, Y., Hinnov, L., Wu, H., Chen, Z.-Q., & Zou, Z. (2019). Paleoclimate proxies for cyclostratigraphy: Comparative analysis using a Lower Triassic marine section in South China. *Earth-Science Reviews*, *189*, 125–146. doi:10.1016/j.earscirev.2019.01.011

Lim, J., Lee, J. Y., Hong, S. S., Park, S., Lee, E., & Yi, S. (2019). Holocene coastal environmental change and ENSO-driven hydroclimatic variability in East Asia. *Quaternary Science Reviews*, *220*, 75–86. doi:10.1016/j.quascirev.2019.07.041

Limpens, J., Berendse, F., Blodau, C., Canadell, J. G., Freeman, C., Holden, J., & Schaepman-Strub, G. (2008). Peatlands and the carbon cycle: From local processes to global implications–a synthesis. *Biogeosciences*, *5*(5), 1475–1491. doi:10.5194/bg-5-1475-2008

Lind, L., Schuler, M. S., Hintz, W. D., Stoler, A. B., Jones, D. K., Mattes, B. M., & Relyea, R. A. (2018). Salty fertile lakes: How salinization and eutrophication alter the structure of freshwater communities. *Ecosphere*, *9*(9), e02383. doi:10.1002/ecs2.2383

Ling, M. A., King, S., Mapendembe, A., & Brown, C. A. (2018). Review of Ecosystem Service Valuation Progress and Approaches by the Member States of the European Union. UNEP-WCMC.

Ling, J. E., Casanova, M. T., Shannon, I., & Powell, M. (2018). Development of a wetland plant indicator list to inform the delineation of wetlands in New South Wales. *Marine and Freshwater Research*, *70*(3), 322–344. doi:10.1071/MF18114

Lionello, P., Malanotte-Rizzoli, P., Boscolo, R., Alpert, P., Artale, V., Li, L., Luterbacher, J., May, W., Trigo, R., Tsimplis, M., Ulbrich, U., & Xoplaki, E. (2006). The Mediterranean climate: An overview of the main characteristics and issues. Mediterranean Climate Variability. Developments in Earth and Environmental Sciences, 4, 1-26.

Lionello, P. (2012). *The Climate of the Mediterranean Region. From the Past to the Future*. Elsevier.

Liu, G.F., He, J., Yang, Y.Z., & Han, S.Q. (2015). Article. *Huanjingkexue= Huanjingkexue, 36*(8), 2887–2894.

Liu, X.Q., Huang, T.L., Li, N., Yang, S.Y., Li, Y., Xu, J., &Wang, H.Y. (2019). Article. *Huanjingkexue= Huanjingkexue,*40(5), 2258–2264.

Liu, D., Duan, H., Loiselle, S., Hu, C., Zhang, G., Li, J., Yang, H., Thompson, J. R., Cao, Z., Shen, M., Ma, R., Zhang, M., & Han, W. (2020). Observations of water transparency in China's lakes from space. *International Journal of Applied Earth Observation and Geoinformation*, *92*, 102187. doi:10.1016/j.jag.2020.102187

Liu, X. P., Kelin, W., & Geli, Z. (2004). Perspectives and policies: Ecological industry substitutes in wetland restoration of the Middle Yangtze. *Wetlands*, *24*, 633–641.

Li, X., Rensing, C., Vestergaard, G., Arumugam, M., Nesme, J., Gupta, S., Brejnrod, A. D., & Sørensen, S. J. (2021). Metagenomic evidence for co-occurrence of antibiotic, biocide and metal resistance genes in pigs. *Environment International*, *158*, 106899. Advance online publication. doi:10.1016/j.envint.2021.106899 PMID:34598063

Li, Y., Meng, J., Zhang, C., Ji, S., Kong, Q., Wang, R., & Liu, J. (2020). Bottom-up and top-down effects on phytoplankton communities in two freshwater lakes. *PLoS One*, *15*(4), e0231357. doi:10.1371/journal.pone.0231357 PMID:32271852

Loreau, M. (2000). Biodiversity and ecosystem functioning: Recent theoretical advances. *Oikos*, *91*(1), 3–17. doi:10.1034/j.1600-0706.2000.910101.x

Lou, Y., Pan, Y., Gao, C., Jiang, M., Lu, X., & Xu, Y. J. (2016). Response of plant height, species richness and aboveground biomass to flooding gradient along vegetation zones in floodplain wetlands, Northeast China. *PLoS One*, *11*(4), e0153972. doi:10.1371/journal.pone.0153972 PMID:27097325

Lu, X., Zhang, X., Li, L. Y., & Chen, H. (2014). Assessment of metals pollution and health risk in dust from nursery schools in Xi'an, China. *Environmental Research*, *128*, 27–34. doi:10.1016/j.envres.2013.11.007 PMID:24407476

MacGregor, S. (Ed.). (2017). *Routledge handbook of gender and environment*. Taylor & Francis. doi:10.4324/9781315886572

Macreadie, P. I., Anton, A., Raven, J. A., Beaumont, N., Connolly, R. M., Friess, D. A., Kelleway, J. J., Kennedy, H., Kuwae, T., Lavery, P. S., Lovelock, C. E., Smale, D. A., Apostolaki, E. T., Atwood, T. B., Baldock, J., Bianchi, T. S., Chmura, G. L., Eyre, B. D., Fourqurean, J. W., ... Duarte, C. M. (2019). The future of Blue Carbon science. *Nature Communications*, *10*(1), 3998. doi:10.103841467-019-11693-w PMID:31488846

Madera-Parra, C. A., Peña, M. R., Peña, E. J., & Lens, P. N. L. (2015). Contamination status of *Salmonella* spp., *Shigella* spp. and *Campylobacter* spp. in surface and groundwater of the Kelani river basin, Sri Lanka. *Water (Basel)*, *12*(8), 2187.

Madrange, L., Chaboury, P., Ferrandon, O., Mazet, M., & Rodeaud, J. (1993). Study of the formation and stability of chemical foam on the Vienne river between Limoges and Confolens [Etude de la formation et de la stabilite des mousses chimiques de surface de la Vienne]. *Revue des sciences de l'eau/journal of water science*, *6*(3), 315-334.

Mahmoud, E. K., & Ghoneim, A. M. (2016). Effect of polluted water on soil and plant contamination by heavy metals in El-Mahla El-Kobra, Egypt. *Solid Earth*, *7*(2), 703–711. doi:10.5194e-7-703-2016

Mahyuni, M., & Syahrin, M. N. (2021). Sister City Cooperation Model In Wetland Environmental Governance: Case Study In Banjarmasin, South of Kalimantan. *International Journal of Politic. Public Policy and Environmental Issues*, *1*(02), 115–125.

Maiga, Y., von Sperling, M., & Mihelcic, J. R. (2017). Constructed Wetlands. In Water and Sanitation for the 21st Century: Health and Microbiological Aspects of Excreta and Wastewater Management (Global Water Pathogen Project). Part 4: Management Of Risk from Excreta and Wastewater - Section: Sanitation System Technologies, Pathogen Reduction in Sewered System Technologies). Michigan State University. doi:10.14321/waterpathogens.66

Mairs, D. F. (1966). A total alkalinity atlas for marine lake waters. *Limnology and Oceanography*, *11*(1), 68–72. doi:10.4319/lo.1966.11.1.0068

Majumdar, D. (2013). Biogeochemistry of N2O uptake and consumption in submerged soils and rice fields and implications in climate change. *Critical Reviews in Environmental Science and Technology*, *43*(24), 2653–2684. doi:10.1080/10643389.2012.694332

Malanotte-Rizzoli, P., Artale, V., Borzelli-Eusebi, G. L., Brenner, S., Crise, A., Gacic, M., Kress, N., Marullo, S., Ribera d'Alcalà, M., Sofianos, S., Tanhua, T., Theocharis, A., Alvarez, M., Ashkenazy, Y., Bergamasco, A., Cardin, V., Carniel, S., Civitarese, G., D'Ortenzio, F., ... Triantafyllou, G. (2014). Physical forcing and physical/biochemical variability of the Mediterranean Sea: A review of unresolved issues and directions for future research. *Ocean Science*, *10*(3), 281–322. doi:10.5194/os-10-281-2014

Malik, M., & Rai, S. C. (2019). Drivers of land use/cover change and its impact on Pong Dam Wetland. *Environmental Monitoring and Assessment*, *191*(4), 203–217. doi:10.100710661-019-7347-x PMID:30834470

Malini, B. H., Rao, K. S., & Rao, K. N. (1993) Evolution and dynamics of the Chilka Lake. In *Proceedings of the eighth symposium on Coastal and Ocean Management*. American Society of Civil Engineers.

Malone, T. C., & Newton, A. (2020). The globalization of cultural eutrophication in the coastal ocean: Causes and consequences. *Frontiers in Marine Science*, *7*, 670. doi:10.3389/fmars.2020.00670

Maltby, E., & Barker, T. (2009). *The Wetlands Handbook*. Blackwell Science. doi:10.1002/9781444315813

Manna, M. C., & Ghosh, A. (2021). Soil Organic Carbon Dynamics, Stabilization, and Environmental Implication. In *Soil Science: Fundamentals to Recent Advances* (pp. 13–33). Springer. doi:10.1007/978-981-16-0917-6_2

Mansfeldt, C., Deiner, K., Mächler, E., Fenner, K., Eggen, R., Stamm, C., Schönenberger, U., Walser, J. C., & Altermatt, F. (2020). Microbial community shifts in streams receiving treated wastewater effluent. *The Science of the Total Environment*, *709*, 135727. doi:10.1016/j.scitotenv.2019.135727 PMID:31887504

Manuel, J. (2014). Nutrient pollution: A persistent threat to waterways. *Environmental Health Perspectives*, *122*(11), A304–A309. doi:10.1289/ehp.122-A304 PMID:25360879

Margalef, R. (1958) Temporal succession and spatial heterogeneity in phytoplankton. In Perspectives in Marine biology (pp. 323-347). University of California Press. doi:10.1525/9780520350281-024

Marquer, L., Pomel, S., Abichou, A., Schulz, E., Kaniewski, D., & Van Campo, E. (2008). Late Holocene high resolution palaeoclimatic reconstruction inferred from Sebkha Mhabeul, southeast Tunisia. *Quaternary Research*, *70*(2), 240–250. doi:10.1016/j.yqres.2008.06.002

Martin, H. (1997). From Care to Action making a Sustainable World, IUCN and Earthscan Publication Ltd.

Martinez-Harms, M. J., Bryan, B. A., Balvanera, P., Law, E. A., Rhodes, J. R., Possingham, H. P., & Wilson, K. A. (2015). Making decisions for managing ecosystem services. *Biological Conservation*, *184*, 229–238. doi:10.1016/j.biocon.2015.01.024

Martinez, J., Garcia, A., Beaucourt, N., & Nuñez, E. (2002). Combined seasonal and longitudinal variations of element concentrations in two aquatic mosses (*Fontinalis antipyretica* and *Fontinalis squamosa*). *Nova Hedwigia*, *74*(3-4), 349–364. doi:10.1127/0029-5035/2002/0074-0349

Martinez-Ruiz, F., Kastner, M., Gallego-Torres, D., Rodrigo-Gámiz, M., Nieto-Moreno, V., & Ortega-Huertas, M. (2015). Paleoclimate and paleoceanography over the past 20,000 yr in the Mediterranean Sea Basins as indicated by sediment elemental proxies. *Quaternary Science Reviews*, *107*, 25–46. doi:10.1016/j.quascirev.2014.09.018

Marty, J. T. (2005). Effects of cattle grazing on diversity in ephemeral wetlands. *Conservation Biology*, *19*(5), 1626–1632.

Marx, S. K., McGowan, H. A., Kamber, B. S., Knight, J. M., Denholm, J., & Zawadzki, A. (2014). Unprecedented wind erosion and perturbation of surface geochemistry marks the Anthropocene in Australia. *Journal of Geophysical Research. Earth Surface*, *119*(1), 45–61. doi:10.1002/2013JF002948

Mateo-Sagasta, J., MarjaniZadeh, S., Turral, H., & Burke, J. (2017). *Water pollution from agriculture: a global review.* Food and Agriculture Organization of the United Nations. Retreived from: https://www.fao.org/3/i7754e/i7754e.pdf

Mathuthu, A. S., Mwanga, K., & Simoro, A. (1997). *Impact assessment of industrial and sewage effluents on water quality of the receiving Marimba River in Harare. Lake Chivero: A Polluted Lake*. University of Zimbabwe Publications.

Mayewski, P. A., Rohling, E. E., Stager, J. C., Karlén, W., Maasch, K. A., Meeker, L. D., Meyerson, E. A., Gasse, F., van Kreveld, S., Holmgren, K., Lee-Thorp, J., Rosqvist, G., Rack, F., Staubwasser, M., Schneider, R. R., & Steig, E. J. (2004). Holocene climate variability. *Quaternary Research*, *62*(3), 243–255. doi:10.1016/j.yqres.2004.07.001

McCravy, K. W. (2018). A Review of Sampling and Monitoring Methods for Beneficial Arthropods in Agroecosystems. *Insects*, *9*(4), 170. doi:10.3390/insects9040170 PMID:30477100

McDermott, F., Mattey, D. P., & Hawkesworth, C. (2001). Centenial scale Holocene climate variability revealed by a high-resolution speleothem y18O record from SW Ireland. *Science*, *294*(5545), 1328–1331. doi:10.1126cience.1063678 PMID:11701925

McGrane, S. J. (2016). Impacts of urbanisation on hydrological and water quality dynamics, and urban water management: A review. *Hydrological Sciences Journal*, *61*(13), 2295–2311. doi:10.1080/02626667.2015.1128084

McInnes, R. J., & Everard, M. (2017). Rapid assessment of wetland ecosystem services (RAWES): An example from Colombo, Sri Lanka. *Ecosystem Services*, *25*, 89–105. doi:10.1016/j.ecoser.2017.03.024

McKown, I. D., Gregg, E., Moore, I. D., Payne, A. R., Natalie, A. W., & Gibson, J. L. (2021). Successional dynamics of a 35 year old freshwater mitigation wetland in southeastern New Hampshire. *PLoS One*, *16*(5), e0251748. doi:10.1371/journal.pone.0251748 PMID:33999932

McManus, J. F., Oppo, D. W., & Cullen, J. L. (1999). A 0.5-million-year record of millennial scale climate variability in the North Atlantic. *Science*, *283*(5404), 971–975. doi:10.1126cience.283.5404.971 PMID:9974387

Meehl, G. A., Arblaster, J. M., Matthes, K., Sassi, F., & Van Loon, H. (2009). Amplifying the Pacific Climate System Response to a Small 11-Year Solar Cycle Forcing. *Science*, *325*(5944), 1114–1118. doi:10.1126cience.1172872 PMID:19713524

Meena, A. K., & Sharma, C. (2019). *Wetlands in India: Ecosystem Benefits*. Threats and Management Strategies.

Meena, R., Sathishkumar, P., Ameen, F., Yusoff, A., & Gu, F. L. (2018). Heavy metal pollution in immobile and mobile components of lentic ecosystems-a review. *Environmental Science and Pollution Research International*, *25*(5), 4134–4148. doi:10.100711356-017-0966-2 PMID:29247419

Meese, D. A., Gow, A. J., Grootes, P., Mayewski, P. A., Ram, M., Stuiver, M., Taylor, K. M., Waddington, K. C., & Zielinski, G. A. (1994). The accumulation record from the GISP2 record as an indicator of climate change throughout the Holocene. *Science*, *266*(5191), 1680–1682. doi:10.1126cience.266.5191.1680 PMID:17775628

Megonigal, J. P., Hines, M. E., & Visscher, P. T. (2004). Anaerobic metabolism: Linkages to trace gases and aerobic processes. *Biogeochemistry*.

Mekonnen, B. (2019). Assessment of the Effect of Solid Waste Dump Site on Surrounding Soil and River Water Quality in Tepi Town, Southwest Ethiopia. *Journal of Environmental and Public Health*. PMID:32587623

Meli, P., Rey Benayas, J. M., Balvanera, P., & Martínez Ramos, M. (2014). Restoration enhances wetland biodiversity and ecosystem service supply, but results are context-dependent: A meta-analysis. *PLoS One*, *9*(4), e93507. doi:10.1371/journal.pone.0093507 PMID:24743348

Melosi Martin, V. (2005). *Garbage in the cities, refuse, reform and the environment* (revised edition). University of Pittsburgh Press.

Men'endez, P., Losada, I. J., Torres-Ortega, S., Narayan, S., & Beck, M. W. (2020). The global flood protection benefits of mangroves. *Scientific Reports*, *10*(1), 1–11. doi:10.103841598-020-61136-6 PMID:32157114

Mendelsohn, R., & Olmstead, S. (2009). The economic valuation of environmental amenities and disamenities: Methods and applications. *Annual Review of Environment and Resources*, *34*(1), 325–347. doi:10.1146/annurev-environ-011509-135201

Meng, L., Roulet, N., Zhuang, Q., Christensen, T. R., & Frolking, S. (2016). Focus on impact of climate cane on wetland ecosystems and carbon dynamics. *Environmental Research Letters*, *11*(10), 100201. doi:10.1088/1748-9326/11/10/100201

Meybeck, M. (2001). River basin under Anthropocene conditions. In *Science and Integrated Basin Management Dahlem workshop series* (pp. 275–294). Wiley.

Meybeck, M., Kuusisto, E., Makela, A., & Malkki, E. (1996). *Water quality monitoring: a practical guide to the design and implementation of freshwater quality studies and monitoring programmes* (J. Bartram & R. Ballance, Eds.). CRC Press.

Michalak, A. M., Anderson, E. J., Beletsky, D., Boland, S., Bosch, N. S., Bridgeman, T. B., Chaffin, J. D., Cho, K., Confesor, R., Daloglu, I., Depinto, J. V., Evans, M. A., Fahnenstiel, G. L., He, L., Ho, J. C., Jenkins, L., Johengen, T. H., Kuo, K. C., Laporte, E., ... Zagorski, M. A. (2013). Record-setting algal bloom in Lake Erie caused by agricultural and meteorological trends consistent with expected future conditions. *Proceedings of the National Academy of Sciences of the United States of America*, *110*(16), 6448–6452. doi:10.1073/pnas.1216006110 PMID:23576718

Middleton, B. (2016). Succession in Wetlands. In *The Wetland Book*. Springer. doi:10.1007/978-94-007-6172-8_1-2

Middleton, B. A. (1999). Succession and herbivory in monsoonal wetlands. *Wetlands Ecology and Management*, *6*(4), 189–202. doi:10.1023/A:1008495121557

Miles, A., & Ebrey, J. (2017). The village in the city: Participation and cultural value on the urban periphery. *Cultural Trends*, *26*(1), 58–69. doi:10.1080/09548963.2017.1274360

Millar, J., Robinson, W., Baumgartner, L., Homsombath, K., Chittavong, M., Phommavong, T., & Singhanouvong, D. (2019). Local perceptions of changes in the use and management of floodplain fisheries commons: The case of Pak Peung wetland in Lao PDR. *Environment, Development and Sustainability*, *21*(4), 1835–1852. doi:10.100710668-018-0105-3

Millennium Ecosystem Assessment (MEA). (2005). *Ecosystems and human well-being: a framework for assessment. Wetlands and water*. World Resources Institute.

Ministry of Environment & Forest. (2006). *National Environment Policy 2006*. Government of India. http://www.indiaenvironmentportal.org.in/files/nep2006e.pdf

Ministry of Environment and Forests. (2007). *Conservation of wetlands in India: a profile (approach and guidelines). MoEF*. Government of India.

Ministry of Environment, Forest, and Climate Change. (2020). *Guidelines for implementing Wetlands (Conservation and Management) Rules, 2017*. Government of India. http://www.indiaenvironmentportal.org.in/files/file/Wetlands%20(Conservation%20and%20Management)%20Rules,%202017.pdf

Mirza, A., Forbes, J. D., Zhu, F., Bernstein, C. N., Van Domselaar, G., Graham, M., Waubant, E., & Tremlett, H. (2020). The multiple sclerosis gut microbiota: A systematic review. *Multiple Sclerosis and Related Disorders*, *37*, 101427. doi:10.1016/j.msard.2019.101427 PMID:32172998

Mitsch, W. L. (2013). Wetland creation and restoration. Encyclopedia of Biodiversity, 7, 367-383. doi:10.1016/B978-0-12-384719-5.00221-5

Mitsch, W. J. (1994). *Global wetlands - old world and new*. Elsevier.

Mitsch, W. J., Bernal, B., & Hernandez, M. E. (2015). Ecosystem services of wetlands. *The International Journal of Biodiversity Science, Ecosystem Services & Management, 11*(1), 1–4. doi:10.1080/21513732.2015.1006250

Mitsch, W. J., Bernal, B., Nahlik, A. M., Mander, Ü., Zhang, L., Anderson, C. J., Jørgensen, S. E., & Brix, H. (2013). Wetlands, carbon, and climate change. *Landscape Ecology, 28*(4), 583–597. doi:10.100710980-012-9758-8

Mitsch, W. J., & Gosselink, J. G. (2000a). The Value of Wetlands: Importance of Scale and Landscape Setting. *Ecological Economics, 35*(1), 25–33. doi:10.1016/S0921-8009(00)00165-8

Mitsch, W. J., & Gosselink, J. G. (2007). *Wetlands* (4th ed.). John Wiley and Sons.

Mitsch, W. J., Marchio, D. A. Jr, Savarese, M., & Bovard, B. (2016). Carbon Sequestration and Sedimentation in Mangrove Swamps Influenced by Hydrogeomorphic Conditions and Urbanization in Southwest Florida. *Forests, 7*(12), 116. doi:10.3390/f7060116

Mitsch, W. J., Zhang, L., Stefanik, K. C., Nahlik, A. M., Anderson, C. J., Bernal, B., Hernandez, M., & Song, K. (2012). Creating wetlands: Primary succession, water quality changes, and self-design over 15 years. *Bioscience, 62*(3), 237–250. doi:10.1525/bio.2012.62.3.5

Miyake, F., Suzuki, A., Masuda, K., Horiuchi, K., Motoyama, H., Matsuzaki, H., Motizuki, Y., Takahashi, K., & Nakai. Y. (2015). Cosmic ray event of AD 774–775 shown in quasi-annual 10Be data from the Antarctic Dome Fuji ice core. *Geophysical Research Letters, 42*(1), 84-9.

MOEF. (1990). Wetlands of India - A directory. Ministry of Environment and Forests, Govt. of India.

Mokany, A., Wood, J. T., & Cunningham, S. A. (2008). Effect of shade and shading history on species abundances and ecosystem processes in temporary ponds. *Freshwater Biology, 53*(10), 1917–1928. doi:10.1111/j.1365-2427.2008.02076.x

Molina, V., Eissler, Y., Fernandez, C., Cornejo-D'Ottone, M., Dorador, C., Bebout, B. M., & Hengst, M. (2021). Greenhouse gases and biogeochemical diel fluctuations in a high-altitude wetland. *The Science of the Total Environment, 768*, 144370. doi:10.1016/j.scitotenv.2020.144370 PMID:33454466

Mondal, B., Dolui, G., Pramanik, M., Maity, S., Biswas, S. S., & Pal, R. (2017). Urban expansion and wetland shrinkage estimation using a GIS-based model in the East Kolkata Wetland, India. *Ecological Indicators, 83*, 62–73. doi:10.1016/j.ecolind.2017.07.037

Mondal, S., & … . (2021). A Pipeline for Assessment of Pathogenic Load in the Environment Using Microbiome Analysis. In M. Nath, D. Bhatt, P. Bhargava, & D. K. Choudhary (Eds.), *Microbial Metatranscriptomics Belowground*. Springer. doi:10.1007/978-981-15-9758-9_23

Montague, C. L., Zale, A. V., & Percival, H. F. (1987). Ecological effects of coastal marsh impoundments: A review. *Environmental Management, 11*(6), 743–756.

Moomaw, W. R., Chmura, G. L., Davies, G. T., Finlayson, C. M., Middleton, B. A., Perry, J. E., & Sutton-Grier, A. E. (2018). The relationship between wetlands and a changing climate. *Wetlands, 38*, 183–205. doi:10.100713157-018-1023-8

Moreno, A., Cacho, I., Canals, M., Grimalt, J. O., Sánchez Goñi, M. F., Shackleton, N., & Sierro, F. J. (2005). Links between marine and atmospheric processes oscillating on a millennial time-scale. A multi-proxy study of the last 50,000 yr from the Alboran Sea (Western Mediterranean Sea). *Quaternary Science Reviews*, *24*(14-15), 1623–1636. doi:10.1016/j.quascirev.2004.06.018

Morris, J. T., Sundareshwar, P. V., Nietch, C. T., Kjerfve, B., & Cahoon, D. R. (2002). Responses of coastal wetlands to rising sea level. *Ecology*, *83*(10), 2869–2877. doi:10.1890/0012-9658(2002)083[2869:ROCWTR]2.0.CO;2

Morrison, K. D., Hammer, E., Boles, O., Madella, M., Whitehouse, N., Gaillard, M. J., Bates, J., Vander Linden, M., Merlo, S., Yao, A., Popova, L., Hill, A. C., Antolin, F., Bauer, A., Biagetti, S., Bishop, R. R., Buckland, P., Cruz, P., Dreslerová, D., ... Zanon, M. (2021). Mapping past human land use using archaeological data: A new classification for global land use synthesis and data harmonization. *PLoS One*, *16*(4), e0246662. doi:10.1371/journal.pone.0246662 PMID:33852578

Moseley, K. (2018). *Wetland Ecology- Basic Principles. Regional ESS.* Soil Science Division.

Moseman-Valtierra, S. (2012). Reconsidering the climatic role of marshes: Are they sinks or sources of greenhouse gases. In Marshes: Ecology, Management and Conservation. New York: Nova Science.

Moseman-Valtierra, S., Gonzalez, R., Kroeger, K. D., Tang, J., Chao, W. C., Crusius, J., & Shelton, J. (2011). Short-term nitrogen additions can shift a coastal wetland from a sink to a source of N2O. *Atmospheric Environment*, *45*(26), 4390–4397. doi:10.1016/j.atmosenv.2011.05.046

Moyle, J. B. (1946). Some chemical factors infl uencing the distribution of aquatic plants in Minnesota. *American Midland Naturalist*, *34*(2), 402–420. doi:10.2307/2421128

Mulder, O. J., Mulder, K. P., Kubiszewski, I., Anderson, S. J., Costanza, R., & Sutton, P. (2020). The value of coastal wetlands for storm protection in Australia. *Ecosystem Services*, *46*, 101205. doi:10.1016/j.ecoser.2020.101205

Müller, F., & Burkhard, B. (2012). The indicator side of ecosystem services. *Ecosystem Services*, *1*(1), 26–30. doi:10.1016/j.ecoser.2012.06.001

Myers-Smith, I. H., Harden, J. W., Wilmking, M., Fuller, C. C., McGuire, A. D., & Chapin Iii, F. S. (2008). Wetland succession in a permafrost collapse: Interactions between fire and thermokarst. *Biogeosciences*, *5*(5), 1273–1286. doi:10.5194/bg-5-1273-2008

Nagendra, H., Bai, X., Brondizio, E. S., & Lwasa, S. (2018). The urban south and the predicament of global sustainability. *Nature Sustainability*, *1*(7), 341–349. doi:10.103841893-018-0101-5

Naggar, Y. A., Naiem, E., Mona, M., Giesy, J. P., & Seif, A. (2014). Metals in agricultural soils and plants in Egypt. *Toxicological and Environmental Chemistry*, *96*(5), 730–742. doi:10.1080/02772248.2014.984496

Nascimento, A. L., Souza, A. J., Andrade, P., Andreote, F. D., Coscione, A. R., Oliveira, F. C., & Regitano, J. B. (2018). Sewage Sludge Microbial Structures and Relations to Their Sources, Treatments, and Chemical Attributes. *Frontiers in Microbiology*, *9*, 1462. doi:10.3389/fmicb.2018.01462 PMID:30018612

National Park Service. (2016). *Why are wetlands important.* https://www.nps.gov/subjects/wetlands/why.htm

National Research Council (NRC). (1995). *Wetlands: Characteristics and Boundaries.* National Academy Press.

National Research Council (NRC). (2001). *Compensating for Wetland Losses Under the Clean Water Act.* National Academy Press.

National Research Council. (2001). *Compensating for wetland losses under the Clean Water Act.* DC, National Academy Press.

Nengroo, Z. A., Shah, A. H., & Bhat, M. S. (2017). Dynamics of land use change in rural-urban fringe: A case study of Srinagar city. *Environmental Science: An Indian Journal, 13*(4), 142.

New Delhi Ministry of Environment and Forests. (2011). *National wetland inventory and assessment report. MoEF*. Government of India.

New Delhi Ministry of Environment and Forests. (2012). *Annual report 2011–12. MoEF*. Government of India.

Newell, J. P., Goldstein, B., & Foster, A. (2019). A 40-year review of food–energy–water nexus literature and its application to the urban scale. *Environmental Research Letters, 14*(7), 073003. doi:10.1088/1748-9326/ab0767

Ngigi, A. N., Magu, M. M., & Muendo, B. M. (2019). Occurrence of antibiotics residues in hospital wastewater, wastewater treatment plant, and in surface water in Nairobi County, Kenya. *Environmental Monitoring and Assessment, 192*(1), 18. doi:10.100710661-019-7952-8 PMID:31820117

Nielsen, D. L., Brock, M. A., Crosslé, K., Harris, K., Healey, M., & Jarosinski, I. (2003). The effects of salinity on aquatic plant germination and zooplankton hatching from two wetland sediments. *Freshwater Biology, 48*(12), 2214–2223. doi:10.1046/j.1365-2427.2003.01146.x

Nielsen, S., & Bruun, E. W. (2015). Sludge quality after 10–20 years of treatment in reed bed systems. *Environmental Science and Pollution Research International, 22*(17), 12885–12891. doi:10.100711356-014-3815-6 PMID:25422113

Niering, W. A. (1987). Vegetation dynamics (succession and climax) in relation to plant community management. *Conservation Biology, 1*(4), 287–295. doi:10.1111/j.1523-1739.1987.tb00049.x

Nieuwenhuijse, D. F., Oude Munnink, B. B., Phan, M., Hendriksen, R. S., Bego, A., Rees, C., Neilson, E. H., Coventry, K., Collignon, P., Allerberger, F., Rahube, T. O., Oliveira, G., Ivanov, I., Sopheak, T., Vuthy, Y., Yost, C. K., Tabo, D., Cuadros-Orellana, S., Ke, C., ... Koopmans, M. P. G. (2020). Setting a baseline for global urban virome surveillance in sewage. *Scientific Reports, 10*(1), 13748. doi:10.103841598-020-69869-0 PMID:32792677

Niu, Z., Zhang, H., Wang, X., Yao, W., Zhou, D., Zhao, K., & Gong, P. (2012). Mapping wetland changes in China between 1978 and 2008. *Chinese Science Bulletin, 57*(22), 2813–2823. doi:10.100711434-012-5093-3

Njagi, D. M., Routh, J., Olago, D., & Gayantha, K. (2021). A multi-proxy reconstruction of the late Holocene climate evolution in the Kapsabet Swamp, Kenya (East Africa). *Palaeogeography, Palaeoclimatology, Palaeoecology, 574*, 110475. doi:10.1016/j.palaeo.2021.110475

Nkwunonwo, U. C., Odika, P. O., & Onyia, N. I. (2020). A Review of the Health Implications of Heavy Metals in Food Chain in Nigeria. *The Scientific World Journal*. doi:10.1155/2020/6594109 PMID:32351345

Noe, G. B., & Hupp, C. R. (2005). Carbon, nitrogen, and phosphorus accumulation in floodplains of Atlantic Coastal Plain rivers, USA. *Ecological Applications, 15*(4), 1178–1190. doi:10.1890/04-1677

NWIA National Wetland Atlas. (2018). *Ministry of Environment and Forests, Government of India; As a part of the project on National Wetland Inventory and Assessment (NWIA)*. https://www.un.org/en/events/citiesday/assets/pdf/the_worlds_cities_in_2018_data_booklet.pdf)

Nyanda, M. (2000). Report on the study of agrochemical use and handling in the Lake Zone. Academic Press.

Nyman, J. A. (2011). Ecological Functions of Wetlands. *Wetlands*, 115–128. doi:10.1007/978-94-007-0551-7_6

O'Neill, B. J., Rogers, D. C., & Thorp, J. H. (2016). Flexibility of ephemeral wetland crustaceans: Environmental constraints and anthropogenic impacts. *Wetlands Ecology and Management, 24*(3), 279–291.

Odum, E. P. (1979). The value of wetlands: a hierarchical approach. *Vegetatio, 118*, 3–16.

Odum, W. E., Odum, E. P., & Odum, H. T. (1995). Nature's pulsing paradigm. *Estuaries*, *18*(4), 547–555. doi:10.2307/1352375

Oertli, B. D. A., Joye, D. A., Castella, E., Juge, R., Cambin, D., & Lachavanne, J.-B. (2002). Does size matter? The relationship between pond area and biodiversity. *Biological Conservation*, *104*(1), 59–70. doi:10.1016/S0006-3207(01)00154-9

Oldfield, F. (2015). Can the magnetic signatures from inorganic fly ash be used to mark the onset of the Anthropocene? *The Anthropocene Review*, *2*(1), 3–13. doi:10.1177/2053019614534402

Oluwole, O., & Cheke, R. A. (2009). Health and environmental impacts of pesticide use practices: A case study of farmers in Ekiti State, Nigeria. *International Journal of Agricultural Sustainability*, *7*(3), 153–163.

Osei, J., Nyame, F. K., Armah, T. K., Osae, S. K., Dampare, S. B., Fianko, J. R., Adomako, D., & Bentil, N. (2010). Application of multivariate analysis for identification of pollution sources in the Densu delta wetland in the vicinity of a landfill site in Ghana. *Journal of Water Resource and Protection*, *2*(12), 1020–1029. doi:10.4236/jwarp.2010.212122

Ozturk, A., & Arici, O. K. (2021). Carcinogenic-potential ecological risk assessment of soils and wheat in the eastern region of Konya (Turkey). *Environmental Science and Pollution Research International*, *28*(12), 15471–15484. https://doi.org/10.1007/s11356-020-11697-w

Palanisami, K., Meinzen-Dick, R., & Giordano, M. (2010). Climate change and water supplies: Options for sustaining tank irrigation potential in India. *Economic and Political Weekly*, *45*(26-27), 183–190.

Pal, S., & Debanshi, S. (2021). Developing wetland landscape insecurity and hydrological security models and measuring their spatial linkages. *Ecological Informatics*, *66*, 101461. doi:10.1016/j.ecoinf.2021.101461

Pandey, P. K., Kass, P. H., Soupir, M. L., Biswas, S., & Singh, V. P. (2014). Contamination of water resources by pathogenic bacteria. *AMB Express*, *4*, 51. https://doi.org/10.1186/s13568-014-0051-x

Pandey, S. (1993). Changes in waterbird diversity due to the construction of Pong Dam Reservoir, Himachal Pradesh, India. *Biological Conservation*, *66*, 125–130.

Pandya, I. Y. (2018). Pesticides and their applications in agriculture. *Asian Journal of Applied Science and Technology*, *2*(2), 894–900.

Pant, N., & Verma, R.K. (2010). *Tanks in Eastern India: A Study in Exploration International Water Management Institute*. TATA Water Policy Research Program and Centre for Development Studies.

Paris, S. (2004). *Bekämpfung von Schwimmschlamm, verursacht durch Microthrix parvicella* (Doctoral dissertation). Technische Universität München.

Parolin, P., & Wittmann, F. (2010). Struggle in the flood: Tree responses to flooding stress in four tropical floodplain systems. *AoB Plants*.

Parvez, S., & Bhat, S. U. (2014). Searching for water quality improvement in Dal lake, Srinagar, Kashmir. *Journal of Himalayan Ecology and Sustainable Development*, *9*, 51–64.

Patel, J. G., Murthy, T. V. R., Singh, T. S., & Panigrahy, S. (2009). Analysis of the Distribution Pattern of Wetlands in India in Relation to Climate Change. *Archives XXXVIII8W3 Workshop Proceedings Impact of Climate Change on Agriculture*, 282–287. www.MAweb.org

Pazvakawambwa, G. T. (2018). *Water resources governance in the upper Swakop basin of Namibia* (Doctoral dissertation). University of Namibia.

Pearce-Higgins, J. W., Stephen, L., Langston, R. H. A., & Bright, J. A. (2008). Assessing the cumulative impacts of wind farms on peatland birds: A case study of golden plover *Pluvialis apricaria* in Scotland. *Mires and Peat*, *4*, 1–13.

Peierls, B. L., & Paerl, H. W. (1997). Bioavailability of atmospheric organic nitrogen deposition to coastal phytoplankton. *Limnology and Oceanography*, *42*(8), 1819–1823. doi:10.4319/lo.1997.42.8.1819

Pejrup, M. (1988). *The triangular diagram used for classification of estuarine sediments: a new approach. Tide-influenced sedimentary environments and facies*. Reidel.

Pena, L. D., Francés, G., Diz, P., Esparza, M., Grimalt, J. O., Nombela, M. A., & Alejo, I. (2010). Climate fluctuations during the Holocene in NW Iberia: High and low latitude linkages. *Continental Shelf Research*, *30*(13), 1487–1496. doi:10.1016/j.csr.2010.05.009

Pepi, M., & Focardi, S. (2021). Antibiotic-Resistant Bacteria in Aquaculture and Climate Change: A Challenge for Health in the Mediterranean Area. *International Journal of Environmental Research and Public Health*, *18*(11), 5723. doi:10.3390/ijerph18115723 PMID:34073520

Pérez-Asensio, J. N., Frigola, J., Pena, L. D., Sierro, F. J., Reguera, M. I., Rodríguez-Tovar, F. J., & Cacho, I. (2020). Changes in western Mediterranean thermohaline circulation in association with a deglacial Organic Rich Layer formation in the Alboran Sea. *Quaternary Science Reviews*, *228*, 106075. doi:10.1016/j.quascirev.2019.106075

Perl, S. M., Celestian, A. J., Cockell, C. S., Corsetti, F. A., Barge, L. M., Bottjer, D., & Melwani Daswani, M. (2021). A proposed geobiology-driven nomenclature for astrobiological in situ observations and sample analyses. *Astrobiology*, *21*(8), 954–967. doi:10.1089/ast.2020.2318 PMID:34357788

Perthuisot, J. P. (1974). *Les dépôts salins de la sebkha El Melah de Zarzis: conditions et modalites de la sedimentation evaporitique*. Academic Press.

Perthuisot, J. P., & Floridia, S. (1973). *Carte géologique de la Sebkha el Melah et de ses bordures 1: 50,000*. Laboratoire de géologie de l'Ecole normale supérieure.

Perthuisot, J. P. (1971). Recent polyhalite from Sebkha el Melah (Tunisia). *Nature. Physical Science (London)*, *232*(35), 186–187. doi:10.1038/physci232186a0

Pessarakli, M. (1991). Formation of saline and sodic soils and their reclamation. *Journal of Environmental Science and Health. Part A, Environmental Science and Engineering & Toxic and Hazardous Substance Control*, *26*(7), 1303–1307.

Peters, S. C. (2008). Arsenic in groundwaters in the Northern Appalachian Mountain belt: A review of patterns and processes. *Journal of Contaminant Hydrology*, *99*(1-4), 8–21.

Petrescu, A. M. R., Lohila, A., Tuovinen, J. P., Baldocchi, D. D., Desai, A. R., Roulet, N. T., & Cescatti, A. (2015). The uncertain climate footprint of wetlands under human pressure. *Proceedings of the National Academy of Sciences of the United States of America*, *112*(15), 4594–4599. doi:10.1073/pnas.1416267112 PMID:25831506

Philips, S., Laanbroek, H. J., & Verstraete, W. (2002). Origin, causes and effects of increased nitrite concentrations in aquatic environments. *Reviews in Environmental Science and Biotechnology*, *1*(2), 115–141. doi:10.1023/A:1020892826575

Phillips, A. A., Speth, D. R., Miller, L. G., Wang, X. T., Wu, F., Medeiros, P. M., & Sessions, A. L. (2021). *Microbial succession and dynamics in meromictic Mono Lake*. Geobiology. doi:10.1111/gbi.12437

Piddock, L. J. (1990). Techniques used for the determination of antimicrobial resistance and sensitivity in bacteria. *The Journal of Applied Bacteriology*, *68*(4), 307–318. doi:10.1111/j.1365-2672.1990.tb02880.x PMID:2190965

Pielou, E. C. (1966). The measurement of diversity in different types of biological collections. *Journal of Theoretical Biology*, *13*, 131–144. doi:10.1016/0022-5193(66)90013-0

Pierson, D. C. (2012). Light and Primary Production in Lakes. In L. Bengtsson, R. W. Herschy, & R. W. Fairbridge (Eds.), *Encyclopedia of Lakes and Reservoirs. Encyclopedia of Earth Sciences Series.* Springer. doi:10.1007/978-1-4020-4410-6_261

Pirazzoli, P. A. (2005). A review of possible eustatic, isostatic and tectonic contributions in eight late-Holocene relative sea-level histories from the Mediterranean area. *Quaternary Science Reviews, 24*(18-19), 1989–2001. doi:10.1016/j.quascirev.2004.06.026

Pittock Jet al. (2015). Managing freshwater, river, wetland and estuarine protected areas. 569–608. In G. L. Worboys, M. Lockwood, A. Kothari, S. Feary, & I. Pulsford (Eds.), *Protected Area Governance and Management.* ANU Press.

Poffenbarger, H. J., Needelman, B. A., & Megonigal, J. P. (2011). Salinity influence on methane emissions from tidal marshes. *Wetlands, 31*(5), 831–842. doi:10.100713157-011-0197-0

Pomel, S., Abichou, A., & Schulz, E. (2006). La signification paléo-impluviométrique des sédiments de la sebkha Mhabeul (Sud-Est de la Tunisie). *L'érosion entre Société, Climat et Paléoenvironnement: Table ronde en l'honneur du Professeur René Neboit-Guilhot*, 435-442.

Poppe, L. J., Eliason, A. H., & Hastings, M. E. (2004). A Visual Basic program to generate sediment grain-size statistics and to extrapolate particle distributions. *Computers & Geosciences, 30*(7), 791–795. doi:10.1016/j.cageo.2004.05.005

Post, V. E. A. (2005). Fresh and saline groundwater interaction in coastal aquifers: Is our technology ready for the problems ahead? *Hydrogeology Journal, 13*(1), 120–123.

Pramod, A., Kumara, V., & Gowda, R. (2011). A study on Physico-Chemical Characteristics of Water in Wetlands of Hebbe Range in Bhadra Wildlife Sanctuary, Mid-Western Ghat Region. *Indian Journal of Experimental Sciences, 2*(10), 9–15.

Preisner, M. (2020). Surface water pollution by untreated municipal wastewater discharge due to a sewer failure. *Environmental Processes, 7*(3), 767–780.

Press, W. H., Teukolsky, S. A., Flannery, B. P., & Vetterling, W. T. (1992). Numerical recipes in Fortran 77: volume 1, volume 1 of Fortran numerical recipes: the art of scientific computing. Cambridge University Press.

Press, W. H., Teukolsky, S. A., Vetterling, W. T., & Flannery, B. P. (1992). *Numerical Recipes in C.* Cambridge University Press.

Press, W. H., Teukolsky, S. A., Vetterling, W. T., & Flannery, B. P. (1992). Numerical Recipes. Cambridge University Press.

Primavera, J. H. (1991). Intensive prawn farming in the Philippines: Ecological, social, and economic implications. *Ambio, 20*, 28–33.

Pueyo-Ros, J., Ribas, A., & Fraguell, R. M. (2019). A cultural approach to wetlands restoration to assess its public acceptance. *Restoration Ecology, 27*(3), 626–637. doi:10.1111/rec.12896

Qian, J., Tang, S., Wang, P., Lu, B., Li, K., Jin, W., & He, X. (2020). From source to sink: Review and prospects of microplastics in wetland ecosystems. *The Science of the Total Environment.*

Quan, N. H., Toan, T. Q., Dang, P. D., Phuong, N. L., Anh, T. T. H., Quang, N. X., Quoc, D. P., Quoi, L. P., Hanington, P., & Sea, W. B. (2018). Conservation of the Mekong Delta wetlands through hydrological management. *Ecological Research, 33*(1), 87–103. doi:10.100711284-017-1545-1

Quintela-Baluja, M., Abouelnaga, M., Romalde, J., Su, J. Q., Yu, Y., Gomez-Lopez, M., Smets, B., Zhu, Y. G., & Graham, D. W. (2019). Spatial ecology of a wastewater network defines the antibiotic resistance genes in downstream receiving waters. *Water Research, 162*, 347–357. doi:10.1016/j.watres.2019.06.075 PMID:31295654

Rabalais, N. N. (2002). Nitrogen in aquatic ecosystems. *Ambio, 31*(2), 102–112. doi:10.1579/0044-7447-31.2.102 PMID:12077998

Radwan, G. A. E. (2020). Physical properties of inland lakes and their interaction with global warming: A case study of Lake Nasser, Egypt. *The Egyptian Journal of Aquatic Research, 46*(2), 103–115.

Ragavan, P., Kathiresan, K., Mohan, P. M., Ravichandran, K., Jayaraj, R. S. C., & Rana, T. S. (2020). Ensuring the adaptive potential of Coastal wetlands of India-the need of the hour for sustainable management. *Wetlands Ecology and Management*, 1–12.

Rai, N. D., Bhasme, S., & Balaji, P. (2018). Power, inequality and rights: A political ecology of forest restoration. In *Forest landscape restoration* (pp. 47–62). Routledge. doi:10.4324/9781315111872-4

Ramachandra, T. V. (2009). *Conservation and management of urban wetlands: Strategies and challenges*. Academic Press.

Ramachandra, T. V., & Ahalya, N. (2001). Wetlands Restoration and Conservation. In *Proceedings of National Conference on 'Control of Industrial Pollution and Environmental Degradation'.* Department of Civil Engineering, PSG College of Technology.

Ramachandra, T. V., Mahapatra, D. M., Vinay, S., Varghese, S., Asulabha, K. S., Bhat, S. P., & Aithal, B. H. (2017). *Bellandur and Varthur Lakes rejuvenation blueprint*. ENVIS Technical Report 116, Environmental Information System, CES, Indian Institute of Science.

Ramachandra, T. V., & Kumar, U. (2008). Wetlands of greater Bangalore, India: Automatic delineation through pattern classifiers. *Electronic Green Journal, 1*(26), 2. doi:10.5070/G312610729

Ramsar Convention on Wetlands. (2012). *The annotated Ramsar list: India*. The Secretariat of the Convention on Wetlands.

Ramsar Convention on Wetlands. (2018). *Global Wetlands Outlook: State of the World's Wetlands and their services to people*. Ramsar Convention Secretariat.

Ramsar Convention on Wetlands. (n.d.). *The Global Wetland Outlook, About.* https://www.global-wetland-outlook.ramsar.org/

Ramsar Convention. (2013). *The Ramsar Convention Bureau*. www.ramsar.org

Ramsar Convention. (2018). *Urban wetlands: prized land, not wasteland.* https://www.ramsar.org/sites/default/files/documents/library/wwd18_infographic_engl.pdf

Ramsar Secretariat. (2013). *The List of Wetlands of International Importance*. The Secretariat of the Convention on Wetlands.

Rasse, D. P., Peresta, G., & Drake, B. G. (2005). Seventeen years of elevated CO2 exposure in a Chesapeake Bay Wetland: Sustained but contrasting responses of plant growth and CO2 uptake. *Global Change Biology, 11*(3), 369–377. doi:10.1111/j.1365-2486.2005.00913.x

Rawat, K., & Yadav, A. K. (2021). Sustainable Practices for Conservation of Wetland Ecosystem. *Wetlands Conservation: Current Challenges and Future Strategies*, 129-139.

Reale, O., & Dirmeyer, P. (2000). Modeling the effects of vegetation on Mediterranean climate during the Roman Classical Period. Part I: Climate history and model sensitivity. *Global and Planetary Change, 25*(3-4), 163–184. doi:10.1016/S0921-8181(00)00002-3

Reddy, K. R., & D'angelo, E. M. (1997). Biogeochemical indicators to evaluate pollutant removal efficiency in constructed wetlands. *Water Science and Technology, 35*(5), 1–10. doi:10.2166/wst.1997.0152

Reddy, K. R., & DeLaune, R. D. (2008). *Biogeochemistry of wetlands: science and applications*. CRC Press. doi:10.1201/9780203491454

Reineck, H.E., & Siefert, W. (1980). Faktoren der Schlickbildung im Sahlenburger Watt und Neuwerker Watt. *Die KuK ste, 35*, 26-51.

Rejmánková, E., Komárek, J., Dix, M., Komárková, J., & Girón, N. (2011). Cyanobacterial blooms in lake Atitlan, Guatemala. *Limnologica, 41*(4), 296–302. doi:10.1016/j.limno.2010.12.003

Richards, D. R., Moggridge, H. L., Warren, P. H., & Maltby, L. (2020). Impacts of hydrological restoration on lowland river floodplain plant communities. *Wetlands Ecology and Management, 28*(3), 403–417. doi:10.100711273-020-09717-0

Richardson, C. J. (1994). Ecological functions and human values n wetlands: A framework for assessing forestry impacts. *Wetlands, 14*(1), 1–9. doi:10.1007/BF03160616

Rietbergen-McCracken, J., & Abaza, H. (2000). *Environmental Valuation: A Worldwide Compendium of Case Studies*. Earthscan.

Rimoldi, S. G., Stefani, F., Gigantiello, A., Polesello, S., Comandatore, F., Mileto, D., Maresca, M., Longobardi, C., Mancon, A., Romeri, F., Pagani, C., Cappelli, F., Roscioli, C., Moja, L., Gismondo, M. R., & Salerno, F. (2020). Presence and infectivity of SARS-CoV-2 virus in wastewaters and rivers. *The Science of the Total Environment, 744*, 140911. doi:10.1016/j.scitotenv.2020.140911 PMID:32693284

Riordan, B., Verbyla, D., & Mc Guire, A. D. (2006). Shrinking ponds in subarctic Alaska based on 1950 2002 remotely sensed images. *Journal of Geophysical Research, III*(G4), G04002. doi:10.1029/2005JG000150

Ripple, W. J., Wolf, C., Newsome, T. M., Galetti, M., Alamgir, M., Crist, E., & Laurance, W. F. (2017). World Scientists' Warning to Humanity: A Second Notice. *Bioscience, 67*(12), 10–1093. doi:10.1093/biosci/bix125

Rodbell, D. T., Seltzer, G. O., Anderson, D. M., Abbott, M. B., Enfield, D. B., & Newman, J. H. (1999). An 15,000-year record of El Nino-driven alluviation in southwestern Ecuador. *Science, 283*(5401), 516–520. doi:10.1126cience.283.5401.516 PMID:9915694

Rodgers, J. (1950). The nomenclature and classification of sedimentary rocks. *American Journal of Science, 248*(5), 297–311. doi:10.2475/ajs.248.5.297

Rodrigo-Gámiz, M., Martínez-Ruiz, F., Rodríguez-Tovar, F. J., Jiménez-Espejo, F. J., & Pardo-Igúzquiza, E. (2014). Millennial-to centennial-scale climate periodicities and forcing mechanisms in the westernmost Mediterranean for the past 20,000 yr. *Quaternary Research, 81*(1), 78–93. doi:10.1016/j.yqres.2013.10.009

Rodrigues, A., Sardinha, R. A., & Pita, G. (2021). Fundamentals of Global Carbon Budgets and Climate Change. In *Fundamental Principles of Environmental Physics* (pp. 267–308). Springer. doi:10.1007/978-3-030-69025-0_8

Romshoo, S. A., & Rashid, I. (2014). Assessing the impacts of changing land cover and climate on Hokersar wetland in Indian Himalayas. *Arabian Journal of Geosciences, 7*(1), 143–160. doi:10.100712517-012-0761-9

Root-Bernstein, M., & Frascaroli, F. (2016). Where the fish swim above the birds: Configurations and challenges of wetland restoration in the Po Delta, Italy. *Restoration Ecology, 24*(6), 773–784. doi:10.1111/rec.12369

Roshnath, R., & Sashikuma, C. (2019). Conservation challenges of the heronries in Kerala. *Journal of the Bombay Natural History Society, 116*, 63–67.

Rosli, F. A., Lee, K. E., Goh, C. T., Mokhtar, M., Latif, M. T., Goh, T. L., & Simon, N. (2017). The Use of Constructed Wetlands in Sequestrating Carbon: An Overview. *Nature Environment & Pollution Technology, 16*(3), 813–819.

Roy-Basu, A., Bharat, G. K., Chakraborty, P., & Sarkar, S. K. (2020). Adaptive co-management model for the East Kolkata wetlands: A sustainable solution to manage the rapid ecological transformation of a peri-urban landscape. *The Science of the Total Environment, 698*, 134203. doi:10.1016/j.scitotenv.2019.134203 PMID:31783445

RStudio Team. (2020). *RStudio: Integrated Development for R.* RStudio. https://www.rstudio.com/

Ruiz, C. A., & Aguirre, J. (2003). Las comunidades de briófitos y su relación con la estructura de la vegetación fanerogamica, en el gradiente altitudinal de la Serranía del Perijá (Cesar–Colombia). *Bryophyte Diversity and Evolution, 24*, 101–113.

Ruoso, L. E., Plant, R., Maurel, P., Dupaquier, C., Roche, P. K., & Bonin, M. (2015). Reading ecosystem services at the local scale through a territorial approach: The case of peri-urban agriculture in the Thau Lagoon, Southern France. *Ecology and Society, 20*(3), art11. doi:10.5751/ES-07694-200311

Russell, J. M., & Johnson, T. C. (2005). A high-resolution geochemical record from Lake Edward, Uganda Congo and the timing and causes of tropical African drought during the late Holocene. *Quaternary Science Reviews, 24*(12-13), 1375–1389. doi:10.1016/j.quascirev.2004.10.003

Russi, D., ten Brink, P., Farmer, A., Badura, T., Coates, D., Förster, J., Kumar, R., & Davidson, N. (2013). *The Economics of Ecosystems and Biodiversity for Water and Wetlands.* IEEP.

Ruzicka, K., Gabriel, O., Bletterie, U., Winkler, S., & Zessner, M. (2009). Cause and effect relationship between foam formation and treated wastewater effluents in a transboundary river. *Physics and Chemistry of the Earth Parts A/B/C, 34*(8-9), 565–573. doi:10.1016/j.pce.2009.01.002

Sadrin, S., Sennoune, S., Gout, B., Marque, S., Moreau, J., Zinoune, K., Grillasca, J. P., Pons, O., & Maixent, J. M. (2020). A 2-strain mixture of Lactobacillus acidophilus in the treatment of irritable bowel syndrome: A placebo-controlled randomized clinical trial. *Digestive and Liver Disease, 52*(5), 534–540. doi:10.1016/j.dld.2019.12.009

Saitou, N., & Nei, M. (1987). The neighbor-joining method: A new method for reconstructing phylogenetic trees. *Molecular Biology and Evolution, 4*(4), 406–425. doi:10.1093/oxfordjournals.molbev.a040454 PMID:3447015

Salata, S., Ronchi, S., & Arcidiacono, A. (2017). Mapping air filtering in urban areas. A land use regression model for ecosystem services assessment in planning. *Ecosystem Services, 28*, 341–350. doi:10.1016/j.ecoser.2017.09.009

Salimi, S., Almuktar, S. A. A. N., & Scholz, M. (2021). Impact of climate change on wetland ecosystems: A critical review of experimental wetlands. *Journal of Environmental Management, 286*, 112160. doi:10.1016/j.jenvman.2021.112160 PMID:33611067

Sandi, S. G., Rodriguez, J. F., Saco, P. M., Saintilan, N., & Riccardi, G. (2021). Accelerated Sea-Level Rise Limits Vegetation Capacity to Sequester Soil Carbon in Coastal Wetlands: A Study Case in South-Eastern Australia. *Earth's Future.*

Sant, K., Mandic, O., Rundić, L., Kuiper, K. F., & Krijgsman, W. (2018). Age and evolution of the Serbian Lake System: Integrated results from middle Miocene Lake Popovac. *Newsletters on Stratigraphy, 51*(1), 117–143. doi:10.1127/nos/2016/0360

Sarkar, A., & Upadhyay, B. (2013). Assessment of the Variations in Physico-Chemical Characteristics of Water Quality of the Wetlands in District Manipuri (UP), India. *International Journal of Geology, Earth and Environmental Sciences, 3*(1), 95–103.

Saubade, A. M., & Zaouali, L. J. (1986). Malacofaune subfaussile de la Bahira el Bibane, Tunisie méridionale. *Bull. Inst. Géol. Bassin d'Aquitaine, Bordeaux, 40*, 23–33.

Saunois, M., Bousquet, P., Poulter, B., Peregon, A., Ciais, P., Canadell, J. G., & Zhu, Q. (2016). The global methane budget 2000–2012. *Earth System Science Data*, *8*(2), 697–751. doi:10.5194/essd-8-697-2016

Savari, M., Damaneh, H. E., & Damaneh, H. E. (2020). Factors influencing local people's participation in sustainable forest management. *Arabian Journal of Geosciences*, *13*(13), 1–13. doi:10.100712517-020-05519-z

Savenije, H. H. G., & Pages, J. (1992). Hypersalinity: A dramatic change in the hydrology of Sahelian estuaries. *Journal of Hydrology (Amsterdam)*, *135*, 157–174.

Savijärvi, H. I., & Harri, A. M. (2021). Water vapor adsorption on Mars. *Icarus*, *357*, 114270. doi:10.1016/j.icarus.2020.114270

Sawyer, C., & McCarthy, P. (1967). *Chemical and Sanitary Engineering* (2nd ed.). McGraw-Hill.

Sawyer, S., & Tapia, A. (2005). The Sociotechnical Nature of Mobile Computing Work: Evidence from a Study of Policing in the United State. *International Journal of Technology and Human Interaction*, *1*(3), 1–14. doi:10.4018/jthi.2005070101

Schilling, K., & Zessner, M. (2011). Foam in the aquatic environment. *Water Research*, *45*(15), 4355–4366. doi:10.1016/j.watres.2011.06.004 PMID:21757217

Schneider, D., Aßmann, N., Wicke, D., Poehlein, A., & Daniel, R. (2020). Metagenomes of Wastewater at Different Treatment Stages in Central Germany. *Microbiology Resource Announcements*, *9*(15), e00201–e00220. doi:10.1128/MRA.00201-20 PMID:32273362

Schraufnagel, F.H. (1967). Pollution associated with chemical de-icing. *Highway Research Record*, *193*, 22–33.

Schulz, E., Smykatz-Kloss, W., & Abichou H., (1995). Zaderg environmental history of the semidesert region in southernTunisia. *Zentralblatt fur Geologie und Paleontology*, *1*(3-4), 423-440.

Schulz, E., Abichou, A., Hachicha, T., Pomel, S., Salzmann, U., & Zouari, K. (2002). Sebkhas as ecological archives and the vegetation and landscape history of southeastern Tunisia during the last two millennia. *Journal of African Earth Sciences*, *34*(3-4), 223–229. doi:10.1016/S0899-5362(02)00021-0

Schulze, E. D., & Mooney, H. A. (1993). *Biodiversity and ecosystem function*. Springer-Verlag. doi:10.1007/978-3-642-58001-7

Schwarzenbach, R. P., Egli, T., Hofstetter, T. B., Von Gunten, U., & Wehrli, B. (2010). Global water pollution and human health. *Annual Review of Environment and Resources*, *35*, 109–136.

Scott, D. A., & Pole, C. M. (1989). *A Status Overview of Asian Wetlands*. Asian Wetlands Bureau.

Segarra, K. E. A., Schubotz, F., Samarkin, V., Yoshinaga, M. Y., Hinrichs, K. U., & Joye, S. B. (2015). High rates of anaerobic methane oxidation in freshwater wetlands reduce potential atmospheric methane emissions. *Nature Communications*, *6*(1), 1–8. doi:10.1038/ncomms8477 PMID:26123199

Sejrup, H. P., Haflidason, H., & Andrews, J. T. (2011). A Holocene North Atlantic SST record and regional climate variability. *Quaternary Science Reviews*, *30*(21-22), 3181–3195. doi:10.1016/j.quascirev.2011.07.025

Sellami, I., Charmantier, G., Naceur, H. B., Kacem, A., & Lorin-Nebel, C. (2020a). Osmoregulatory performance and immunolocalization of Na+/K+-ATPase in the branchiopod Artemia salina from the Sebkha of Sidi El Hani (Tunisia). *Tissue & Cell*, *63*, 101340. doi:10.1016/j.tice.2020.101340 PMID:32223951

Sellami, I., Naceur, H. B., & Kacem, A. (2020c). Study of Cysts Biometry and Hatching Percentage of the Brine Shrimp Artemia salina (Linnaeus, 1758) from the Sebkha of Sidi El Hani (Tunisia) According to Successive Generations. *Aquaculture Studies*, *21*(1), 41–46. doi:10.4194/2618-6381-v21_1_05

Sellami, I., Naceur, H. B., & Kacem, A. (2021b). Reproductive performance in successive generations of the brine shrimp Artemia salina (Crustacea: Anostraca) from the Sebkha of Sidi El Hani (Tunisia). *Animal Reproduction Science, 225*, 106692. doi:10.1016/j.anireprosci.2021.106692 PMID:33450437

Serwecińska, L. (2020). Antimicrobials and Antibiotic-Resistant Bacteria: A Risk to the Environment and to Public Health. *Water (Basel), 12*(12), 3313. doi:10.3390/w12123313

Shah, J. A., Pandit, A. K., & Shah, G. M. (2019). Physico-chemical limnology of a shallow lake in the floodplains of western Himalaya from last four decades: Present status. *Environmental Systems Research, 8*(1), 1–10. doi:10.118640068-019-0136-2

Shakeri, A., & Moore, F. (2010). The impact of an industrial complex on freshly deposited sediments, Chener Rahdar river case study, Shiraz, Iran. *Environmental Monitoring and Assessment, 169*(1), 321–334.

Sharma, S. D. (2020). Risk assessment via oral and dermal pathways from heavy metal polluted water of Kollerulake - A Ramsar wetland in Andhra Pradesh, India. *Environmental Analysis. Health and Toxicology, 35*(3), e2020019.

Sharma, S. S., Dietz, K. J., & Mimura, T. (2016). Vacuolar compartmentalization as indispensable component of heavy metal detoxification in plants. *Plant, Cell & Environment, 39*(5), 1112–1126. doi:10.1111/pce.12706 PMID:26729300

Sharma, S., & Bhattacharya, A. (2017). Drinking water contamination and treatment techniques. *Applied Water Science, 7*, 1043–1067.

Sharma, S., Rattan, R., Jamwal, P. S., Kumar, R., & Shukla, S. (2014). *Wetlands: Strategy for Natural Resource Management through Wetlands Conservation in India. Bharti Publications.*

Shelef, O., Gross, A., & Rachmilevitch, S. (2013). Role of Plants in a Constructed Wetland: Current and New Perspectives. *Water (Basel), 5*, 405–419.

Shepard, F. P. (1954). Nomenclature based on sand-silt-clay ratios. *Journal of Sedimentary Petrology, 24*, 151–154.

Shimizu, M. H., Sampaio, G., Venancio, I. M., & Maksic, J. (2020). Seasonal changes of the South American monsoon system during the Mid-Holocene in the CMIP5 simulations. *Climate Dynamics, 54*(5), 2697–2712. doi:10.100700382-020-05137-1

Shoko, A. P., Limbu, S. M., Mrosso, H. D., & Mgaya, Y. D. (2014). A comparison of diurnal dynamics of water quality parameters in Nile tilapia (Oreochromis niloticus, Linnaeus, 1758) monoculture and polyculture with African sharp tooth catfish (Clarias gariepinus, Burchell, 1822) in earthen ponds. *International Aquatic Research., 6*(1), 1–13. doi:10.100740071-014-0056-8

Short, F. T., Kosten, S., Morgan, P. A., Malone, S., & Moore, G. E. (2016). Impacts of climate change on submerged and emergent wetland plants. *Aquatic Botany, 135*, 3–17. doi:10.1016/j.aquabot.2016.06.006

Sieben, E. J. J., Glen, R. P., Deventer, H. V., & Dayaram, A. (2021). The contribution of wetland flora to regional floristic diversity across a wide range of climatic conditions in southern Africa. *Biodiversity and Conservation, 30*(3), 1–22.

Sievers, F., Wilm, A., Dineen, D., Gibson, T. J., Karplus, K., Li, W., Lopez, R., McWilliam, H., Remmert, M., Söding, J., Thompson, J. D., & Higgins, D. G. (2011). Fast, scalable generation of high-quality protein multiple sequence alignments using Clustal Omega. *Molecular Systems Biology, 7*(1), 539. doi:10.1038/msb.2011.75 PMID:21988835

Silver, W. L., Perez, T., Mayer, A., & Jones, A. R. (2021). The role of soil in the contribution of food and feed. *Philosophical Transactions of the Royal Society B, 376*(1834).

Simboura, N., & Reizopoulou, S. (2007). A comparative approach of assessing ecological status in two coastal areas of Eastern Mediterranean. *Ecological Indicators*, *7*(2), 455–468. doi:10.1016/j.ecolind.2006.05.003

Singh, B. P. (1994). Biodiversity conservation in mangrove environment. *Indian Forester*, *120*, 827–836.

Singh, M., Ganguli, S., & Ghosh, M. M. (2019). Comparative metagenomic dataset of hospital effluent microbiome from rural and urban hospitals in West Bengal. *Data in Brief*, *25*, 104264. doi:10.1016/j.dib.2019.104264 PMID:31440549

Singh, S. P. (1995). India's population problem, a global issue of environment and sustainability: An appeal for cooperation. *Bulletin of the Ecological Society of America*, *76*, 55–59.

Sizo, A., Noble, B., & Bell, S. (2015). Futures analysis of urban land use and wetland cane in Saskatoon, Canada. An application in strategic environmental assessment. *Sustainability*, *7*(1), 811–830. doi:10.3390u7010811

Slavov, A. K. (2017). General Characteristics and Treatment Possibilities ofDairy Wastewater-A Review. *Food Technology and Biotechnology*, *55*(1), 14–28. https://doi.org/10.17113/ftb.55.01.17.4520

Smalling, K.L., Anderson, C.W., Honeycutt, R.K., Cozzarelli, I.M., Preston, T., & Hossack, B.R. (2019). Associations between environmental pollutants and larval amphibians in wetlands contaminated by energy-related brines are potentially mediated by feeding traits. *Environmental Pollution*, *248*, 260–268.

Smith, R. D., Ammann, A., Bartoldus, C., & Brinson, M.M. (1995). *An approach for assessing wetland functions using hydrogeomorphic classification, reference wetlands, and functional indices: Vicksburg, Miss., U.S. Army Engineers Waterways Experiment Station.* Technical Report TRWRP-DE, 10, 100.

Smith, T.J., HI. (1987). Effects of light and intertidal position on seedling survival and growth in tropical tidal forests. *Journal of Experimental Marine Biology and Ecology*, *110*, 133–146.

Sok, K., Méral, P., Pillot, D., & Defossez, S. (2019). Ecosystem Services from Tonle Sap Flood Pulse: Spatial and Economic Analysis in Aek Phnom and Sangkae Districts of Battambang Province, Cambodia. In *Water and Power* (pp. 123–151). Springer. doi:10.1007/978-3-319-90400-9_8

Solanki, V., & Joshi, A. (2017). Disappearing wetland: A study of Basai wetlands, Haryana (India). *International Journal of Economic Research*, *14*(21), 159–169.

Sorensen, J. (1993). The management of enclosed coastal waterbodies: the need for a framework for international information exchange. In *Proceedings of the Eighth Symposium on Coastal and Ocean Management*. American Society of Civil Engineers.

South Asia Network on Dams, Rivers, and People (SANDRP). (2020). *World Wetlands Day 2020: Positive Stories from India*. https://sandrp.in/2020/02/01/world-wetlands-day-2020-positive-stories-from-india/

South Asia Network on Dams, Rivers, and People (SANDRP). (2021). *India's Wetlands 2016: Encroached and Polluted*. https://sandrp.in/2017/02/02/indias-wetlands-2016-encroached-and-polluted/

Southwood, T. R. E. (1978). *Ecological Methods*. Chapman and Hall.

Staley, Z. R., Chuong, J. D., Hill, S. J., Grabuski, J., Shokralla, S., Hajibabaei, M., & Edge, T. A. (2018). Fecal source tracking and eDNA profiling in an urban creek following an extreme rain event. *Scientific Reports*, *8*(1), 14390. doi:10.103841598-018-32680-z PMID:30258068

Steerenberg, P., Van Amelsvoort, L. G. P. M., Colosio, C., Corsini, E., Fustinoni, S., Vergieva, T., & Van Loveren, H. (2008). Toxicological evaluation of the immune function of pesticide workers, a European wide assessment. *Human and Experimental Toxicology*, *27*(9), 701–707.

Stengele, F., & Smykatz-Kloss, W. (1995). Mineralogical and geochemical study of Holocene sebkha sediments in southeastern Tunisia. *Chemie der Erde-Geochemistry*, *55*(4), 241–256.

Steven, A. D. L., Vanderklift, M. A., & Bohler-Muller, N. (2019). A new narrative for the Blue Economy and Blue Carbon. *Journal of the Indian Ocean Region*, *15*(2), 123–128. doi:10.1080/19480881.2019.1625215

Strohschön, R., Wiethoff, K., Baier, K., Lu, L., Bercht, A.L., Wehrhahn, R., & Azzam, R. (2013). Land use and Water Quality in Guangzhou, China: A survey of ecological and Social Vulnerability in Four Urban Units of the Rapidly Developing Megacity. *International Journal of Environmental of Research*, *7*, 343–358.

Stuiver, M., & Brazuinas, T. (1993). Modeling atmospheric 14C influences and 14C ages of marine samples to 10,000 bc. In M. Stuiver, A. Long, & R. S. Kra (Eds.), *Calibration, Radiocarbon*. doi:10.1017/S0033822200013874

Su, J., Friess, D. A., & Gasparatos, A. (2021). A meta-analysis of the ecological and economic outcomes of mangrove restoration. *Nature Communications*, *12*(1), 1–13. doi:10.103841467-021-25349-1 PMID:34413296

Sun, F., & Carson, R. T. (2020). Coastal wetlands reduce property damage during tropical cyclones. *Proceedings of the National Academy of Sciences of the United States of America*, *117*(11), 5719–5725. doi:10.1073/pnas.1915169117 PMID:32123063

Sun, F., Xu, Z., & Fan, L. (2021). Response of heavy metal and antibiotic resistance genes and related microorganisms to different heavy metals in activated sludge. *Journal of Environmental Management*, *300*, 113754. Advance online publication. doi:10.1016/j.jenvman.2021.113754 PMID:34543965

Sur, S., & Mandal, J. (2019). Deepor Beel: A Weeping and Blemishing Ramsar Site. *Jalaplavit*, *9*(3), 48–58.

Surya, B., Syafri, S., Sahban, H., & Sakti, H. H. (2020). Natural resource conservation based on community economic empowerment: Perspectives on watershed management and slum settlements in Makassar City, South Sulawesi, Indonesia. *Land (Basel)*, *9*(4), 104. doi:10.3390/land9040104

Suthar, A. M., Tatu, K., Gujar, R., & Kamboj, R. D. (2019). A Comparative Account of Diversity of Hydrophytes in Some Inland Wetlands (Pariej, Kanewal and Wadhwana) of Central Gujarat. *Research & Reviews. Journal of Life Science*, *9*(2), 39–43.

Sutherland, W. J. (2006). *Ecological Census Techniques: A Handbook*. Cambridge University Press. doi:10.1017/CBO9780511790508

Syakila, A., Kroeze, C., & Slomp, C. P. (2010). Neglecting sinks for N2O at the earth's surface: Does it matter? *Journal of Integrative Environmental Sciences*, *7*(sup1, S1), 79–87. doi:10.1080/1943815X.2010.497492

Syvitski & Kettner. (2011). Sediment flux and the Anthropocene. *Philosophical Transactions of the Royal Society A: Mathematical, Physical and Engineering Sciences, 369*(1938), 957-975.

Tahmiscioğlu, M. S., Anul, N., Ekmekçi, F., & Durmuş, N. (2007, March). Positive and negative impacts of dams on the environment. In *International Congress on River Basin Management* (pp. 759-769). Academic Press.

Talukdar, J. K., & Rajbongshi, M. K. (2018). Ichthyofaunal Diversity and Conservation Status of Puthimari Beel of Barpeta, Assam, India. Academic Press.

Tamiazzo, J., Breschigliaro, S., Salvato, M., & Borin, M. (2015). Performance of a wall cascade constructed wetland treating surfactant-polluted water. *Environmental Science and Pollution Research International*, *22*(17), 12816–12828.

Tanabe, Y., Hori, M., Mizuno, A. N., Osono, T., Uchida, M., Kudoh, S., & Yamamuro, M. (2019). Light quality determines primary production in nutrient-poor small lakes. *Scientific Reports*, *9*(1), 1–8. doi:10.103841598-019-41003-9 PMID:30874599

Tchobanoglous, G., Burton, F. L., & Stensel, H. D. (2003). Metcalf & Eddy wastewater engineering: treatment and reuse. International Edition. McGraw-Hill.

Tchobanoglous, G. (1993). *Integrated Solid Waste Management: Engineering Principles and Management Issues*. McGraw-Hill.

Tedijanto, C., Olesen, S. W., Grad, Y. H., & Lipsitch, M. (2018). Estimating the proportion of bystander selection for antibiotic resistance among potentially pathogenic bacterial flora. *Proceedings of the National Academy of Sciences of the United States of America*, *115*(51), E11988–E11995. doi:10.1073/pnas.1810840115 PMID:30559213

TEEB. (2010). *The Economics of Ecosystems and Biodiversity: Ecological and Economic Foundations* (P. Kumar, Ed.). Earthscan.

TEEB. (2013). The economics of ecosystems and biodiversity for water and wetlands. Institute for European Environmental Policy (IEEP) & Ramsar Secretariat.

Ten Brink, P., Badura, T., Farmer, A., & Russi, D. (2012). *The Economics of Ecosystem and Biodiversity for Water and Wetlands: A Briefing Note*. Institute for European Environmental Policy.

Thai, P. K., Ky, L. X., Binh, V. N., Nhung, P. H., Nhan, P. T., Hieu, N. Q., Dang, N., Tam, N., & Anh, N. (2018). Occurrence of antibiotic residues and antibiotic-resistant bacteria in effluents of pharmaceutical manufacturers and other sources around Hanoi, Vietnam. *The Science of the Total Environment*, *645*, 393–400. doi:10.1016/j.scitotenv.2018.07.126 PMID:30029118

Thapa, S., Wang, L., Koirala, A., Shrestha, S., Bhattarai, S., & Aye, W. N. (2020). Valuation of ecosystem services from an important wetland of Nepal: A Study from Begnas watershed system. *Wetlands*, *40*(5), 1071–1083. doi:10.100713157-020-01303-7

Tharpar, R. (1963). *A History of India* (Vol. 1). Penguin Books.

The Center for Disease. (2020). *Dynamics Economics & Policy. Resistance Map: Antibiotic resistance*. https://resistancemap.cddep.org/Antibiotic Resistance. php

Thoendel, M. (2020). Targeted Metagenomics Offers Insights into Potential Tick-Borne Pathogens. *Journal of Clinical Microbiology*, *58*(11), e01893–e20. doi:10.1128/JCM.01893-20 PMID:32878948

Thomas, S. T., Thomas, E. T., McLean, M., & Titus, T. T. (2021). Paving the way to achieving the United Nations Sustainable Development Goals for women from Indigenous communities: Lessons from Attappady, India. *Discover Sustainability*, *2*(1), 1–15. doi:10.100743621-021-00009-y

Thormann, M. N. (2006). Diversity and function of fungi in peatlands: A carbon cycling perspective. *Canadian Journal of Soil Science*, *86*(Special Issue), 281–293. doi:10.4141/S05-082

Timothy, P. R., & Swindles, G. T. (2014). Influence of ocean–atmospheric oscillations on lake ice phenology in eastern North America. *Climate Dynamics*, *45*(9), 2293–2308. doi:10.100700382-014-2415-y

Tiner, R. W. (1991). The concept of a hydrophyte for wetland identification. *Bioscience*, *41*, 236–247.

Tiner, R. W. (2009). *Ecology of Wetlands: Classification Systems. In Encyclopedia of Inland Waters*. Academic Press.

Tipping, E., Vincent, C. D., Lawlor, A. J., & Lofts, S. (2008). Metal accumulation by stream bryophytes, related to chemical speciation. *Environmental Pollution, 156*, 936–943.

Tiseo, I. (2020). *Generation of municipal solid waste worldwide in 2017, by select country (in million metric tons).* Statista. Retrieved from https://www.statista.com/statistics/916749/global-generation-of-municipal-solid-waste-by-country/

Todd, M. J., Smith, K., & Bannister, P. (2006). Supervising a Social Science Dissertation: Staff Experiences and Perceptions. *Teaching in Higher Education, 11*(2), 161–173. doi:10.1080/13562510500527693

Tomasini-Ortiz, C., Moeller Chávez, G., Sánchez Chávez, J., & Bravo Inclán, L. (2012). Cianobacterias y cianotoxinas en el lago de Pátzcuaro, Michoacán, México. *Revista AIDIS de Ingeniería y Ciencias Ambientales: Inestigación, desarrollo y práctica, 5*(2), 93-101.

Tootoonchi, M., & Gettys, L. A. (2019). Testing salt stress on aquatic plants: Effect of salt source and substrate. *Aquatic Ecology, 53*(3), 325–334. doi:10.100710452-019-09692-6

Tornabene, B. J., Breuner, C. W., & Hossack, B. R. (2020). Relative Toxicity and Sublethal Effects of NaCl and Energy-Related Saline Wastewaters on Prairie Amphibians. *Aquatic Toxicology (Amsterdam, Netherlands), 228*, 105626. doi:10.1016/j.aquatox.2020.105626 PMID:32992088

Trepel, M. (2000). Quantifizierung der Stickstoffdynamik von Ökosystemen auf Niedermoorböden mit dem Modellsystem WASMOD. *EcoSys, 29*, 1–140.

Tulchinsky, T. H. (2018). John Snow, Cholera, the Broad Street Pump; Waterborne Diseases Then and Now. *Case Studies in Public Health*, 77–99. doi:10.1016/B978-0-12-804571-8.00017-2

Turetsky, M. R., Wieder, R. K., Vitt, D. H., Evans, R. J., & Scott, K. D. (2008). The disappearance of relict permafrost in boreal North America: Effects On peatland carbon storage and fluxes. *Global Change Biology, 13*(9), 1922–1934.

Turner, R. K., van der Bergh, J. C. J. M., Soderqvist, T., Barendregt, A., van der Straaten, J., Maltby, E., & van Ierland, E. C. (2000). Ecological-economic analysis of wetlands: Scientific integration for management and policy. *Ecological Economics, 35*(1), 7–23. doi:10.1016/S0921-8009(00)00164-6

Turney, C. S. M., Jones, R. T., Fogwill, C., Hatton, J., Williams, A. N., Hogg, A., Thomas, Z. A., Palmer, J., Mooney, S., & Reimer, R. W. (2016). A 250-year periodicity in Southern Hemisphere westerly winds over the last 2600 years. *Climate of the Past, 12*(2), 189–200. doi:10.5194/cp-12-189-2016

Twiddle, C. (2012). Pollen Analysis: Not Just a Qualitative Tool. *Geo-Morphological Techniques,* 1-8.

U.S. Fish & Wildlife Service. (2021). *Classification and types of wetlands.* https://www.epa.gov/wetlands/classification-and-types-wetlands#undefined

Ulluwishewa, R. (1995). Traditional practices of inland fishery resources management in the dry zone of Sri Lanka: Implications for sustainability. *Environmental Conservation, 22*, 127–3.

UN WWAP (United Nations World Water Assessment Programme). (2017). *The United Nations World Water Development Report (2017). Wastewater: The Untapped Resource*. UNESCO.

UNESCO World Water Assessment Programme (WWAP). (2017). *The United Nations world water development report 2017: wastewater: the untapped resource; facts and figures*. https://unesdoc.unesco.org/ark:/48223/pf0000247553

UN-Habitat. (2016). https://unhabitat.org/un-habitat-launches-the-world-cities-report 2016#:~:text=With%20the%20United%20Nations%20Conference,to%20prosperity%20and%20equitable%20growth

United Nations Environment Program (UNEP). (2007). *Global Environment Outlook GEO.* Retrieved June 24, 2021, from https://www.unep.org/resources/global-environment-outlook-4

United States Environment Protection Agency. (n.d.). *Why are Wetlands Important?* EPA. https://www.epa.gov/wetlands/why-are-wetlands-important

United States Environmental Protection Agency (USEPA). (1995). *America's Wetlands: Our Vital Link Between Land and Water.* Office of Water, Office of Wetlands, Oceans, and Watersheds.

United States Environmental Protection Agency. (1979). *National Secondary Drinking Water Regulations EPA 570/9-76-000.* US EPA.

United States Environmental Protection Agency. (n.d.). *Wetlands Protection and Restoration.* Retrived June 24, 2021, from: https://www.epa.gov/wetlands

UN-Water. (2015a). https://www.unwater.org/publications/wastewater-management-un-water-analytical-brief/

Urvashi. (2006). *Sustainable Solid Waste Management: Issues Policies and Structures.* Academic Foundation.

USEPA. (2016). *Water quality assessment and TMDL information.* United States Environmental Protection Agency.

Valach, A. C., Kasak, K., Hemes, K. S., Anthony, T. L., Dronova, I., Taddeo, S., & Baldocchi, D. D. (2021). Productive wetlands restored for carbon sequestration quickly become net CO2 sinks with site-level factors driving uptake variability. *PLoS One, 16*(3), e0248398. doi:10.1371/journal.pone.0248398 PMID:33765085

Vallès, V., & Bourgeat, F. (1989). Détermination géochimique de l'exigence de plâtre d'une culture des sols sodiques. I. Le développement du modèle thermodynamique GYPSOL simulant l'interaction chimique de l'eau d'irrigation-sol. *Recherche sur les sols aides et de readaptation, 44*(3), 157-163.

Van den Wyngaert, I. J. J., & Bobbink, R. (2006). The influence of vertebrate herbivory on ecological dynamics in wetland ecosystems. In E. Maltby (Ed.), *The wetlands handbook.* Blackwell Scientific.

Van der Valk, A. (1981). Succession in wetlands: A Gleasonian approach. *Ecology, 62*, 688–696.

van der Valk, A. G. (2005). The prairie potholes of North America. *Hydrobiologia, 539*, 171–188.

Van Kranendonk, M. J., Djokic, T., Baumgartner, R., Bontognali, T. R., Sugitani, K., Kiyokawa, S., & Walter, M. R. (2021). Life analog sites for Mars from early Earth: diverse habitats from the Pilbara Craton and Mount Bruce Supergroup, Western Australia. In Mars Geological Enigmas (pp. 357-403). Elsevier.

Van Oirschot, D., Wallace, S., & Van Deun, R. (2015). Wastewater treatment in a compact intensified wetland system at the Badboot: A floating swimming pool in Belgium. *Environmental Science and Pollution Research International, 22*(17), 12870–12878.

Vann, C. D., & Megonigal, J. P. (2003). Elevated CO 2 and water depth regulation of methane emissions: Comparison of woody and non-woody wetland plant species. *Biogeochemistry, 63*(2), 117–134. doi:10.1023/A:1023397032331

vanRaamsdonk, L., van der Zande, M., Koelmans, A. A., Hoogenboom, R., Peters, R., Groot, M. J., Peijnenburg, A., & Weesepoel, Y. (2020). Current insights into monitoring, bioaccumulation, and potential health effects of microplastics present in the food chain. *Foods, 9*(1), 72. doi:10.3390/foods9010072

Vargas, R., & Morales, E. (2014). Hepáticas del Parque Natural Municipal "Robledales de Tipacoque", Boyacá-Colombia. *Universitas Scientiarum, 19*, 201–211.

Vasiliu, D., Bucse, A., Lupascu, N., Ispas, B., Gheablau, C., & Stanescu, I. (2020). Assessment of the metal pollution in surface sediments of coastal Tasaul Lake (Romania). *Environmental Monitoring and Assessment*, *192*(12), 749. doi:10.100710661-020-08698-0 PMID:33155147

Vásquez, C., Calva, J., Morocho, R., Donoso, D. A., & Benítez, Á. (2019). Bryophyte communities along a tropical urban river respond to heavy metal and arsenic pollution. *Water (Basel)*, *11*, 813.

Vasudevan, P., Nair, M. K. M., Annamalai, T., & Venkitanarayanan, K. S. (2003). Phenotypic and genotypic characterization of bovine mastitis isolates of Staphylococcus aureus for biofilm formation. *Veterinary Microbiology*, *92*(1-2), 179–185. doi:10.1016/S0378-1135(02)00360-7 PMID:12488081

Vedogbeton, H., & Johnston, R. J. (2020). Commodity consistent meta-analysis of wetland values: An illustration for coastal marsh habitat. *Environmental and Resource Economics*, *75*(4), 835–865. doi:10.100710640-020-00409-0

Verma, M. (2001). *Economic Valuation of Bhoj Wetlands for Sustainable Use* [EERC Working Paper Series: WB-9]. Indian Institute of Forest Management.

Vicziany, M., Chattopadhyay, D., & Bhattacharyya, S. (2017). Food from sewage: Fish from the east Kolkata wetlands and the limits of traditional knowledge. *South Asia*, *40*(3), 619–644. doi:10.1080/00856401.2017.1341038

Viehweger, K. (2014). How plants cope with heavy metals. *Botanical Studies (Taipei, Taiwan)*, *55*(1), 35. doi:10.1186/1999-3110-55-35 PMID:28510963

Vijayan, V. S., Prasad, S. N., Vijayan, L., & Muralidharan, S. (2004). Inland Wetlands of India: Conservation Priorities. Sàlim Ali Centre for Ornithology and Natural History.

Vikingstad, A. K. (2006). *Static and dynamic studies of foam and foam-oil interactions. Department of Chemistry*. University of Bergen.

Visconti, G. (2021). Geobiology. In *Climate, Planetary and Evolutionary Sciences* (pp. 317–329). Springer. doi:10.1007/978-3-030-74713-8_9

Voigt, A. M., Zacharias, N., Timm, C., Wasser, F., Sib, E., Skutlarek, D., Parcina, M., Schmithausen, R. M., Schwartz, T., Hembach, N., Tiehm, A., Stange, C., Engelhart, S., Bierbaum, G., Kistemann, T., Exner, M., Faerber, H. A., & Schreiber, C. (2020). Association between antibiotic residues, antibiotic resistant bacteria and antibiotic resistance genes in anthropogenic wastewater - An evaluation of clinical influences. *Chemosphere*, *241*, 125032. doi:10.1016/j.chemosphere.2019.125032 PMID:31622887

Votteler, T. H., & Muir, T. A. (2002). *Wetland Protection Legislation*. Wetland Management and Research. https://water.usgs.gov/nwsum/WSP2425/legislation.html

Vuori, K. M., & Helisten, H. (2010). The use of aquatic mosses in assessment of metal pollution: Appraisal of type-specific background concentrations and inter-specific differences in metal accumulation. *Hydrobiologia*, *656*, 99–106.

Vymazal, J., Brix, H., Cooper, P. F., Green, M. B., & Haberl, R. (1998). *Constructed wetlands for wastewater treatment in Europe*. Backhuys Publishers.

Wakode, H. B., Baier, K., Jha, R., & Azzam, R. (2018). Impact of urbanization on groundwater recharge and urban water balance for the city of Hyderabad, India. *International Soil and Water Conservation Research*, *6*, 51–6261.

Walker, L. R., Wardle, D. A., Bardgett, R. D., & Clarkson, B. D. (2010). The use of chronosequences in studies of ecological succession and soil development. *Journal of Ecology*, *98*(4), 725–736. doi:10.1111/j.1365-2745.2010.01664.x

Wallace, J., & Waltham, N. J. (2021). On the potential for improving water quality entering the Great Barrier Reef lagoon using constructed wetlands. *Marine Pollution Bulletin, 170*, 112627. doi:10.1016/j.marpolbul.2021.112627 PMID:34182303

Wang, J., Chen, J., Wen, Y., Fan, W., Liu, Q., & Tarolli, P. (2021). Monitoring the coastal wetlands dynamics in Northeast Italy from 1984 to 2016. *Ecological Indicators, 129*, 107906. doi:10.1016/j.ecolind.2021.107906

Wang, J., Chu, L., Wojnárovits, L., & Takács, E. (2020). Occurrence and fate of antibiotics, antibiotic resistant genes (ARGs) and antibiotic resistant bacteria (ARB) in municipal wastewater treatment plant: An overview. *The Science of the Total Environment, 744*, 140997. doi:10.1016/j.scitotenv.2020.140997 PMID:32755790

Wang, L., & Liang, T. (2015). Distribution characteristics of phosphorus in the sediments and overlying water of Poyang lake. *PLoS One, 10*(5), e0125859.

Wang, N., & Wang, C. (2018). Effects of microcystin-LR on the tissue growth and physiological responses of the aquatic plant *Iris pseudacorus* L. *Aquatic Toxicology (Amsterdam, Netherlands), 200*, 197–205.

Wang, X., Lan, B., Fei, H., Wang, S., & Zhu, G. (2021). Heavy metal could drive co-selection of antibiotic resistance in terrestrial subsurface soils. *Journal of Hazardous Materials, 411*, 124848. doi:10.1016/j.jhazmat.2020.124848 PMID:33858075

Wang, X., Sun, Y., Li, S., & Wang, H. (2019). Spatial distribution and ecological risk assessment of heavy metals in soil from the Raoyanghe Wetland, China. *PLoS One, 14*(8), e0220409. doi:10.1371/journal.pone.0220409 PMID:31398209

Wang, Y., Pan, C. Q., & Xing, H. (2019). Advances in Gut Microbiota of Viral Hepatitis Cirrhosis. *BioMed Research International, 9726786*, 1–9. Advance online publication. doi:10.1155/2019/9726786 PMID:31886272

Wang, Z., Wu, J., Madden, M., & Mao, D. (2012). China's wetlands: Conservation plans and policy impacts. *Ambio, 41*(7), 782–786. doi:10.100713280-012-0280-7 PMID:22457078

Wan, J., Wang, X., Yang, T., Wei, Z., Banerjee, S., Friman, V. P., Mei, X., Xu, Y., & Shen, Q. (2021). Livestock manure type affects microbial community composition and assembly during composting. *Frontiers in Microbiology, 12*, 621126. https://doi.org/10.3389/fmicb.2021.621126

Wanner, H., Beer, J., Bütikofer, J., Crowley, T. J., Cubasch, U., Flückiger, J., & Widmann, M. (2008). Mid-to Late Holocene climate change: An overview. *Quaternary Science Reviews, 27*(19-20), 1791–1828. doi:10.1016/j.quascirev.2008.06.013

Warren, J. K. (2006). *Evaporites: sediments, resources and hydrocarbons*. Springer Science & Business Media. doi:10.1007/3-540-32344-9

Wasli, A. (2016). *Holocene cyclostratigraphy within the sedimentary record of exoreic (Boujmel) and endorheic (Mhabeul) saline systems: Eustatic and climatic controls* (Master Thesis). Higher Institute of Applied Sciences and Technology of Gabes.

Waters, C. N., Zalasiewicz, J. A., Williams, M., Ellis, M. A., & Snelling, A. M. (2014). A stratigraphical basis for the Anthropocene? *Geological Society, 395*(1), 1-21.

Waters, C. N., Zalasiewicz, J., Summerhayes, C., Barnosky, A. D., Poirier, C., Gałuszka, A., & Jeandel, C. (2016). The Anthropocene is functionally and stratigraphically distinct from the Holocene. *Science, 351*(6269).

Waters, C. N., Zalasiewicz, J., Summerhayes, C., Fairchild, I. J., Rose, N. J., Loader, W., ... Edgeworth, M. (2018). Global Boundary Stratotype Section and Point (GSSP) for the Anthropocene Series: Where and how to look for potential candidates (2018). *Earth-Science Reviews, 178*, 379–42.

Watson, S. B. (2003). Cyanobacterial and eukaryotic algal odour compounds: Signals or by-products? A review of their biological activity. *Phycologia*, *42*(4), 332–350. doi:10.2216/i0031-8884-42-4-332.1

Werners, S. E., Bos, E. J., Civic, K., Hlásny, T., Hulea, O., Jones-Walters, L. M., & Zsuffa, I. J. (2015). *Climate change vulnerability and ecosystem-based adaptation measures in the Carpathian region: final report-Integrated assessment of vulnerability of environmental resources and ecosystem-based adaptation measures* (No. 2572). Wageningen-UR.

West, A. (2015). Core concept: Ecosystem services. *Proceedings of the National Academy of Sciences of the United States of America*, *112*(24), 7337–7338. doi:10.1073/pnas.1503837112 PMID:26082537

Weston, N. B., Neubauer, S. C., Velinsky, D. J., & Vile, M. A. (2014). Net ecosystem carbon exchange and the greenhouse gas balance of tidal marshes along an estuarine salinity gradient. *Biogeochemistry*, *120*(1), 163–189. doi:10.100710533-014-9989-7

Wetlands (Conservation and Management) Rules. (2010). *Ministry of Environment and Forests, Gazette of India*. https://thc.nic.in/Central%20Governmental%20Rules/Wetlands%20(Conservation%20And%20Management)%20Rules,%202010.pdf

Wetlands (Conservation and Management) Rules. (2017). *Ministry of Environment and Forests, and Climate Change. Gazette of India*. http://www.indiaenvironmentportal.org.in/content/447822/wetlands-conservation-and-management-rules-2017/

Wetzel, R. G. (2001). *Limnology: lake and river ecosystems*. Academic Press.

Wetzel, R. G., & Likens, G. E. (2000). Light and Temperature. In *Limnological Analyses*. Springer., doi:10.1007/978-1-4757-3250-4_2

White, E. E. Jr, Ury, E. A., Bernhardt, E. S., & Yang, X. (2021). Climate Change Driving Widespread Loss of Coastal Forested Wetlands Throughout the North American Coastal Plain. *Ecosystems (New York, N.Y.)*, 1–16. doi:10.100710021-021-00686-w

WHO Media Centre. (2017). *News Release. WHO publishes list of bacteria for which new antibiotics are urgently needed.* Available from: https://www.who.int/mediacentre/news/releases/2017/bacteria-antibiotics-needed/en/

WHO. (2015). *Connecting Global Priorities: Biodiversity and Human Health*. https://apps.who.int/iris/bitstream/handle/10665/174012/9789241508537_eng.pdf

Wilcox, D. A. (2004). Implications of hydrologic variability on the succession of plants in Great Lakes wetlands. *Aquatic Ecosystem Health & Management*, *7*(2), 223–231. doi:10.1080/14634980490461579

Wild, D., Von Schulthess, R., & Gujer, W. (1995). Structured modelling of denitrification intermediates. *Water Science and Technology*, *31*(2), 45–54. doi:10.2166/wst.1995.0070

Wiles, G. C., Barclay, D. J., Calkin, P. E., & Lowell, T. V. (2008). Century to millennial-scale temperature variations for the last two thousand years indicated from glacial geologic records of Southern Alaska. *Global and Planetary Change*, *60*(1-2), 115–125. doi:10.1016/j.gloplacha.2006.07.036

Wilkinson, I.P., Poirier, C., Head, M. J., Sayer, C. D., & Tibby, J. (2014). Microbiotic signatures of the Anthropocene in marginal marine and freshwater palaeoenvironments (2014). *Geological Society, 395*(1), 185-219.

Williamson, S., Ball, A., & Pretty, J. (2008). Trends in pesticide use and drivers for safer pest management in four African countries. *Crop Protection (Guildford, Surrey)*, *27*(10), 1327–1334.

Wilson, D. N., Hauryliuk, V., Atkinson, G. C., & O'Neill, A. J. (2020). Target protection as a key antibiotic resistance mechanism. *Nature Reviews. Microbiology*, *18*(11), 637–648. doi:10.103841579-020-0386-z PMID:32587401

Wu, Q. (2018). GIS and Remote Sensing Applications in Wetland Mapping and Monitoring. In B. Huang (Ed.), *Comprehensive Geographic Information Systems* (Vol. 2, pp. 140–157). Elsevier. doi:10.1016/B978-0-12-409548-9.10460-9

Wurts, W. A. (2003). Daily pH cycle and ammonia toxicity. *World Aquaculture*, *34*(2), 20–21.

Wurts, W. A., & Durborow, R. M. (1992). Interactions of pH, carbon dioxide, alkalinity and hardness in fish ponds. Southern Regional Aquaculture Centre. *Fact Sheet*, (464), 1–4.

WWF. (1993). *Directory of Indian Wetlands*. World Wide Fund for Nature.

Xagoraraki, I., & O'Brien, E. (2020). Wastewater-Based Epidemiology for Early Detection of Viral Outbreaks. In Women in Water Quality (pp. 75-97). Springer. doi:10.1007/978-3-030-17819-2_5

Xie, P. (2011). Ammonium, microcystins, and hypoxia of blooms in eutrophic water cause oxidative stress and C-N imbalance in submersed and floating-leaved aquatic plants in Lake Taihu, China. *Chemosphere*, *82*(3), 329–339.

Xi, H., Feng, Q., Zhang, L., Si, J., Chang, Z., Yu, T., & Guo, R. (2016). Effects of water and salinity on plant species composition and community succession in Ejina Desert Oasis, northwest China. *Environmental Earth Sciences*, *75*(2), 138. doi:10.100712665-015-4823-7

Xing, W., Wu, H., Hao, B., Huang, W., & Liu, G. (2013). Bioaccumulation of heavy metals by submerged macrophytes: Looking for hyperaccumulators in eutrophic lakes. *Environmental Science & Technology*, *47*, 4695–4703. doi:10.1021/es303923w

Xi, Y., Peng, S., Ciais, P., & Chen, Y. (2021). Future impacts of climate change on inland Ramsar wetlands. *Nature Climate Change*, *11*(1), 45–51. doi:10.103841558-020-00942-2

Xu, T., Weng, B., Yan, D., Wang, K., Li, X., Bi, W., Li, M., Cheng, X., & Liu, Y. (2019). Wetlands of International Importance: Status, Threats, and Future Protection. *International Journal of Environmental Research and Public Health*, *16*(10), 1818. doi:10.3390/ijerph16101818 PMID:31121932

Xu, X., Jiang, B., Tan, Y., Costanza, R., & Yang, G. (2018). Lake-wetland ecosystem services modeling and valuation: Progress, gaps and future directions. *Ecosystem Services*, *33*, 19–28. doi:10.1016/j.ecoser.2018.08.001

Yan, C. P., & Ma, Y. J. (2010). Research progress on impacts of human industrial activities on wetland environment. *Wetland Science*, *8*(1), 98–104.

Yang, H., Mirsepasi-Lauridsen, H. C., Struve, C., Allaire, J. M., Sivignon, A., Vogl, W., Bosman, E. S., Ma, C., Fotovati, A., Reid, G. S., Li, X., Petersen, A. M., Gouin, S. G., Barnich, N., Jacobson, K., Yu, H. B., Krogfelt, K. A., & Vallance, B. A. (2020). Ulcerative Colitis-associated E. coli pathobionts potentiate colitis in susceptible hosts. *Gut Microbes*, *12*(1), 1847976. doi:10.1080/19490976.2020.1847976 PMID:33258388

Yang, S. L., Zhang, J., Zhu, J., Smith, J. P., Dai, S. B., Gao, A., & Li, P. (2005). Impact of dams on Yangtze River sediment supply to the sea and delta intertidal wetland response. *Journal of Geophysical Research. Earth Surface*, *110*(F3).

Yang, Y., Li, B., Zou, S., Fang, H. H., & Zhang, T. (2014). Fate of antibiotic resistance genes in sewage treatment plant revealed by metagenomic approach. *Water Research*, *62*, 97–106. doi:10.1016/j.watres.2014.05.019 PMID:24937359

Yardi, K. D., Bharucha, E., & Girade, S. (2019). Post-restoration monitoring of water quality and avifaunal diversity of Pashan Lake, Pune, India using a citizen science approach. *Freshwater Science*, *38*(2), 332–341. doi:10.1086/703440

Yasir, M. (2020). Analysis of Microbial Communities and Pathogen Detection in Domestic Sewage Using Metagenomic Sequencing. *Diversity (Basel)*, *13*(1), 6. doi:10.3390/d13010006

Yechieli, Y., & Wood, W. W. (2002). Hydrogeologic processes in saline systems: Playas, sabkhas, and saline lakes. *Earth-Science Reviews*, *58*(3-4), 343–365. doi:10.1016/S0012-8252(02)00067-3

Yoo, K., Yoo, H., Lee, J., Choi, E. J., & Park, J. (2020). Exploring the antibiotic resistome in activated sludge and anaerobic digestion sludge in an urban wastewater treatment plant via metagenomic analysis. *Journal of Microbiology (Seoul, Korea)*, *58*(2), 123–130. doi:10.100712275-020-9309-y PMID:31875929

Young, H. S., McCauley, D. J., Galetti, M., & Dirzo, R. (2016). Patterns, causes, and consequences of anthropocene defaunation (2016). *Annual Review of Ecology, Evolution, and Systematics*, *47*(1), 333–358. doi:10.1146/annurev-ecolsys-112414-054142

Yuan, H., Liu, E., & Shen, J. (2015). The accumulation and potential ecological risk of heavy metals in microalgae from an eutrophic lake (Taihulake, China). *Environmental Science and Pollution Research International*, *22*(21), 17123–17134.

Yuan, Q. B., Zhai, Y. F., Mao, B. Y., & Hu, N. (2018). Antibiotic resistance genes and intI1 prevalence in a swine wastewater treatment plant and correlation with metal resistance, bacterial community and wastewater parameters. *Ecotoxicology and Environmental Safety*, *161*, 251–259. doi:10.1016/j.ecoenv.2018.05.049 PMID:29886312

Yusoff, K. (2013). Geologic life: Prehistory, climate, futures in theAnthropocene. *Environment and Planning. D, Society & Space*, *31*(5), 779–795. doi:10.1068/d11512

Zahedi, A., Monis, P., Deere, D., & Ryan, U. (2021). Wastewater-based epidemiology—Surveillance and early detection of waterborne pathogens with a focus on SARS-CoV-2, *Cryptosporidium* and *Giardia. Parasitology Research*, 1–22.

Zaïbi, C., Kamoun, F., Carbonel, P., & Montacer, M. (2011). Distribution des ostracodes dans les sédiments de subsurface de la Sebkha El-Guettiate (Skhira, golfe de Gabès). Intérêt pour la reconstitution des paléo-environnements de l'Holocène. *Carnets de géologie*.

Zaïbi, C., Carbonel, P., Kamoun, F., Fontugne, M., Azri, C., Jedoui, Y., & Montacer, M. (2012). Evolution of the sebkha Dreîaa (South-Eastern Tunisia, Gulf of Gabes) during the late Holocene: Response of ostracod assemblages. *Revue de Micropaleontologie*, *55*(3), 83–97. doi:10.1016/j.revmic.2012.03.003

Zedler, J. B. (2005). Progress in wetland restoration. *Trends in Ecology & Evolution*, *10*, 402–407. PMID:10998517

Zedler, J. B., & Kercher, S. (2005). Wetland resources: Status, trends, ecosystem services, and restorability. *Annual Review of Environment and Resources*, *30*(1), 39–74. doi:10.1146/annurev.energy.30.050504.144248

Zehr, J. P., & Ward, B. B. (2002). Nitrogen cycling in the ocean: New perspectives on processes and paradigms. *Applied and Environmental Microbiology*, *68*(3), 1015–1024. https://doi.org/10.1128/AEM.68.3.1015-1024.2002

Zeng, H., & Wu, J. (2013). Heavy metal pollution of lakes along the mid-lower reaches of the Yangtze River in China: Intensity, sources and spatial patterns. *International Journal of Environmental Research and Public Health*, *10*(3), 793–807. doi:10.3390/ijerph10030793 PMID:23442559

Zeng, M., Soric, A., & Roche, N. (2015). Modeling partial nitrification and denitrification in a hybrid biofilm reactor: Calibration by retention time distribution and respirometric tests. *Environmental Science and Pollution Research International*, *22*(17), 12849–12860.

Zhang, D., Ling, H., Huang, X., Li, J., Li, W., Yi, C., Zhang, T., Jiang, Y., He, Y., Deng, S., Zhang, X., Wang, X., Liu, Y., Li, G., & Qu, J. (2020). Potential spreading risks and disinfection challenges of medical wastewater by the presence of Severe Acute Respiratory Syndrome Coronavirus 2 (SARS-CoV-2) viral RNA in septic tanks of Fangcang Hospital. *The Science of the Total Environment*, *741*, 140445. doi:10.1016/j.scitotenv.2020.140445 PMID:32599407

Zhang, G., Bai, J., Tebbe, C. C., Zhao, Q., Jia, J., Wang, W., Wang, X., & Yu, L. (2021). Salinity controls soil microbial community structure and function in coastal estuarine wetlands. *Environmental Microbiology*, *23*(2), 1020–1037. doi:10.1111/1462-2920.15281 PMID:33073448

Zhang, K., Li, A., Liu, X., Chen, M. T., Lu, J., Zhang, J., & Wang, H. (2021). Heavy mineral record from the east China sea inner shelf: Implications for provenance and climate changes over the past 1500 years. *Continental Shelf Research*, *226*, 104488. doi:10.1016/j.csr.2021.104488

Zhang, L., Tu, D., Li, X., Lu, W., & Li, J. (2020). Impact of long-term industrial contamination on the bacterial communities in urban river sediments. *BMC Microbiology*, *20*(1), 1–16. doi:10.118612866-020-01937-x PMID:32795344

Zhang, P., Grutters, B., van Leeuwen, C., Xu, J., Petruzzella, A., van den Berg, R. F., & Bakker, E. S. (2019). Effects of Rising Temperature on the Growth, Stoichiometry, and Palatability of Aquatic Plants. *Frontiers in Plant Science*, *9*, 1947. doi:10.3389/fpls.2018.01947 PMID:30671079

Zhao, Q., Liu, S., & Dong, S. (2010). Effect of dam construction on spatial-temporal change of land use: A case study of Manwan, Lancang River, Yunnan, China. *Procedia Environmental Sciences*, *2*, 852–858. doi:10.1016/j.proenv.2010.10.096

Zhong, X., Cui, P., Jiang, J., Ning, C., Liang, B., Zhou, J., Tian, L., Zhang, Y., Lei, T., Zuo, T., Ye, L., Huang, J., & Chen, H. (2021). *Streptococcus*, the Predominant Bacterium to Predict the Severity of Liver Injury in Alcoholic Liver Disease. *Frontiers in Cellular and Infection Microbiology*, *11*, 649060. doi:10.3389/fcimb.2021.649060 PMID:33816353

Zhou, Y., Wylie, K. M., El Feghaly, R. E., Mihindukulasuriya, K. A., Elward, A., Haslam, D. B., Storch, G. A., & Weinstock, G. M. (2016). Metagenomic Approach for Identification of the Pathogens Associated with Diarrhea in Stool Specimens. *Journal of Clinical Microbiology*, *54*(2), 368–375. doi:10.1128/JCM.01965-15 PMID:26637379

Zou, Q., Lin, G., Jiang, X., Liu, X., & Zeng, X. (2018, September 18). (2018). Sequence clustering in bioinformatics: An empirical study. *Briefings in Bioinformatics*. Advance online publication. doi:10.1093/bib/bby090 PMID:30239587

Zou, Y., Duan, X., Xue, Z., Mingju, E., Sun, M., Lu, X., & Yu, X. (2018). Water use conflict between wetland and agriculture. *Journal of Environmental Management*, *224*, 140–146. doi:10.1016/j.jenvman.2018.07.052 PMID:30036808

About the Contributors

Ashok K. Rathoure holds a doctorate degree in Bioremediation/Biotechnology from Central University (HNBGU Srinagar UK India) and M.Tech. Biotechnology. Currently, he is serving as CEO, Akone Services. He served as Managing Director Biohm Consultare Pvt. Ltd. Surat India. Previously, he was associated with Eco Group of Companies Surat Gujarat; Vardan Environet Gurgaon; En-vision Group Surat (En-vision Environmental Services and En-vision Enviro Engineers Pvt. Ltd.) for EIA studies; Himachal Institute of Life Sciences Paonta and Beehive College of Ad. Studies Dehradun for teaching to Biotechnology, Microbiology, Biochemistry, and other biosciences subjects. His area of research is environmental biotechnology and publication includes 100 full length research papers in international and national journals of repute, more than 45 books from reputed publishers in India & abroad and more than 25 Book chapters. He is also associated as Editor-in-Chief for Octa Journal of Environmental Research, Managing Editor for Octa Journal of Biosciences and Executive Editor for Scientific India Magazine. He also holds PG Diploma in Human Resource Management.

* * *

Debolina Banerjee is a guest faculty in Kidderpore College, Kolkata from February 2020 in the Environmental Studies Department. She completed her bachelor's on B.Sc. Botany Honors from Charuchandra College, Calcutta University (2014-2017) and completed M.Sc. in Environmental Science from Asutosh College, Calcutta University (2017-2019). She was a summer trainee under the same project for 3 months (September 2019-November, 2019). She is interning in the "Fishing Cat Project" under the Zoology Department of Ballygunge Science College, University of Calcutta (December 2019- Continuing). She was also OPA Intern in WWF-India (WBSO) for October 2019- April 2020. Debolina is a regular active volunteer since February 2018 and got the best Volunteer prize 2018-2019 and 2019-2020 from WWF-India (West Bengal, Kolkata). She was Intern in Summer Intern in CSIR Summer Research Training Program NEIST (June- August, 2020).She is an Editorial Member (FOE/EN/00113) of FRIENDS OF ENVIRONMENT (3rd January, 2021- contd.), also the member of Member (CON-1042808) at British Ecological Society (13th July, 2021- contd.) and Member (ID: 83) at International Primatological Society (27th June, 2021- contd.). Ms. Banerjee has many publications in many scientific journals and sites. Areas of interests are Ecology, Wildlife conservation and behaviour ecology.

Priya Banerjee is presently working as Assistant Professor, Environmental Studies, Centre for Distance and Online Education, Rabindra Bharati University, Kolkata. She has completed her PhD research from the Department of Environmental Science, University of Calcutta, Kolkata. Her work has been published in various internationally reputed journals like Carbohydrate Polymers, Desalination, Chemosphere, Journal of Environmental Management, Ecotoxicology and Environmental Safety, etc. and has been broadly cited in recent studies. She has published 35 research articles, 25 book chapters and a book to her credit. She has also participated in various national and international seminars till date and has received the best poster presentation award at the 100th Indian Science Congress, Kolkata and Excellent Paper Award at the 6th International Conference on Solid Waste Management, Kolkata. She is also an active reviewer for reputed journals like Journal of Cleaner Production, Heylion, Process Safety and Environmental Protection, Nature Scientific Reports, etc. Her current research interests include integrated waste management strategies, bioremediation, nanotechnology, and toxicology. She is actively involved in research collaborations several other national and international institutes.

Sana Bedoui is a professor in Higher Institute of Water Sciences and Techniques of Gabes, University of Gabes, Tunisia.

Chandani Bhattacharjee is a professor in HR College of Commerce, Mumbai, Maharashtra, India.

Brij Bhushan has been involved in studies related to recovery of resources by integration of membrane technology from wastewater. His other field of expertise is fabrication of adsorbents for use in wastewater treatment, environmental remediation, and preservation. He is currently associated with Department of Chemistry, Graphic Era University, Dehradun.

Vidhi Chaudhary is currently working as an Assistant Professor at the Department of Botany, Daulat Ram College, University of Delhi, Delhi – 110007, India. She has obtained her Ph. D. and M. Phil. degrees in Botany from the Department of Botany, University of Delhi. Her research areas include Plant-Microbe Interactions and Microbial Biotechnology. She has been actively involved in teaching different Botany papers including Environmental Science. She has worked in several research projects funded by various funding agencies like ICAR AMAAS and Delhi University Innovation Project. She has published more than 20 research publications in the journals of international repute, review papers, book chapters and popular articles etc. Her work has been cited as "one of the most requested articles of the journal" for the year 2008 for a paper published in the Journal of Basic Microbiology. She has been an active member of the Reviewer Board for Acta Scientific Journal. She is one of the Editors of Proceedings of the Symposium on Environment and Sustainable Development: New Perspectives. She has been a life member of Prestigious Societies such as the Association of Microbiologists of India. She has total of 1141 citations with an h-index of 13 and an i-10 index of 16. She has been awarded the "Young Scientist International Women Empowerment Award" by Sardar Vallabh Bhai Patel Science and Technology Organization. She is also the recipient of several Best Poster Presentation Awards at national and international conferences. She has received the Best Paper Presentation Award in Rhizosphere3 International Conference at the University of Western Australia, Perth.

Shahid Ahmad Dar is a Senior Research Fellow, Department of Environmental Science, University of Kashmir. He has his Master's and Doctorate degree in Environmental Science from University of Kashmir. He has to his credit more than 10 research articles in various national and international journals.

Soumi Datta is currently working in one of the top FMCG companies based in Delhi as a Principal Scientist and has vivid industrial experience over 15 years in the field of plant biotechnology, DNA fingerprinting, nursery, and greenhouse techniques, hydroponics, palynology (honey analysis through microscopy), Corporate Social Responsibility (CSR) activities, development of Standard Operating Procedures (SOPs) and Good Laboratory Practices (GLPs). Presently she is also pursuing her Ph. D program from the Department of Pharmacognosy & Phytochemistry, School of Pharmaceutical Education & Research (SPER) Jamia Hamdard, Hamdard Nagar, New Delhi-1100062. She completed her B.Sc (Botany Hons) from Hindu College, University of Delhi, North Campus, Delhi, followed by M.Sc (Botany); specialization in Plant Molecular Biology & Crop Genetics and M. Phil (Botany); specialization in Bioinformatics from Department of Botany, North Campus, Delhi-110007. She also holds a PGDM degree in International Trade Management from Narsee Monjee Institute of Management Studies (NMIMS Global Access-School for Continuing Education), Mumbai. She has completed a certified course on Intellectual Property Rights (IPR) from WIPO (World Intellectual Property Organization) and a certified course on "Writing in sciences" from Standford online. She has organized and participated in many national and international conferences/ workshops; presented posters in conferences, and have papers publications in peer-reviewed journals. She has mentored more than 100 interns in their masters (M.Tech/M.Sc Biotechnology) dissertation project work and helped them to document their research findings.

Rashmi Gupta is an energy and environmental lawyer with a Master's in energy & clean technology law, and environmental law from the University of California, Berkeley. Her research focuses on clean energy, climate change, and human rights laws. She has represented many stakeholders including governments and industries in environmental issues ranging from the Delhi Pollution case to the Taj Mahal Trapezium Zone case. She has authored/co-authored research papers on residential electrification, carbon neutrality, and clean energy.

Soumaya Hajji is an Engineer in Geology from National Engineering School in Sfax (ENIS), Tunisia in 2000. She obtained her PhD in Modeling in Hydrogeology in 2015. She obtained her HDR in Environmental Hydrogeology and geo-Informatic in 2021. She is a researcher at geochemistry and hydrogeology, modelling and geostatistics at 'Water-Energy-Environment' Research Laboratory (LR3E) in Engineering School Sfax, Tunisia. Her research activities deal with geochemistry, modeling, analysis, optimisation, simulation of groundwater resources, groundwater quality and. She is an Associate Professor in Geology Department in Faculty of Sciences of Sfax, Tunisia.

Anwesha Haldar is an Assistant Professor of Geography at East Calcutta Girls' College, Kolkata, India. She has also taught at the postgraduate level in West Bengal State University, University of Calcutta and Rabindra Bharati University, Kolkata. She graduated from Loreto College and was a UGC-SRF in the Department of Geography, University of Calcutta, from where she received her Ph.D. in Science. She has edited a few books and journals, published many research articles in reputed journals and edited volumes, and presented papers in the fields of Anthropocene geomorphology, climatology, and environmental geography.

Somnath Kar is presently working as an Assistant Professor, Department of Botany, Holy Cross College, Agartala. He has completed his graduation with Botany Major from MBB College, Under Tripura University. He received his Master's degree in Botany with a specialization in Plant taxonomy and Biodiversity from Tripura University. He has received his Ph.D. degree in Botany from Tripura University. He has research experience of more than 9 years. He also worked as UGC BSR Fellow. He has teaching experience of M.Sc. theory and practical classes in the Department of Botany, Tripura University. He has also teaching experience in both theory and practical classes of B.Sc. in the Department of Botany, Ram Thakur College, Agartala, Tripura. He has guided more than 12 M. Sc and 20 B.Sc. students in their projects, His area of research interest are Plant Taxonomy and Biodiversity, Ethno-Botany, Ethno-Medicine, Plant Reproductive Biology, Plant Developmental Biology, Plant Breeding, Ecology. He has received Prof. R. C. Srivastava Memorial Award (Paper Presentation-First) in the National seminar on "Plant Science Researches in North-East India" held on 7th June 2018. Also received the prize for best poster at "Assam botanical congress (ABC-01) &International Conference on Plant science" held on 4th -6th February 2019. He is a Life member of the East Himalayan Society for Spermatophyte Taxonomy (Taxo Club), The Society of Plant Reproductive Biologists, The Indian Botanical Society, The Botanical Society of Bengal, and InSC. He has published 12 research papers published in different peer-review journals and 2 book chapters having ISBN numbers. He has presented 10 research articles in different national and international conferences and also participated in more than 40 conferences and 100 webinars.

Rupsha Karmakar is very passionate towards exploring new avenues. Currently, she is working on evaluating wastewater samples from West Bengal. She had an experience on snail culture. She has an ardent interest towards research and would like to contribute into social welfare.

Manoj Kumar Karnena received the B.S. degree in Environmental Science and the M.S. degree in Environmental science from GITAM (Deemed to be) University, India. Later, he pursued his Ph.D. in Water and Waste Water treatment at the same institution. He was the university topper and gold medalist in bachelors and post-graduation. He holds an Environmental Law degree from National Law University, Delhi, and Industrial Safety Degree from Annamalai University, Tamilnadu. Since 2019, he has been working as a teaching assistant with the Department of Environmental Science, GITAM Institute of Science, GITAM (Deemed to be) University, India. His greatest strengths are his research and communication skills. He started teaching and working in technical work in 2019 in various backgrounds. Manoj Kumar Karnena has written more than 50 papers since he started writing in 2017. He has written several book chapters and books on the topic within his research interest and holds two patents. His research interest is in Water and Wastewater Treatment, Natural Coagulants, Nanocatalysis, Storage Technologies, and membrane technology. Manoj Kumar Karnena had won several engineering designs and environmental protection awards in water and wastewater treatment.

Nisha Khatik is currently working as a Guest Faculty in the Department of Botany, Maharshi Dayanand Saraswati University, Ajmer, Rajasthan -305009, India. She has obtained her Ph.D and M. Phil. degrees in Botany from the Department of Botany, Samrat Prithviraj Chauhan Government College, Maharshi Dayanand Saraswati University, Ajmer. She has been actively involved in research and teaching in the field of Botany & major research interests include plant tissue culture and molecular biology. She has published several book chapters and research papers in the journals of national and international repute and has also participated in several National and International conferences. Dr. Khatik is the Lifetime member in Indian Science Congress Association (ISCA) and The Indian Botanical Society, Kolkata. & Member of Editorial board of many reputed journals.

Madhavi Konni holds a doctorate in Nanoscience and Nanotechnology in the Department of Chemistry from GITAM (Deemed to be University), Visakhapatnam, India. Currently, she is working as Associate Professor at Dadi Institute of Engineering and Technology, Anakapalle. Her research interests include synthesizing Metal-Organic Frameworks, Material Chemistry, Carbon Nano Materials, Hydrogen Storage Technologies, and Nano Catalysis. She published many research articles and book chapters in reputed international journals and holds two patents. Dr. Madhavi is a fellow and received HEAM Scholar Award (2018), the Young Scientist Award (December 2017 and March 2017), and few other awards from different societies.

Rituparna Kundu Chaudhuri studied Botany at the University of Calcutta, West Bengal, India, and graduated and ranked 3rd as Master of Science in 2001. She was awarded GATE, NET (Lectureship and CSIR) and joined the research group of Prof.Timir Baran Jha, at erstwhile Presidency College, now known as Presidency University, Kolkata. She received her Ph.D. degree in 2009 from the University of Calcutta entitled 'Tissue Culture Studies and Biochemical Analysis of An Indigenous Medicinal Plant Swertia chirata". During her Ph.D. tenure, she worked at INRA, Versailles, France as visiting researcher for three months under the supervision of Prof. David Tepfer. She obtained the position as an Assistant Professor in UG and PG Dept. of Botany at Bethune College, Kolkata, India, in the year 2008. She joined Krishnagar Govt. College during 2015 on transfer. With specialization in Plant Physiology, Biochemistry, and Molecular biology, her research interest is mainly focused on Plant Tissue Culture and conservation of medicinally important plants along with analysis and scaling up of secondary metabolites. Also, she pursues an interest in the genetic improvement of commercially important legume crops. To date, she has published 21 research articles in peer-reviewed journals, contributed to several book chapters in edited books published by International and National publication houses. She has also published in National and International seminar proceedings. Currently, she is associated with research activities in Crop-Plant Biotechnology.

Nibedita Maji is pursuing PhD from Department of Zoology, Chandernagore College, Chandannagar, West Bengal.

Pallavi Mitra is presently working as Assistant Professor in Environmental Studies at Centre for Distance and Online Education, Rabindra Bharati University, Kolkata. She was awarded her Ph.D. degree in Environmental Science in 2015 from University of Calcutta. She has several publications in peer reviewed journals and books. One of her research works entitled "Ceramic Glazing Pigment from Chrome-Iron Sludge" was granted India Patent in 2018.

Aniruddha Mukhopadhayay is presently Professor at Department of Environmental Science, University of Calcutta and has been teaching for more than the last two decades. After completion of his graduation and master's from University of Calcutta he was awarded the D. N. Roy Chowdhury Gold Medal for his results in post graduation. He had also completed his M. Phil. in Environmental Science from University of Calcutta. He had qualified UGC NET (Life Science) in 1986 and pursued his Doctoral Studies from University of Calcutta. He had carried out his post doctoral research in USA.He has been elected as FELLOW of West Bengal Academy of Science and Technology (FAScT). He has supervised doctoral students and been in research collaborations with research institutes and other Universities both at national and international level. He is a member of different committees instituted by the Government Agencies. His research interests include areas like environmental biotechnology, toxicology and ecology.

Arunima Nayak is a researcher cum academician involved in the development and advancement in the knowledge, tools and processes in the field of different bio refinery and valorisation strategies for wastes and wastewater treatment. Future research interest is to address the emerging challenges in the field of waste management and water treatment technologies via activated carbon adsorption, anaerobic digestion and extraction of antioxidants.

Moumit Roy Goswami currently working as Assistant Professor in Environmental Science at Netaji Nagar College for Women, Kolkata and has been teaching for more than a decade. His research interests include areas of Environmental Biology and Environmental Toxicology.

Dipanwita Sarkar (Paria) has a an excellent academic career as Gold Medalist in both B.Sc. and M.Sc. in Zoology from the University of kalyani.. She has a vast research experience of about 21 years in the field of Ecology, Ecological Engineering, Wetlands, BIodiversity and Fisheries. She has Published a number of research papers in the Journals of national and International repute. She has teaching experience of 11 years in Postgraduate and Undergraduate courses.

Nabeela Siddiqui's research revolves around Water Conflicts. She has completed her B.A.LL.B (Hons) from Aligarh Muslim University, Aligarh, India. She was awarded the prestigious President of India, Dr. Shankar Dayal Sharma Gold medal, The Innes Prize, The Carmicheal Prize, and The V. Suresham Memorial Award for her Masters (LL.M.) from University of Madras, Chennai, India. She has previously worked as a Research cum Teaching Assistant at National Law University, Jabalpur, and as a Research Assistant at the National Green Tribunal, Principal Bench, New Delhi. She is an active member of The Global Network for Human Rights and the Environment (GNHRE) and The International Society for Therapeutic Jurisprudence (ISTJ).

Anjana Singh is currently working as an Assistant Professor in the Department of Botany, Deshbandhu College, University of Delhi, Kalkaji Main Rd, Block H, Kalkaji, Delhi- 110019, India. She has completed M.Phil and M. Sc. from Department of Botany, University of Delhi, Delhi-110007, in addition, M. Sc (Ecology and Environment) from Sikkim Manipal University (distance learning) and B.Sc (Environmental Science) from Maitreyi College, University of Delhi, Chanakyapuri, New Delhi-110021. She has more than 15 years of teaching experience. She is actively involved in the examination, admission and cultural events in college and in particular the botany department, various other responsibilities during SABRANG, the annual cultural festival of Deshbandhu College, in organising workshops, webinars, conferences and also in other assignments etc. Her area of research interests includes the study of diversity and distribution of Bryophytes, the study of the growth patterns in relation to the changing climatic conditions, accumulation of metals especially the pollutants by bryophytes. She is the author of a number of papers in peer-reviewed journals and has authored a number of chapters in books of international repute. She has actively organized and participated in many national and international conferences/workshops; presented posters in conferences. She has qualified in UGC-NET (LS) in Environmental Sciences and CSIR-UGC-NET (LS) in Life Sciences; agricultural Scientists Recruitment Board (ICAR) National Eligibility Test for Lectureship and many has been awarded with many other prestigious awards and allocations.

Dwaipayan Sinha is presently a faculty member under the West Bengal Education Service in the Department of Botany Government General Degree College, Mohanpur, Paschim Medinipur. He was previously posted in the Department of Botany, Bidhannagar College, Salt Lake, Kolkata under the same capacity. He has done is graduation with honors in Botany from the prestigious Presidency College, Kolkata securing First Class and ranking in the University. He has done his masters in Botany from the Department of Botany, Delhi University, North Campus with a specialization in Plant Molecular Biology and Crop Genetics. He subsequently qualified for CSIR NET under the JRF category and joined the National Institute of Oceanography, Goa as a research fellow, and finally moved to CSIR-IITR. He completed his Ph.D. from CSIR-IITR and registered in the Department of Botany, University of Lucknow. His broad research area includes antioxidant properties of Gymnosperms. He has more than 15 years of Research experience and 12 years of teaching experience. He is actively involved in research activities and presents research interests include the exploration of phytodiversity in various blocks of West Bengal. He is also working on sacred groves and ethnomedicinal uses of plants of West Bengal. He is the author of a number of papers in peer reviewed journals and has authored a number of chapters in books of international repute.

Mohamed Ali Tagorti is a professor in Geochemistry at University of Sousse, Tunisia.

Vinaya Tari has completed her Ph.D. in Environmental Science from the University of Mumbai. She did under graduation studies in Zoology, M.Sc. in Environmental Science, and a Diploma in Biotechnology (2 years) from the University of Mumbai and B.Ed. from S.N.D.T. University. She has qualified State Level Eligibility Test for Assistant Professorship valid in the states of Maharashtra and Goa, India. She has been awarded the 'Young Scientist Award' from Deccan Environmental Research Organization (DERO). She has attended around 34 conferences/ workshops/ webinars/ faculty development programme. She has published more than 22 research items in reputed journals, Magazines/e-Newsletters, and conferences. She is Associate Researcher at Australian Centre for Sustainable Development, Research and Innovation (ACSDRI) and South Asian Institute of Sustainable Development (SAISD), Australia. Also appointed as a Scientific reviewer and Writer (Part time) at 'The BrainAura', United Kingdom based organization. Her current research interests are Environmental economics, sustainable crop protection practices, botanical pesticide formulations & Environmental pollution. She has also engaged as a reviewer for various reputed journals. She has co-authored book chapters with several reputed publishers. She is a certified mentor on 'Expertrons', for guiding students and professionals on their career path. She has collaborated with abroad countries such as Thailand, Malaysia, Turkey, Qatar, Indonesia, Australia, etc.

Saritha Vara is presently Associate Professor in the Department of Environmental Science, GITAM (Deemed to be University), Visakhapatnam and has been teaching for about 16 years. After completion of her graduation and master's from Andhra University, she pursued her M.Tech (Environmental Management) and Ph.D in Environmental Science and Technology from JNTU, Hyderabad. She has published around eighty national and international research publications and nearly 8 book chapters to her credit. She has supervised doctoral students. Her research interests include areas like water and wastewater treatment, environmental toxicology, epidemiological studies, and corrosion studies.

Index

A

Allogenic 131
Anthropogenic activities 34, 52, 57, 99, 107, 115, 138-139, 186-187, 189-190, 238, 291, 297
Antibiotic Resistant Bacteria (ARB) 72, 90
Aquatic 1-3, 8-10, 12, 14, 18, 20, 27-29, 31-37, 39, 52, 54, 56-57, 59-65, 72, 93-95, 106, 129, 131, 136, 140, 150, 155, 157, 159-160, 162, 290-291
Autogenic 131

B

benefits 1-2, 8, 25, 93-94, 97, 101, 109, 124, 139-140, 154, 167, 177, 188, 281, 283, 293, 296
Biodiversity 2-3, 10, 17, 19, 21, 27-28, 30, 58, 96-97, 99, 101-102, 106-107, 124, 129, 137, 139-141, 150-152, 154-155, 165-167, 188, 191, 283, 290, 294

C

chemical parameters 51-52, 59, 252
Civil society 97, 99-100, 295
classification 3, 5, 28, 129, 151, 197-198, 202, 204-206, 208-209, 213
Climate change 1-2, 10, 21, 31, 39, 83, 92-94, 96-103, 106, 137, 139, 152, 173, 182-185, 190, 216, 241, 256, 280, 283, 290, 292, 294
climatic phase 226, 234, 241, 255
climatic variability 96, 218, 238, 252-253, 256, 259-260, 267, 272
Climax 30, 140, 254
Conservation 13-14, 17-19, 21, 23-24, 38-39, 52, 94-95, 97-103, 107, 110, 122-124, 129, 137, 140, 152, 154, 166-167, 173-174, 177-178, 188-190, 270, 280-282, 285, 291, 294-297
Contaminated Wastewater 71
Contingent valuation 138
Controlling 95, 246, 280
Culture Dependent 75, 77, 81
Culture Independent 71, 75, 83
cyclostratigraphy 216, 226, 230-234, 236, 245-247, 253, 259-260

D

Deforestation 186, 290-291
Disease Network 71, 73, 78, 80
Disturbance 30, 183, 188
Diversity index 164
Drought 30, 93, 106, 130, 137-138, 140

E

Ecological restoration 19, 152, 167, 294
Ecosystem services 3, 51, 97, 100-101, 140, 151, 166-167, 188, 280-281, 283, 285, 296
Ecotourism 94, 124, 138, 281
Environmental management 154, 178-179, 183, 282, 294
Exoreic Saline System 216, 229-230

F

Field Testing Kits 51-52
Flame Photometer 218, 229, 235, 254
Flooding 8-9, 11-12, 33, 93, 136-140, 151, 233, 238, 253
Flora 27-29, 35, 39, 62, 96, 129-130, 137, 140, 155, 162, 167, 173-174
Freshwater 2-4, 6, 12, 24, 28, 30, 39, 93-94, 96, 129, 140, 155, 190, 216, 280, 291
functions 1, 3-4, 8, 12, 18, 95-96, 100, 106-107, 109, 111, 116, 123, 151, 154, 166-167, 189

G

geobiology 267-268
Geochemistry 215, 218, 222, 229, 236, 238, 241, 268

grain size distribution 197-198, 201, 204-205, 207-208, 210, 213, 256, 272
Greenhouse gases 94, 183, 185

H

Habitat 3, 9-12, 28, 39, 107, 109, 115, 129, 131, 151-152, 155, 157, 159-162, 281, 296
Heavy Metals 33, 35-36, 38, 72, 78, 90, 133, 176, 231, 241
holocene 216, 218-219, 226, 230-236, 245-247, 252-253, 257, 259-260, 267
Holocene Cyclostratigraphy 216, 226, 230, 233-234, 236, 245-247, 253, 259-260
Hospital Effluent 73, 90

I

Illumina Sequencing 76, 90
Impact 10, 31, 33-36, 83, 92-94, 96-97, 99, 106, 115, 128, 133, 137-139, 171, 174, 176, 178, 186, 231, 281, 290, 296
Inclusive Graphic Statistics 197-198, 201-204
In-Situ Measurements 51-52, 66

L

Laboratory Analysis 66
Land cover 52, 115, 139, 166
Land use 14, 52, 106-107, 115, 117, 122-124, 129, 166

M

Magnetic Susceptibility 220-223, 231-232, 234, 238, 240-241
Mchiguig 252-257, 259-260
Metagenomic Analysis 75
Microcystin 37
Mineral 3, 35, 38, 229, 236, 255-256
Mineralogy 215, 226, 229, 236, 243
Monitoring 3, 24, 35, 39, 51-52, 58, 64, 66, 71, 73-76, 82-83, 91, 98-99, 102, 141, 167, 179, 283
MSW 172-173, 175, 177

O

Outbreak Response 91

P

parameters 51-52, 59, 66, 131, 163, 197, 201-202, 204-205, 208, 222, 230, 232, 234-235, 238-239, 241, 244, 252-253, 255-257, 259-260, 274, 283, 285, 294, 296
Pathogenic Load 71, 73, 75-76, 78, 90
Physico-chemical 51, 102
Plants 1, 3-4, 8-9, 11-14, 17-18, 28-30, 35-37, 54-55, 61-62, 72-74, 93, 95-96, 136-137, 150-151, 162, 174, 178, 183
Pollution 2, 12, 14, 19, 27-33, 35-39, 52-54, 60, 94-95, 102, 128-129, 131-133, 140-141, 171, 175-179, 186, 231, 241, 290
preservation 12-14, 17-18, 23-24, 66, 102, 140, 178, 184, 290
Protection 3, 8-10, 13-14, 17-18, 21-24, 93, 97, 100, 102-103, 112, 123-124, 137, 139-140, 166-167, 173-174, 178-180, 183, 190, 290, 295
Public policy 122, 296

Q

Quadrat method 155, 162

R

Ramsar Convention 2-3, 13-14, 18, 21, 23-24, 94, 97, 99-100, 102, 106, 129, 140-141, 150, 152, 174, 281-283, 291, 293, 296
Restoration 3, 10, 13-14, 17-22, 24-25, 39, 52, 92, 97, 99-103, 124, 152, 167, 171, 174-175, 183, 187-190, 280, 282-283, 291-297

S

Salinity 3, 12, 29-30, 58, 66, 128, 130, 138-139, 273-274, 277
sebkha 197, 204, 208, 216-220, 224-226, 229-235, 238-239, 241-247, 252-257, 260, 268-269, 272-275
Sebkha El Melah 216-220, 224-226, 229, 232
sebkha Mchiguig 253-257, 260
sedimentary 139, 197-198, 208, 217, 222, 230-232, 234, 238-239, 241, 252-255, 257, 267-268
Sidi El Hani 197-201, 204, 208, 211, 213, 253, 267-269, 272-273, 275, 277
Solid Waste 171-174, 176-177, 179, 291

Spectral Analysis 215, 218-224, 226, 229, 235, 243-244, 247, 252-254, 258-260
Sustainable development 13, 106-107, 124, 152, 180, 281
Sustainable Development Goals 152, 180, 281

T

terrestrial analogue 267
threats 1, 3, 32-33, 35, 39, 75, 100, 102, 106, 114, 116, 152, 167
Toxicity 133
Transect Method 160, 162
Turbulence 130

U

Upper Ganga Ramsar site 92, 97, 101
Urbanization 2, 12, 27, 33, 36, 39, 52, 94, 100, 106-107, 115, 129, 171, 290

V

variability 2-3, 19, 96, 197, 215-216, 218-220, 223, 231-232, 235, 238-239, 241, 243-244, 246-247, 252-254, 256, 259-260, 267, 272, 277
Vegetation 2, 8-10, 12-13, 21, 28, 30, 33, 39, 115, 130-131, 137, 139-141, 150-151, 155, 161-163, 166, 171, 177, 216, 274

W

Waste Management 172, 175, 179
Wastewater Monitoring 74-76, 91
Water bodies 2, 9-10, 27-28, 31-37, 73, 75, 93, 95, 98-99, 102, 115, 132-133, 137, 141, 171-174, 291, 293, 295
water pollution 12, 27-28, 31-33, 35-37, 39, 95, 176
Water Quality 2-3, 8-10, 12, 17, 29-30, 34, 51-52, 58, 64, 66, 71, 78, 93, 102, 139, 174, 283, 290, 297
Wetland Management 14, 38, 52, 98, 282, 285, 291, 294, 297
wetlands 1-4, 8-9, 11-14, 17-23, 25, 27-32, 35-36, 38-39, 51-55, 57-59, 61-66, 92-93, 95-103, 106-118, 122-124, 128-131, 137-141, 150-152, 166-167, 171-177, 179, 182-187, 241, 267, 280-283, 285, 290-297

X

X-Ray Diffraction 225-226, 229, 234

www.ingramcontent.com/pod-product-compliance
Lightning Source LLC
LaVergne TN
LVHW060406300926
849162LV00006B/1358